COMPUTER AIDED LOGICAL DESIGN
WITH EMPHASIS ON VLSI

COMPUTER AIDED LOGICAL DESIGN
WITH EMPHASIS ON VLSI

FOURTH EDITION

Frederick J. Hill
UNIVERSITY OF ARIZONA

Gerald R. Peterson

JOHN WILEY & SONS, INC.
NEW YORK ■ CHICHESTER ■ BRISBANE ■ TORONTO ■ SINGAPORE

ACQUISITIONS EDITOR	Steven Elliot
MARKETING MANAGER	Susan Elbe
PRODUCTION SUPERVISOR	Elizabeth Austin
DESIGNER	Ann Marie Renzi
MANUFACTURING MANAGER	Andrea Price
COPY EDITING SUPERVISOR	Deborah Herbert
ILLUSTRATION	Anna Melhorn

This book was set in Times Roman by Acu-Type, Inc. and printed and bound by Hamilton Printing.

Recognizing the importance of preserving what has been written, it is a policy of John Wiley & Sons, Inc. to have books of enduring value published in the United States printed on acid-free paper, and we exert our best efforts to that end.

Library of Congress Cataloging in Publication Data:

Hill, Frederick J.

Computer aided logical design with emphasis on VLSI / Frederick J. Hill, Gerald R. Peterson.—4th ed.

p. cm.

Rev. ed. of: Introduction to switching theory and logical design. 3rd ed. © 1981.

Includes index.

ISBN 0-471-57527-5 (cloth)

1. Logic design—Data processing. 2. Integrated Circuits—Very large scale integration—Design and construction—Data processing. 3. Computer-aided design. I. Peterson, Gerald R. II. Hill, Frederick J. Introduction to switching theory and logical design. III. Title.

TK7888.4.H54 1993 92-20489
621.39′5—dc20 CIP

Printed in the United States of America

10 9 8 7 6 5 4 3 2 1

This book must be dedicated to my coauthor Gerald R. Peterson, whose untimely death in April 1990 could not blot out the significant impact he made on its contents.

PREFACE

Considerable change has taken place in electrical engineering and computer engineering curricula during the tenure of the third edition of our book *Introduction to Switching Theory and Logical Design*, which appeared in 1980. Now most curricula include a required sophomore or junior course in digital logic which is often followed by a senior elective in digital VLSI design. This book is not a fourth edition of the previous work. It does, however, freely appropriate material from that book and is intended to serve a similar audience.

The introduction of a digital VLSI course on most campuses was spurred by the publication by Meade and Conway of *Introduction to VLSI Systems* in 1980. At first these courses were supported only by layout editors and low-level simulators. The subject was always circuit design and layout as well as logic design. As schematic capture programs supported by standard cell libraries became available, the emphasis in many of these courses moved away from circuit design back in the direction of logic design. This book supports the new emphasis on VLSI design. To the extent that circuit design issues impact logic design, they are within the scope of this book. The assumption is, however, that most readers will be content to allow software tools to translate design to layout.

This book is tied to no particular set of computer-aided logic design tools. The instructor may set up a supporting laboratory using whatever tools are available or are most convenient. We include support of layout synthesis from description in a register transfer level language as well as from design capture. This will not be an obstruction to the reader who wishes to rely exclusively on design capture tools.

The similarity in titles of this book and the one it replaces is no coincidence. The popularity of the latter, over 24 years and three editions, convinced us to retain many of its key chapters. A stand-alone digital VLSI course will take advantage of the same careful introduction to Boolean algebra, Karnaugh maps, and sequential circuits found in the earlier work. Treating CMOS and NMOS design has made it necessary to lengthen the treatment of combinational logic. Some of these topics may be skipped in a stand-alone course.

It is anticipated that this book will often be used in a course with introductory logic design as a prerequisite. In that case the course beginning is probably Chapter 6. From that point there remains more than enough material for one semester.

The introduction to truth tables and logic operators has been moved to Chapter 1. The material on codes has been merged with the treatment of binary arithmetic in Chapter 2. A slight expansion of the familiar approach to Boolean algebra is now presented in Chapter 3. A substantially updated treatment of switching circuits is now presented in Chapter 4. The time-honored introduction to Karnaugh maps comes one chapter earlier as Chapter 5. Tabular and multiple output minimization has been expanded by incorporating the discussion of PLAs and a brief encounter with Espresso. The result is now Chapter 6. Chapter 7 is devoted to the special issues associated with logic design in NMOS and CMOS.

Chapter 8 explores standard parts and cells, applicable to design with MSI packages as well as VLSI. The classic treatment of design of clockmode sequential circuits in Chapter 10 is almost unchanged.

Chapter 11 has been adjusted to better support layout synthesis based on register transfer level descriptions. A probing of the fundamental principles of the spectrum of computer-aided logic design tools is included next, as Chapter 12, so that it will fall naturally into every course syllabus. Which and how many of the remaining chapters to incorporate in the course may depend on the background of entering students. Chapters 13 and 14, are carried over from the present book's predecessor. Although less important in the VLSI environment, the previous book's careful treatment of incompletely specified and level-mode sequential circuits has not been left out.

Chapter 15 offers a comprehensive treatment of test generation for VLSI. This has been my area of research and academic interest for many years. It is hoped that the blend of new and traditional approaches will bring the carefully selected set of topics to life. Chapter 16 is a shortened version of the similar chapter of the earlier book. A section using the iterative network analogy as the basis for the most self-evident approach to test generation for sequential circuits has been added as a conclusion to the chapter. This ordering of material has that topic following sufficiently close to the other test generation topics in Chapter 15, while allowing the introduction of iterative networks to be postponed until Chapter 16.

It is expected that this book will always be supported by some set of design capture, simulation, and layout tools. If synthesis from register transfer descriptions is stressed, some CALD tools that were developed under my direction may prove helpful. (A function level simulator HPSIM has for several years been available to support our book, *Digital Systems: Hardware Organization and Design* (Hill and Peterson, 1987). Also available are a pair of programs that compile AHPL descriptions to EDIF netlists. These netlists may serve as input to automatic place and route programs and gate level simulators, which are available from traditional CAD tool vendors. Specifications and availability of these tools may be obtained by writing to me at the Department of Electrical and Computer Engineering, University of Arizona, Tucson, AZ 85721 or by electronic mail using hill @ ece.arizona.edu.

I would like to extend thanks to all who participated at the various stages of the review process. Included are Wade Shaw, Air Force Institute of Technology; Louis Weinberg, Professor Emeritus, City College; Professor, Graduate Center CUNY; Don Bouldin, University of Tennessee; R. G. Deshmukh, Florida Institute of Technology; Darrow Dawson, University of Missouri, Rolla; Ron DeVries, University of New Mexico; and John Robinson, University of Iowa.

Frederick J. Hill

ACKNOWLEDGMENT

Professor Ronald DeVries of the University of New Mexico, the first Ph.D. student of Dr. Peterson, selflessly provided a comprehensive review of this book. His sincere and sound criticism resulted in the rewriting of several sections.

F. J. Hill

CONTENTS

CHAPTER 1

LOGICAL ANALYSIS

1.1 LOGIC DESIGN WITH AND WITHOUT CAD TOOLS

In *Introduction to Switching Theory and Logical Design*, which first appeared in 1968, the authors considered it essential to introduce the concept of a digital system. It was then necessary to argue that such systems could be implemented by using digital logic elements. Today the only issue is how to approach the design of systems incorporating digital logic. One might take any of the following four approaches when confronted by some significant intellectual task:

1. Accomplish the task directly.
2. Program a computer to accomplish the task.
3. Design a digital system to accomplish the task.
4. Program a computer to design a digital system to accomplish the task.

The choice between these alternatives will, of course, depend on the nature and complexity of the task, the number of times the task will be repeated, and the number of potential customers for a digital system that can accomplish the task. For most readers alternatives 2 and 4 might best be modified to read: **use a computer software tool to**. The intention of this volume is to address alternatives 3 and 4. As will be amply justified in the next section, the act of designing a digital system is usually called logic design. When software tools are included, we have chosen to call the process *computer-aided logic design* or CALD. We thus distinguish the subject from the more general term CAD (computer-aided design) whose scope reaches back to elementary drafting. Logic design and computer-aided logic design are jointly the subject of this book.

At the optimistic extreme we might interpret **use a computer software tool to design a digital system** to imply that we may freely describe a potential digital hardware item in a high-level language such as Pascal and expect the tool to do the rest. Indeed, this is the approach of some tool developers who envision a design process following the form of Fig. 1.1*a*. As yet, this goal is not quite reality.

The following might be a fragment of a hardware design:

<pre>
              if not signal then
                      incrementcounter
          else
                      decrementcounter
</pre>

By itself this fragment could be translated to an optimal (in terms of economics and VLSI* chip area) realization of the hardware element symbolized in Fig. 1.1*b*. The block-labeled optimizer, as it appears in Fig. 1.1*a*, is deceptively simple.

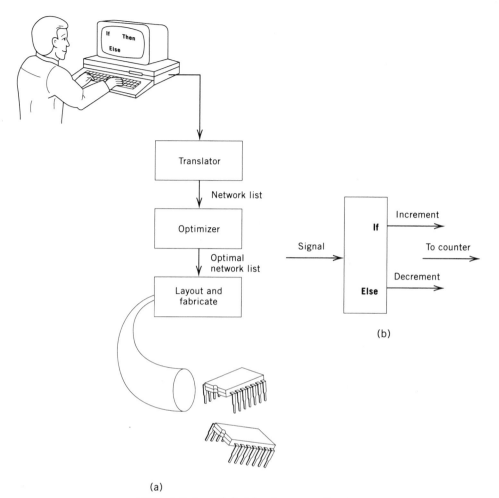

Figure 1.1   Digital hardware synthesis.

The chapters to follow should amply demonstrate that the optimization of digital circuits is composed of several difficult subtasks, whose complexity and computation time increase steeply with the complexity of the target design. The burden on the optimizer can be eased significantly by controls on the translator and on descriptions accepted by the translator.

The algorithms used in the software tools making up the optimization process are often extensions of those used by the logic designer who did not have access to these tools. Treatment of classic logic design techniques is not neglected in this volume. Just as arithmetic remained in the elementary school curriculum through the age of the calculator, it is important that reliance on software tools not rob logic designers of the understanding of their craft.

In a few instances such as the Quine–McCluskey algorithm to be encountered in Chapter 6, we have placed renewed emphasis on formal procedures that might serve as the basis for algorithms implemented in software tools. More often our approach has been to consider refinement, extension, and performance evaluation of algorithms for software implementation to be beyond the scope of the book. Whenever the topic is computer-aided logic design, the emphasis will be on understanding what may be expected of available software tools.

1.2 TRUTH FUNCTIONS

In 1938 a mathematician working for the Bell Telephone Laboratories, Claude Shannon, proposed a switching algebra for use in analyzing and designing discrete binary systems. Shannon's switching algebra was an outgrowth of an algebra of logic devised nearly 100 years earlier by George Boole, an English mathematician and logician. AND, OR, and NOT building blocks are the most convenient physical implementations of this algebra. Although switching algebra is only a special case of Boole's algebra, most people simply use the term Boolean algebra to refer to the mathematics used in dealing with digital systems and circuits.

Boole developed his algebra as a means of describing the complex logical relationships of natural language. This connection to natural language is very important, as it is the basis for a technique for moving from a verbal description of the function of the desired circuit to an unambiguous mathematical description. This procedure is very important, in that the designer is forced to continue consideration of his description until he or she is certain it actually describes what he intends to build. It is surprising how much engineering time is wasted by attempts to bypass this step and to proceed from some vague mental picture of the circuit function directly to wiring circuits. Such a process rarely works and never produces good designs. With the following example we shall illustrate the algebraic formulation of a combinational logic problem.

■ EXAMPLE 1.1

An elevator door control is to function as follows. When the elevator stops at a floor, the door will open and a timer will turn on a signal that will remain on for 10 sec to allow time for passengers to get on or off. In addition, there will be an electric eye to ensure that the doors do not close on someone in the doorway. After the 10 sec are up, if there is no one in the doorway, the doors will close if a call button has been pressed on another floor or a passenger in the elevator has pressed a button for another floor. Write an expression of logically related statements that will evaluate to **TRUE** whenever the elevator door is to close.

SOLUTION As a beginning, let us delete extraneous information to reduce the preceding paragraph to a more concise logical statement.

The elevator door will close if and only if the timer signal is not on and the electric eye circuit is closed and a call button at another floor has been pressed or a floor button in the elevator has been pressed.

Next we observe that this compound statement is made up of simple declarative statements and assign a symbol to represent each statement.

The door will close	D
The timer is on	TM
The electric eye circuit is closed	E
A call button has been pushed	C
A floor button has been pushed	FL

Each statement is true or false. The letters representing the statements are *logic variables*, which will be used as the names of binary signals in a logic circuit. A logic variable will have the value **F**, if the corresponding statement is false, or the value **T**, if the statement is true. For example, the E signal is a signal derived from the electric eye circuit, taking on the value **T**, if the beam is closed, and the value **F**, if the beam is interrupted by a passenger.

One drawback of the English language is that when a number of AND's, OR's, and NOT's appear in a single sentence, the order in which they are to be applied is not always clear. Before writing the final logical expression, it is convenient to add some parentheses to the above statement to form a quasi-English expression emphasizing order.

[(the timer is NOT on) AND the electric eye signal is closed] AND (a call button has been pressed OR a floor button has been pushed)

The statement "the timer is not on" can be symbolized by $\overline{TM}$. The statement $\overline{TM}$ is true whenever TM is false and $\overline{TM}$ is false whenever TM is true. Replacing this and the remaining simple statements with logic variables and symbols to be defined results in Equation 1.1, a Boolean algebraic expression for D. We shall see shortly that this expression can be implemented as a combinational logic circuit with output D.

$$D \equiv (\overline{TM} \wedge E) \wedge (C \vee FL) \tag{1.1}$$

We have represented the phrase *if and only if* by the symbol $\equiv$, which has the same meaning when the variables are restricted to the values **T** and **F**.   ∎

By definition, Equation 1.1 is a statement that is always true. Such a statement is called a *tautology*. A *statement variable* was assigned to each statement in Example 2.1. The **T** or **F** values that may be assumed by these statements are *truth values*. In general, we define a *statement* as any declarative sentence that may be classified as true or false. The classical mathematics of manipulating statement variables and assigning truth values is formally known as *truth-functional calculus*.

Consider, for example, the statement $\overline{TM} \wedge E$ from within Equation 1.1. This is a *truth-functional compound*, a compound statement for which the truth value is determined by the truth values of the component statements. The relationship between the truth of the component statements and the truth of the compound statement is determined by the *connective* relating the components. In this case, the connective is **AND**, indicating that the compound statement is true if and only if both component statements are true. Since AND is a common relationship between statements, we have assigned a standard symbol $\wedge$. There are only four possible combinations of truth values of two component statements, so that we can completely define $A \wedge B$ by listing its truth values in a four-row table like that shown in Fig. 1.2. A table of this form is called a *truth table*.

This table also defines the **OR** connective, symbolized by $A \vee B$. The compound statement $A \vee B$ is true if A is true, OR if B is true, OR if both are true.

A	B	$A \equiv B$	$A \oplus B$	$A \wedge B$	$A \vee B$
F	**F**	**T**	**F**	**F**	**F**
F	**T**	**F**	**T**	**F**	**T**
T	**F**	**F**	**T**	**F**	**T**
T	**T**	**T**	**F**	**T**	**T**

Figure 1.2

Common usage of the word *OR* is not always consistent with the definition of $A \vee B$. Consider this sentence:

"Joe will be driving either a green sedan or a blue convertible."

Clearly, the statement is not intended to mean that Joe might be driving both cars simultaneously. To express that statement A OR statement B but not both are true, we write $A \oplus B$. The truth values for $A \oplus B$ are also tabulated in Fig. 1.2.

Note that problem specifications can be expressed as truth functional compounds only if statements are related by connectives that are completely defined by the tabulation of truth values. Not all declarative sentences are subject to such expression. Consider the following sentence:

"*The door did not close because* a call button was not pressed."

This is not a truth-functional compound, since the truth of the statement cannot be determined solely from the truth of the component statements. Even if both components are true, the compound statement may not be true. The electric eye beam rather than the call button may have been the cause of the door closing. Such sentences cannot be handled by truth-functional calculus. In the next section, it will become clear that the usefulness of electric circuits in making logical decisions is limited to situations in which truth-functional calculus is applicable. The designer must avoid formulating a design statement in such a way that it does not constitute a truth-functional compound.

1.3 BINARY CONNECTIVES

The connectives defined by the truth table of Fig. 1.2 are *binary*, since they relate only two component statements. Each binary connective corresponds to a unique assignment of truth values to the four rows of the truth table, corresponding to the four possible combinations of truth values of the two component statements. There are $2^4 = 16$ possible ways of arranging T's and F's on four rows, so there are 16 possible binary connectives, as tabulated in Fig. 1.3. Each connective is numbered, and symbols are shown for connectives of particular interest. Connectives 1,6,7,9 correspond to common relationships in natural language, as already discussed. Connectives 3 and 5 are simple identities, and connectives 10 and 12 correspond to the negation of either component.

Connectives 8 and 14 do not appear to conform to any simple relationships in natural language. However, if we compare them directly to connectives 1 and 7, as is done in Fig. 1.4, we see that 14 is the negation of 1, that is, the truth values of 14 are precisely opposite those of 1. Similarly, connective 8 is the negation of 7. These relations are expressed in Equations 1.2 and 1.3:

		∧		A		B	⊕	∨	↓	≡	$\bar{B}$		$\bar{A}$		⊃	↑
A	B	0	1	2	3	4	5	6	7	8	9	10	11	12	13	14	15
F	F	F	F	F	F	F	F	F	F	T	T	T	T	T	T	T	T
F	T	F	F	F	F	T	T	T	T	F	F	F	F	T	T	T	T
T	F	F	F	T	T	F	F	T	T	F	F	T	T	F	F	T	T
T	T	F	T	F	T	F	T	F	T	F	T	F	T	F	T	F	T

Figure 1.3

A	B	$\wedge$ 1	$\uparrow$ 14	$\vee$ 7	$\downarrow$ 8
F	F	F	T	F	T
F	T	F	T	T	F
T	F	F	T	T	F
T	T	T	F	T	F

(a)

A	B	$\equiv$ 9	$\supset$ 13
F	F	T	T
F	T	F	T
T	F	F	F
T	T	T	T

(b)

Figure 1.4

$$A \uparrow B = \overline{A \wedge B} \tag{1.2}$$
$$A \downarrow B = \overline{A \vee B} \tag{1.3}$$

Because of these relationships, connective 14 is commonly known as NAND (not AND), and connective 8 is commonly known as NOR (not OR). The importance of these two connectives will become clear in later sections.

It is also instructive to compare connectives 9 and 13. Connective 13 is written symbolically

$$A \supset B$$

corresponding to the truth-functional compound

IF A is true, **THEN** B is true.

The *if-then* connective is important in truth-functional calculus, but is not used in logical design, for very important reasons. In Fig. 1.4b we note that *if-then* differs from *if-and-only-if* only in the second row. A is **F** and B is **T** is not a contradiction to

IF A is true, **THEN** B is true,

since this statement says nothing about B when A is false. By contrast,

B is true **IF-AND-ONLY-IF** A is true

specifies that B must be false if A is false.

This distinction must be carefully noted when translating possibly imprecise natural language to truth function form. Consider the statement, "The car interior light will be on if the driver's door is open." If it is intended that the light will be turned on only by opening the driver's door, then the statement should be written, "The light will be on if and only if the driver's door is open." If the light is to be on by opening either the driver's or passenger's door, the first statement is true, but is only a partial description of the desired circuit. The complete specification should be stated, "The light will be on if and only if the driver's door is open or the passenger's door is open."

1.4   EVALUATION OF TRUTH FUNCTIONS

Before returning to consideration of the elevator, let us consider compounds of two components, related by more then one connective. Suppose, for example, we are interested in the truth value of

$$A \vee (A \wedge \bar{B}) \tag{1.4}$$

for the case in which both A and B are false. We replace the statements A and B in the expression by the truth value **F**, yielding

A	B	$\bar{B}$	$A \wedge \bar{B}$	$A \vee (A \wedge \bar{B})$	A
F	F	T	F	F	F
F	T	F	F	F	F
T	F	T	T	T	T
T	T	F	F	T	T

Figure 1.5

$$\mathbf{F} \vee (\mathbf{F} \wedge \bar{\mathbf{F}}) \equiv \mathbf{F} \vee (\mathbf{F} \wedge \mathbf{T}) \equiv \mathbf{F} \vee \mathbf{F} \equiv \mathbf{F} \tag{1.5}$$

Successive steps of the evaluation are connected by $\equiv$, indicating that the truth value is the same for the expressions on either side of the symbol. This expression is evaluated for the three remaining combinations of values of A and B in Fig. 1.5.

We observe that the resulting truth value in each row is the same as A. We may therefore write

$$A \vee (A \wedge \bar{B}) = A \tag{1.6}$$

Note also that the placement of parentheses makes a difference. Suppose the parenthesis had been misplaced, as given by Equation 1.7:

$$(A \vee A) \wedge \bar{B} \tag{1.7}$$

We will leave it to the reader to verify that

$$(A \vee A) \wedge \bar{B} = A \wedge \bar{B} \tag{1.8}$$

which is not the same as A.

Actually, the parenthesis could have been omitted altogether in this case. The normal precedence convention is that, in the absence of parentheses, all AND operations are executed before OR operations. We will observe in Chapter 3 that this convention results in a significant reduction in the length of expressions.

1.5 MANY-STATEMENT COMPOUNDS

Consider a statement Z, which can be symbolized by

$$Z \equiv (H \wedge \bar{R}) \supset D \tag{1.9}$$

A truth-table analysis of this sort of function is distinguished from those in the previous section only in that the truth tables have more than four rows. In this case, an eight-row truth table is required to provide for all possible combinations of truth or falsity of H, R, and D. This table, with a derivation of truth values for expression 1.9, appears in Fig. 1.6. By reference to Fig. 1.3, the reader can verify that

$$A \supset B \equiv \bar{A} \vee B \tag{1.10}$$

It then follows immediately that

$$Z \equiv (H \wedge \bar{R}) \supset D \equiv \overline{(H \wedge \bar{R})} \vee D \tag{1.11}$$

The validity of expression 1.11 is also demonstrated in Fig. 1.6. Note that the right side of 1.11 employs only $\wedge$, $\vee$, and NOT. We will see that any truth-functional compound, of any number of statements, may be expressed in terms of these three binary connectives. A proof of this assertion must await methods to be developed in Chapter 5.

H	R	D	$\bar{R}$	$H \wedge \bar{R}$	$(H \wedge \bar{R}) \supset D$	$\overline{H \wedge \bar{R}}$	$(\overline{H \wedge \bar{R}}) \vee D$
F	F	F	T	F	T	T	T
F	F	T	T	F	T	T	T
F	T	F	F	F	T	T	T
F	T	T	F	F	T	T	T
T	F	F	T	T	F	F	F
T	F	T	T	T	T	F	T
T	T	F	F	F	T	T	T
T	T	T	F	F	T	T	T

Figure 1.6

With this background, we can return to the elevator control problem. The truth-functional compound (Equation 1.1), expressing control of the elevator door, is repeated as Equation 1.12:

$$D \equiv (\overline{TM} \wedge E) \wedge (C \vee FL) \qquad (1.12)$$

The truth table for this expression is shown in Fig. 1.7.

1.6   PHYSICAL REALIZATIONS

At this point, a brief consideration of the physical realizations of binary connectives may help clarify the relationships between truth-functional calculus and logic circuits. Although any of the connectives of Fig. 1.3 might theoretically be realized physically, in practice only AND, OR, NAND, NOR, exclusive-OR, and negation (NOT) are implemented in logic circuits. Circuits realizing the first five connectives are known as *gates* and circuits realizing NOT as *inverters*. Logic gates can be realized in a great variety of physical forms. For example, hydraulic gates are occasionally encountered and optical logic is envisioned in the future. However, in the vast majority of cases, electronic gates are used, with two discrete voltage levels representing the TRUE and FALSE values. Such devices have one output and an arbitrary number of inputs. The voltage with respect to ground at

TM	E	C	FL	$\overline{TM}$	$X = \overline{TM} \wedge E$	$Y = C \vee FL$	$D = X \wedge Y$
F	F	F	F	T	F	F	F
F	F	F	T	T	F	T	F
F	F	T	F	T	F	T	F
F	F	T	T	T	F	T	F
F	T	F	F	T	T	F	F
F	T	F	T	T	T	T	T
F	T	T	F	T	T	T	T
F	T	T	T	T	T	T	T
T	F	F	F	F	F	F	F
T	F	F	T	F	F	T	F
T	F	T	F	F	F	T	F
T	F	T	T	F	F	T	F
T	T	F	F	F	F	F	F
T	T	F	T	F	F	T	F
T	T	T	F	F	F	T	F
T	T	T	T	F	F	T	F

Figure 1.7   Truth table for elevator door function.

any of these inputs or outputs may take on one of only two values. One of these values will represent TRUE throughout the system, and the other will represent FALSE. All that is necessary is that there be two voltages sufficiently distinct that the possibility of confusing one with the other is minimal. Whatever these levels, there is no guarantee of freedom from error. Just as the sound of jungle drums may be obscured by wind in the trees, any difference in voltages may be obscured by noise resulting from electronic disturbances.

In an AND gate, the output will be at the voltage representing TRUE if all the inputs are at that voltage or at the voltage representing FALSE if any of the inputs are at that voltage. For an OR gate, the output will be TRUE if any of the inputs are TRUE and FALSE only if all inputs are FALSE. For other types of gates, the input and output voltages must correspond in the same manner to the truth values given in Fig. 1.3 for the corresponding connective. An inverter has one input and one output, with the output being FALSE when the input is TRUE and vice versa. A discussion of the circuit aspects of these logic elements will be deferred until Chapter 4.

The most common pictorial representation of logic circuits is the *block diagram*, in which the logic elements are represented by standard symbols. The standard logic symbols specified by IEEE Standard No. 91 (ANSI Y 32.14, 1973) are shown in Figs. 1.8a and 1.8b. Although the symbols are shown with two inputs for the gates, AND, OR, NAND, and NOR gates may have any number of inputs. Exclusive-OR, however, is defined only for two inputs.

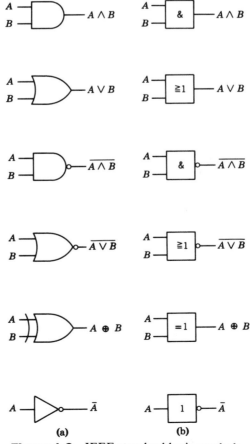

(a)   (b)

Figure 1.8 IEEE standard logic symbols.

Figure 1.9   Example of dot negation.

The reader will note that there are two types of symbols: *uniform-shape* and *distinctive-shape*. The uniform-shape symbols are those established by the International Electrotechnical Commission (IEC Publication 117-15) and are widely used in Europe. The IEEE has included these symbols in its standard, but the distinctive-shape symbols remain the standard of preference in the United States and have found wide acceptance in other parts of the world. The distinctive-shape symbols will be used in this volume.

There are two ways of indicating inversion or negation. One is the inverter symbol, as shown in Fig. 1.8. The second is a small circle on the signal line where it enters or leaves a logic symbol, sometimes referred to as *dot negation*. This system is used for NAND and NOR symbols, as shown in Fig. 1.8, and Fig. 1.9 gives an example of a more complex use of this notation. Generally, the dot negation should be used only if the inversion is actually accomplished within the circuitry of the associated gate. If the inverter is a physically distinct element, it is preferable to use the separate symbol.

We may now use the distinctive-shape logic elements of Fig. 1.8 to construct a circuit realization that will activate the closing of an elevator door as characterized in Example 1.1. We do this by substituting the appropriate gate for each connective in Equation 1.12. The ordering of operations is reflected by the interconnecting wires. The resulting logic circuit is given in Fig. 1.10.

The reader should be able to verify that the output of the circuit of Fig. 1.10 will correspond to TRUE for those combinations of inputs indicated in the truth table of Fig. 1.7. As will be illustrated presently, it is also possible to determine a logical expression for a truth-functional compound directly from its truth table. With this in mind, we may phrase the design procedure for decision circuits as follows.

Step 1   Obtain a statement of the design problem that can be symbolized as a truth-functional compound or translated directly to a truth table.

Step 2   Obtain an expression for the output of the problem in terms of the connectives AND, OR, and NOT.

Step 3*   Obtain the logical expression, equivalent to that obtained in Step 2, that will result in the most economical physical realization.

Step 4   Construct the physical realization corresponding to Step 3. This procedure will be illustrated in the following example.

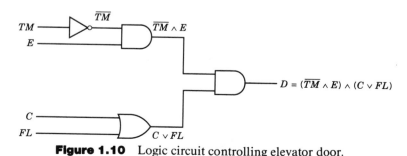

Figure 1.10   Logic circuit controlling elevator door.

---

*This step is the topic of Chapters 5 and 6.

■ **EXAMPLE 1.2**

The flight of a certain satellite may be controlled in any one of three ways: by ground control, manually by the astronaut, or automatically by a computer aboard the satellite. The ground station, which has ultimate authority, will cause a voltage source S aboard the satellite to assume a true level when

"Control is aboard the satellite": S

is true. The astronaut then chooses between manual and automatic control by means of a switch M corresponding to

"Control is manual": M

A check signal C generated within the computer will remain at the true level as long as the computer is functioning properly and is able to assume or remain in control.

There are panel lights at both the ground station and aboard the satellite that should go out, sounding an alarm, if the computer fails to function properly at a time when it should be in control. Letting the variable L correspond to the lights being lit, the above conditions may be represented by

$$\bar{L} = (S \wedge \bar{M}) \wedge \bar{C} \tag{1.13}$$

Alternatively, we may state that the lights should be lit if the computer is functioning or if it is not supposed to be in control, that is,

$$L = (\overline{S \wedge \bar{M}}) \vee C \tag{1.14}$$

A physical implementation of the right side of Equation 1.14 is shown in Fig. 1.11. The reader should verify that the lamp will light for the proper combinations of input truth values. ■

1.7 FUNCTIONALLY COMPLETE SETS OF CONNECTIVES

We saw in the previous section that certain of the binary connectives in Fig. 1.3 could be expressed in terms of various other connectives. The question arises: Can a smaller set of connectives be found in terms of which all 16 functions can be expressed? Let us search for such a smaller set.

For example, consider the set of connectives containing only $\wedge$ and $\vee$. The case in which A and B are both false is critical. Clearly, $A \wedge B$ and $A \vee B$ must both be false. By way of induction, visualize a set of statements, arrived at by relating false statements A and B by $\wedge$ and $\vee$ in a fashion of arbitrary complexity, but such that

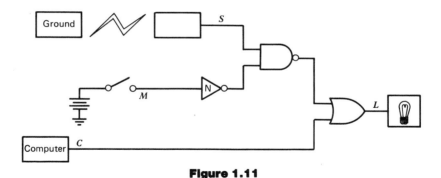

Figure 1.11

Table 1.1

$f_0 \equiv A \wedge \bar{A}$	$f_8 \equiv \overline{A \vee B}$
$f_1 \equiv A \wedge B$	$f_9 \equiv (\bar{A} \wedge \bar{B}) \vee (A \wedge B)$
$f_2 \equiv A \wedge \bar{B}$	$f_{10} \equiv \bar{B}$
$f_3 \equiv A$	$f_{11} \equiv A \vee \bar{B}$
$f_4 \equiv B \wedge \bar{A}$	$f_{12} \equiv \bar{A}$
$f_5 \equiv B$	$f_{13} \equiv \bar{A} \vee B$
$f_6 \equiv (\bar{A} \wedge B) \vee (A \wedge \bar{B})$	$f_{14} \equiv \overline{A \wedge B}$
$f_7 \equiv A \vee \bar{B}$	$f_{15} \equiv A \vee \bar{A}$

they are all false. Now relate any two of these statements, or relate one of them to A or B, utilizing $\wedge$ or $\vee$. By definition, the result must be another false statement. Because the set initially contained only false statements A and B, induction tells us that the set will always include only false statements. In Fig. 1.3, however, note that functions 8 through 15 must be true when A and B are both false. Therefore, these eight functions cannot be realized by any expression written in terms of only the connectives $\wedge$ and $\vee$.

Suppose the NOT operation is added to our set of connectives. The reader who intuitively expects that all functions can be written in terms of these three connectives is quite correct. That each of the functions in Fig. 1.3 can be expressed in terms of $\wedge$, $\vee$, and $^-$ may be verified by determining truth values for the corresponding expression listed in Table 1.1. This task is left to the reader.

A subset of the connectives sufficient for realizing all the connectives and, by extension, any function expressed in terms of any of the connectives, is said to be *functionally complete*. The reader may wonder whether any list of fewer than three connectives might be functionally complete. The answer is yes. Either AND and NOT or OR and NOT is a functionally complete set. As will be apparent in Chapter 4, statements can usually be expressed most conveniently by using all three.

At this point, we pause briefly in our discussion of functionally complete sets of connectives to verify, for truth-functional calculus, two very important theorems called *DeMorgan's theorems*. We will prove these theorems again, algebraically, in Chapter 3. Because they facilitate our discussion, we introduce them here.

THEOREM 1.1

1. $A \downarrow B \equiv \overline{A \vee B} \equiv \bar{A} \wedge \bar{B}$
2. $A \uparrow B \equiv \overline{A \wedge B} \equiv \bar{A} \vee \bar{B}$

PROOF   As elsewhere in the chapter, the proof consists of a truth-table analysis as given in Fig. 1.12.

A	B	$A \vee B$	$\overline{A \vee B}$	$\bar{A}$	$\bar{B}$	$\bar{A} \wedge \bar{B}$	$A \wedge B$	$\overline{A \wedge B}$	$\bar{A} \vee \bar{B}$
F	F	F	T	T	T	T	F	T	T
F	T	T	F	T	F	F	F	T	T
T	F	T	F	F	T	F	F	T	T
T	T	T	F	F	F	F	T	F	F

Figure 1.12   Truth-functional proof of DeMorgan's theorems.

Using DeMorgan's theorems, we see that $\wedge$, $\vee$, and NOT can be expressed in terms of $\uparrow$ and $\downarrow$. Clearly, $X \wedge X \equiv X$, so we have an expression of $\bar{X}$ in Equation 1.15. Using the obvious

$$\bar{X} \equiv \overline{X \wedge X} \equiv X \uparrow X \tag{1.15}$$

theorem $\bar{\bar{A}} \equiv A$ and DeMorgan's theorems, we have expressions for $X \wedge Y$ and $X \vee Y$ in Equations 1.16 and 1.17.

$$X \wedge Y \equiv \overline{X \uparrow Y} \equiv (X \uparrow Y) \uparrow (X \uparrow Y) \tag{1.16}$$

$$X \vee Y \equiv \overline{\overline{X \vee Y}} \equiv \overline{\bar{X} \wedge \bar{Y}} \equiv \bar{X} \uparrow \bar{Y} \equiv (X \uparrow X) \uparrow (Y \uparrow Y) \tag{1.17}$$

We similarly obtain expressions 1.18, 1.19, and 1.20 in terms of $\downarrow$:

$$\bar{X} \equiv \overline{X \vee X} \equiv X \downarrow X \tag{1.18}$$

$$X \wedge Y \equiv \overline{\overline{X \wedge Y}} \equiv \overline{\bar{X} \vee \bar{Y}} \equiv \bar{X} \downarrow \bar{Y} \equiv (X \downarrow X) \downarrow (Y \downarrow Y) \tag{1.19}$$

$$X \vee Y \equiv \overline{X \downarrow Y} \equiv (X \downarrow Y) \downarrow (X \downarrow Y) \tag{1.20}$$

We now conclude that any truth-functional compound may be expressed in terms of a single connective, which may be either $\downarrow$ or $\uparrow$. An $\wedge$, $\vee$, $^-$ expression may be converted to a $\uparrow$ expression by replacing the former connectives and the statements to which they apply by the appropriate expression out of Equations 1.15, 1.16, and 1.17. For instance, f_{13} in Fig. 1.3 becomes

$$\bar{A} \vee B \equiv (A \uparrow A) \vee B \equiv [(A \uparrow A) \uparrow (A \uparrow A)] \uparrow [B \uparrow B] \tag{1.21}$$

If the "obvious" theorem $\bar{\bar{A}} \equiv A$ (easily verified by truth-table analysis) is applied, a much simpler expression will result. That is,

$$[(A \uparrow A) \uparrow (A \uparrow A)] \uparrow (B \uparrow B) \equiv (\bar{A} \uparrow \bar{A}) \uparrow (B \uparrow B)$$
$$\equiv \bar{\bar{A}} \uparrow (B \uparrow B)$$
$$\equiv A \uparrow (B \uparrow B) \tag{1.22}$$

Certainly, other theorems might be used to simplify complex logical expressions. A systematic exposition of such theorems in the form of Boolean algebra is the topic of Chapter 3.

The observation that all logical expressions can be expressed in terms of either $\downarrow$ or $\uparrow$ previews an important notion that will be developed in Chapter 4. That is, all logic circuits can be constructed using only NOR or NAND gates. More formally, we may say that either NAND or NOR is functionally complete.

PROBLEMS

1.1 Symbolize the truth-functional compound describing the following design problem. Determine the truth values of the compound.

The flow of water into the brine solution to be used in a chemical process will be turned off if and only if (1) the tank is full, or (2) the output is shut off, the salt concentration does not exceed 2.5%, and the water level is not below a designated minimum level.

1.2 Symbolize the following statement and determine its truth values:

If it is hot in Arizona and it is raining outside or demonstrators are in the streets, then it is hot in Arizona, demonstrators are in the streets, and it is snowing in Argentina.

1.3   A burglar alarm system for a bank is to be operative only if a master switch at the police station has been turned on. Subject to this condition, the alarm will ring if the vault door is disturbed in any way, or if the door to the bank is opened unless a special switch is first operated by the security guard's key. The vault door will be equipped with a vibration sensor that will cause a switch to close if the vault door is disturbed, and a switch will be mounted on the bank door in such a way that it will close whenever the bank door is opened.

   Symbolize the above system as a truth-functional compound and construct the corresponding logic block diagram.

1.4   In many cars, the seat-belt alarm buzzer is also used to warn against leaving the key in the ignition or leaving the lights on when the car is unoccupied. The following statement describes how such a system might operate:

   The alarm is to sound if the key is in the ignition when the door is open and the motor is not running, or if the lights are on when the key is not in the ignition, or if the driver seat belt is not fastened when the motor is running, or if the passenger seat is occupied and the passenger seat belt is not fastened when the motor is running.

   Symbolize the above system as a truth-functional compound and construct the corresponding logic block diagram. You may assume that switch contacts are available to indicate the occurrence of the various individual conditions.

1.5   Using a truth table, verify that

$$(A \supset B) \supset [\overline{A \vee (\overline{A \wedge B})}] \equiv A \wedge \bar{B}$$

1.6   Express $X \oplus Y$ in terms of only the connective $\downarrow$.

1.7   Prove that no connective in Fig. 1.3 other than $\downarrow$ or $\uparrow$ will serve as the only binary connective in expressions for every function in that figure.

1.8   Tabulate the truth values for Example 1.2.

1.9   Suppose three one-digit binary numbers, a, b, and c, are to be added together to form a two-digit number whose digits are denoted by $s_1 s_2$. For each of the above five binary digits, define the corresponding capital letter to be a statement variable that is TRUE whenever the small letter digit is 1 and FALSE otherwise. Determine a truth-functional expression for S_2 such that S_2 will be true whenever $s_2 = 1$. Determine a similar expression for S_1. Construct the diagram of logic circuits realizing S_2 and S_1.

1.10   An argument consists of several compound statements regarded as *premises* with one additional compound statement called the conclusion, which is asserted to follow logically from the premises. We say that the argument is valid if the conclusion is true whenever the conjunction of the premises is true. Determine by using truth tables the validity of each of the following arguments:

   (a)   If I work, I earn money, and if I don't work, I enjoy myself. Therefore, if I don't earn money, I enjoy myself.

   (b)   The governor of California hails from either Los Angeles or San Francisco. Jones doesn't come from San Francisco. Therefore, if Jones is not from Los Angeles, Jones is not the governor of California.

(c) If Harvard wins the Ivy League football championship, then Princeton will be second, and if Princeton is second, then Dartmouth will place in second division. Harvard will win or Dartmouth will not finish among the top four teams. Therefore, Princeton will not take second place.

BIBLIOGRAPHY

1. Hill, F. J. and G. R. Peterson. *Digital Systems: Hardware Organization and Design*, 3rd ed. Wiley, New York, 1987.
2. Copi, I. M. *Symbolic Logic*. Macmillan, New York, 1954.
3. Quine, W. V. *Mathematical Logic*. Harvard University Press, Cambridge, 1955.

CHAPTER 2

NUMBER SYSTEMS AND CODES

2.1 INTRODUCTION

Familiarity with the binary number system is necessary in using this book for two reasons. First, binary numbers appear frequently as a notational device. Digital computers make use of binary numbers, and many of the best examples of combinational and sequential circuits are drawn from subsystems within computers. Readers who already possess a working knowledge of the binary number system may wish to skip to Section 2.5 to consider coding or proceed directly to Chapter 3.

To begin, let us review what is involved in the definition of a number system. We write, for example,

$$7419$$

as a shortened notation for

$$7 \cdot 10^3 + 4 \cdot 10^2 + 1 \cdot 10^1 + 9 \cdot 10^0$$

Because the decimal number system is almost universal, the powers of 10 are implied without further notation. We say that 10 is the *radix* or *base* of the decimal number system. The fact that 10 was chosen by man as the base of the number system is commonly attributed to the 10 fingers on two hands. Most any other relatively small number (a large number of symbols would be awkward) would have done as well.

Consider the number

$$N = 110101$$

and assume that the base of N is 2. The number system with base 2 is called the *binary* number system. We may rewrite N as follows:

$$N = 1 \cdot 2^5 + 1 \cdot 2^4 + 0 \cdot 2^3 + 1 \cdot 2^2 + 0 \cdot 2^1 + 1 \cdot 2^0$$
$$= 32 + 16 + 0 + 4 + 0 + 1 = 53 \text{ (decimal)}$$

thus obtaining the decimal equivalent. Note that the binary number system requires only two symbols, 0 and 1, whereas the familiar 10 symbols are employed in the decimal system.

In contrast to men, digital computers do not have 10 fingers and are not biased toward the decimal system. They are constructed of physical devices that operate much more reliably when required to switch between only two stable operating points. Some examples are switches, relays, data transmission lines, transistors, etc. One of the stable operating points may be assigned to represent 1 and the other 0. Thus, base 2 must play some role within any digital system.

Many examples throughout the book will be problems in digital computer design. This, however, is not the only reason for introducing binary numbers at this point. We will find the binary number system indispensable as a notational device in the next few chapters.

The reader who is already familiar with arithmetic in other bases may skip rapidly over the next two sections.

2.2 CONVERSION BETWEEN BASES

Suppose that some number N is expressed in base s. It may be converted to base r by the following sequence of divisions carried out in base s. The digits N_i are the quotients and A_i the remainders from each division, so that $A_i < r$.

Remainders

$r \lfloor N$
$\quad r \lfloor N_1 \qquad\qquad A_0$
$\qquad r \lfloor N_2 \qquad\qquad A_1$

.

.

.

$\qquad r \lfloor N_n \qquad\qquad A_{n-1}$
$\qquad\quad \lfloor 0 \qquad\qquad A_n$

Alternatively, this division may be written as follows:

$$N = r \cdot N_1 + A_0$$
$$N_1 = r \cdot N_2 + A_1$$

.

.

.

$$N_n = r \cdot 0 + A_n \tag{2.1}$$

or as

$$N = r(rN_2 + A_1) + A_0$$
$$= r^2 N_2 + rA_1 + A_0$$
$$= r^2(rN_3 + A_2) + rA_1 + A_0$$

.

.

.

$$= A_n r^n + A_{n-1} r^{n-1} + \cdots + A_1 r + A_0 \tag{2.2}$$

■ EXAMPLE 2.1

Convert 653_{10} (the subscript indicates base 10) to (a) base 2 and (b) base 5.

(a)

```
2 | 653
2 | 326    1
2 | 163    0
2 |  81    1
2 |  40    1
2 |  20    0
2 |  10    0
2 |   5    0
2 |   2    1
2 |   1    0
        0    1    ← MSB (most significant bit)
```

Check
$1010001101_2 = 512 + 128 + 8 + 4 + 1$
$\qquad\qquad = 653_{10}$

(b)

$$
\begin{array}{r|l}
5 & 653 \\
5 & 130 \quad 3 \\
5 & 26 \quad 0 \\
5 & 5 \quad 1 \\
5 & 1 \quad 0 \\
& 0 \quad 1
\end{array}
$$

$$10103_5 = 1 \times 5^4 + 1 \times 5^2 + 3 \times 5^0$$
$$= 1 \times 625 + 1 \times 25 + 3$$
$$= 653_{10}$$

∎

■ EXAMPLE 2.2

Convert the number 1606_{10} to base 12: The decimal numbers 10_{10} and 11_{10} become one-digit numbers in base 12, which may be represented by α and β, respectively.

$$
\begin{array}{r|l}
12 & 1606 \\
12 & 133 \quad 10 = \alpha \\
12 & 11 \quad 1 \\
& 0 \quad 11 = \beta
\end{array}
$$

Check
$$\beta 1\alpha_{12} = 11(144) + 1(12) + 10$$
$$= 1606_{10}$$

∎

In each of the above examples, arithmetic was carried out in base 10. This is convenient because we are familiar with base 10 arithmetic. If a number is to be converted from base r ($r \neq 10$) to some base b, we have the choice of doing the arithmetic in base r or else first converting to base 10 (as shown in Section 2.1) and then to base b. Arithmetic in bases other than 10 is the subject of the next section.

The numbers we have discussed so far have been integers. Suppose it is desired to convert numbers with fractional parts from one number system to another. Consider a number

$$N = N_I + N_F = A_n r^n + \cdots + A_1 r + A_0 + A_{-1} r^{-1} + A_{-2} r^{-2} \cdots \tag{2.3}$$

where N_I and N_F are the integral and fractional parts, respectively. The fractional part in one base will always lead to the fractional part in any other base. Therefore, N_I may be converted as before and N_F converted separately as follows:

$$N_F = A_{-1} r^{-1} + A_{-2} r^{-2} + A_{-3} r^{-3} \cdots \tag{2.4}$$

To determine the coefficients A_{-1}, A_{-2}, A_{-3}, etc., for the base r, we note that each of these coefficients is itself an integer. We first multiply by r.

$$rN_F = A_{-1} + A_{-2} r^{-1} + A_{-3} r^{-2} \tag{2.5}$$

Thus, the integral part of rN_F is A_{-1}. We subtract A_{-1} and again multiply by r

$$r(rN_F - A_{-1}) = A_{-2} + A_{-3} r^{-1} + A_{-4} r^{-2} \tag{2.6}$$

thus determining A_{-2}. This process is continued until as many coefficients as desired are obtained. The process may not terminate.

Note that we have avoided the terms *decimal point* or *decimal places*. For base 2, for example, the appropriate terms are *binary point* and *binary places*.

■ EXAMPLE 2.3

Convert the number 653.61_{10} to base 2.

$$
\begin{array}{ll}
2 \cdot (0.61) = 1.22 & A_{-1} = 1 \\
2 \cdot (0.22) = 0.44 & A_{-2} = 0 \\
2 \cdot (0.44) = 0.88 & A_{-3} = 0
\end{array}
$$

Check
$$653.61_{10} = 1010001101.1001110 \cdots$$
$$= 653 + \tfrac{1}{2} + \tfrac{1}{16} + \tfrac{1}{32} + \tfrac{1}{64} \cdots$$

$$2 \cdot (0.88) = 1.76 \quad A_{-4} = 1$$
$$2 \cdot (0.76) = 1.52 \quad A_{-5} = 1$$
$$2 \cdot (0.52) = 1.04 \quad A_{-6} = 1$$
$$2 \cdot (0.04) = 0.08 \quad A_{-7} = 0$$

$$= 653 + 0.5 + 0.0625 + 0.03125$$
$$+ 0.015625$$
$$= 653.609375$$

$\vdots$

∎

A disadvantage of the binary number system is the fact that so many bits are needed for numbers of even modest magnitude. For example,

$$35465$$

a number easy enough to interpret in decimal is equivalent to

$$111010101011001$$

Not only is it virtually impossible to get any sense of the magnitude of this string of 0's and 1's, you are very likely to make a mistake if you try to copy it. Because of such problems, it is common to convert binary numbers to hexadecimal (base 16) form. The conversion is simple because 16 is a power of 2. The process used in converting the binary number to base 16 may be informally derived as follows. Let N be a number expressed in 12-bit binary form. Therefore,

$$N = d_{11}2^{11} + d_{10}2^{10} + d_9 2^9 + d_8 2^8 + d_7 2^7 + d_6 2^6 + d_5 2^5 + d_4 2^4 + d_3 2^3$$
$$+ d_2 2^2 + d_1 2^1 + d_0 2^0 \tag{2.7}$$
$$= (2^4)^2(d_{11}2^3 + d_{10}2^2 + d_9 2^1 + d_8) + 2^4(d_7 2^3 + d_6 2^2 + d_5 2^1 + d_4)$$
$$+ 2^0(d_3 2^3 + d_2 2^2 + d_1 2^1 + d_0)$$
$$= (d_{11}2^3 + d_{10}2^2 + d_9 2^1 + d_8) \cdot 16^2 + (d_7 2^3 + d_6 2^2 + d_5 2^1 + d_4) \cdot 16^1$$
$$+ (d_3 2^3 + d_2 2^2 + d_1 2^1 + d_0) \tag{2.8}$$

Each of the expressions in parentheses represents a four-bit binary number in the range 0–15 (decimal). We noted in the last section that a number system requires a number of symbols equal to the radix. Thus, we will need 16 symbols to represent the terms in parentheses. Standard practice for hexadecimal is to use the 10 decimal symbols, plus the first six letters of the alphabet, as shown in Fig. 2.1.

Based on Fig. 2.1, we see that all we need to do to convert a binary number to hexadecimal is to divide the number into groups of four bits, starting at the least significant bit, and convert each group of four bits to the equivalent hexadecimal digit given in Fig. 2.1.

Binary	Hex	Decimal	Binary	Hex	Decimal
0000	0	0	1000	8	8
0001	1	1	1001	9	9
0010	2	2	1010	A	10
0011	3	3	1011	B	11
0100	4	4	1100	C	12
0101	5	5	1101	D	13
0110	6	6	1110	E	14
0111	7	7	1111	F	15

Figure 2.1 Hexadecimal equivalents of binary and decimal numbers.

■ **EXAMPLE 2.4**

Convert 10101010110011101 to hex.

SOLUTION

0001	0101	0101	1001	1101
1	5	5	9	D

1559D

2.3   ARITHMETIC WITH BASES OTHER THAN 10

Our ability to perform arithmetic in base 10 depends on a set of basic addition and multiplication relations that are committed to memory. These operations become so familiar that we tend to forget the significance of these tables, particularly the addition table. Doing arithmetic in some other base similarly requires a certain familiarization with the counterparts of these tables for that base. We see in Fig. 2.2 the addition and multiplication tables for base 2 and base 5.

All further rules of arithmetic, such as the carry and borrow operations, apply in precise analogy for bases other than 10. We are now ready to consider nondecimal arithmetic.

■ **EXAMPLE 2.5**

Add 321 to 314 and multiply 142×32 in base 5.

```
      1   1   ← Carry
      3 1 4                    1 4 2
    +3 2 1                   ×  3 2
    -------                  -------
    1 1 4 0                        1   ← Carry
                                2 3 4
                            1 2 1       ← Carry
                              3 2 1
                            ---------
                            1 1 1 4 4
```
■

■ **EXAMPLE 2.6**

Using base-5 arithmetic, convert 431_5 to base 2

```
2 | 431
2 | 213   0
2 | 104   0
2 |  24   1
2 |  12   0
2 |   3   1
2 |   1   1
      0   1

    N₂ = 1110100
```

Check:
$$N = 1 \cdot 2^{11} + 1 \cdot 2^{10} + 1 \cdot 2^4 + 0 \cdot 2^3$$
$$+ 1 \cdot 2^2 + 0 \cdot 2 + 0 \cdot 1$$
$$= (2^3)^2 + 2^2 \cdot 2^3 + 2^2 \cdot 2^2 + 2^2$$
$$= (13)^2 + 4(13) + 4 \cdot 4 + 4$$
$$= 224 + 112 + 31 + 4 = 431$$

■

(a) Base 2

+	0	1
0	0	1
1	1	10

×	0	1
0	0	0
1	0	1

(b) Base 5

+	0	1	2	3	4
0	0	1	2	3	4
1	1	2	3	4	10
2	2	3	4	10	11
3	3	4	10	11	12
4	4	10	11	12	13

×	0	1	2	3	4
0	0	0	0	0	0
1	0	1	2	3	4
2	0	2	4	11	13
3	0	3	11	14	22
4	0	4	13	22	31

(*Note:* Where two digits are included in a square, the most significant may be regarded as a carry.)

Figure 2.2 Addition and multiplication tables. (*a*) Base 2. (*b*) Base 5.

It is not difficult to acquire a facility for base-2 arithmetic because of the simplicity of the base-2 addition and multiplication tables. We will see in succeeding chapters that, for this same reason, physical implementation of base-2 arithmetic is economical.

In the decimal number system, the multiplication of a number by a power of 10 requires only a shifting operation. The same thing is true for powers of 2 in base 2. Consider, for example,

$$25_{10} \times 2^4_{10} = (11001 \times 10000)_2 = 110010000$$

2.4 NEGATIVE NUMBERS

Negative numbers in base 2 may be treated in much the same way as in base 10. For example, in adding a positive number to a negative number, the number smaller in magnitude is subtracted from the larger, and the sign of the larger is assigned to the result. Thus, the mind must carry on a sequence of logical manipulations independent of the actual arithmetic. We shall see that these logical operations may be implemented in the circuitry of an electronic digital computer.

If a slightly different approach is used, the computer need not distinguish between addition and subtraction. That is, the circuitry employed in the operation of addition may also be used for subtraction.

Usually, negative numbers are stored in a computer in *two's complement form*. That is, the number $-M$ is represented by TWOSCOMPLEMENT(M). If numbers are stored in n-bit binary form,

$$\text{TWOSCOMPLEMENT}(M) = 2^n - M, \quad \text{where } M < 2^{n-1} \tag{2.9}$$

A slight modification of Equation 2.9 results in a convenient mechanism for computing the two's complement of a number.

$$\text{TWOSCOMPLEMENT}(M) = [(2^n - 1) - M] + 1 \tag{2.10}$$

We recognize that $2^n - 1$ is merely a string of n 1's. Thus, the subtraction $(2^n - 1) - M$ may be accomplished by merely complementing every bit of the binary M. The computation of TWOSCOMPLEMENT(M) using Equation 2.10 is accomplished in two steps. First, $(2^n - 1) - M$ is computed as suggested and then 1 is added as a separate step.

■ EXAMPLE 2.7

Determine the two's complement representation of -25 to be stored in eight-bit binary form.

SOLUTION   We first convert 25 to binary 00011001. To illustrate the process, we express the subtraction explicitly.

$$
\begin{array}{r}
2^8 - 1 = \quad 11111111 \\
- \ 00011001 \\
\hline
11100110 \\
+ \ 1 \\
\hline
11100111 = \text{TWOSCOMPLEMENT (25)}
\end{array}
$$

It was not really necessary to write out the subtraction, as 11100110 was determined by merely complementing each bit of 00011001.   ■

We note that the most significant bit of the two's complement number 11100111 is 1. This bit, usually called the sign bit, must always be 1 to distinguish the two's complement from a positive number. Thus, when negative numbers are stored in a computer memory in two's complement form, the magnitude of all numbers, positive and negative, must be $\leq 2^{n-1} - 1$ (-2^{n-1} is often allowed). If at any point in an arithmetic computation, this assertion is violated, *overflow* is said to have occurred.

2.5   CODING OF DECIMAL NUMBERS

Although virtually all digital systems are binary in the sense that all signals within the systems may take on only two values, some nevertheless perform arithmetic in the decimal system. In some circumstances, the identity of decimal numbers is retained to the extent that each decimal digit is individually represented by a binary code. There are 10 decimal digits, so four binary bits are required in each code element. The most obvious choice is to let each decimal digit be represented by the corresponding four-bit binary number, as shown in Fig. 2.3. This form of representation is known as the BCD (binary-coded decimal) representation or code.

In computer systems, most input or output is in the decimal system because this is the system most convenient for human users. On input, the decimal numbers are converted into some binary form for processing, and this process is reversed on output. In a "straight binary" computer, a decimal number, such as 75, would be converted into the binary equivalent, 1001011. In a computer using the BCD system, 75 would be converted into 0111 0101. If the BCD form is used, it must be preserved in arithmetic processing. For example, addition in a BCD machine would be carried out as shown below.

$$
\begin{array}{rr}
37 & 0011\ 0111 \\
+24 & +0010\ 0100 \\
\hline
61 & 0110\ 0001 \\
\text{(Decimal)} & \text{(BCD)}
\end{array}
$$

Decimal Digit	Binary Representation $X_3\ X_2\ X_1\ X_0$
0	0000
1	0001
2	0010
3	0011
4	0100
5	0101
6	0110
7	0111
8	1000
9	1001

Figure 2.3 Binary-coded decimal digits.

The principal advantage of the BCD system is the simplicity of input/output conversion; the principal disadvantage is the complexity of arithmetic processing. The choice (between binary and BCD) depends on the type of problems the system will be handling. Computers are built both ways, and the proper choice is a subject of some debate.

The BCD code is not the only possible coding for decimal digits. Any set of at least 10 distinct combinations of at least four binary bits could theoretically be used. Indeed, some writers use the term *BCD* to refer to any set of 10 binary codes used to represent the decimal digits, but we will use BCD only to refer to the code of Fig. 2.3.

Figure 2.4 includes a listing of three other possible codes. The excess-3 code is particularly useful when it is desired to perform arithmetic by the method of complements. The nine's complement of a decimal digit a is defined as $9-a$. The nine's complement of a decimal digit expressed in excess-3 code may be obtained by complementing each bit individually. This fact is easily verified by the reader.

Not all codes of decimal digits use only four bits. The 2-out-of-5 code represents each decimal digit by one of the 10 possible combinations of two 1's and three 0's. The third code in Fig. 2.4 utilizes the center 10 characters of a four-bit Gray code. The distinguishing feature of the Gray code is that successive coded characters never differ in more than one bit. The basic Gray code configuration is shown in Fig. 2.5a. A three-bit Gray code may be obtained by merely reflecting

Decimal Digit	Excess-3 $X_3\ X_2\ X_1\ X_0$				2-out-of-5 $X_4\ X_3\ X_2\ X_1\ X_0$					Gray Code $X_3\ X_2\ X_1\ X_0$			
0	0	0	1	1	0	0	0	1	1	0	0	1	0
1	0	1	0	0	0	0	1	0	1	0	1	1	0
2	0	1	0	1	0	0	1	1	0	0	1	1	1
3	0	1	1	0	0	1	0	0	1	0	1	0	1
4	0	1	1	1	0	1	0	1	0	0	1	0	0
5	1	0	0	0	0	1	1	0	0	1	1	0	0
6	1	0	0	1	1	0	0	0	1	1	1	0	1
7	1	0	1	0	1	0	0	1	0	1	1	1	1
8	1	0	1	1	1	0	1	0	0	1	1	1	0
9	1	1	0	0	1	1	0	0	0	1	0	1	0

Figure 2.4

```
0 0        0 0 0      0 0 0 0
0 1        0 0 1      0 0 0 1
1 1        0 1 1      0 0 1 1
1 0        0 1 0      0 0 1 0
--------------
           1 1 0      0 1 1 0
           1 1 1      0 1 1 1
           1 0 1      0 1 0 1
           1 0 0      0 1 0 0
-----------------
                      1 1 0 0
                      1 1 0 1
                      1 1 1 1
                      1 1 1 0
                      1 0 1 0
                      1 0 1 1
                      1 0 0 1
                      1 0 0 0
```

(a)   (b)   (c)

Figure 2.5

the two-bit code about an axis at the end of the code and assigning a third bit as 0 above the axis and as 1 below the axis. This is illustrated in Fig. 2.5*b*. By reflecting the three-bit code, a four-bit code may be obtained as in Fig. 2.5*c*, etc. The Gray code is particularly useful in minimizing errors in analog-digital conversion.

For a display of decimal digits, the standard form is the seven-segment display, shown in Fig. 2.6*a*. Each of the seven segments is a separate light-emitting diode (LED) or crystal that can be turned on or off individually. It should be apparent to the reader that each decimal digit can be formed by lighting some subset of the seven segments. To control this display, we must generate a seven-bit code to indicate whether each segment should be on or off. If we let 0 correspond to OFF and 1 to ON, the *seven-segment code* for the decimal digits will be as shown in Fig. 2.6*b*. Since the seven-segment code is highly redundant, it would be unsuitable for internal use in a computer. Thus, the use of this form of display will require logic to convert from some internal code, such as BCD, to the seven-segment code.

2.6 PARITY

Avoiding error is of major interest in any data-handling system. Among the major causes of error are component failure and intermittent signal deviation resulting from additive noise. Errors of the first type may be reduced in frequency by using duplicate or redundant circuits. Random errors due to noise occur with greater frequency in some parts of a digital system than in others. Errors are particularly likely when transmission of information from one system or subsystem to another is involved. The longer and noisier the transmission path, the greater the likelihood of error. Noise will generally have less effect in an integral subsystem, such as an arithmetic unit, where information is primarily in the form of levels rather than pulses.

If it were possible to determine that data received from a transmission line or sampled from a memory were erroneous, it might be possible to effect a retransmission or resampling of the data. It is possible to reduce the probability of an error going undetected. The simplest approach is called a *parity check*. Suppose,

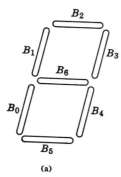

Digit	Seven-Segment Code						
	B_6	B_5	B_4	B_3	B_2	B_1	B_0
0	0	1	1	1	1	1	1
1	0	0	1	1	0	0	0
2	1	1	0	1	1	0	1
3	1	1	1	1	1	0	0
4	1	0	1	1	0	1	0
5	1	1	1	0	1	1	0
6	1	1	1	0	0	1	1
7	0	0	1	1	1	0	0
8	1	1	1	1	1	1	1
9	1	0	1	1	1	1	0

(a)   (b)

Figure 2.6   Seven-segment display and code.

for example, that information is to be stored on a magnetic tape in characters of seven binary bits each. Let us add an eighth bit to each character in such a way that the number of 1 bits in the character will always be even. We say that the character is coded with even parity. After a character has been read from the magnetic tape, a parity check is made to see that the number of 1 bits is still even.

■ EXAMPLE 2.8

Establish even parity by adding a bit to each of the seven-bit characters 1101001 and 0101111. These even-parity characters are to be transmitted through a medium in which an error occurs once in every 10^4 bits (one out of 10^4 bits is received in complemented form). Determine the probability that an erroneous character will be developed that will go undetected by a parity check at the receiver.

SOLUTION   Determination of the parity bits is immediate.

		Data Character								Parity-Bit-Coded Character					
X_6	X_5	X_4	X_3	X_2	X_1	X_0		X_7	X_6	X_5	X_4	X_3	X_2	X_1	X_0
1	1	0	1	0	0	1		0	1	1	0	1	0	0	1
0	1	0	1	1	1	1		1	0	1	0	1	1	1	1

For the above, or any even-parity character, complementing two bits will result in a character that still has even parity, as will the complementation of any even number of bits. The parity checker at the receiver would be unable to distinguish such characters from valid coded characters. There are

$$_8C_2 = \frac{8!}{6! \cdot 2!}$$

combinations of two of the eight bits. For each of these combinations, the probability that those two bits are in error and the remaining six are correct is $10^{-8}(1-10^{-4})^6$. Proceeding similarly for four, six, and eight bits, we arrive at

$$P(\text{even number of bits in error}) = \frac{8!}{6! \cdot 2!} 10^{-8}(1 - 10^{-4})^6$$

$$+ \frac{8!}{4! \cdot 4!} 10^{-16}(1 - 10^{-4})^4 + \frac{8!}{6! \cdot 2!} 10^{-24}(1 - 10^{-4})^2 + 10^{-32} \qquad (2.11)$$

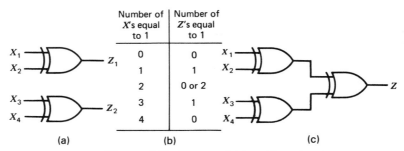

Figure 2.7   Four-variable odd-parity checker.

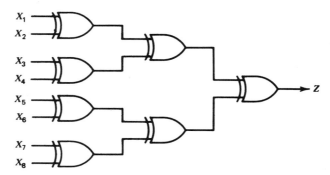

Figure 2.8   Eight-variable odd-parity checker.

as the expression for the probability of an undetectable erroneous character. Neglecting the obviously insignificant terms yields the close approximation.

$$P(\text{even number of bits in error}) = 28(10^{-8}) \qquad (2.12)$$

This is very significantly less than $7 \cdot 10^{-4}$, which is the approximate probability of a single error in a seven-bit character transmitted without a parity bit.   ∎

The circuit for detecting odd parity over two bits is simply exclusive-OR. In Fig. 2.7a we show two exclusive-OR gates with two distinct pairs of inputs. If exactly one of the four variables X_1, X_2, X_3, and X_4 is 1, then exactly one of the outputs, Z_1 or Z_2, must be 1. This must also be true if three of the four X's are 1. If all of the X's are 0 or all are 1, clearly $Z_1 = Z_2 = 0$. If the two inputs to one gate are 1 while the two inputs to the other gate are 0, both Z_1 and Z_2 must be 0. If one input to each gate is 1, then $Z_1 = Z_2 = 1$. This discussion is summarized in the table of Fig. 2.7b.

From the table, the number of outputs equal to 1 is odd if and only if an odd number of inputs are 1. Thus, a check of odd parity may be accomplished by adding a third exclusive-OR gate, as shown in Fig. 2.7c. Extending the argument to eight variables leads to the circuit of Fig. 2.8, which may be realized by using 21 NAND gates and 14 inverters. We could have arrived at this same result by verifying that $X_1 \oplus X_2$ is an associative operation.

2.7   ERROR-DETECTING AND CORRECTING CODES*

Noting the decrease in probability of error resulting from the addition of a single parity bit, we are led naturally to consider the possibility of adding more than

*This material is *not* essential to subsequent chapters.

one redundant bit. This approach will lead to a further reduction in the probability of error and, as we shall see, can even facilitate error correction without retransmission.

Let us suppose that the four coded characters in Fig. 2.9 are the only ones that will be transmitted over some communications system. Note that each of the four coded characters differs from the other coded characters in at least three of the five bits. We say that the *minimum distance* of the code is 3. In general, the minimum distance M of a code is defined as the minimum number of bits in which any two characters of a code differ.

Because there are four distinct two-bit coded characters, we see that $M = 3$ has been achieved at the price of adding three redundant bits. Suppose that character D is transmitted as shown but is received as 11000. Although the last two bits are in error, 11000 will not be confused with any of the coded characters A, B, and C. Indeed, two erroneous bits in any character of the code will not distort it into any other character of the code. It should be apparent to the reader that we may generalize our observation to the statement: Errors in two or fewer bits may be detected in any minimum-distance-3 code.

Errors in three or more bits cannot always be detected in a minimum-distance-3 code. Note, for example, that errors in the first three bits of B in Fig. 2.9 will distort this character into the code for A.

The probability of an error in two bits may be sufficiently small in some applications to be disregarded. In this case, it is possible to employ a minimum-distance-3 code as an error-correcting code. Suppose that the character 11000 is received in a communications system employing the code of Fig. 2.9. If it is assumed that no more than one bit will be in error, we may deduce that the transmitted character was 11100. Thus, circuitry at the receiving end may be provided to convert 11000 directly to 11100. In general, we may say that a received character with a single-bit error will differ in only one bit from the transmitted character but in at least two bits from any other character in the code. Thus, circuitry may be provided to correct all single-bit errors without retransmission. The above is true for any minimum-distance-3 code.

Suppose a two-bit error actually occurs in a character of a minimum-distance-3 code set up to correct single-bit errors. The correction process is still carried out on the erroneous character, but it will be corrected to the wrong coded character. Thus, the two-bit error will in effect go undetected.

We have discussed above the special case of the general relation

$$M - 1 = C + D, \quad C \le D \tag{2.13}$$

where C is the number of erroneous bits that will be corrected and D the number of errors that will be detected. For a specified minimum distance M, various combinations of values of C and D will satisfy Equation 2.13. A partial list is given in Fig. 2.10. The single-parity-bit code and the 2-out-of-5 code are examples of minimum-distance-2 codes for detecting single-bit errors. As discussed above, a minimum-distance-3 code may be used either for detecting two-bit errors or for detecting and correcting single-bit errors. It is necessary to choose between rejecting all erroneous characters and requiring retransmission or correcting single-

$$
\begin{array}{cccccc}
A & 0 & 0 & 0 & 0 & 0 \\
B & 1 & 1 & 1 & 0 & 0 \\
C & 0 & 0 & 1 & 1 & 1 \\
D & 1 & 1 & 0 & 1 & 1 \\
\end{array}
$$

Figure 2.9

M	1	2	3	4	5	6
D	0	1	2 1	3 2	4 3 2	5 4 3
C	0	0	0 1	0 1	0 1 2	0 1 2

Figure 2.10

bit errors on the spot and passing two-bit errors. We cannot have the benefit of both modes of operation.

A minimum-distance-4 code may be used for detecting triple errors. Alternatively, it may be assumed that triple errors are sufficiently improbable to be ignored. A two-bit error will result in a character differing in two bits from the correct and in only two bits from at least one other character in the code. Thus, two-bit errors cannot be corrected. Single-bit errors, however, may be corrected. This correction process will alter characters with three-bit errors to the wrong coded character. Two-bit errors, however, will not be mistaken for single-bit errors. Thus, it is again necessary to choose between the two modes of operation listed in Fig. 2.10.

The solutions of Equation 2.13 for still larger M may be similarly justified.

2.8   HAMMING CODES

One of the most convenient minimum-distance-3 codes was devised by R. W. Hamming [5]. We number the bit positions in sequence from left to right. Those positions numbered as a power of 2 are reserved for parity check bits. The remaining bits are information bits.

The seven-bit code is shown in Fig. 2.11, where P_1, P_2, and P_4 indicate parity bits. The bits X_3, X_5, X_6, and X_7 make up the information character to be transmitted. From these bits, the parity bits are determined as illustrated in the following example.

$$\begin{array}{ccccccc} 1 & 2 & 3 & 4 & 5 & 6 & 7 \\ P_1 & P_2 & X_3 & P_4 & X_5 & X_6 & X_7 \end{array}$$

Figure 2.11

■ EXAMPLE 2.9

Determine the Hamming-coded character corresponding to the information character $X_3X_5X_6X_7 = 1010$.

SOLUTION   Where $a \oplus b = 1$ when $a \neq b$, we may write

$$P_1 = X_3 \oplus X_5 \oplus X_7 \qquad (2.14)$$

The above extension of the exclusive-OR notation is defined to mean that P_1 will be 1 if there are an odd number of 1's among X_3, X_5, and X_7 and that P_1 will be zero otherwise. Thus,

$$P_1 \oplus X_3 \oplus X_5 \oplus X_7 = 0 \qquad (2.15)$$

to establish even parity over bits 1, 3, 5, and 7. In this case,

C_4	(odd parity over 4 5 6 7)	0	0	0	0	1	1	1	1
C_2	(odd parity over 2 3 6 7)	0	0	1	1	0	0	1	1
C_1	(odd parity over 1 3 5 7)	0	1	0	1	0	1	0	1

Erroneous bit   none 1 2 3 4 5 6 7

Figure 2.12

$$P_1 = 1 \oplus (0 \oplus 0) = 1 \oplus 0 = 1 \qquad (2.16)$$

Similarly,

$$P_2 = X_3 \oplus X_6 \oplus X_7 = 1 \oplus (1 \oplus 0) = 0$$

and

$$P_4 = X_5 \oplus X_6 \oplus X_7 = 0 \oplus (1 \oplus 0) = 1 \qquad (2.17)$$

The final coded character is 1011010. ■

The correction process at the receiving end is very convenient in the case of a Hamming code. Since we must assume that only one bit is in error, it is only necessary to locate that bit. This may be accomplished by checking for odd parity over the same three combinations of bits for which even parity was established at the transmitting end as follows:

$$C_1 = P_1 \oplus X_3 \oplus X_5 \oplus X_7$$
$$C_2 = P_2 \oplus X_3 \oplus X_6 \oplus X_7 \qquad (2.18)$$
$$C_4 = P_4 \oplus X_5 \oplus X_6 \oplus X_7$$

If $C_1 = 1$, there must have been an error in one of the four bits, 1, 3, 5, and 7, etc.

The erroneous bit may be determined directly from Fig. 2.12. If all three check bits indicate 0 or even parity, then no single bit is in error. This was the situation upon transmission, and we again remind ourselves that we have assumed there are no two-bit errors.

If $C_1 = 1$ but $C_2 = C_4 = 0$, we may conclude that one of the bits 1, 3, 5, or 7 is in error but that 4, 5, 6, and 7 as well as 2, 3, (6, 7) are all correct. Thus, bit 1 must be in error as indicated in the second column of Fig. 2.12. Should $C_4 = C_2 = C_1 = 1$, we may conclude that bit 7 is in error because this is the only bit influencing all three parity checks. The entries at the bottom of the remaining columns of Fig. 2.12 may be similarly determined.

■ EXAMPLE 2.10

Suppose the character $P_1P_2X_3P_4X_5X_6X_7 = 1101101$ is received in a communications system employing the Hamming code. In this case,

$$C_1 = 1 \oplus 0 \oplus 1 \oplus 1 = 1$$
$$C_2 = 1 \oplus 0 \oplus 0 \oplus 1 = 0$$
$$C_4 = 1 \oplus 1 \oplus 0 \oplus 1 = 1$$

Thus, bit 5 is in error and the correct character is 1101001. ■

If the seven-bit Hamming code is employed for single-bit detection and correction, the probability of an undetected error is the same as the original probability of error in two or more of the seven bits. This is approximately*

$$P_2 = \frac{7!}{5! \, 2!} (P)^2 (1 - P)^5 \tag{2.19}$$

where P is the probability of an error in any one bit.

By rejecting and calling for the retransmission of any impossible characters, the seven-bit Hamming code may be used for double-error detection. In this case, the probability of an undetected error would be the probability of a 3-or-more-bit error, which is approximately

$$P_3 = \frac{7!}{4! \, 3!} P^3 (1 - P)^4 \tag{2.20}$$

If $P = 10^{-4}$, then $P_2 = 20.9(10^{-8})$ and $P_3 = 34.9(10^{-12})$.

Minimum-distance-3 Hamming codes are possible for any $2^n - 1$ bits, where $2^n - n - 1$ of these are information bits. In these cases, all bits numbered as a power of 2 will be parity-check bits. The groups of bits checked for parity continue in the same binary pattern. For example, in the 15-bit code, even parity is achieved over the last eight bits through P_8. The bit P_4 falls in the group consisting of every other series of four bits, beginning at the left, etc. Each of these codes may be extended to minimum-distance 4 by adding one more bit to provide even parity over the entire character.

PROBLEMS

2.1 Convert the following base-10 numbers to base 2 and check.
(a) 13   (b) 94   (c) 356

2.2 Repeat Problem 2.1, converting to base 5.

2.3 Repeat Problem 2.1 for base 16.

2.4 Convert the following base-10 numbers to base 2.
(a) 0.00625   (b) 43.32   (c) 0.51

2.5 Find the hexidecimal equivalent of the following base-2 numbers.
(a) 10111100101   (b) 1101.101   (c) 1.0111

2.6 Express (a) 734_8 and (b) 41.5_8 in base 2.

2.7 Express the base-2 numbers of Problem 2.5 in base 4 without employing division.

2.8 Construct addition and multiplication tables for base 4.

2.9 Perform the indicated arithmetic operations in base 5.

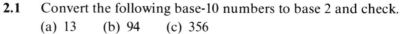

(a) 143_5 (b) 124.22_5 (c) 413_5 (d) 404_5
   $\underline{+221_5}$ $\underline{+\ 21.34_5}$ $\underline{\times\ 32_5}$ $\underline{-213_5}$

2.10 Convert the base-5 number $4{,}433{,}214_5$ directly to base 12.

---

*This expression neglects the probability that more than three bits may be in error, which is considerably smaller still.

2.11 Express the following negative numbers in eight-bit two's complement form.

(a) -103 (b) -99 (c) -131

In which case, if any, did overflow occur?

2.12 Convert the arguments of the following arithmetic operations to 16-bit binary numbers with negative numbers in two's complement form and then perform the indicated operations.

(a) 695_{10} (b) 695_{10} (c) 272_{10} (d) 131_{10}
  $+272_{10}$ -272_{10} $\times\ 23_{10}$ -219_{10}

2.13 Carry out the operations of Problem 2.11 in BCD; that is, convert the operands to BCD and express the results in BCD.

2.14 The method of subtracting by adding complements may also be employed in the decimal number system.

(a) Determine the ten's complement of 373.

(b) Accomplish the subtraction $614 - 373$ by adding the ten's complement.

2.15 Determine a circuit that checks for odd parity over seven bits using exclusive-OR gates.

2.16 Encode the information character 0 1 1 0 1 1 1 0 1 0 1 according to the 15-bit Hamming code.

2.17 Determine which bit, if any, is in error in the Hamming-coded character 1 1 0 0 1 1 1.

2.18 Determine the exact versions of the approximate expressions given by Equations 2.18 and 2.19.

BIBLIOGRAPHY

1. Ware, W. H. *Digital Computer Technology and Design*, Vol. I. Wiley, New York, 1963.
2. Flores, I. *The Logic of Computer Arithmetic*. Prentice-Hall, Englewood Cliffs, N.J., 1962.
3. Kline, R. M. *Digital Computer Design*. Prentice-Hall, Englewood Cliffs, N.J., 1977.
4. Hwang, K. *Computer Arithmetic*. Wiley, New York, 1979.
5. Hamming, R. W. "Error Detecting and Error Correcting Codes," *BST J.*, **29**:2, 147–160 (April 1950).

CHAPTER 3

BOOLEAN ALGEBRA

3.1 POSTULATES

Associated with the physical realization of any logic element is a cost. Similarly, a cost may be assigned to any logic circuit equal to the total cost of all the logic elements in the circuit. When it is necessary to construct a physical realization of any truth function, it is often desirable that this be done at minimum cost. It is possible to verify that a given circuit realizes the proper truth function by checking its output against the truth table for each combination of inputs. Similarly, two functions may be shown equivalent by this method. Truth-functional calculus, however, offers no convenient way for finding the most economical realization of a given function.

For this purpose, we introduce Boolean algebra. Perhaps the most convenient set of postulates for a Boolean algebra is the one set forth by Huntington in 1904 [1]. The first postulate can be thought of as establishing the system under study.

 I. *There exists a set* K *of objects or elements, subject to an equivalence* relation, denoted "$=$," which satisfies the principle of substitution.*

By substitution, it is meant that if $a = b$, a may be substituted for b in any expression involving b without affecting the validity of the expression.

The remainder of Huntington's postulates are:

 IIa. *A rule of combination "$+$" is defined such that* a + b *is in* K *whenever both* a *and* b *are in* K.

 IIb. *A rule of combination "$\cdot$" is defined such that* a $\cdot$ b *(abbreviated* ab*) is in* K *whenever both* a *and* b *are in* K.

 IIIa. *There exists an element 0 in* K *such that, for every* a *in* K, $a + 0 = a$.

 IIIb. *There exists an element 1 in* K *such that, for every* a *in* K, a $\cdot$ 1 = a.

 IVa. a + b = b + a } (commutative laws).
 IVb. a $\cdot$ b = b $\cdot$ a

 Va. a + (b $\cdot$ c) = (a + b) $\cdot$ (a + c) } (distributive laws)
 Vb. a $\cdot$ (b + c) = (a $\cdot$ b) + (a $\cdot$ c)

 VI. *For every element* a *in* K *there exists an element* $\bar{a}$ *such that*

$$a \cdot \bar{a} = 0$$

 and

$$a + \bar{a} = 1$$

 VII. *There are at least two elements* x *and* y *in* K *such that* x $\neq$ y.

---

*A definition of an equivalence relation may be found in Section 3.6.

Note that nothing has been said that would fix the number or type of elements that make up K. As a matter of fact, many systems satisfy these postulates. A few will be illustrated later.

The reader may have observed a similarity between these postulates and those of ordinary algebra. Note, however, that the first distributive law—that is, distribution over addition—does not hold for ordinary algebra. Also, there is no $\bar{a}$ defined in ordinary algebra.

If a set of postulates is to be of any use, it must be consistent. That is, none of the postulates in the set may contradict any other postulate of the set. To verify consistency, we might attempt a postulate-by-postulate examination to ascertain that no postulate contradicted any possible group of postulates. This is rather awkward; in fact, a much easier way is available. It is merely necessary to find an example of a Boolean algebra that is known independently to be consistent. If such a consistent system satisfies all of Huntington's postulates, then the postulates themselves must be consistent.*

The simplest example of a Boolean algebra consists of only two elements, 1 and 0, defined to satisfy

$$\bar{1} = 0 \qquad \bar{0} = 1$$
$$1 \cdot 1 = 1 + 1 = 1 + 0 = 0 + 1 = 1$$

and

$$0 + 0 = 0 \cdot 0 = 1 \cdot 0 = 0 \cdot 1 = 0$$

We see that Postulates I, II, III, and VII are satisfied by definition. Consider, for example, IIIb. If $a = 1$, we have

$$a \cdot 1 = 1 \cdot 1 = 1$$

If $a = 0$, we have

$$a \cdot 1 = 0 \cdot 1 = 0$$

The satisfaction of both of the above equations results from our definition of the " $\cdot$ " function for all possible combinations of arguments. Satisfaction of the commutative laws is evident, and verification of the distributive laws requires only a truth-table listing of the values of each side of both equations for all combinations of values of a, b, and c. This task is left to the reader.

We observe also the satisfaction of Postulate VI upon letting a take on the values 1 and 0. That is, letting $a = 1$, we obtain

$$a \cdot \bar{a} = 1 \cdot \bar{1} = 1 \cdot 0 = 0$$
$$a + \bar{a} = 1 + \bar{1} = 1 + 0 = 1 \tag{3.1}$$

and letting $a = 0$ yields

$$a \cdot \bar{a} = 0 \cdot \bar{0} = 0 \cdot 1 = 0$$
$$a + \bar{a} = 0 + \bar{0} = 0 + 1 = 1 \tag{3.2}$$

In addition to consistency, the question of *independence* of the postulates has been of some interest. By independence, it is meant that none of the postulates

---

*This matter of the consistency of a set of postulates is a subtle mathematical concept. The interested reader should consult Huntington's paper [1] for a full discussion.

can be proved from the others. The postulates presented here are, in fact, independent. However, a demonstration of this fact would be lengthy and is not essential to our discussion.

It is not necessary that we begin with an independent set of postulates. In fact, some authors save effort by including some of the theorems that we will develop later as postulates. It is our opinion, however, that the proof of these theorems constitutes the best possible introduction to the manipulation of Boolean algebra.

3.2 TRUTH-FUNCTIONAL CALCULUS AS A BOOLEAN ALGEBRA

There is a one-to-one correspondence between the truth-functional calculus of Chapter 1 and the two-element algebra introduced in the previous section. Table 3.1 interprets each truth value or logical connective as an element or a rule of combination, respectively, in a Boolean algebra. These interpretations imply the truth-table correspondence shown in Fig. 3.1.

A precise distinction between the two columns of Table 3.1 is not always made. Generally, " · " is identified in the literature as AND, whereas "+" is called OR. We will follow this convention until it is necessary to discuss logic and arithmetic in the same context. In Chapter 11 the symbols of truth functional calculus will be used for AND and OR.

In the remainder of this chapter, Boolean algebra will be developed into a convenient manipulative tool. As this is accomplished, the results will be immediately applicable to truth functions and therefore to logic circuits.

3.3 DUALITY

Notice that Huntington's postulates are presented in pairs. A closer look reveals that in each case one postulate in a pair can be obtained from the other by interchanging 0 and 1 along with + and " · " symbols. For example,

Table 3.1

Truth Calculus	Boolean Algebra
$\wedge$ $\longleftrightarrow$	$\cdot$
$\vee$ $\longleftrightarrow$	$+$
F $\longleftrightarrow$	0
T $\longleftrightarrow$	1
$\bar{A}$ $\longleftrightarrow$	$\bar{A}$

$A\,B$	ab	$A \wedge B$ $a \cdot b$	$A \vee B$ $a + b$
FF $\longrightarrow$ 00		F $\longrightarrow$ 0	F $\longrightarrow$ 0
FT $\longrightarrow$ 01		F $\longrightarrow$ 0	T $\longrightarrow$ 1
TF $\longrightarrow$ 10		F $\longrightarrow$ 0	T $\longrightarrow$ 1
TT $\longrightarrow$ 11		T $\longrightarrow$ 1	T $\longrightarrow$ 1

Figure 3.1

$$a + 0 = a$$
$$\downarrow \quad \downarrow$$
$$a \cdot 1 = a \tag{3.3}$$

and

$$a + (b \cdot c) = (a + b) \cdot (a + c)$$
$$\downarrow \quad \downarrow \qquad \downarrow \quad \downarrow \quad \downarrow$$
$$a \cdot (b + c) = (a \cdot b) + (a \cdot c) \tag{3.4}$$

Every theorem that can be proved for Boolean algebra has a dual that is also true. That is, every step of a proof of a theorem may be replaced by its dual, yielding a proof of the dual of the theorem. In a sense, this doubles our capacity for proving theorems.

3.4 FUNDAMENTAL THEOREMS OF BOOLEAN ALGEBRA

In this section, we prove the theorems necessary for the convenient manipulation of Boolean algebra. Some of these are labeled *theorems*, whereas others are labeled *lemmas*. Our criterion for making this distinction is that only equalities that are valuable tools for working problems are designated as theorems. Intermediate results necessary in the proof of theorems and results, which are included only for the sake of logical completeness, are called lemmas.

LEMMA 3.1   The 0 and 1 elements are unique.

PROOF   By way of contradiction, assume that there are two zero elements, 0_1 and 0_2. For any elements a_1 and a_2 in K, we have

$$a_1 + 0_1 = a_1 \quad \text{and} \quad a_2 + 0_2 = a_2 \qquad \text{(Post. IIIa)}$$

Now let $a_1 = 0_2$ and $a_2 = 0_1$. Therefore,

$$0_2 + 0_1 = 0_2 \quad \text{and} \quad 0_1 + 0_2 = 0_1$$

Thus, using the first commutative law and transitive property of equality, we have

$$0_1 = 0_2$$

As an example of duality, let us make the following change of symbols in our proof:

$$a_1 + 0_1 = a_1 \qquad a_2 + 0_2 = a_2$$
$$\downarrow \downarrow \qquad\qquad \downarrow \downarrow$$
$$a_1 \cdot 1_1 = a_1 \qquad a_2 \cdot 1_2 = a_2$$

and

$$0_2 + 0_1 = 0_2 \qquad 0_1 + 0_2 = 0_1$$
$$\downarrow \downarrow \downarrow \quad \downarrow \qquad \downarrow \downarrow \downarrow \quad \downarrow$$
$$1_2 \cdot 1_1 = 1_2 \qquad 1_1 \cdot 1_2 = 1_1$$

and

$$0_1 = 0_2$$
$$\downarrow \qquad \downarrow$$
$$1_1 = 1_2$$

Because we were aware that the dual of each postulate exists, it was unnecessary to cite postulates as the dual of each step was compiled. In fact, we could have merely stated that the dual of the theorem was true by the principle of duality. This approach will be followed in the remaining theorems of this section.

LEMMA 3.2   For every element a in K, $a + a = a$ and $a \cdot a = a$.

PROOF

$a + a = (a + a) \cdot 1$	(Post. IIIb)
$a + a = (a + a)(a + \bar{a})$	(Post. VI)
$a + a = a + a\bar{a}$	(Post. Va)
$a + a = a + 0$	(Post. VI)
$a + a = a$	(Post. IIIa)
$a \cdot a = a$	(Duality)

LEMMA 3.3   For every a in K, $a + 1 = 1$ and $a \cdot 0 = 0$.

PROOF

$a + 1 = 1 \cdot (a + 1)$	(Post. IIIb)
$a + 1 = (a + \bar{a})(a + 1)$	(Post. VI)
$a + 1 = a + \bar{a} \cdot 1$	(Post. Va)
$a + 1 = a + \bar{a}$	(Post. IIIb)
$a + 1 = 1$	(Post. VI)
$a \cdot 0 = 0$	(Duality)

LEMMA 3.4   The elements 1 and 0 are distinct and $\bar{1} = 0$.

PROOF   Let a be any element in K.

$a \cdot 1 = a$	(Post. IIIb)
$a \cdot 0 = 0$	(Lemma 3.3)

Now assume that $1 = 0$. In this case, the above expressions are satisfied only if $a = 0$. Postulate VII, however, tells us that there are at least two elements in K. The contradiction can be resolved only by concluding that $1 \neq 0$. To prove the second assertion, we need only write

$\bar{1} = \bar{1} \cdot 1$	(Post. IIIb)
$= 0$	(Post. VI)

LEMMA 3.5   For every pair of elements a and b in K, $a + ab = a$ and $a(a + b) = a$.

PROOF

$a + ab = a \cdot 1 + ab$	(Post. IIIb)
$= a(1 + b)$	(Post. Vb)
$= a \cdot 1$	(Lemma 3.3)
$= a$	(Post. IIIb)

and

$a(a + b) = a$	(Duality)

LEMMA 3.6   The $\bar{a}$ defined by Postulate VI for every a in K is unique.

PROOF   By contradiction. Assume that there are two distinct elements, $\bar{a}_1$ and $\bar{a}_2$, that satisfy Postulate VI, that is, assume that

$$a + \bar{a}_1 = 1, \qquad a + \bar{a}_2 = 1, \qquad a\bar{a}_1 = 0, \qquad a\bar{a}_2 = 0$$

$$
\begin{aligned}
\bar{a}_2 &= 1 \cdot \bar{a}_2 && \text{(Post. IIIb)} \\
&= (a + \bar{a}_1)\bar{a}_2 && \text{Hypothesis} \\
&= a\bar{a}_2 + \bar{a}_1\bar{a}_2 && \text{(Post. Vb)} \\
&= 0 + \bar{a}_1\bar{a}_2 && \text{Hypothesis} \\
&= a\bar{a}_1 + \bar{a}_1\bar{a}_2 && \text{Hypothesis} \\
&= (a + \bar{a}_2)\bar{a}_1 && \text{(Post. Vb)} \\
&= 1 \cdot \bar{a}_1 && \text{Hypothesis} \\
&= \bar{a}_1 && \text{(Post. IIIb)}
\end{aligned}
$$

LEMMA 3.7   For every element a in K, $a = \bar{\bar{a}}$.

PROOF   Let $\bar{\bar{a}} = x$. Therefore,

$$\bar{a}x = 0 \qquad \text{and} \qquad \bar{a} + x = 1 \qquad\qquad \text{(Post. VI)}$$

but

$$a\bar{a} = 0 \qquad \text{and} \qquad \bar{a} + a = 1 \qquad\qquad \text{(Post. VI)}$$

Thus, both x and a satisfy Postulate VI as the complement of $\bar{a}$. Therefore, by Lemma 3.6,

$$x = a$$

LEMMA 3.8 $a[(a + b) + c] = [(a + b) + c]a = a.$

PROOF
$$
\begin{aligned}
a[(a + b) + c] &= a(a + b) + ac && \text{(Post. Vb)} \\
&= a + ac && \text{(Lemma 3.5)} \\
&= a = [(a + b) + c]a && \text{(Post. IVb, Lemma 3.5)}
\end{aligned}
$$

THEOREM 3.9   For any three elements $a, b,$ and c in $K, a + (b + c) = (a + b) + c$ and $a \cdot (bc) = (ab) \cdot c$.

The reader will recognize Theorem 3.9 as the associative laws from ordinary algebra. Some authors include these laws among the postulates, although, as we will see, this is unnecessary.

PROOF   Let

$$
\begin{aligned}
Z &= [(a + b) + c] \cdot [a + (b + c)] \\
&= [(a + b) + c]a + [(a + b) + c] \cdot (b + c) \\
&= a + [(a + b) + c] \cdot (b + c) && \text{(Lemma 3.8)} \\
&= a + \{[(a + b) + c]b + [(a + b) + c] \cdot c\} && \text{(Post. Vb)} \\
&= a + \{b + [(a + b) + c] \cdot c\} && \text{(Lemma 3.8, Post. IVb)} \\
&= a + (b + c) && \text{(Lemma 3.5)}
\end{aligned}
$$

However, we may also write

$$Z = (a + b)[a + (b + c)] + c[a + (b + c)] \qquad \text{(Post. Vb)}$$
$$= (a + b)[a + (b + c)] + c \qquad \text{(Lemma 3.8)}$$
$$= \{a[a + (b + c)] + b[a + (b + c)]\} + c \qquad \text{(Post. Vb)}$$
$$= \{a[a + (b + c)] + b\} + c \qquad \text{(Lemma 3.8)}$$
$$= (a + b) + c \qquad \text{(Lemma 3.5)}$$

Therefore, by transitivity

$$a + (b + c) = (a + b) + c$$

and

$$(a \cdot b)c = a(b \cdot c) \qquad \text{(Duality)}$$

Now that we have established the associative laws, certain expressions may be simplified by omitting parentheses as follows:

$$(a + b) + c = a + b + c \qquad (3.5)$$

and

$$(a \cdot b) \cdot c = abc \qquad (3.6)$$

This format may be extended to Boolean sums and products of any number of variables.

THEOREM 3.10 For any pair of elements a and b in K, $a + \bar{a}b = a + b$; $a(\bar{a} + b) = ab$.

PROOF

$$a + \bar{a}b = (a + \bar{a})(a + b) \qquad \text{(Post. Va)}$$
$$a + \bar{a}b = a + b \qquad \text{(Post. VI, Post. IIIb)}$$
$$a \cdot (\bar{a} + b) = a \cdot b \qquad \text{(Duality)}$$

THEOREM 3.11 For every pair of elements a and b in K, $\overline{a + b} = \bar{a} \cdot \bar{b}$ and $\overline{a \cdot b} = \bar{a} + \bar{b}$.

The expressions in Theorem 3.11 are the two forms of the very important DeMorgan's law, which we introduced in Chapter 1. The second form is the dual of the first, but for the sake of instruction, we will arrive at the second form in another way.

$$(a + b) + \bar{a} \cdot \bar{b} = [(a + b) + \bar{a}] \cdot [(a + b) + \bar{b}] \qquad \text{(Post. Va)}$$
$$(a + b) + \bar{a} \cdot \bar{b} = [\bar{a} + (a + b)] \cdot [\bar{b} + (b + a)] \qquad \text{(Post. IVa)}$$
$$(a + b) + \bar{a} \cdot \bar{b} = 1 \cdot 1 = 1 \qquad \text{(Theorem 3.9, Lemma 3.3)}$$
$$(a + b) \cdot (\bar{a} \cdot \bar{b}) = a(\bar{a} \cdot \bar{b}) + b(\bar{b} \cdot \bar{a}) \qquad \text{(Post. Vb, Post. IVb)}$$
$$(a + b) \cdot (\bar{a} \cdot \bar{b}) = 0 + 0 = 0 \qquad \text{(Theorem 3.9, Lemma 3.3)}$$

Both requirements of Postulate VI have been satisfied, so $a + b$ is the unique complement of $\bar{a} \cdot \bar{b}$. Therefore, we may write

$$a + b = \overline{\bar{a} \cdot \bar{b}} \qquad \text{or} \qquad \overline{a + b} = \bar{a} \cdot \bar{b}$$

The above holds equally well for $\bar{a}$ and $\bar{b}$ in place of a and b, so we can write

$$\overline{\bar{a} + \bar{b}} = \bar{\bar{a}} \cdot \bar{\bar{b}} = a \cdot b$$

or

$$\overline{\bar{a} + \bar{b}} = \overline{a \cdot b} \qquad\qquad \text{(Lemma 3.6, 3.7)}$$

The classic development of Boolean algebra based on Huntington's postulates as given above has the advantage of illustrating the same techniques required in the application of the algebra.

3.5 SET THEORY AS AN EXAMPLE OF BOOLEAN ALGEBRA

We have already discussed truth-functional calculus as one example of Boolean algebra. Since this example satisfies all of Huntington's postulates, it must also satisfy each of the theorems and lemmas proved so far.

Set theory is a second example of a Boolean algebra. A *set* may be regarded as any collection of objects. In order to talk of an algebra of sets, we must limit the objects that make up these sets with some meaningful criteria. We define a *universal set* to include every object that satisfies these criteria. For a reason that will soon be apparent, the universal set must necessarily contain at least one object. The universal set may, in fact, contain a finite or an infinite number of objects. The collection of points in a plane or in a finite region of a plane make interesting examples of a universal set. When we say that R is a subset of S, we mean that every object in R is also found in S. This is symbolized by $R \subset S$. When referring to individual elements, we use $p \in S$ to indicate that p is a member of S. Only those sets that are subsets of the universal set are material to the development of an algebra of sets. A second set of importance is the *null set*, or the set containing no objects. The universal set and the null set will be symbolized by S_U and S_Z, respectively. The *union* of two sets, R and S, is the set that contains all objects contained in either R or S. This is symbolized by $R \cup S$. The *intersection* of two sets, R and S, is that set containing these elements found in both sets R and S. The intersection is designated by $R \cap S$. The *complement* of a set S, designated $C(S)$, contains those objects found in the universal set but not in set S.

A helpful illustration of a universal set is the collection of points in a rectangle, such as shown in Fig. 3.2. The two sets R and S contain those points of the universal set found within the respective circles.* A representation of this type is known as a *Venn diagram*.

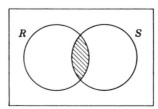

Figure 3.2 $R \cap S$.

*The student of advanced calculus may be concerned about the status of points on the boundaries of the circles, a problem not critical to this development. One possible resolution of the conflict is to imagine the universal set to include only a large but finite number of points in the rectangle. In this way, boundaries can be deliberately arranged to avoid points in the universal set.

The darkened area in Fig. 3.2 is the intersection of sets R and S. Similarly, $R \cup S$ and $C(S)$ are darkened in Figs. 3.3 and 3.4, respectively.

Recalling the postulates in Section 3.1, let us choose a set of K objects to be those sets containing no points other than points found in the universal set. Such sets, one of which is the null set, are subsets of the universal set. If equality is taken to relate only identical sets, then Postulate I is satisfied. We will proceed to show that, in light of the correspondences in Table 3.2, the remaining postulates will also be satisfied.

The sets $A \cap B$ and $A \cup B$ are subsets of S_U, so Postulate II is satisfied. Because $R \cup S_Z$ and $R \cap S_U$ contain precisely the same objects as R, Postulate III is satisfied. Satisfaction of Postulate IV, the commutative laws, is evident. Under the set theory interpretation, the distributive laws become

$$A \cup (B \cap C) = (A \cup B) \cap (A \cup C) \tag{3.7}$$

and

$$A \cap (B \cup C) = (A \cap B) \cup (A \cap C) \tag{3.8}$$

The fact that Equations 3.7 and 3.8 are satisfied is best illustrated by using the Venn diagram. Note in Fig. 3.5a that set $B \cap C$ is shaded with vertical lines, whereas set A is shaded by horizontal lines. The set $A \cup (B \cap C)$ consists of the areas darkened in either manner. In Fig. 3.5b, the set $A \cup B$ is indicated by horizontal lines, and $A \cup C$ is shaded by vertical lines. The intersections of these two sets, $(A \cup B) \cap (A \cup C)$ is therefore given by the criss-crossed area in Fig. 3.5b. Note that the areas representing both sides of Equation 3.7 are identical. The validity of Equation 3.8 can be similarly demonstrated by using the Venn diagram. By the definition of the complement, it follows immediately that

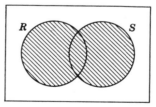

Figure 3.3 $R \cup S$.

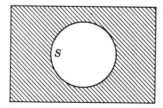

Figure 3.4 $C(S)$.

Table 3.2

$\cap$	$\longleftrightarrow$	$\cdot$
$\cup$	$\longleftrightarrow$	$+$
S_Z	$\longleftrightarrow$	0
S_U	$\longleftrightarrow$	1
$C(S)$	$\longleftrightarrow$	$\bar{S}$

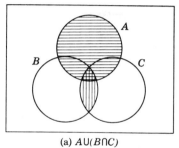 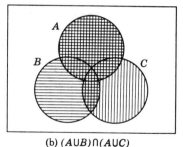

(a) $A \cup (B \cap C)$ (b) $(A \cup B) \cap (A \cup C)$

Figure 3.5   (a) $A \cup (B \cap C)$. (b) $(A \cup B) \cap (A \cup C)$.

$$R \cap \bar{R} = S_Z \tag{3.9}$$

and

$$R \cup \bar{R} = S_U \tag{3.10}$$

Thus, Postulate VI is satisfied. The fact that S_U and S_Z are always defined and distinct guarantees the satisfaction of Postulate VII.

3.6 ALTERNATIVE DEFINITION OF A BOOLEAN ALGEBRA

Mathematicians did not ignore the topic of Boolean algebra after Huntington and Sheffer published their sets of postulates in 1904 and 1913, respectively. To some it is preferable to define a Boolean algebra as a distributive complemented lattice, thereby tying it more closely to other mathematical structures. To support Section 12.8 and provide for this alternative, we proceed in developing the concept of a lattice.

If S is a set of n elements, there are n^2 ordered pairs of elements (a,b), where both a and $b \in S$.

Definition 3.1   A *relation* R on set S is a subset of the set of all ordered pairs (a,b), where a and $b \in S$. The notation a R b is equivalent to $(a,b) \in$ R.

There are two classes of relations that are of particular interest: (1) equivalence relations and (2) partial orderings. An *equivalence relation* on S is a relation that satisfies the following three properties for any a, b, and c in S.

Reflexive: a R a

Symmetric: If a R b, then b R a.

Transitive: If a R b and b R c, then a R c.

We shall find an important application of equivalence relations when we consider sequential circuits in Chapter 10, where we will demonstrate that an equivalence relation on S partitions S into a set of disjoint classes. It is the definition of a partial ordering that leads us to a lattice.

Definition 3.2   A *partial ordering* $\leq$ on S is a relation that satisfies the following properties for any elements a, b, and c in S.

Reflexive: $a \leq a$

Antisymmetric: If $a \leq b$ and $b \leq a$, then $a = b$.

Transitive: If $a \leq b$ and $b \leq c$, then $a \leq c$.

A diagram called a *Hasse* or ordering diagram that shows point *a* above point *b* if *b* ≤ *a* is often a convenient graphical representation of a partial ordering. Edges are included in the diagram so that there is a path of upward directed edges from *b* to *a* if and only if *b* ≤ *a*.

■ EXAMPLE 3.1

Consider a relation **R** = {(0,*b*), (*b*,*a*), (*a*,1)} defined on set *S* = {0,*a*,*b*,1}. The antisymmetric property is satisfied by the omission of (*b*,0), (*a*,*b*), and (1,*a*). What pairs must be added to the relation to form a partial ordering? Construct a Hasse diagram of the resulting relation.

SOLUTION   The pairs {(0,0), (*a*,*a*), (*b*,*b*), (1,1)} must be added to satisfy the reflexive property. For transitivity we must have 0 ≤ *a*, 0 ≤ 1, and *b* ≤ 1. The Hasse diagram is shown in Fig. 3.6*a*.   ■

■ EXAMPLE 3.2

Consider a reflexive relation on *S* that includes the additional pairs {(0,*a*), (0,*b*), (0,1), (*a*,1), (*b*,1)}. The relation is a partial ordering with a Hasse diagram as given in Fig. 3.6*b*.   ■

Let *P* be a subset of a partially ordered set *S*. Then an element *u* ∈ *S* is an upper bound of *P*, if and only if *p* ≤ *u* for every *p* ∈ *P*. Likewise, *l* is a lower bound of *P*, if and only if *l* ≤ *p* for every *p* ∈ *P*. An element u_l ∈ *S* is said to be a *least upper bound* (lub) of *P* if and only if u_l ≤ *u*, where *u* is any upper bound of *P*. Similarly, l_g is the *greatest lower bound* (glb) of *P*.

Definition 3.3   A partially ordered set *S* in which every pair of elements has a glb and a lub is a *lattice*.

Using induction in conjunction with Definition 3.3, we can show that every lattice *S* has a least element 0 such that 0 ≤ *s* for every element *s* ∈ *S*. Similarly, a lattice has a greatest element 1. For convenience let us denote the glb of *a* and *b* as *a* · *b* and the lub of *a* and *b* as *a* + *b*.

THEOREM 3.12   A distributive complemented lattice (a lattice that satisfies Huntington's postulates V and VI) is a Boolean algebra.

PROOF   The operators of Postulate II have just been defined as the lub and glb, respectively. Because 0 is the least element, the lub(*a*,0) = *a* + 0 = *a*. Similarly, *a* · 1 = *a*, thus satisfying Postulate III. The lub and glb of two elements do not depend on the argument order of the two elements, so the commutative laws of Postulate

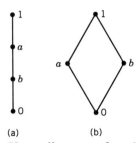

(a)   (b)

Figure 3.6   Hasse diagrams of partial orderings.

IV are satisfied. Therefore, with the inclusion of Postulates V and VI in the hypothesis, the lattice is a Boolean algebra.

■ EXAMPLE 3.3

Are the partially ordered sets of Fig. 3.6 lattices? Are they Boolean algebras?

SOLUTION   Both partial orderings are lattices. Enumeration of all possible pairs of elements will verify that each pair has a glb and an lub. In the lattice of Fig. 3.6a, a and b do not have complements. In the lattice of Fig. 3.6b, $0+1 = 1, 0 \cdot 1 = 0$, $a + b = 1$, and $a \cdot b = 0$. That is, every element has a unique complement.

It is easy to see that the lattice in Fig. 3.6a is not a Boolean algebra. Neither a nor b has a complement.

The lattice of Fig. 3.6b is complemented because $a = \bar{b}, b = \bar{a}$, and $\bar{0} = 1$. We can show that this lattice is distributive by resorting to enumeration. It is easy to show that $x + yz$ and $(x+y)(x+z)$ are identical and that $x(y+z)$ and $xy + xz$ are identical, if any of the three literals is 1 or 0. It is now only necessary to consider all combinations of assignments of a and $\bar{a}$ to x, y, and z. Suppose $y = z = a$ (similar for $y = z = \bar{a}$). In this case, $x + yz = x + a$ and $(x+y)(x+z) = (x+a)(x+a) = x + a$. Similarly, $x(y+z) = xa$ and $xy + xz = xa + xa = xa$.

The only remaining alternative is for $y = \bar{z}$, for example $y = a$ and $z = \bar{a}$. Now $x + yz = x + a\bar{a} = x + 0 = x$ and $(x+y)(x+z) = (x+a)(x+\bar{a})$, which is $\bar{a}$ if $x = \bar{a}$ and a if $x = a$.   ■

Lemmas 3.1 through 3.5 and the associative laws follow easily from the definitions of lub and glb. The development might have been shortened somewhat by using the statement of Theorem 3.12 as the definition of a Boolean algebra. However, the development as given in Section 3.4 is complete and entirely algebraic.

3.7   EXAMPLES OF BOOLEAN SIMPLIFICATION

We are now ready to utilize Boolean algebra in the simplification of some meaningful examples. The first two examples are a problem in set theory and a needlessly complex logic circuit.

■ EXAMPLE 3.4

Scholars A, B, and C collect old manuscripts. Mr. A is a collector of English political works and foreign-language novels. Ms. B is a collector of all political works, except English novels, and English works that are not novels. Mr. C collects nonfictional items that are English works or foreign-language political works. Determine those books for which there is competition—that is, those desired by two or more collectors.

SOLUTION   Let us first define the various sets involved in the problem.

> A set of books collected by Mr. A
> B set of books collected by Ms. B
> C set of books collected by Mr. C
> E set of all English language books
> N set of all novels
> P set of all political works

The set of books collected by two or more persons can be symbolized by

$$Z = (A \cap B) \cup (A \cap C) \cup (B \cap C) \tag{3.11}$$

Translating the problem statements into set theoretic expressions gives

$$A = (E \cap P) \cup (\bar{E} \cap N)$$
$$B = (P \cap \overline{E \cap N}) \cup (E \cap \bar{N})$$
$$C = [E \cup (\bar{E} \cap P)] \cap \bar{N}$$

Note that it was more convenient to use the Boolean algebra symbol than the set theoretic symbol for the complement. In order to utilize Boolean algebra in obtaining a solution, the symbols for intersection and union will also be replaced by the corresponding "." and "+" symbols. Before substituting into Equation 3.11, let us write the expression for A, B, and C in Boolean form and make whatever simplifications are possible.

$$A = EP + \bar{E}N$$
$$B = P(\overline{E\bar{N}}) + E\bar{N} = P(\bar{E} + \bar{N}) + E\bar{N} \qquad \text{(DeMorgan's theorem)}$$
$$= P\bar{E} + P\bar{N}(E + \bar{E}) + E\bar{N}$$
$$= P\bar{E} + P\bar{N}\bar{E} + PE\bar{N} + E\bar{N}$$
$$= P\bar{E}(1 + \bar{N}) + (P + 1)E\bar{N} = P\bar{E} + E\bar{N}$$
$$C = (E + \bar{E}P)\bar{N} = (E + P)\bar{N} \qquad \text{(Theorem 3.10)}$$

Now writing Equation 3.11 as a Boolean expression and substituting, we get

$$Z = (EP + \bar{E}N)(P\bar{E} + E\bar{N}) + (EP + \bar{E}N)(E\bar{N} + P\bar{N})$$
$$+ (P\bar{E} + E\bar{N})(E\bar{N} + P\bar{N})$$

or

$$Z = (EP + \bar{E}N)(P\bar{E} + E\bar{N} + E\bar{N} + P\bar{N})$$
$$+ (P\bar{E} + E\bar{N})(E\bar{N} + P\bar{N}) \qquad \text{(Distributive laws)}$$
$$= (EP + \bar{E}N)(P\bar{E} + E\bar{N} + P\bar{N}) + (P\bar{E} + E\bar{N})(E\bar{N} + P\bar{N})$$
$$= (EP + \bar{E}N)(P\bar{E} + E\bar{N} + P\bar{E}\bar{N} + PE\bar{N}) + (P\bar{E} + E\bar{N})(E\bar{N} + P\bar{N})$$
$$= (EP + \bar{E}N)(P\bar{E}(1 + \bar{N}) + (P + 1)E\bar{N}) + (P\bar{E} + E\bar{N})(E\bar{N} + P\bar{N})$$
$$= (EP + \bar{E}N)(P\bar{E} + E\bar{N}) + (P\bar{E} + E\bar{N})(E\bar{N} + P\bar{N})$$
$$= (P\bar{E} + E\bar{N})(EP + \bar{E}N + E\bar{N} + P\bar{N}) \qquad \text{(Distributive law)}$$
$$= E\bar{N} + P\bar{E}(PE + \bar{E}N + P\bar{N}) \qquad \text{(Distributive law)}$$
$$= E\bar{N} + P\bar{E}E + P\bar{E}N + P\bar{E}\bar{N}$$
$$= E\bar{N} + P\bar{E}(N + \bar{N}) = E\bar{N} + P\bar{E}$$

Thus, we conclude that more than one person is interested in English nonfiction and foreign-language political works.

An alternative solution of this problem constitutes an interesting application of the Venn diagram. Note that sets A, B, and C are represented by the darkened areas in Figs. 3.7a, 3.7b, and 3.7c, respectively.

By inspection, we see that the area darkened in at least two of these three diagrams is that shown in Fig. 3.8. This can be expressed as

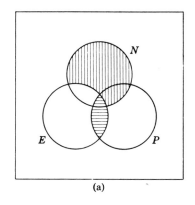

(a)

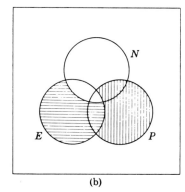

(b)

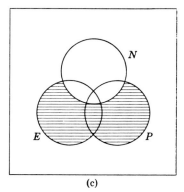

(c)

Figure 3.7

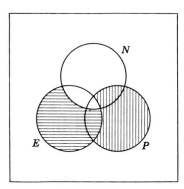

Figure 3.8

$$(E \cap \bar{N}) \cup (P \cap \bar{E})$$

which is the same result obtained by the algebraic method. ∎

The reader may have noticed that expressions similar to

$$xy + \bar{x}z + yz \qquad (3.12)$$

appeared several times in Example 3.4. Each time, a simplification involving several lines of algebraic manipulation was made. The following theorem makes it possible to shortcut this manipulation in future examples.

THEOREM 3.13 (Consensus Theorem)   For any three elements a, b, and c in K,

$$ab + \bar{a}c + bc = ab + \bar{a}c$$

and

$$(a + b)(\bar{a} + c)(b + c) = (a + b)(\bar{a} + c)$$

PROOF
$$ab + \bar{a}c + bc = ab + \bar{a}c + bc(a + \bar{a})$$
$$= ab + abc + \bar{a}c + \bar{a}cb$$
$$= ab(1 + c) + \bar{a}c(1 + b)$$
$$= ab + \bar{a}c$$
$$(a + b)(\bar{a} + c)(b + c) = (a + b)(\bar{a} + c) \qquad \text{(Duality)}$$

Theorem 3.13 is illustrated by the Venn diagram in Fig. 3.9. Note that the outlined area bc is partially covered by ab and $\bar{a}c$. The theorem may be applied whenever two products jointly cover a third in this manner.

■ EXAMPLE 3.5

The logic circuit shown in Fig. 3.10 was designed by trial and error. To simplify the figure, some of the actual connections to the inputs and outputs of the inverters are not shown. Let us utilize Boolean algebra to design an equivalent circuit free from unnecessary hardware.

SOLUTION   It is first necessary to determine the function realized by this circuit. To this end, the output functions for the various logic circuits are shown in Fig. 3.10. The output function is given by

$$F(x, y, z) = y\bar{z}(\bar{z} + \bar{z}x) + (\bar{x} + \bar{z})(\bar{x}y + \bar{x}z) \qquad (3.13)$$
$$= y\bar{z} + (\bar{x} + \bar{z})(\bar{x}y + \bar{x}z) \qquad \text{(Lemma 3.5)}$$
$$= y\bar{z} + (\bar{x} + \bar{z})\bar{x}(y + z)$$
$$= \bar{z}y + \bar{x}z + \bar{x}y \qquad \text{(Lemma 3.5)}$$

Now if $\bar{z}$ is identified as a, y as b, and $\bar{x}$ as c in Theorem 3.13, we have

$$F(x, y, z) = \bar{z}y + \bar{x}z \qquad \text{(Theorem 3.13)}$$

We see that the considerably simplified and therefore cheaper circuit in Fig. 3.11 realizes the same function as the one in Fig. 3.10.   ■

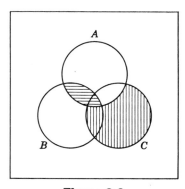

Figure 3.9

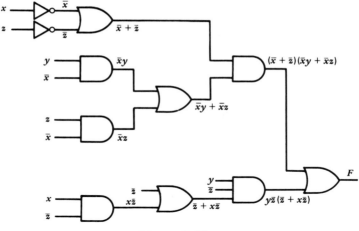

Figure 3.10

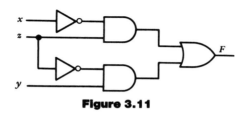

Figure 3.11

■ **EXAMPLE 3.6**

Simplify the following Boolean expression, which represents the output of a logical decision circuit:

$$F(w, x, y, z) = x + xyz + yz\bar{x} + wx + \bar{w}x + \bar{x}y$$
$$= x + yz(x + \bar{x}) + \bar{x}y + x(w + \bar{w})$$
$$= (x + \bar{x}y) + yz + x$$
$$= x + y + x + yz$$
$$= x + (y + yz)$$
$$= x + y$$

Another approach to the problem involves the use of Lemma 3.7. That is,

$$\overline{F(w, x, y, z)} = \overline{x + xyz + yz\bar{x} + \bar{x}y + wx + \bar{w}x}$$
$$= \bar{x}(\overline{xyz})(\overline{\bar{x}yz})(\overline{xy})(\overline{wx})(\overline{\bar{w}x})$$
$$= [\bar{x}(\bar{x} + \bar{y} + \bar{z})](x + \bar{y} + \bar{z})(x + \bar{y})(\bar{w} + \bar{x})(w + \bar{x})$$
$$= [\bar{x} \cdot (\bar{w} + \bar{x})(w + \bar{x})](x + \bar{y})(\bar{y} + \bar{z} + x)$$
$$= [\bar{x}](x + \bar{y})$$
$$\overline{F(w, x, y, z)} = \bar{x} \cdot \bar{y}$$

Therefore, by Lemma 3.7 we have

$$F(w, x, y, z) = \overline{\overline{F(w, x, y, z)}} = \overline{\bar{x} \cdot \bar{y}} = x + y.$$

■

3.8 REMARKS ON SWITCHING FUNCTIONS

The reader may have wondered why the Venn diagram that was so helpful in Example 3.4 was not utilized in the two subsequent examples. It should be recalled that this device has so far been introduced only as an illustration of set theory. The Boolean algebra of sets has many elements, not just two, as is the case with truth values and logic voltage levels. The Venn diagram can, however, be made useful in the simplification of switching circuits and truth functions if the following definitions and interpretations are kept in mind.

First, it is convenient to distinguish Boolean functions of only two elements by calling them *switching* functions. We then define a *switching variable* as a letter (usually chosen from the beginning or end of the alphabet, excluding F) that may take on either of the element values 0 or 1. In the case of logic circuits, a fixed number of variables, say n, serve as inputs to a circuit under consideration. The situation is similar in the case of truth functions.

There are 2^n possible ways of assigning values to n variables. The three-variable case is illustrated in Fig. 3.12. If the eight question marks are replaced by any combination of 0's and 1's, a specific *function* of $x_1, x_2,$ and x_3 is defined. There are 2^8, or 2^{2^3}, ways of replacing the eight question marks by 0's and 1's, so there are 2^{2^3} *switching functions* of three variables. The value of F for a particular row of the truth table is known as the functional value for the corresponding combination of input values. Formally, we have the following definition.

Definition 3.4 A *switching function* of n variables is any one particular assignment of functional values (1's or 0's) for all 2^n possible combinations of values of the n variables.

In general, then, there are 2^{2^n} distinct switching functions of n variables.

We have seen that there are many *expressions* for a given switching function of truth-table assignment. In the two-element algebra, many distinct expressions are equal to each of the two elements 1 and 0. For example,

$$1 \cdot (1 + 0) = 1$$

Boolean algebra provides a means of determining whether such expressions are equal to 1 or 0.

Let us consider now the possibility of an extended Boolean algebra whose *elements* are *all possible switching functions* of n variables. Such an algebra will provide us with a means of determining which *function* is represented by a given *expression* in the n variables. For example, we can determine that

x_1	x_2	x_3	F
0	0	0	?
0	0	1	?
0	1	0	?
0	1	1	?
1	0	0	?
1	0	1	?
1	1	0	?
1	1	1	?

Figure 3.12

$$x + xy + y + yz$$
$$x + \bar{x}y + x(w + \bar{w})$$

and

$$x + y$$

are all expressions for the same function, specifically that function defined by the truth table of Fig. 3.13.

The theorems we have proved so far, when specialized for a two-valued algebra, express equality between switching functions or, if you prefer, between elements of the extended algebra. Thus, as in ordinary algebra, two switching functions are identically equal if their evaluations are equal for every combination of variable values. This notion of equality satisfies Postulate I. It remains to be shown that the proposed algebra of 2^{2^n} elements satisfies the remainder of the postulates and is indeed a Boolean algebra. Functional expressions may certainly be related by the connectives AND, OR, and NOT in the same way variables are related. Therefore, Postulates II, IV, V are satisfied. The zero element is the function that is identically 0, and the unit element is the function that is identically 1. The definition of the negation of F must certainly be $\bar{F}$. If a function h is to satisfy the expression

$$h(x_1, x_2, ..., x_n) = \bar{F}(x_1, x_2, ..., x_n) \tag{3.14}$$

then h must be 0 whenever F is evaluated as 1 and vice versa. Satisfaction of Postulate VII is immediate.

We will now see that a Venn diagram provides a convenient illustration of the algebra whose elements are the 2^8 functions of three variables. In Fig. 3.14, areas are assigned to each of the eight combinations of values. A function can then be indicated by darkening those areas corresponding to combinations of values for

x	y	F
0	0	0
0	1	1
1	0	1
1	1	1

Figure 3.13

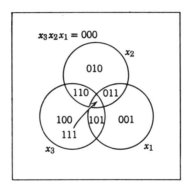

Figure 3.14

which the function is 1. Every possible area configuration stands for a distinct function. For example, the function

$$F(x, y, z) = \bar{z}y + \bar{x}z + \bar{x}y = \bar{z}y + \bar{x}z \tag{3.15}$$

resulting from Example 3.5 is depicted in Fig. 3.15. The differently lined regions indicate the simplification resulting from the application of Theorem 3.14.

Almost every problem in Boolean algebra will be found to be some variation of the following statement:

Given one of the 2^{2^n} functions of n variables, determine from the large number of equivalent expressions of this function one which satisfies some criteria for simplicity.

If the problem concerns an electronic switching circuit, the simplicity criteria will result from the desirability of using the cheapest possible circuit that provides the desired performance. The Venn diagram is a means of solving this type of problem in which three or fewer variables are involved.

For more than three variables, the basic illustrative form of the Venn diagram is inadequate. The most convenient extension, the Karnaugh map, will be discussed in Chapter 5.

3.9 SUMMARY

In this chapter, we have developed a formal algebra for the representation and manipulation of switching functions. We have also laid the necessary foundation for the development in later chapters of systematic procedures for simplification of algebraic expressions and the corresponding circuits.

Following is a list of the important postulates, lemmas, and theorems for ready reference in problems and proofs:

PIIIa $a + 0 = a$	PIIIb $a \cdot 1 = a$
PIVa $a + b = b + a$	PIVb $ab = ba$
PVa $\quad a + bc = (a + b)(a + c)$	PVb $\quad a(b + c) = ab + ac$
PVIa $a \cdot \bar{a} = 0$	PVIb $a + \bar{a} = 1$
L 2a $a + a = a$	L 2b $a \cdot a = a$
L 3a $a + 1 = 1$	L 3b $a \cdot 0 = 0$
L 4 $\quad 1 = 0$	
L 5a $a + ab = a$	L 5b $a(a + b) = a$
L 7 $\quad \bar{\bar{a}} = a$	

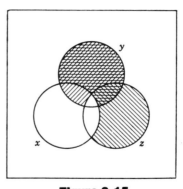

Figure 3.15

T 9a $a + (b + c) = (a + b) + c$
$\qquad\qquad = a + b + c$
T10a $a + \bar{a}b = a + b$
T11a $\overline{a + b} = \bar{a} \cdot \bar{b}$
T12a $ab + \bar{a}c + bc = ab + \bar{a}c$

T 9b $a(bc) = (ab)c = abc$

T10b $a(\bar{a} + b) = a \cdot b$
T11b $\overline{a \cdot b} = \bar{a} + \bar{b}$
T12b $(a + b)(\bar{a} + c)(b + c)$
$\qquad\qquad = (a + b)(\bar{a} + c)$

PROBLEMS

3.1   (a) Verify the distributive law

$$a(b + c) = ab + ac$$

for the two-valued Boolean algebra.

(b) Repeat (a) for $a + bc = (a + b)(a + c)$.

3.2   Write out a proof of the second associative law.

$$(a \cdot b) \cdot c = a(b \cdot c)$$

3.3   Verify by a truth-table analysis (using 1 and 0 instead of **T** and **F**) that the following theorems of Boolean algebra hold for the two-valued truth-functional calculus.

(a) Lemma 3.2

(b) Lemma 3.5

(c) Lemma 3.7

(d) Theorem 3.9

(e) Theorem 3.10

(f) Theorem 3.13

3.4   Write in its simplest possible form the Boolean function realized by the logic network of Fig. P3.4.

3.5   Use the Venn diagram to illustrate the validity of Equation 3.8.

3.6   Verify the following by Boolean algebraic manipulation. Justify each step with a reference to a postulate or theorem.

(a) $(X + \bar{Y} + XY)(X + \bar{Y})\bar{X}Y = 0$

(b) $(X + \bar{Y} + X\bar{Y})(XY + \bar{X}Z + YZ) = XY + \bar{X}\bar{Y}Z$

(c) $(AB + C + D)(\bar{C} + D)(\bar{C} + D + E) = AB\bar{C} + D$

3.7   Using the postulates and theorems of Boolean algebra, simplify the following to a form with as few occurrences of each variable as possible.

(a) $(X + \bar{Y})[XYZ + \bar{Y}(Z + X)] + XY\bar{Z}(X + \bar{X}Y)$

(b) $(X + \bar{Y}\bar{X})[XZ + X\bar{Z}(Y + \bar{Y})]$

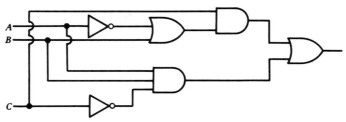

Figure P3.4

3.8 Simplify the following as far as possible by manipulation of Boolean algebra.

(a) $X\bar{Z}Y + (X\bar{Z}Y + Z\bar{X})[Y(Z + X) + \bar{Y}Z + \bar{Y}X\bar{Z}]$

(b) $[(a \downarrow b)] \uparrow [(a + (\bar{a} + xz)\bar{a})(\bar{z} + \bar{a})]$

3.9 Repeat Problem 3.8, utilizing a Venn diagram.

3.10 Prove that $(a + b)(\bar{a} + c)(b + c) = (a + b)(\bar{a} + c)$.

3.11 By manipulation of Boolean algebra verify that

$$[\bar{X}_1\bar{X}_2(X_3X_1 + \bar{X}_2)] + (X_1 + X_2)\overline{(X_1\bar{X}_2\bar{X}_3)} \downarrow (\overline{X_1X_2X_3})$$
$$= \bar{X}_2\bar{X}_3 + \bar{X}_1X_3$$

3.12 Let the dual of a Boolean function f be designated by f^D. Prove the theorem $f^D(X_1, X_2, ..., X_n) = \bar{f}(\bar{X}_1, \bar{X}_2, ..., \bar{X}_n)$ for all functions of one, two, or three variables. Assume that the variables may take on only the values 1 and 0.

3.13 A Boolean function F is called self-dual if the dual of the function is equal to the function itself ($F = F^D$). Show that there are $2^{2^{n-1}}$ self-dual functions of n variables.

3.14 Prove that

$$f(X_1, X_2, ..., X_n) = X_1 f(1, X_2, ..., X_n) + \bar{X}_1 f(0, X_2, ..., X_n)$$

where the variables may take on only the values 1 and 0.

3.15 State and prove the dual of the expression in Problem 3.14.

3.16 Prove that exclusive-OR is associative, that is,

$$A \oplus (B \oplus C) = (A \oplus B) \oplus C$$

3.17 Prove that AND is distributive over exclusive-OR, that is,

$$A \cdot (B \oplus C) = A \cdot B \oplus A \cdot C$$

3.18 Consider the possibility of a Boolean algebra of three elements, say, 0, 1, and 2. Show that a Boolean algebra cannot be defined for three elements.

3.19 Honest John, the used-car dealer, has some two-toned, eight-cylinder cars. He also has in stock some air-conditioned cars and some cars with fewer than eight cylinders, all of which are overpriced. Another dealer, Insane Charlie, has in his lot some cars without air-conditioning or eight cylinders that are not overpriced. He also has some air-conditioned, eight-cylinder models and some overpriced solid-color cars. Write the simplest possible Boolean expression for the categories of cars currently stocked by both dealers.

BIBLIOGRAPHY

1. Huntington, E. V. "Sets of Independent Postulates for the Algebra of Logic," *Trans. Am. Math. Soc.*, **5**:288–309 (1904).

2. Sheffer, H. M. "A Set of Five Independent Postulates for Boolean Algebras, with Applications to Logical Constants," *Trans. Am. Math. Soc.*, **14**:481–488 (1913).

3. Boole, G. *An Investigation of the Laws of Thought.* Dover Publications, New York, 1854.

4. Shannon, C. E. "Symbolic Analysis of Relay and Switching Circuits," *Trans. AIEE*, **57:**713–723 (1938).
5. Carroll, L. *The Complete Works*. The Nonesuch Press, London, 1939.
6. Stone, H. S. *Discrete Mathematical Structures and Their Applications*. SRA Inc., Palo Alto, Calif., 1973.
7. Tremblay, J. P. and R. Manochar. *Discrete Mathematical Structures and Applications to Computer Science*. McGraw-Hill, New York, 1975.

CHAPTER 4

LOGIC ELEMENT REALIZATION

4.1 OVERVIEW

The ideal logic elements of Section 1.6 were defined independent of implementation. The goal of this chapter is to consider alternative physical realizations of these elements and to determine under what conditions they will function consistent with the ideal elements. At this writing CMOS and NMOS are the most widely used very large-scale integrated circuit (VLSI) technologies. Consequently, the most refined computer-aided logic design tools are available for these technologies. MOS is the acronym for *metal oxide semiconductor*. Metal layers continue to play an important role in MOS and other integrated circuit technologies, but metal is no longer part of the active devices, as it was when the term MOS came into use.

Because the MOS device is closest to a perfect switch, we shall begin our treatment of physical logic elements in the next section with circuits composed of ideal MOS devices or perfect switches. Other technologies do enjoy some advantages over MOS (usually speed) and cannot be neglected. Consideration of other logic families will follow in Section 4.3, beginning with the introduction of the diode as the principal component of several families of logic elements. Finally, the discussion will return to consider physical MOS devices in more detail. We shall see that these second-order device properties are important in NMOS logic design and in the modeling of both NMOS and CMOS logic elements for gate- and switch-level simulation (Chapter 12).

4.2 LOGIC ELEMENT REALIZATION WITH IDEAL SWITCHES

The simplest switching device is the switch itself. A switch is any mechanical device by means of which two (or more) electrical conductors may be conveniently connected or disconnected. The simplest form of a switch consists of two strips of spring metal on which are mounted electrical contacts. A lever or push-button controls whether the switch is *open* (contacts separated) or *closed* (contacts touching). The manner in which switches may be used to implement logic functions is illustrated in Fig. 4.1. In the circuit of Fig. 4.1a, the light bulb will be lit only if both switches A AND B are closed. In Fig. 4.1b, the light bulb will be lit if either switch A OR B is closed.

The MOS logic family is based almost entirely on a single device, a MOS field effect transistor (FET), that very closely approximates an ideal switch. Shown in Fig. 4.2a is the standard symbol for the NMOS transistor with the accepted names for its three terminals. Actually, the device is symmetric, and the source and drain can be readily interchanged. The gate may be regarded as the input

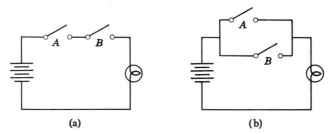

Figure 4.1   Switch realizations of AND or OR.

line. As shown in Fig. 4.2*b*, the device is a close approximation of an open switch, when the gate voltage V_x is close to 0. When V_x is a positive voltage, typically 1 to 5 V, the switch is closed and behaves as a very small electrical resistance. This situation is depicted in Fig. 4.2*c*. In the logic circuits to be described below, we shall let 0 V represent logical 0 and the positive voltage (for convenience 5 V) represent logical 1. In Section 4.6 we shall encounter a similar device, the PMOS switch that is closed by a negative gate voltage.

The fundamental MOS logic element is the inverter shown in Fig. 4.3*a*. The upper (pull-up) device is actually a depletion mode transistor designed to function as a relatively large resistor when connected as given and the lower transistor switch is closed. For the purpose of this simplified discussion, you should simply regard it as a large resistor as shown in Fig. 4.3*b*. Now when *x* is logical 0 and V_x is 0 V as in Fig. 4.3*c*, the lower transistor (pull-down) switch is open. Because no current is present in the resistor *R*, the output voltage is +5 V and the logical output *z* is 1. When *x* = 1 and V_x is positive, the lower transistor switch is closed, as depicted in Fig. 4.3*d*. Now, the output is connected to 0 V and *z* = 0. These values are tabulated in Fig. 4.3*e*. We observe from this table that the circuit of Fig. 4.3*a* implements the logical NOT operation or is indeed an inverter, as asserted at the beginning of the paragraph. The standard symbol for the inverter is repeated in Fig. 4.3*f*.

Next consider the circuit of Fig. 4.4*a* in which a second pull-down transistor has been added in series. Now both inputs must be logical 1, if the output is to be pulled down to 0 V. If either or both devices has a 0 input, the output will be +5 V or logical 1. The tabulation of these values in Fig. 4.4*b* is consistent with those of the logical NAND gate, the standard symbol for which is repeated in Fig. 4.4*c*. Figure 4.5 shows a similar device in which the two pull-down transistors are connected in parallel. Now the output *z* will be 0, if either input is 1. It will be logical 1, if both inputs are 0. This device is a NOR gate. An AND gate may be realized by adding an inverter to the output of a NAND gate, as shown in Fig. 4.6. This may be verified by considering all possible combinations of input values as given in

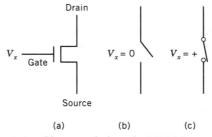

Figure 4.2   Characteristics of a NMOS transistor.

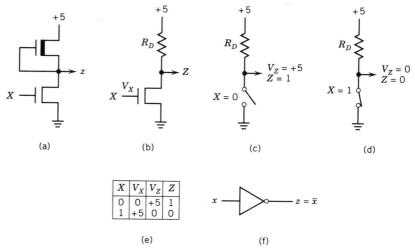

(a) (b) (c) (d)

X	V_X	V_Z	Z
0	0	+5	1
1	+5	0	0

(e) (f)

$z = \bar{x}$

Figure 4.3 Ideal NMOS inverter.

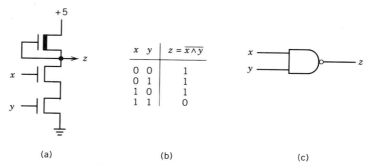

x	y	$z = \overline{x \wedge y}$
0	0	1
0	1	1
1	0	1
1	1	0

(a) (b) (c)

Figure 4.4 NMOS NAND gate.

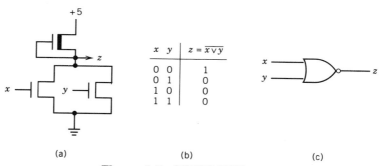

x	y	$z = \overline{x \vee y}$
0	0	1
0	1	0
1	0	0
1	1	0

(a) (b) (c)

Figure 4.5 NMOS NOR gate.

Fig. 4.6*b*. The standard symbol for an AND gate is repeated in Fig. 4.6*c*. Adding an inverter to the NOR yields an OR gate.

Looking again at Fig. 4.6*a* will reveal a capacitor depicted at the gate input of the second inverter. This capacitor is not a separate component, but instead represents an inherent property of the MOS device. The gate of a MOS transistor will draw no steady-state current. The input resistance is infinite. However, when the value of the line connected to the device input changes, time is required for the charging or discharging of the capacitor to the new value.

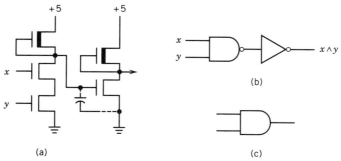

(a)

(b)

(c)

Figure 4.6 AND gate.

4.3 BIPOLAR DIODE-TRANSISTOR LOGIC

The earliest electronic logic elements included vacuum tubes and diodes. When vacuum tubes gave way to bipolar transistors in the late 1950s, diodes remained. The symbol for a diode is shown in Fig. 4.7a. Ideally, a diode offers zero resistance to current from anode to cathode (forward direction) and infinite resistance in the reverse direction. The ideal current voltage characteristic is shown in Fig. 4.7b.

Circuits for the two forms of diode gates are shown in Fig. 4.8. There are two possible input voltages V_+ and V_- that may differ slightly from the supply voltage V_H and 0, respectively. In Fig. 4.8a, $e_0 = V_+$ only if both inputs are V_+. If one or more inputs is V_-, then $e_0 = V_-$. This behavior is summarized in Fig. 4.8b. In the circuit of Fig. 4.8d, $e_0 = V_-$ only if both inputs are V_-. If one of the inputs is V_+, then $e_0 = V_+$. This behavior is summarized in Fig. 4.8e. Letting V_H be logical 1 and V_1 be logical 0 (positive logic) allows the translation of the voltage tabulations to the truth tables of Figs. 4.8c and 4.8e. Given positive logic, Fig. 4.8a is an AND gate and Fig. 4.8d an OR gate.

A reversal of the logical assignment of V_+ and V_- is *negative logic* and causes a reversal of the significance of the two logic networks given in Fig. 4.8. There are physical reasons for using negative logic on occasion. The natural inclination, however, is to let the voltage nearest to 0 V represent logical 0. Such will be the practice throughout this book.

If the diodes were ideal, there would be no limit on the number of inputs on a single gate (*fan-in*), or on the number of other gates a single gate could drive (*fan-out*), or on the number of gates that could be connected in series one after another (levels of gating). However, the voltage across conducting diodes and the finite reverse current of nonconducting diodes result in a continual degradation signal level as the circuit complexity increases. For example, even in a single gate, the

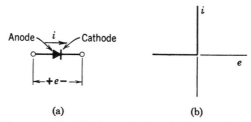

(a)

(b)

Figure 4.7 Diode symbol and characteristics.

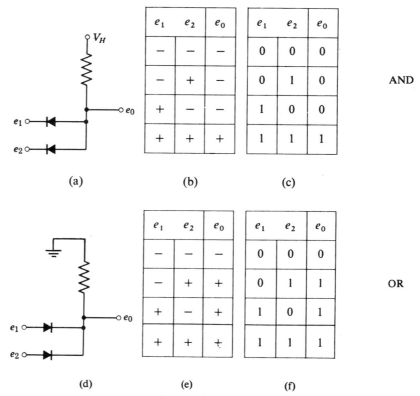

(a)

e_1	e_2	e_0
–	–	–
–	+	–
+	–	–
+	+	+

(b)

e_1	e_2	e_0
0	0	0
0	1	0
1	0	0
1	1	1

AND

(c)

(d)

e_1	e_2	e_0
–	–	–
–	+	+
+	–	+
+	+	+

(e)

e_1	e_2	e_0
0	0	0
0	1	1
1	0	1
1	1	1

OR

(f)

Figure 4.8 Positive logic diode AND and OR gates.

output voltage will always be slightly different from the input voltages because of the drops across the conducting diodes.

The signal degradation problem of diode gates is usually addressed by following each diode network with a bipolar transistor inverter, which serves as an amplifier. A positive voltage on the base of the NPN transistor depicted in Fig. 4.9a causes current to flow from collector to emitter, just as a positive gate voltage on a NMOS device establishes a path from the drain to source. A bipolar transistor is a poorer approximation of an ideal switch than the MOS device for two reasons. (1) When the device of Fig. 4.9a is "on," current flows into the base. The input impedance is not infinite. (2) The collector and emitter terminals are not interchangeable. Current must flow from collector to emitter, **accounting for the term bipolar**.

A typical supply and logical 1 voltage for the elementary inverter of Fig. 4.9b is 5 V. If e_i is 5 V, the base will be positive, and the maximum or *saturation current* will flow from collector to emitter. The output voltage e_0 will be in the range of 0.1–0.5 V. When e_i is less than some threshold (typically 1 V), e_0 will be 5 V. A diode AND network and an inverter can be merged to form a diode-transistor (DTL) NAND gate as depicted in Fig. 4.10a. The relative voltage values are shown in Fig. 4.10b and the logical values in Fig. 4.10c. A NOR gate is similarly composed of an OR gate and inverter.

The output of a diode network without an inverter cannot be expected to drive or fan-out to more than one similar network. A bipolar inverter, NAND, or NOR gate may drive several gates, but there is a fan-out limit. Consider the situation shown in Fig. 4.11, in which one NAND gate drives another. When the output of

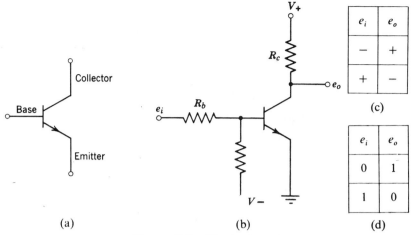

Figure 4.9 Bipolar inverter.

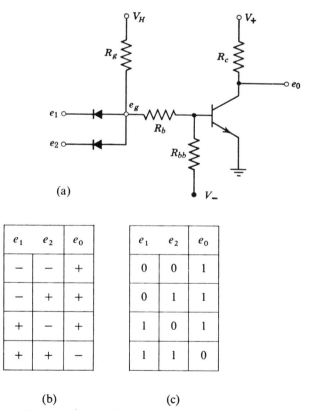

Figure 4.10 Diode-transistor NAND gate.

the first gate is at the low level (transistor in saturation), a load current I_L will flow as shown. If other gates are connected in parallel, each will contribute a similar I_L. However, there is a limit, set by base current and current gain, to the amount of current that can flow into the transistor if it is to remain in saturation. If so many gates are connected that this limit is exceeded, the transistor will come out of saturation and the output voltage will rise above the low logic level.

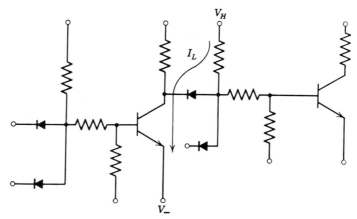

Figure 4.11   Loading a NAND gate.

4.4   SPEED AND DELAY IN LOGIC CIRCUITS

Faster operation generally means greater computing power, so there is a continual search for faster and faster logic circuits. The development of faster circuits is the province of the electronic circuit engineer, not the logic designer, and the actual speed of the electronics is of only indirect interest to the logic designer. However, the qualitative nature of the transitions and delays in logic circuits plays an important role in the logical theory of sequential circuits.

Figure 4.12 shows a grounded-emitter inverter and the response to a positive pulse at the base, taking the transistor from cutoff to saturation and back. A number of complex physical factors enter into the determination of this response, of which we can consider only a few. First, from the time of the start of the base pulse, there is a delay time t_d before the collector current starts to rise. This delay is primarily caused by the effective base-to-emitter capacitance of the transistor (shown dotted in Fig. 4.12a). This capacitance must be charged to a positive level before the transistor starts to turn on. After the transistor starts to conduct, there is

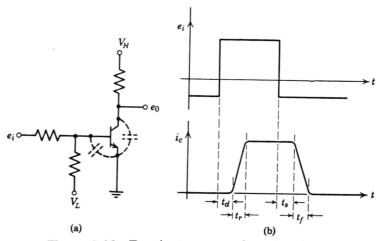

Figure 4.12   Transient response of transistor inverter.

a finite rise time t_r primarily controlled by the collector capacitance. On the trailing edge of the base pulse, there is again a delay t_s before the collector current starts to drop. This delay is due to the base-to-emitter capacitance, as well as the storage of charge in the base region during saturation. The delay due to the storage effect t_s is longer than t_d. Finally, there is a finite fall time t_f, again due to the collector capacitance.

In terms of specifying delay times through gates, the times t_d and t_s shown in Fig. 4.12*b* are meaningful only if the input rise and fall times are negligible compared to those at the output. This is rarely the case, since logic circuits are usually driven by other logic circuits of the same type, so that rise and fall times are comparable at input and output and are generally of the same order of magnitude as the delay times. It is thus necessary to specify just what points on the input and output waveforms are to be used in specifying delay times.

In this regard, it is important to note that the transistors in logic circuits are operating in the switching mode. Even though the input may vary over the full logic range (0–3.6 for transistor–transistor logic—TTL), only a very small change in input voltage is required to cause switching. For TTL the output switches fully from one logic level to the other as the input passes through a narrow range of about 0.1 V, nominally centered at 1.5 V. Unless the input change is so slow that it takes longer than the nominal rise time of the circuit to pass through this 0.1-V range, which is most unlikely, the output rise and fall times will be totally determined by the internal parameters of the logic circuit.

On the basis of this switching at a nominal voltage of 1.5 V, delay time is normally measured from the time the input waveform passes through 1.5 V until the corresponding output change passes through 1.5 V. A typical test setup for measuring delay time is shown in Fig. 4.13, along with the corresponding waveforms. As shown, the gate under test is driven by another gate of the same type, which is in turn driven by a drive pulse having rise and fall times of the same order of magnitude as those of the gates.

As shown, the delay times are measured from the 1.5-V points of the input transitions to the 1.5-V points of the corresponding output transitions. Because one transition involves the transistor turning on, the other the transistor turning off, the delays, as noted above, are not equal. The notation for these delays is unfortunately not standardized among manufacturers. The most common notations are t_{pd0} or t_{pd-} for the negative-going output transition and t_{pd1} or t_{pd+} for the positive-going output transition. The average gate delay t_{pd} is the arithmetic mean of these two delays. The transition voltages will be different for other logic families. The user should consult the manufacturer's data sheet for details.

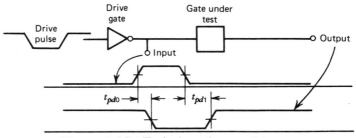

Figure 4.13   Typical test for gate delay.

4.5 TWO MORE BIPOLAR LOGIC FAMILIES

In the mid–1960s logic circuits constructed from discrete components, resistors, transistors, diodes, and the like began to disappear and the integrated circuit, an entire circuit fabricated as a unit on the same chip, soon dominated. First to appear were small-scale integrated circuits, SSI, which consisted of one to four logic gates or memory elements packaged as a unit, often in the dual-in-line package (DIP) shown in Fig. 4.14. There remain applications in which levels of integration higher than SSI and MSI (medium-scale integration) cannot be conveniently applied. MSI parts are usually packaged subsystems near enough to the low end of the complexity scale so that fixed configurations will have many applications.

Over the years, the most popular SSI and MSI logic family has been TTL. We shall introduce TTL by relating it to the already discussed DTL. The integrated circuit version of the DTL NAND gate is given in Fig. 4.15. The extra transistor adds amplification to increase fan-out. The extra diode provides biasing for the output transistor, thereby avoiding a separate negative power supply.

The first change from DTL to TTL is based on the fact that, in integrated circuits, it is easier to fabricate transistors than diodes. When diodes are required, it is usual to use the base-emitter junction of a transistor, with the base serving as an anode, the emitter as a cathode, and the collector tied to the base. In the circuit of Fig. 4.15, we note that the anodes of the input diodes are common, so that these

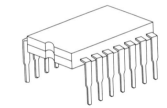

Figure 4.14 Dual-in-line package.

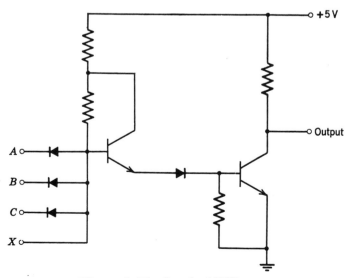

Figure 4.15 Standard DTL gate.

diodes can be realized in the form of a multiple-emitter transistor, as shown in Fig. 4.16*a*.

The remaining changes in the evolution from DTL to TTL are made to achieve increased speed. When the circuit of Fig. 4.16*a* is to switch from the low-output condition to the high-output condition, transistor $Q1$ must go from saturation to cutoff. As noted earlier, this requires removal of charge from the base of $Q1$. This switching action is brought about by transistor $Q2$ going from saturation to cutoff, so that the only path for discharging the base of $Q1$ is through resistor $R1$, and the speed of this discharge is limited by the time constant of this circuit. To obtain faster switching, we take advantage of the fact that the input diodes are realized in terms of a transistor and change the circuit to that shown in Fig. 4.16*b*. Now the charge on the base of $Q1$ is removed through transistor $Q3$, resulting in a considerable reduction in the storage delay time t_s.

Another factor limiting the transition speed is the collector capacitance, which must be charged as the output voltage switches from the low to high value. In the circuit of Fig. 4.16*b*, the only path for charging this capacitance is via the collector resistance $R2$. The charging speed can be increased by reducing $R2$, but this will result in increased power dissipation. An output circuit of this form is

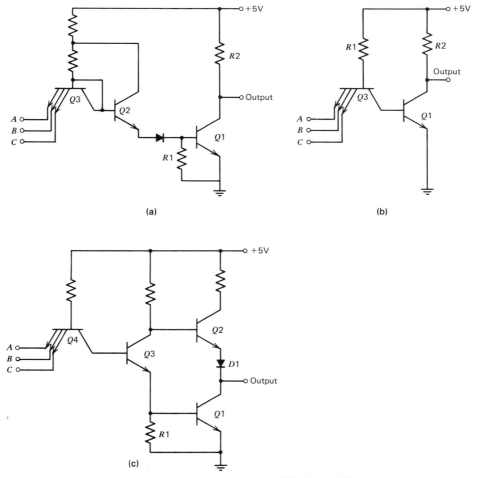

Figure 4.16 Evolution of TTL from DTL.

known as a *passive pull-up* circuit, since the output capacitance is "pulled up" via the passive element *R*2. An alternative that provides faster charging without increased power dissipation is the *active pull-up* circuit shown in Fig. 4.16*c*. In the low-output condition, the *phase-splitter* transistor *Q*3 is in saturation and the saturation current through *R*1 results in a sufficiently positive voltage to saturate output transistor *Q*1. Because of the diode *D*1, the emitter of output transistor *Q*2 is more positive than the collector of *Q*3, so that *Q*2 is cut off. When the gate goes to the high-output condition, transistor *Q*3 cuts off, so that the base of *Q*1 goes to 0, the base of *Q*2 goes sharply positive, *Q*1 cuts off, and *Q*2 saturates. Thus, the output circuit essentially acts as a two-pole switch, switching the output between ground and the supply voltage. Because of voltage drops in the output circuit, the actual output voltages, with V_{CC} = 5*v*, are 0.2 (low) and 3.4 V (high). In terms of speed, the result of these changes in circuitry is that the typical TTL gate is approximately three times as fast as the corresponding DTL gate. TTL fan-out is typically 8 to 10, expressed as the number of similar gate inputs that can be driven.

One important advantage of TTL is that the speed and power can be varied over quite a wide range simply by changing the size of the resistors. Increasing the size of the resistors decreases the speed and power; decreasing the size of the resistors increases the speed and power. Whatever may be done to the resistance values, the speed of TTL is ultimately limited by the time required to pull the output transistors out of saturation. In *Schottky* TTL (STTL), the transistors are kept out of saturation by Schottky barrier diodes connected between the base and collector. Standard STTL (74S series) has typical gate delays and power dissipation of 3 nsec and 20 mW/gate, and low-power STTL (74LS series) has typical gate delays and power dissipation of 10 nsec and 2 mW/gate. The 74LS series is the current industry-standard form of TTL.

The fastest form of logic currently available is ECL (emitter-coupled logic), in which a totally different circuit configuration is utilized to ensure that all transistors are operated in the *active* region between saturation and cutoff. The circuit of the standard ECL gate is shown in Fig 4.17. The theory of operation will not be discussed, because it is rather complex, but the reader will note that the gate provides both OR and NOR outputs. This is a distinct advantage, because it virtually eliminates the need for separate inverters. The logic levels are typically −0.75 and

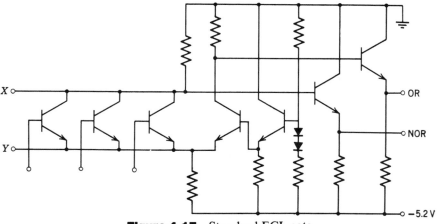

Figure 4.17   Standard ECL gate.

-1.75 V. The inputs per gate and gates per packages are somewhat less than for TTL, but the fan-out is considerably increased, typically 20–25. Expansion, to a maximum fan-in of 20, is possible by connecting additional input transistors between points X and Y. Several types of ECL circuits are available, but the two most commonly used provide average gate delays of 2 and 0.7 nsec, with power dissipations of 25 and 50 mW/gate, respectively.

4.6 COMPUTER-AIDED DESIGN OF VLSI

Integrated circuits offer significant advantages over discrete circuits in three primary areas: size, power, and cost. A number of complete gate circuits can be deposited on a single chip about the size of the head of a common pin. Practical problems of connecting these minute circuits may require mounting them in much larger packages, but the size advantage relative to discrete circuits is nevertheless substantial.

Closely related to the size advantage is the reduction in power dissipation. Whatever the size of the individual circuit packages, the density with which large numbers of packages can be mounted is largely determined by their power dissipation because of the problems associated with dissipating the heat generated by large numbers of closely packed circuits. The reduced power requirements also reduce power supply requirements, providing for still further reductions in equipment size and cost. Some families of IC logic require so little power that battery-powered operation of even very complex devices becomes practical, a real advantage when portability is important.

Perhaps most important are the reductions in cost made possible by integrated circuits. The fabrication processes of integrated circuits are basically similar to those of transistors and diodes, and the cost of a complete integrated circuit is generally comparable to the cost of a single transistor. Furthermore, all the costs of interconnecting discrete components are eliminated. As a result, the cost of logic has been reduced dramatically, to the point where the logic cost is a minor part of the cost of most digital systems.

The gate/package upper bounds of SSI and MSI were defined by the expected range of applications of each individual part. Except for special structures such as memories and microprocessors, relatively few fixed parts larger than 100 gates are considered general-purpose. The special-purpose parts in this complexity range have come to be known as ASICS, application-specific integrated circuits. Depending on complexity, ASICS may be considered LSI or VLSI.

When all aspects of each particular VLSI design are open to unconstrained optimization with respect to speed, power consumption, chip size, etc., the design process is called *full custom*. A full custom design consists of a complete set of masks, one for each step in the fabrication process for the particular technology used. The data files required to specify mask sets is created using a layout editor. With a layout editor as the only CAD tool, design costs for a custom VLSI chip can increase as the second power of circuit complexity, as suggested by Fig. 4.18*a*. This design cost/complexity relationship is less severe for designs realized by the interconnection of SSI and MSI parts.

The cost of an integrated circuit will include design costs, which are fixed regardless of the number of copies produced, together with incremental production and testing costs. Figure 4.18*b* depicts design and fabrication costs for some typical digital system, for which more than one approach to realization merits

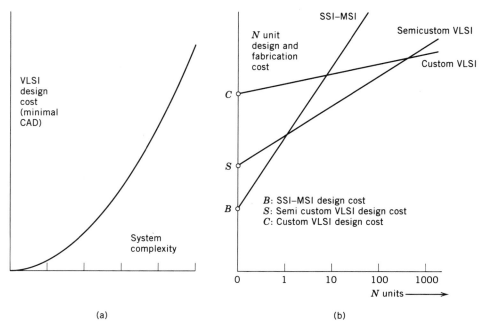

Figure 4.18   Typical design and fabrication cost alternatives.

consideration. At one extreme, it might be realized as a VLSI ASIC. At the other extreme, the system might be constructed by interconnecting SSI and MSI parts on a printed circuit board. Per unit production costs are much less for the VLSI realization, but design costs are greater.

For low-volume parts or when the ultimate production volume is unclear at the onset of design, the design costs suggested by Fig. 4.18 make the full custom procedure an unattractive alternative for ASIC design. Vendors have developed several distinct design methodologies, all aimed at reducing "one-time" costs both for the vendor and user of the ASIC. These include gate arrays, standard cells, programmable logic arrays, and storage logic arrays (SLA's). All these approaches have been adapted to take advantage of available computer-aided logic design techniques to reduce design costs. The design cost characteristic marked semicustom in Fig. 4.18*b* is intended to represent this entire family of approaches. All these approaches constrain and simplify design at the price of increases in chip area and, in some cases, reduced speed for a given design target. Clearly, the points of intersection in Fig. 4.18*b* will vary with the semicustom approach chosen and the individual design task. A major goal of this book is to develop the fundamentals of each semicustom approach, so that the reader may use it with the support of available design tools.

4.7   MOS SWITCH REVISITED

Although a few readers may eventually be involved in the design of LSI digital networks using the bipolar circuits discussed in Sections 4.3 and 4.5, more likely these technologies will be encountered as already realized SSI or MSI building blocks to be integrated at the circuit board level. The MOS technology, to be developed in more depth in this section, is more likely to be encountered in VLSI design. The smallest primitive logic element in TTL or ECL technology is the

gate. Therefore, design may be accomplished while considering only the properties specified for gates. Although these gate properties may depend on internal
circuit properties, the designer need not master the details of the circuits themselves. MOS devices may similarly be used to form gates that may, in turn, be
treated as primitive elements. However, it is also possible to combine MOS
devices into subnetworks that accomplish logic, but are not gates. Therefore,
understanding the properties of the devices themselves becomes more critical.

The fundamental MOS device (actually an n-channel enhancement-mode
device) is symbolized in Fig. 4.19a. As suggested by the open switch at the right,
this device functions very much as a voltage-controlled switch. The controlling
input is the gate, labeled g, in Fig. 4.19a. A very important property of the MOS
device is that the gate draws no current, regardless of the voltage at point g. We say
that the input impedance is infinite. The voltage at the gate input does, however,
control the existence or nonexistence of current between points 1 and 2. Associated
with every MOS device is a threshold voltage V_{th}. Often, this threshold is about
$+0.2V_{dd}$. V_{dd} is the supply voltage. If as in Fig. 4.19a, the voltage between point g
and both of the terminals 1 and 2 is less than V_{th}, then the switch is open as
depicted, and no current flows between points 1 and 2.

Figure 4.19b shows the upper terminal to be somewhat more positive than the
lower terminal, which we label s. Now the gate voltage V_{gs} exceeds V_{th}, so current
flows from the more positive point d (drain) to the less positive point s (source). As
shown at the right of Fig. 4.19b, the switch is closed. Although outside the scope of
this book, it can be demonstrated that for small values of V_{ds}, the current I_{ds} is
approximated by Equation 4.1, where β is a constant.

$$\text{Linear: } V_{ds} < V_{gs} - V_{th} \qquad I_{ds} = \beta(V_{gs} - V_{th} - V_{ds}/2)V_{ds} \qquad (4.1)$$

The maximum I_{ds} for fixed V_{gs} occurs at

$$V_{ds} = V_{gs} - V_{th}$$

Beyond that point, the device is said to be in the saturation region, with I_{ds} not a
function of V_{ds} as given by Equation 4.2.

$$\text{Saturation: } V_{ds} > V_{gs} - V_{th} \qquad I_{ds} = \beta(V_{gs} - V_{th})^2 \qquad (4.2)$$

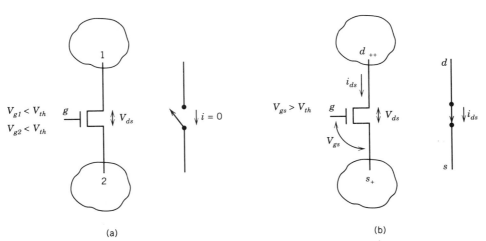

Figure 4.19   NMOS switch.

Figure 4.20 clearly indicates the linear and saturation regions for typical voltage values.

4.8 STRUCTURE OF *n*-CHANNEL AND *p*-CHANNEL DEVICES

The layout of MOS devices on the chip is relevant to their application in logic design. Eventually, the logic designer's interest in layout will be confined to a two-dimensional map of devices on the available chip area. For this to be meaningful, we must first look briefly at a model of the solid-state NMOS enhancement-mode device (the positive gate voltage enhances the density of electrons in the channel) as it appears in three dimensions in Fig. 4.21. The material called the substrate is *p*-type silicon, that is, silicon containing impurity atoms with the result that the majority of available charge carriers are positively charged carriers or holes. The regions marked *n* result from the controlled diffusion of another impurity into the substrate that makes electrons available to move about as majority charge carriers. Conduction takes place only when carrier electrons are attracted into the region marked channel by a positive charge on the gate. The flow of electrons will be from the less positively charged source to the more positively charged drain.

The electrons move the length of the channel, marked *L* in Fig. 4.21. The channel width is *W*. The channel geometry is defined by the combined presence of *n*-type material at each end and a thin overlayer of polysilicon gate material insulated from the channel by an oxide. Although not shown in Fig. 4.21, both polysilicon and diffusion areas together with superimposed metal layers may be used to interconnect points within the overall integrated circuit. The constant β introduced in the previous section may now be written as $\beta = K(W/L)$, where K depends on properties of the channel material and thickness of the oxide. Conduction in the channel resembles conduction in any other resistive area. Conductance is proportional to W/L.

The threshold voltage V_{th} can be reduced by implanting ions in the channel. When the threshold voltage of a device is reduced to about $-0.8V_{dd}$, the result is called a depletion-mode device. The threshold voltage of a depletion device will

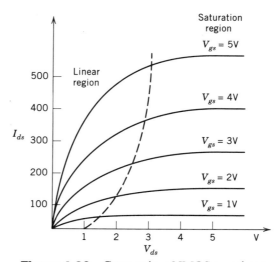

Figure 4.20 Current in a NMOS transistor.

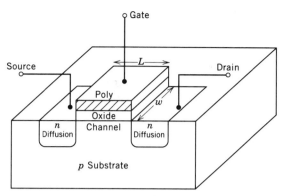

Figure 4.21 Structure of a NMOS transistor.

be referred to as V_{dep}. We shall see in the next section that the depletion device is to be preferred over a resistor as a pull-up or load in a NMOS inverter. A variety of symbols are used in the literature to represent depletion-mode devices. We shall be consistent in the use of the symbol given in Fig. 4.22b. The relation that must be satisfied to turn each of the devices on is also given in Fig. 4.22. Typical values in terms of the positive supply voltage V_{dd} are given to make clear which thresholds are negative.

The remaining symbol in Fig. 4.22 is the p- channel enhancement-mode transistor. Looking back at Fig. 4.21, suppose that n- and p-type regions were everywhere reversed. That is, suppose that the substrate was made of n-type material and that p areas were formed at either end of the channel by diffusion. The result would be the p-channel device shown in Fig. 4.22c. Now conduction takes place when the charge on the gate is sufficiently negative with respect to the p-type source that positive charge carriers or holes are attracted into the channel. The circle included as part of the gate symbol denotes the p-channel device and suggests that V_{gs} must be less than some negative threshold value. PMOS, a technology that depended exclusively on p-channel devices, preceded NMOS in the evolutionary process. Now p-channel devices are used primarily in conjunction with n-channel devices to form a technology called CMOS (complementary MOS) to be described in the next section.

4.9   NMOS AND CMOS LOGIC ELEMENTS

The basic NMOS inverter is shown again in Fig. 4.23a. In NMOS, in contrast to a bipolar technology such as TTL, linear resistors are rarely used as the pull-up element in an inverter. Depletion-mode devices are used as pull-ups to simplify the

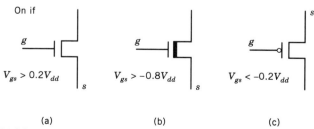

(a)   (b)   (c)

Figure 4.22 MOS device symbols. (a) n-channel enhancement. (b) n-channel depletion. (c) p-channel enhancement.

fabrication process and to provide greater current when the pull-down device is first turned off and a capacitive load must be charged.

In a custom layout of a VLSI digital system in NMOS, not all devices will have the same dimensions. Let us look at the issue of dimensions first in terms of the static characteristic of the inverter. Once the capacitive load is charged when V_{gs} is low, the output z will be approximately V_{dd}, regardless of device dimensions. If V_{gs} is logical 1 (typically V_{dd}), the current will flow in both the enhancement and depletion devices, and the output will assume some voltage V_{lo}. It is important that V_{lo} always be lower than the threshold voltage, so that it may be regarded as logical zero. If this is to be the case, the pull-down device will be operating in the linear region. The voltage across the pull-up device $V_{dd} - V_{lo}$ is large, so that this device is in the saturation region. Substituting $-|V_{dep}|$ for the threshold voltage in Equation 4.2 results in Equation 4.3.

$$I_{pu} = \frac{\beta_{pu}}{2}(0 + |V_{dep}|)^2 \tag{4.3}$$

The current in the pull-down device is given by Equation 4.1, with V_{lo} substituted for V_{ds}. Because the output z will drive the gates of other NMOS devices that have infinite input impedance in the steady state, the currents I_{pu} and I_{pd} must be the same, yielding Equation 4.4.

$$\beta_{pd}(V_{gs} - V_{th} - \frac{V_{lo}}{2}) V_{lo} = \frac{\beta_{pu}}{2}(V_{dep})^2 \tag{4.4}$$

Neglecting $V_{lo}/2$, solving for V_{lo}, letting $V_{dep} = -0.8V_{dd}$, and defining $R = \beta_{pd}/\beta_{pu}$ as the ratio of the β's of the pull-down and pull-up devices results in Equation 4.5.

$$V_{lo} = \frac{1}{2R} \frac{(0.8V_{dd})^2}{(V_{gs} - V_{th})} \tag{4.5}$$

Now we let $V_{gs} = V_{dd}$ and V_{th} assume the typical enhancement-mode threshold value $0.2V_{dd}$ to form the simple relationship between R and V_{lo} given by Equation 4.6.

$$V_{lo} = \frac{1}{2R} \frac{(0.8V_{dd})^2}{0.8V_{dd}} = \frac{0.4V_{dd}}{R} \tag{4.6}$$

If we insist that V_{lo} be $< 0.5V_{th} = 0.1V_{dd}$ to insure that any gate driven by z will be turned off, we arrive at what is typically considered a minimum β ratio, $R = 4$.

$$R = \frac{\beta_{pd}}{\beta_{pu}} = \frac{(W/L)_{pd}}{(W/L)_{pu}} > \frac{0.4V_{dd}}{0.1V_{dd}} = 4$$

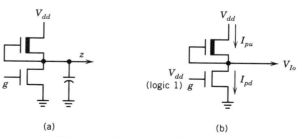

(a) (b)

Figure 4.23 An NMOS inverter.

We have now established a simple but fundamental relationship on the geometries of pull-up and pull-down devices in NMOS logic gates. We shall use and extend this relationship as we consider the unique aspects of NMOS logic design.

Next let us look at an inverter realized in CMOS technology, as symbolized in Fig. 4.24. A three-dimensional model of one approach to the fabrication of this device will be found in Fig. 4.26. We expect the reader who approaches a VLSI design and fabrication experience in CMOS to be more dependent on standard-device geometries provided by the silicon foundry broker than he or she would be for a NMOS design. Indeed, some readers might approach a NMOS design as a full custom project while using only "standard cells," which will be defined in Chapter 7, in a CMOS realization of the same network. Accepting this reality allows us to be less quantitative in our analysis of the more complicated CMOS structure.

Consistent with Fig. 4.22, the p-channel device labeled p in Fig. 4.24 will be turned on if

$$V_{\text{in}} - V_{dd} < -0.2V_{dd} \qquad \text{or} \qquad V_{\text{in}} < 0.8V_{dd} \qquad (4.7)$$

Using Equation 4.7 and the similar relation for the n-channel device allows us to generate the table of input/output values given in Fig. 4.25.

As shown, device p is off and device n is on when $V_{\text{in}} > 0.8V_{dd}$, and the output is 0, as would be expected in an inverter. When $V_{\text{in}} < 0.2V_{dd}$, device p is on, device n is off, and the output once the capacitor is charged is V_{dd}. The fact that under static conditions one of the two devices is always turned off is the principle advantage of CMOS logic. Static logic levels are not a function of device geometries. The static input level to an inverter must always be 0 or V_{dd}, and the output level is always V_{dd} or 0. In most digital systems, only a few elements are in the process of changing values at any given time. In a CMOS implementation, current will con-

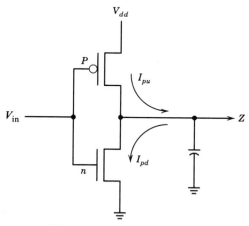

Figure 4.24 CMOS inverter.

Input		p	n	Z
Logic 1	$V_{\text{in}} > 0.8V_{dd}$	Off	On	0 V, logic 0
	$0.2V_{dd} < V_{\text{in}} < 0.8V_{dd}$	On	On	Depends on V_{in}, β_{pd}, β_{pu}
Logic 0	$V_{\text{in}} < 0.2V_{dd}$	On	Off	V_{dd}, logic 1

Figure 4.25 CMOS inverter values.

sequently flow in only a few devices at a given time. The resulting savings in power consumption is enormous. During a transition on input V_{in}, both the pull-up and pull-down devices in the inverter of Fig. 4.24 will conduct and power will be consumed. Now the second row of Fig. 4.25 applies, and device geometries will affect current in both devices and the current available to drive the capacitive load. A detailed analysis of this circumstance is beyond the scope of this book. Often, the smallest geometries that can be fabricated are used in order to minimize chip area for each design.

One approach to the realization of a CMOS circuit is given in Fig. 4.26. The process begins with a substrate of n-type silicon. By diffusion of the appropriate impurity, relatively large sections of the material are transformed to the p-type, to a depth well below the limits of the n and p channels. These areas are called p wells. Now the sources and drains for the n-channel devices are transformed back to n-type material by a second diffusion. One or several n-channel devices may be formed in a single p well. The sources and drains for the p-channel devices are diffused directly on the n-type substrate. The p well and n substrate are electrically isolated, but must be connected to reference voltages via the n and p plugs as shown.

This book is not intended to stand alone as a resource for full custom layout in either NMOS or CMOS technology. For a brief orientation on this subject, we rely on the widely accepted lambda- (λ-) based design rules developed by Mead and Conway [1]. Rather than in microns all dimensions will be specified in terms of λ, which represents the smallest distance separating two distinct areas on the chip. The smallest length or width of any device area, including metal connections, is 2λ. It is this minimum metal width that is conventionally used to identify a process. Therefore, for a 2-μ process, $\lambda = 1$.

To suggest how layout might be accomplished, let us pursue a two-input NMOS NAND gate. A device diagram of the NAND gate is given in Fig. 4.27a. How does the presence of two pull-down devices in series with the pull-up transistor affect the β ratio R and the channel dimensions as expressed in Equation 4.6? The two n channels in series may simply be regarded as two resistors in series. The width of the two channels will be the same. If the dimensions of the depletion device remain unchanged and V_{lo} is to satisfy the criteria that led to Equation 4.6, then channel resistance of the two enhancement-mode devices must be the same as the single device in the inverter.

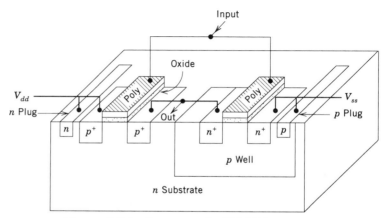

Figure 4.26 Structure of a CMOS inverter.

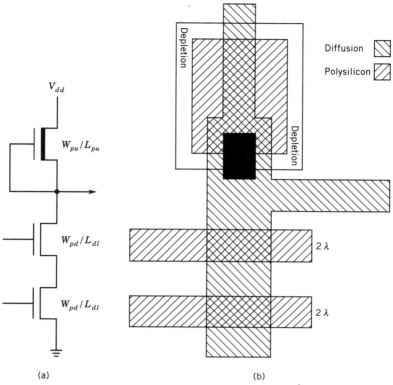

Figure 4.27 A Two-input NMOS NAND gate.

From Equation 4.6, $\beta_{pd}/\beta_{pu} = 4$ may be achieved for the inverter by setting

$$(W/L)_{pd} = 1 \quad \text{and} \quad (W/L)_{pu} = 1/4$$

Now let $W_{pd} = L_{pd} = 2L_{dl}$, where L_{dl} is the length of each of the two channels in series. Using the 2λ minimal dimension, we may satisfy these relations by $W_{pd} = 4\lambda$, $L_{dl} = 2\lambda$, $W_{pu} = 2\lambda$, and $L_{pu} = 8\lambda$.

A representation of the actual chip mask layouts for the NMOS NAND gate is shown in Fig. 4.27b. Dimensions may be deduced by comparison to the 2λ length of the two enhancement-mode channels. The channels are the areas where the polysilicon apparently overlays the diffusion. Depositing the polysilicon on the chip prior to the diffusion step actually prevents diffusion into the channels. The black area represents a buried metal contact that electrically connects the polysilicon to the diffusion area.

4.10 PASS TRANSISTOR

The principal difference between logic design with SSI logic gates and logic design on the MOS VLSI chip is the ability to use enhancement-mode devices as switches (much like relays) to steer current into gate inputs. Devices used in this manner are called *pass transistors* or *transmission gates*. The input signal V_s is steered through a pass transistor to form V'_{in} at the input of a NMOS inverter in Fig. 4.28a. The signal V_s could be the output of an arbitrary gate or switch network. For the pass transistor to be turned on, it is necessary that $V'_{in} < V_{dd} - V_{thb}$.

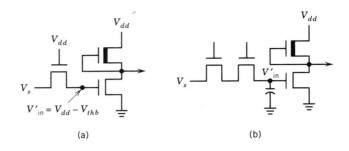

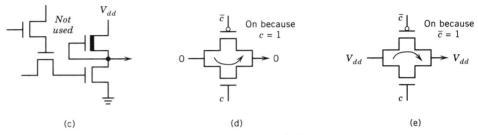

Figure 4.28 Pass transistor and transmission gate networks.

Because the pass transistor is not in a path to ground, a phenomenon known as body effect will increase the threshold V_{thb} from $0.2V_{dd}$ to $0.3V_{dd}$. In order that the output V_{lo} remain the same as for the inverter without a pass transistor, the β ratio R must be adjusted to compensate for the lower gate voltage. Now from Equation 4.5, we get

$$0.1V_{dd} > V_{lo} = \frac{(0.8V_{dd})^2}{2R(V_{dd} - V_{thb} - V_{th})} \tag{4.8}$$

$$R > \frac{(0.8V_{dd})^2}{2(0.1V_{dd})(0.5V_{dd})} = 6.4 \tag{4.9}$$

With the 2-λ minimal channel dimension, the β ratio will usually be doubled from 4 to 8 if a pass transistor network is included at the input of a gate.

Adding additional pass transistors in series, as shown in Fig. 4.28b, does not further reduce the static voltage at the inverter input. Once the gate capacitor is charged, there is no actual voltage V_{ds} across the pass transistors. The channel resistances in series will have a significant impact on the delay in charging or discharging the gate capacitor during transitions in the logic value of the inverter. Driving the gate of a pass transistor with another pass transistor as shown in Fig. 4.28c will reduce the inverter gate voltage by another threshold value. This configuration is never implemented.

It is the logic 1 signal that is reduced by the amount V_{thb} before it reaches the gate of the inverter. If V_s in Fig. 4.28b is 0 and the gates of the pass transistors are V_{dd}, the pass transistors remain turned on (in fact, the gate to source voltage increases) as 0 propagates through the network to V_{in}.

In CMOS even small static deviations of gate voltages from V_{dd} or 0 are intolerable. Fortunately, it is possible to avoid the degradation of the signal at the output of a CMOS pass network, regardless of whether the input is 1 or 0. This is accomplished by forming *a transmission gate* by connecting a *p*-channel device in parallel with an *n*-channel device, as shown in Fig. 4.28d. In this case, both devices will be "on," if $C = 1$. The transmission gate is said to be normally open. If the connections to the *p*- and *n*-channel devices are reversed, the transmission gate is normally closed. That is, it is on, if $C = 0$. Figure 4.28*d* shows the logic 0 or 0 V propagating through the *n*-channel device with no change in value, as was the case in the NMOS network. A V_{dd} logic level will pass unchanged through the *p*-channel device, as suggested by the arrow in Fig. 4.28*e*. We may conclude that in CMOS transmission gate networks have no impact on the design of the gates they drive.

4.11   NOISE MARGIN

The term *noise* refers to spurious signals generated in a system by any of a variety of environmental causes. In digital systems most noise is due to inductive or capacitive coupling between signal lines. For example, if the current in one line changes rapidly, the resultant rapid change in the surrounding magnetic field may generate a voltage in adjacent lines. Whatever the source of the noise, if it is large enough, it may change the signal enough to cause it to take on the opposite logical value. The *noise margin* is a measure of how much noise can be added to the logic inputs of a circuit without the circuit responding improperly.

Figure 4.29 shows the *transfer characteristic* of a TTL inverter, that is, a plot of output voltage V_O as a function of the input voltage V_I. Similar curves can be drawn for any family of logic. Because the use of actual voltages will make the basic concepts easier to understand, we will use the specific example of TTL in the discussion to follow, but the ideas involved apply to any family of logic. With a low-level voltage applied at the input, the output will be at the high level, nominally 3.4 V for TTL. The exact voltage will vary from one gate to another, depending on statistical variations in the characteristics of gates. To place bounds on this variation, the manufacturer specifies a minimum high-level output voltage, $V_{OH_{min}} = 2.4$ V. With this specification, the manufacturer guarantees that if you apply a low-level voltage at the input, the output will never be less than 2.4 V. Similarly, the manufacturer specifies a maximum low-level output voltage, $V_{OL_{max}} = 0.4$ V. If a high-level input is applied, the output is guaranteed not to exceed 0.4 V.

These two values, $V_{OH_{min}}$ and $V_{OL_{max}}$, place bounds on the output voltages, but what about the inputs? We need similar bounds to define the input levels. To define the low-level input, the manufacturer specifies $V_{IL_{max}} = 0.8$ V. This means that the circuit is guaranteed to interpret an input as low-level and produce the corresponding output (high for an inverter) if the input voltage does not exceed

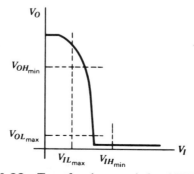

Figure 4.29   Transfer characteristic of TTL inverter.

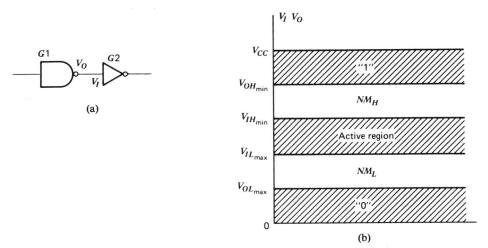

Figure 4.30 Noise margin in logic circuits.

0.8 V. Similarly, an input voltage is guaranteed to be interpreted as a high-level input as long as the voltage is not less than $V_{IH\min} = 2.0$ V.

The manner in which these specifications determine the noise margin can be considered in terms of Fig. 4.30a, showing one TTL gate driving another, together with a graphic interpretation of the relevant levels in Fig. 4.30b. Assume that the inputs to $G1$ are such that the output is at the low level. This voltage is guaranteed not to exceed $V_{OL\max} = 0.4$ V. The gate $G2$ will interpret an input signal as low-level as long as it does not exceed $V_{IL\max} = 0.8$ V. Thus, additive noise up to 0.4 V can appear on the line connecting these two gates, without causing gate $G2$ to respond improperly. We thus define the *low noise margin* as

$$NM_L = V_{IL\max} - V_{OL\max}$$

Similarly, if the output of $G1$ is at the high level, the voltage will be no less than $V_{OH\min} = 2.4$ V. Gate $G2$ will interpret the input as high-level as long as it is not less than $V_{IH\min} = 2.0$ V, so that noise can reduce the voltage from $G1$ by as much as 0.4 V without $G2$ responding improperly. We define the *high noise margin* as

$$NM_H = V_{OH\min} - V_{IH\min}$$

The overall noise margin is defined as the smaller of the two noise margins. For TTL they are the same and $NM = 0.4$ V.

The noise margin provides information as to how a logic family responds to noise. Equally important in considering the overall noise performance is the amount of noise generated by the logic. The noise margin of TTL is good, but TTL also generates a lot of noise, so that its overall noise performance is considered only fair. Of the three types of logic used in SSI and MSI, CMOS has the best noise performance and ECL the worst, with TTL in the middle. In LSI, CMOS again has the best noise performance.

PROBLEMS

4.1 Assume that a power NAND gate, implemented by the simple circuit given in Fig. P4.1, is to be used in a network composed only of NAND gates of the same type. Assume that $R_1 = 2000\ \Omega$, $R_L = 1000\ \Omega$, and β (current gain) = 200. Also assume that the base current = 0.5 ma when the gate is turned on.

(a) Determine a steady-state fan-out limit for the gate subject to these conditions.

(b) Considering delay, why might a lower nominal fan-out be set for the gate.

4.2 Suppose two gates with the specifications given in Problem 4.1 are to be connected in the network shown in Fig. P4.2, where gate 1 drives only gate 2 and the low-current lamp. What is the maximum dc resistance allowed for the lamp, if the circuit is to function properly? **Hint: The lamp will be lighted and $Z = 1$, when $A = B = C = 1$.**

4.3 Suppose the voltage waveforms A and B, shown in Fig. P4.3, were measured at the input and output of an inverter. Determine values for the net turn-on and turn-off delay of the inverter.

4.4 The turn-on delay is 8 nsec and the turn-off 12 nsec for all gates in the network of Fig. P4.4. Suppose that initially $A = B = C = 0$. Determine the time delay before output Z changes following a change on input C from 0 to 1.

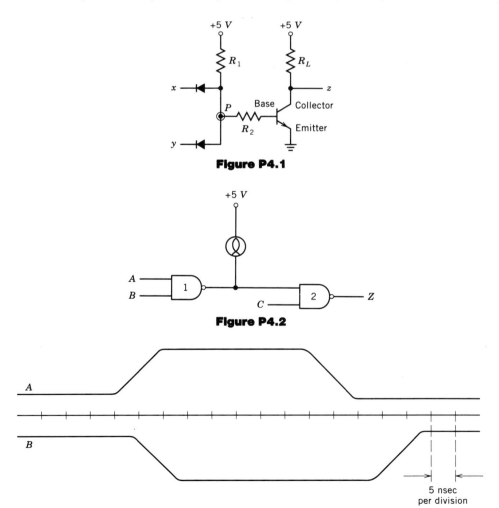

Figure P4.1

Figure P4.2

Figure P4.3

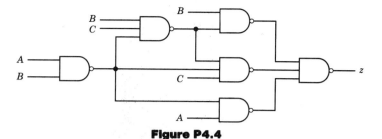

Figure P4.4

4.5 Consider a NMOS inverter with an enhancement-mode pull-down transistor with $V_{th} = 0.2V_{dd}$ and a depletion-mode pull-up device with $V_{dep} = -0.8V_{dd}$. As the input or gate voltage changes value (slowly, so that output current may be neglected), the output voltage will traverse between V_{lo} and V_{dd}. At some point, the input or gate voltage will equal the output voltage. Suppose this happens at $V_{in} = V_{out} = V_{dd}/2$.

   (a) Using the appropriate expressions for current in both devices, determine the value of the β ratio R of the pair of devices. Will the pull-up actually be in saturation if $V_{dep} = -0.8V_{dd}$?

   (b) What would the cross-over point ($V_{in} = V_{out}$) be if the β ratio R were 4, as determined in Equation 4.6. Again, use the saturation expressions for both currents.

4.6 Use the β ratio $R = 4$ determined from Equation 4.6. Determine a more accurate value of V_{lo} by not neglecting $V_{lo}/2$ in Equation 4.4.

4.7 Suppose that β is the same for the n- and p-channel devices in Fig. 4.24 and that $V_{th} = 0.2V_{dd}$ for the n-channel device and $-0.2V_{dd}$ for the p-channel device. Assume a very large load capacitance so that the output voltage changes slowly, as V_{in} goes from V_{dd} to 0.

   (a) If the output voltage is $0.05V_{dd}$ when V_{in} is $0.5V_{dd}$, use Equations 4.1 and 4.2 to determine an expression for the current into the capacitor in terms of β.

   (b) If the output voltage is $0.15V_{dd}$ when V_{in} is 0, use Equations 4.1 and 4.2 to determine an expression for the current into the capacitor in terms of β.

4.8 It is desired to implement a NMOS inverter on a chip to be processed as CMOS. Consequently, n- and p-channel enhancement-mode devices are available, but depletion-mode devices are not. Suppose the thresholds of the n- and p-channel devices are $0.2V_{dd}$ and $-0.2V_{dd}$, respectively. A pull-up device is provided by connecting the gate of a PMOS transistor to 0 V.

   (a) Use Equations 4.1 and 4.2, as appropriate, to determine the minimum β ratio R that will result in $V_{lo} = 0.1V_{dd}$.

   (b) How would this calculation differ, if the gate of the PMOS device were connected to the inverter output, as in the case of a depletion-mode device.

4.9 Construct a mask layout diagram for a NMOS NOR gate similar to that for the NAND gate as given in Fig. 4.27b. It is important that dimensions are such that the β ratio R is at least 4, but not larger than necessary.

4.10 Construct a mask layout diagram of the pass transistor inverter combination of Fig. 4.28a. It is important that channel dimensions be consistent with the assumed logic levels.

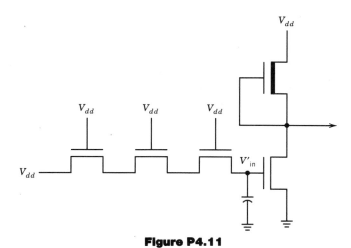

Figure P4.11

4.11 What is the value of V'_{in} in the pass transistor configuration shown in Fig. P4.11?

BIBLIOGRAPHY

1. Mead, C. and L. Conway. *Introduction to VLSI Systems*. Addison-Wesley, Reading, Mass., 1980.
2. Mukherjee, A. *Introduction to NMOS & CMOS VLSI Systems Design*. Prentice-Hall, Englewood Cliffs, N.J., 1986.
3. *The TTL Data Book*. Texas Instruments, Inc., Dallas, Texas, 1984.
4. Dillinger, T. E. *VLSI Engineering*. Prentice-Hall, Englewood Cliffs, N.J., 1988.
5. Heinbuch, D. V. *The CMOS3 Cell Library*. Addison-Wesley, Reading, Mass., 1988.
6. Pucknell, D. A. and K. Eshraghian. *Basic VLSI Design*. Prentice-Hall, Englewood Cliffs, N.J., 1988.
7. Garrett, L. S. "Integrated-Circuit Digital Logic Families," *IEEE Spectrum:* 46–58 (Oct. 1970), 63–72 (Nov. 1970), 30–42 (Dec. 1970).
8. Torero, E. A. "Focus on Fast Logic," *Electronic Design*, **20:**12, 50–57 (June 8, 1972).
9. Kohonen, T. *Digital Devices and Circuits*. Prentice-Hall, Englewood Cliffs, N.J., 1972.
10. Barna, A. and D. I. Porat. *Integrated Circuits in Digital Electronics*. Wiley, New York, 1973.
11. Casasent, D. *Digital Electronics*. Quantum, New York, 1974.
12. Deem, B. R., K. Muchow, and A. Zeppa. *Digital Computer Circuits and Concepts*. Reston, Va., 1977.
13. Williams, G. E. *Digital Technology*. SRA, Chicago, 1977.
14. Taub, H. and D. Schilling. *Digital Integrated Electronics*. McGraw-Hill, New York, 1977.
15. Sandige, R. S. *Digital Concepts Using Standard Integrated Circuits*. McGraw-Hill, New York, 1978.

CHAPTER 5

MANIPULATION OF BOOLEAN EXPRESSIONS

5.1 MOTIVATION

The Karnaugh map (K-map) is a graphical aid to human manipulation of Boolean algebraic expressions. We offer three reasons for persisting with the introduction of a map method in a volume entitled *Computer-Aided Logic Design*. First, confidence in the result provided by any tool is related to the user's own ability to accomplish the same type of task, albeit for much simpler examples. Second, as CAD tools evolve, they do not necessarily shed the need for occasional human assistance. Third, the application of the Karnaugh map in a human manipulation of Boolean functions is broader in scope than any single optimization technique that might be incorporated into a particular tool.

The *best* design is generally the *simplest* design that will get the job done. It follows that our objective in manipulating the mathematical model will usually be simplification or minimization. Because of the myriad possible forms of Boolean expressions, no completely general criteria for the simplest expression have been developed. As we will see, however, it is possible to define a simplest form of the two-level, or minimum-delay-time, circuit.

We have seen that the rules of the algebra can be applied formally to the manipulation and simplification of Boolean expressions, but such methods are far from easy to apply. The algebraic manipulation of Boolean functions is often quite involved, and finding the right line of attack requires considerable ingenuity and, sometimes, just plain luck. It is obvious that standard, systematic methods of minimization would be very useful. Two-level Boolean minimization is an important use of the Karnaugh map. Optimization is the only purpose of the Quine–McCluskey algorithm to be discussed in Chapter 6.

5.2 TWO-LEVEL REALIZATIONS with NAND or NOR GATES

We observed several examples in Chapters 1 and 3 that were most naturally formulated in AND-OR terms, that is, expressions of the general form

$$AB + CD + AD$$

We will see in the next section that any Boolean function can be expressed in AND-OR and also OR-AND form like

$$f = (A + B)(B + C)(C + D)$$

The minimization techniques to be developed in this and the next chapter are applicable only to these forms. As will become clear shortly, realizations of these *two-level* expressions are faster than multilevel realizations, because no signal is required to propagate through more than two gates.

In Chapter 4 we observed that a logical inversion was a natural feature of bipolar and CMOS logic elements. Therefore, it will often be more convenient to realize two-level expressions using NAND and/or NOR gates rather than AND and OR gates. We saw in Chapter 1 that the NAND and NOR operators are functionally complete. Let us apply this result to two-level realizations.

Consider the simple logical circuit of Fig. 5.1*a*, which consists of three NAND gates driving another NAND gate. From DeMorgan's law,

$$\overline{XYZ} = \bar{X} + \bar{Y} + \bar{Z} \qquad \text{(Theorem 3.11)}$$

we see that the final NAND gate can be replaced by an OR gate with inversion on the inputs (Fig. 5.1*b*). From Lemma 3.7,

$$\bar{\bar{X}} = X \qquad \text{(Lemma 3.7)}$$

so that the two successive inversions on the lines between the input and output gates cancel, yielding the circuit of Fig. 5.1*c*. Algebraically, the same result is obtained as follows:

$$
\begin{aligned}
f &= \overline{(\overline{A\bar{B}}) \cdot (\overline{BC}) \cdot (\overline{\bar{A}\,C})} \\
&= (\overline{\overline{A\bar{B}}}) + (\overline{\overline{BC}}) + (\overline{\overline{\bar{A}\,C}}) \qquad \text{(Theorem 3.11)} \\
&= A\bar{B} + BC + \bar{A}\bar{C} \qquad\qquad\quad \text{(Lemma 3.7)}
\end{aligned}
$$

Thus, we see that a two-level NAND circuit is equivalent to a two-level AND-OR circuit. In a similar fashion, we can show that a two-level NOR circuit is equivalent to a two-level OR-AND circuit (Fig. 5.2). The reader should note carefully that these simple relationships hold only for two-level AND-OR or OR-AND forms. When there are more levels of gating, the procedures for finding all-NAND or all-NOR designs are considerably more complex. This topic will be discussed in Chapter 7.

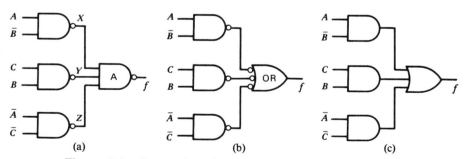

Figure 5.1 Conversion of NAND-NAND to AND-OR circuit.

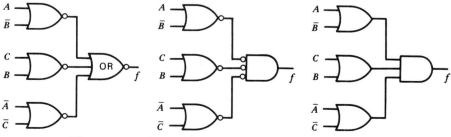

Figure 5.2 Conversion of NOR-NOR to OR-AND circuit.

5.3   STANDARD FORMS OF BOOLEAN FUNCTIONS

A standard, or general, method of analysis would seem to imply a standard starting point and a standard objective. In this section, we will develop two standard forms that may be used as starting points for simplification. The definitions of *literal*, *product term*, *sum term*, and *normal term*, which are given in Table 5.1, will be useful in discussing these standard forms. In a sense, the normal form is the "best" form. Since

$$A \cdot A = A = A + A$$

and   (Lemma 3.2)

$$A + \bar{A} = 1 \quad \text{and} \quad A\bar{A} = 0$$

a multiple occurrence of a variable in a sum or product term is always redundant or results in a trivial function.

As a first step, we will consider the form known as the *sum of products* (SOP). The nature of this form and manner of developing it are illustrated in the following examples.

■ **EXAMPLE 5.1**

Find the sum-of-products form of the function

$$f(A, B, C, D) = (AC + B)(CD + \bar{D})$$

First, let $AC = x$ and $CD = y$. Then

$$
\begin{aligned}
f(A, B, C, D) &= (x + B)(y + \bar{D}) \\
&= (x + B)y + (x + B)\bar{D} &&\text{(Distributive law)} \\
&= xy + By + x\bar{D} + B\bar{D} &&\text{(Distributive law)} \\
&= ACD + BCD + AC\bar{D} + B\bar{D} &&\text{(5.1)}
\end{aligned}
$$

■

■ **EXAMPLE 5.2**

Repeat Example 5.1 for the function

$$
\begin{aligned}
f(A, B, C, D, E) &= (\overline{AC} + \bar{D})(\overline{B + CE}) \\
&= [(\bar{A} + \bar{C}) + \bar{D}](\bar{B} \cdot \overline{CE}) &&\text{(DeMorgan's law)} \\
&= [(\bar{A} + \bar{C})\bar{B} + \bar{B}\bar{D}](\bar{C} + \bar{E}) &&\text{(DeMorgan's law)} \\
& &&\text{(Distributive law)} \\
&= (\bar{A}\bar{B} + \bar{B}\bar{C} + \bar{B}\bar{D})(\bar{C} + \bar{E}) &&\text{(Distributive law)} \\
&= \bar{A}\bar{B}\bar{C} + \bar{B}\bar{C} + \bar{B}\bar{C}\bar{D} + \bar{A}\bar{B}\bar{E} + \bar{B}\bar{C}\bar{E} + \bar{B}\bar{D}\bar{E} &&\text{(5.2)}
\end{aligned}
$$

■

From these examples, the general method to be followed in converting any Boolean function to the sum-of-products form should be evident. If no terms other than single variables are negated, only repeated application of the second distributive law is required. When terms other than single variables are negated, DeMorgan's law must also be applied. Recalling the definition of a normal term,

Table 5.1

Term	Definition	Synonym
Literal	Variable or its complement ($A, \bar{A}, B, \bar{B}$, etc.)	
Product term	Series of literals related by AND, for example, $A\bar{B}D, AC\bar{D}E$, etc.	Conjunction
Sum term	Series of literals related by OR, for example, $A + C + \bar{D}, A + B + \bar{D} + E$, etc.	Disjunction
Normal term	Product or sum term in which no variable appears more than once	

we see that Equation 5.1 or 5.2 could be referred to as a *sum of normal products*, but the shorter notation is preferred.

Continuing with the sum of products, Equation 5.1, we can write

$$f(A, B, C, D) = ACD + BCD + AC\bar{D} + B\bar{D} = ACD(B + \bar{B})$$
$$+ BCD(A + \bar{A}) + AC\bar{D}(B + \bar{B}) + B\bar{D}(A + \bar{A})$$
$$(a + \bar{a} = 1, a \cdot 1 = a)$$
$$= ABCD + A\bar{B}CD + ABCD + \bar{A}BCD + ABC\bar{D}$$
$$+ A\bar{B}C\bar{D} + B\bar{D}A + B\bar{D}\bar{A} \qquad \text{(Distributive law)}$$
$$= ABCD + A\bar{B}CD + \bar{A}BCD + ABC\bar{D} + A\bar{B}C\bar{D}$$
$$+ B\bar{D}A(C + \bar{C}) + B\bar{D}\bar{A}(C + \bar{C})$$
$$(a + a = a, a + \bar{a} = 1, a \cdot 1 = a)$$
$$= ABCD + A\bar{B}CD + \bar{A}BCD + ABC\bar{D} + A\bar{B}C\bar{D}$$
$$+ ABC\bar{D} + AB\bar{C}\bar{D} + \bar{A}BC\bar{D} + \bar{A}B\bar{C}\bar{D}$$
$$\text{(Distributive law)}$$
$$f(A, B, C, D) = ABCD + A\bar{B}CD + \bar{A}BCD + ABC\bar{D} + A\bar{B}C\bar{D}$$
$$+ AB\bar{C}\bar{D} + \bar{A}BC\bar{D} + \bar{A}B\bar{C}\bar{D} \qquad (a + a = a)$$
$$(5.3)$$

Note that Equation 5.3 is a sum of normal products with every product containing as many literals as there are variables in the function. Such products are called *canonic products*, *standard products*, *or minterms*. The terms *standard* or *canonic sum of products* and *full disjunctive normal form* have been used for expressions of the form of Equation 5.3. *Standard sum of products* will be preferred here.

We are led to the following theorem, the validity of which may already be evident to the reader. A proof of this theorem will be presented in the next section.

THEOREM 5.1 Any switching function of n variables $f(x_1, x_2, ..., x_n)$ may be expressed as a standard sum of products.

By virtue of the principle of duality, we expect that a *standard product of sums* (POS), *full conjunctive normal form*, or *product of maxterms* will also exist.

■ **EXAMPLE 5.3**

Express $f(A, B, C, D) = A + C + \bar{B}\bar{D}$ as a standard product of sums.

$$A + (C + \bar{B}\bar{D}) = A + (C + \bar{B})(C + \bar{D}) \qquad \text{(Distributive law)}$$
$$= (A + C + \bar{B})(A + C + \bar{D})$$

Then, since

$$A \cdot \bar{A} = 0 \qquad \text{(Postulate VI)}$$

$$A + 0 = A \qquad \text{(Postulate IIIa)}$$

and

$$(a + b\bar{b}) = (a + b)(a + \bar{b}) \qquad \text{(Distributive law)}$$

we can write

$$
\begin{aligned}
f(A, B, C, D) &= (A + \bar{B} + C + D\bar{D})(A + C + \bar{D} + B\bar{B}) \\
&= (A + \bar{B} + C + D)(A + \bar{B} + C + \bar{D})(A + B + C + \bar{D}) \\
&\quad \cdot (A + \bar{B} + C + \bar{D}) \\
&= (A + \bar{B} + C + D)(A + \bar{B} + C + \bar{D})(A + B + C + \bar{D}) \quad (5.4)
\end{aligned}
$$

∎

Equation 5.4 is seen to consist of a product of normal sums, with each sum containing as many literals as there are variables in the function. Such sums are known as *canonic sums*, or *standard sums*, or *maxterms*.*

Generalizing, we have the dual of Theorem 5.1.

THEOREM 5.2   Any switching function of n variables $f(x_1, x_2, ..., x_n)$ may be expressed as a standard product of sums.

It may seem to the reader that we are moving in the wrong direction, since the standard forms found in the above examples were more complicated than the expressions with which we started. It is often the case that the canonic expressions will be very complex, but our basic purpose in using them is to provide a common form for starting simplification procedures.

5.4   MINTERM AND MAXTERM DESIGNATION OF FUNCTIONS

It may have occurred to the reader, in following the examples of the preceding section, that writing out all the minterms or maxterms of a given function may be laborious. A shorthand notation for switching functions would certainly be useful. A switching function is defined by its truth table, a listing of the function values for all possible input combinations. Therefore, a simple and precise means of designating functions is obtained by numbering the rows of the truth table for a function and then listing the rows (input combinations) for which the function has the value 1, or alternately, the rows for which it has the value 0.

Figure 5.3 shows the truth table of a particular three-variable function, with the rows assigned identifying numbers. The row numbers are simply the decimal equivalents of the input combinations on each row interpreted as binary numbers. For example, the input combination $A = 0, B = 1, C = 1$, interpreted as a binary number, gives us $011_2 = 3_{10}$, so that row is called row 3. Now we can specify the function by listing the rows for which it has the value 1

$$f(A, B\,C) = \sum m(0, 4, 5, 7) \qquad (5.5)$$

or the rows for which it has the value 0

$$f(A, B, C) = \prod M(1, 2, 3, 6) \qquad (5.6)$$

*The reasons for the names *minterm* and *maxterm* will become evident in a later section.

Row. No.	A	B	C	f
0	0	0	0	1
1	0	0	1	0
2	0	1	0	0
3	0	1	1	0
4	1	0	0	1
5	1	0	1	1
6	1	1	0	0
7	1	1	1	1

Figure 5.3 Truth table with row numbers assigned.

The symbols $\sum m$ and $\prod M$ in the above lists are not just arbitrarily chosen but rather indicate a direct correspondence between these lists and the standard forms. To show this, we write out the standard sum-of-products form (Equation 5.5a) and the standard product-of-sums form (Equation 5.6a) for the function of Fig. 5.1.

$$f(A, B, C) = \bar{A}\bar{B}\bar{C} + A\bar{B}\bar{C} + A\bar{B}C + ABC \tag{5.5a}$$

$$f(A, B, C) = (A + B + \bar{C})(A + \bar{B} + C)$$
$$\cdot (A + \bar{B} + \bar{C})(\bar{A} + \bar{B} + C) \tag{5.6a}$$

Consider first the sum-of-products form. From Lemma 3.3 ($a + 1 = 1$), we see that the function will take on the value 1 whenever any one (or more) of the products takes on the value 1. Since these standard products, or minterms, contain all three variables, there is only one combination of inputs for which a given minterm will equal 1. Since each minterm is unique, no two minterms will equal 1 for the same input combination.

For example, for a combination of values $A = 1, B = 0, C = 0$ (row 4) and for no other, the second minterm in Equation 5.5 equals 1. That is,

$$A\bar{B}\bar{C} = 1\,1\,1 = 1$$

Similarly, the other minterms in Equation 5.5a are m_0, m_5, and m_7, and Equation 5.5 may then be interpreted as a list of the minterms in standard sum-of-products form.

Consider next the standard product of sums, Equation 5.6a. From the dual version of Lemma 3.3 ($a \cdot 0 = 0$), we see that this function will be 0 whenever one (or more) of the products is 0. By the dual of the above argument, a given maxterm can be 0 for only one input combination, and vice versa. For example, for input $A = 0, B = 1, C = 0$ (row 2),

$$A + \bar{B} + C = 0 + \bar{1} + 0 = 0 + 0 + 0 = 0$$

Because $A + \bar{B} + \bar{C}$ evaluates to 0 for $A,B,C = 010$, we designate this maxterm as M_2, etc.

Figure 5.4 shows the minterms and maxterms associated with each row of a three-variable truth table. The extension to more variables should be obvious. From this table, we can derive a rule for determining the actual product or sum, given the row number or vice versa. For the minterms, each uncomplemented variable is associated with a 1 in the corresponding position in the binary row number, and each complemented variable is associated with a 0. For the maxterms, the rule is just the opposite.

Row No.	A	B	C	Minterms	Maxterms
0	0	0	0	$\bar{A}\bar{B}\bar{C} = m_0$	$A + B + C = M_0$
1	0	0	1	$\bar{A}\bar{B}C = m_1$	$A + B + \bar{C} = M_1$
2	0	1	0	$\bar{A}B\bar{C} = m_2$	$A + \bar{B} + C = M_2$
3	0	1	1	$\bar{A}BC = m_3$	$A + \bar{B} + \bar{C} = M_3$
4	1	0	0	$A\bar{B}\bar{C} = m_4$	$\bar{A} + B + C = M_4$
5	1	0	1	$A\bar{B}C = m_5$	$\bar{A} + B + \bar{C} = M_5$
6	1	1	0	$AB\bar{C} = m_6$	$\bar{A} + \bar{B} + C = M_6$
7	1	1	1	$ABC = m_7$	$\bar{A} + \bar{B} + \bar{C} = M_7$

Figure 5.4

■ EXAMPLE 5.4

Convert Equation 5.3 to the minterm list form.

$f(A, B, C, D) =$

$ABCD + A\bar{B}CD + \bar{A}BCD + ABC\bar{D} + A\bar{B}C\bar{D} + AB\bar{C}\bar{D} + \bar{A}BC\bar{D} + \bar{A}B\bar{C}\bar{D}$

1111	1011	0111	1110	1010	1100	0110	0100
15	11	7	14	10	12	6	4

$f(A, B, C, D) = \sum m(4, 6, 7, 10, 11, 12, 14, 15)$ ■

■ EXAMPLE 5.5

Convert the following equation to the maxterm list form:

$f(A, B, C, D) =$

$(A + \bar{B} + C + \bar{D})(\bar{A} + B + C + D)(A + B + \bar{C} + \bar{D})(\bar{A} + B + C + \bar{D})$

0	1	0	1	1	0	0	0	0	0	1	1	1	0	0	1

5   8   3   9

$f(A, B, C, D) = \prod M(3, 5, 8, 9)$ ■

Since the truth table provides the complete specification of any switching function and we have now demonstrated a procedure for converting any truth table to a sum of minterms or product of maxterms, Theorems 5.1 and 5.2 have now been proved.

5.5 KARNAUGH MAP REPRESENTATION OF BOOLEAN FUNCTIONS

The *Karnaugh map* [1] is one of the most powerful tools in the repertoire of the logic designer. The power of the Karnaugh map does not lie in its application of any new theorems, but instead in its utilization of the remarkable ability of the human mind to perceive patterns in pictorial representations of data. A K-map may be regarded either as a pictorial form of a truth table or as an extension of the Venn diagram. First, consider a truth table for two variables. We list all four possible input combinations and the corresponding function values, for example, the truth tables for AND and OR (Fig. 5.5).

As an alternative approach, let us set up a diagram consisting of four small boxes, one for each combination of variables. Place a 1 in any box representing a

A	B	$A \cdot B$
0	0	0
0	1	0
1	1	1
1	0	0

A	B	$A + B$
0	0	0
0	1	1
1	1	1
1	0	1

Figure 5.5 Truth tables for AND and OR.

combination of variables for which the function has the value 1. There is no logical objection to putting 0's in the other boxes, but they are usually omitted for clarity. Figure 5.6 shows two forms of Karnaugh maps for AB and $A + B$.

The reader will observe that the tables and maps presented in this section all have combinations of input variables arranged in Gray code order. We shall see shortly that this convention will be very helpful in the interpretation of functions on maps. The diagrams of Fig. 5.6a are perfectly valid Karnaugh maps, but it is more common to arrange the four boxes in a square, as shown in Fig. 5.6b.

For an alternative interpretation, recall the Venn diagram. We interpret the universal set as the set of all 2^n combinations of values of n variables, divide this set into 2^n equal areas, and then darken the areas corresponding to those combinations for which the function has the value 1. We start with the universal set represented by a square (Fig. 5.7a) and divide it in half, corresponding to input combinations in which $A = 1$ and combinations in which $\bar{A} = 1$ (Fig. 5.7b). We then divide it in half again, corresponding to $B = 1$ and $\bar{B} = 1$ (Fig. 5.7c). With this notation, the interpretations of AND as the intersection of sets and OR as the

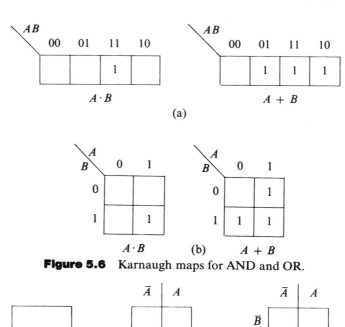

Figure 5.6 Karnaugh maps for AND and OR.

Figure 5.7 Development of maps by Venn-diagram approach.

union of sets make it particularly simple to determine which squares should be darkened in K-maps of these functions (Fig. 5.8).

We note that the shaded areas in Fig. 5.8 correspond to the squares containing 1's in Fig. 5.6. Thus, both interpretations lead to the same result. We might say that the Karnaugh map is essentially a diagrammatic form of a truth table and that the Venn-diagram concepts of union and intersection of areas aid us in setting up or interpreting a Karnaugh map.

Since there must be one square for each input combination, there must be 2^n squares in a Karnaugh map for n variables. Whatever the number of variables, we may interpret the map in terms of the graphical form of the truth table (Fig. 5.9a) or the union and intersection of areas (Fig. 5.9b).

Note particularly that the function mapped in Fig. 5.9a is minterm, m_7. Each minterm is represented by one square. Each one of the eight squares corresponds to one of the eight minterms of three variables. This is the origin of the name *minterm*. A minterm is the form of the Boolean function corresponding to the minimum possible area, other than 0, on a Karnaugh map. A *maxterm*, on the other hand, is the form of the Boolean function corresponding to the maximum

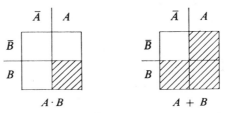

Figure 5.8   Maps of AND and OR (Venn form).

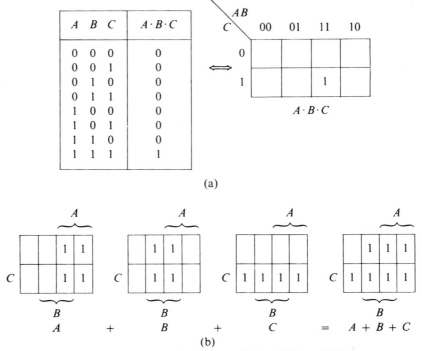

(a)

(b)

Figure 5.9   Maps for three-variable AND and OR.

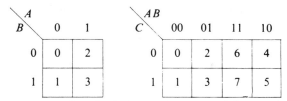

Figure 5.10 Standard K maps for two and three variables.

possible area, other than 1, on a K-map. Figure 5.9b is maxterm M_0 with 1's in the maximum possible area—all the squares but one.

Since each square on a K-map corresponds to a row in a truth table, it is appropriate to number the squares just as we numbered the rows. These standard K-maps are shown in Fig. 5.10 for two and three variables. If a function is stated in the form of the minterm list, all we need to do is enter 1 in the corresponding squares to produce the K-map.

■ **EXAMPLE 5.6**

Develop the K-map of $f(A, B, C) = \sum m(0, 2, 3, 7)$.

SOLUTION

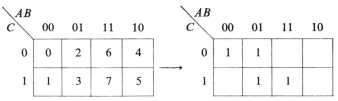

Figure 5.11

■

If a function is stated as a maxterm list, we can enter 0 in the squares listed or 1 in those not listed.

■ **EXAMPLE 5.7**

Develop the K-map of $f(A, B, C) = \prod M(0, 1, 5, 6)$.

SOLUTION

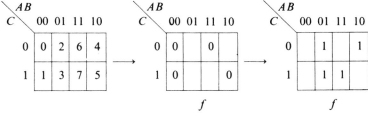

Figure 5.12

■

A map showing the 0's of a function is a perfectly valid K-map, although it is more common to show the 1's.

In developing these basic concepts, we have restricted ourselves to the simple two- and three-variable K-maps, but in practical cases, we will more often be using maps for functions of more variables. The standard map for four variables is shown in Fig. 5.13 in both notations.

We have seen that one requirement for a Karnaugh map is that there must be a square corresponding to each input combination; the maps of Fig. 5.13 satisfy this requirement. Another requirement is that the squares must be so arranged that any pair of squares immediately adjacent to each other (horizontally or vertically) must correspond to a pair of input conditions that are *logically adjacent*, that is, differ in only one variable. For example, squares 5 and 13 on the maps in Fig. 5.13 correspond to input combinations $\bar{A}B\bar{C}D$ and $AB\bar{C}D$, identical except in A. Note that squares at the ends of columns or rows are also logically adjacent.

The standard K-map for five variables is shown in Fig. 5.14. Here we have two four-variable maps placed side by side. They are identical in $BCDE$, but one corresponds to $A = 1$, the other to $A = 0$. The standard four-variable adjacencies apply in each map. In addition, squares in the same relative position on the two maps, for example, 4 and 20, are also logically adjacent. Any maps that satisfy the requirements of 2^n squares in proper adjacency can be considered K-maps. In Figs. 5.15 and 5.16, we give the most common forms of K-maps, for two to six variables, in the two alternative notations.

There is no particular preference between the two types of notation. The notation of Fig. 5.15 makes it simple to determine the number of a square, since the

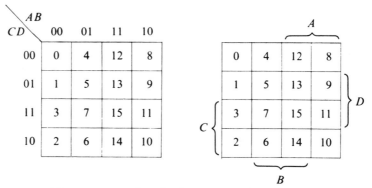

Figure 5.13   Standard K-maps for four variables.

BC DE	A = 0 00	01	11	10		A = 1 00	01	11	10	BC DE
00	0	4	12	8		16	20	28	24	00
01	1	5	13	9		17	21	29	25	01
11	3	7	15	11		19	23	31	27	11
10	2	6	14	10		18	22	30	26	10

Figure 5.14   Standard K-map for five variables.

binary equivalent is directly available. The alternative notation emphasizes the areas associated with each variable and may be preferable when starting from forms other than minterm or maxterm lists. We suggest that the reader try working with both forms and then use whichever seems most convenient.

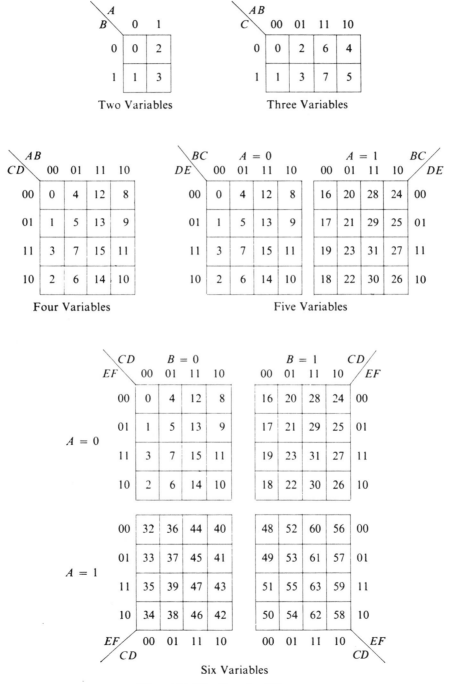

Figure 5.15 Standard K-maps.

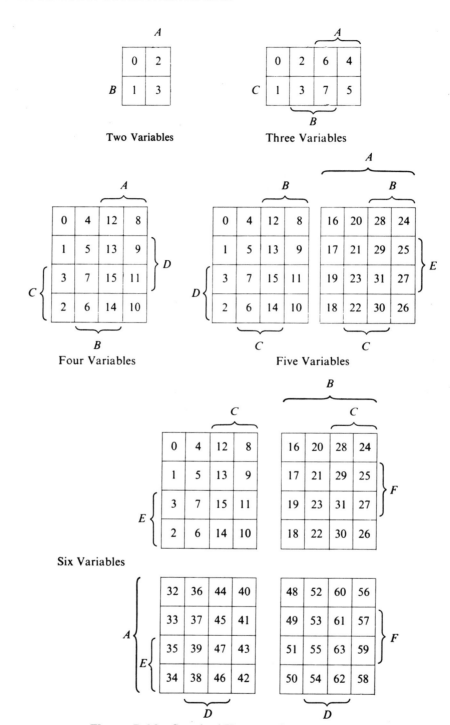

Figure 5.16 Standard K-maps: alternate notation.

■ EXAMPLE 5.8

Find K-maps for the following functions:

(a) $f(V, W, X, Y, Z) = \sum m(9, 20, 21, 29, 30, 31)$.

(b) $f(A, B, C, D, E) = AB + \bar{C}D + DE$.

SOLUTION

(a) For the first function, we enter 1 in the listed squares in Fig. 5.17.

(b) Referring to the standard map for five variables, we easily identify the maps of the individual product terms, as shown in Fig. 5.18.

We then take the union of these three maps to form the final K-map (Fig. 5.19). The minterm list of this function may now be read directly from the map.

$$f(A, B, C, D, E) = AB + \bar{C}D + DE$$

$$= \sum m(2, 3, 7, 10, 11, 15, 18, 19, 23, 24, 25, 26, 27, 28, 29, 30, 31)$$ ∎

In part (a) of Example 5.8, note that the variables do not have to be A, B, C, etc. Obviously, we can call the variables anything we want. The only precaution to be observed is that the variables must appear on the K-map in the proper manner corresponding to the order of their listing in the function statement.

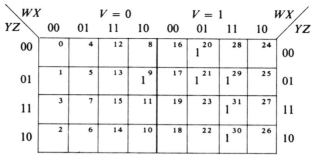

Figure 5.17

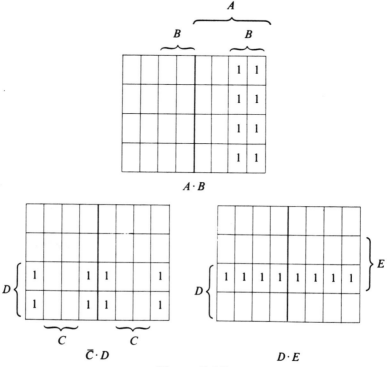

Figure 5.18

BC \ DE	A = 0 00	01	11	10	A = 1 00	01	11	10	BC / DE
00	0	4	12	8	16	20	28 _1_	24 _1_	00
01	1	5	13	9	17	21	29 _1_	25 _1_	01
11	_1_ 3	_1_ 7	_1_ 15	_1_ 11	_1_ 19	_1_ 23	_1_ 31	_1_ 27	11
10	_1_ 2	6	14 _1_	_1_ 10	_1_ 18	22	_1_ 30	_1_ 26	10

Figure 5.19

5.6   SIMPLIFICATION OF FUNCTIONS ON KARNAUGH MAPS

It has taken us awhile to establish the necessary tools, but we are now ready to use them in minimizing functions. As mentioned earlier, we must have some circuit format in mind before a criterion for the simplest circuit can be defined. Let us consider direct realizations of three different forms of a four-variable function as shown in Fig. 5.20.

$$f(A, B, C, D) = (A + B)(C\bar{D}) + (A + B)(\bar{C}D)$$

$$= AC\bar{D} + A\bar{C}D + BC\bar{D} + B\bar{C}D$$

$$= \sum m(5, 6, 9, 10, 13, 14) \tag{5.7}$$

Now, which of these forms is the simplest? Obviously, circuit (c), the realization of the standard sum of products, is the most expensive to build. Comparing circuits (a) and (b), the reader will probably feel that (a) is the simpler of the two. It is simpler, in the sense of having fewer gates, with fewer inputs, but it has one drawback relative to circuit (b). Note that the signals A and B, entering the left gate, pass through three gates, or three *levels of gating*, before reaching the output. By comparison, all the signals into circuit (b) pass through only two levels of gating.

In Chapter 4, it was pointed out that each level of logic adds to the delay in the development of a signal at the circuit output. In high-speed digital systems, it is desirable that this delay be as small as possible.

The choice between a simpler circuit and a faster circuit is generally a matter of engineering judgment. More speed almost invariably costs more money or increases chip area, and one of the most difficult jobs of the design engineer is to decide just how much speed is affordable. For the present, we will assume that speed is the dominant factor and we will design for the fastest possible circuit. Any sum-of-products or product-of-sums expression can be realized by two levels of gating. Whether these be AND-OR, OR-AND, NAND-NAND, or NOR-NOR configurations, they will be referred to as *two-level* circuits.

The other reason for concentrating on two-level circuits is the existence of gate minimization procedures for these forms that are both general and straightforward. Until about 1985 when it became apparent that hardware synthesis in its most general form implied a broad interpretation of minimization, only very special circuit structures (see Section 8.4) were realized using more than two levels of logic.

In assigning a cost to a sum-of-products expression, no cost is assigned for inverting the variables to form-complemented literals. Typically, both com-

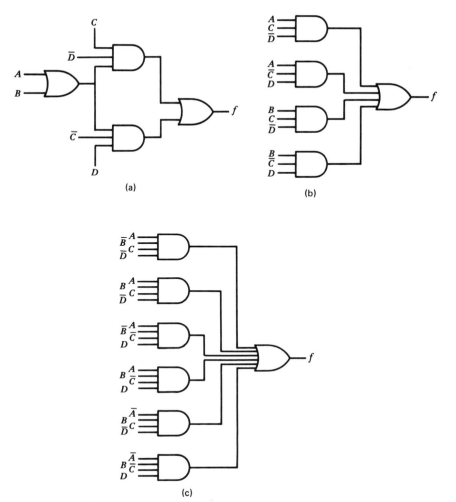

Figure 5.20 Alternate realizations of a function.

plemented and uncomplemented literals are available at the outputs of memory elements and, therefore, available when needed throughout the network. This situation is characterized as the availability of *double-rail inputs*. In VLSI chip design the cost of double-rail routing may exceed the costs of extra inverters. In that case, only the uncomplemented variables or *single-rail inputs* might be available.

In a sum-of-products form, each product corresponds to a gate and each literal to a gate input. The same holds for each sum in a product-of-sums form. The exact ratio between the cost of a gate and the cost of a gate input will depend on the type of gate, but in practically every case, the cost of an additional gate will be several times that of an additional input on an already existing gate. On this basis, the elimination of gates will be the primary objective of any minimization process, leading to the following definition of a minimal expression.

Definition 5.1 A second-order sum-of-products expression will be regarded as a *minimal* expression if there exists (1) no other equivalent expression involving fewer products, and (2) no other equivalent expression involving the same number of products but a smaller number of literals. A minimal product of sums is the same with the word *products* replaced by the word *sums*, and vice versa.

Note that *a* minimal rather than *the* minimal expression is characterized by Definition 5.1. As we will see, there may very well be several distinct but equivalent expressions satisfying this definition and having the same number of both products and literals. For the remainder of this section, only the sum-of-products form will be considered. In the next section minimization of product-of-sums expressions will be discussed.

Simplification of functions on the K-map is based on the fact that sets of minterms that can be combined into simpler product terms either will be adjacent or appear in symmetric patterns on the K-map.

Consider this function

$$f(A, B, C) = \sum m(0, 1, 4, 6)$$
$$= \bar{A}\bar{B}\bar{C} + \bar{A}\bar{B}C + A\bar{B}\bar{C} + AB\bar{C} \qquad (5.8)$$

By algebraic manipulation, we can simplify this function as follows:

$$f(A, B, C) = \bar{A}\bar{B}(C + \bar{C}) + A\bar{C}(B + \bar{B})$$
$$= \bar{A}\bar{B} + A\bar{C}$$

On the map, we have the pattern shown in Fig. 5.21. Note that the minterms that have combined into a simpler term are adjacent on the K-map. This is a general principle. *Any pair of* n-*variable minterms that are adjacent on a K-map may be combined into a single product term of* n − *1 literals.* As we noted earlier, K-maps are so arranged that minterms in adjacent squares are identical, except in one variable. This variable appears in the true form in one minterm and in the complemented (false) form in the other. Thus, the value of the function will be independent of the value of this variable. For example, in the function given by Equation 5.8, if $A = 0$ and $B = 0$, then $f = 1$, regardless of the value of C. Similarly, if $A = 1$ and $C = 0$, then $f = 1$, regardless of the value of B.

Now consider

$$f(A, B, C, D) = \sum m(0, 8, 12, 14, 5, 7)$$
$$= \bar{A}\bar{B}\bar{C}\bar{D} + A\bar{B}\bar{C}\bar{D} + AB\bar{C}\bar{D} + ABC\bar{D} + \bar{A}\bar{B}\bar{C}D + \bar{A}BCD$$
$$= \bar{B}\bar{C}\bar{D}(A + \bar{A}) + AB\bar{D}(C + \bar{C}) + \bar{A}BD(C + \bar{C})$$
$$= \bar{B}\bar{C}\bar{D} + AB\bar{D} + \bar{A}BD \qquad (5.9)$$

which is mapped in Fig. 5.22. Note that here the pairing includes 0 and 8, and 12 and 14, which do not appear to be adjacent. The variables are arranged in "ring" pattern of symmetry, so that these squares would be adjacent if the map were inscribed on a torus (a doughnut-shaped form). If you have difficulty visualizing the map on a torus, just remember that squares in the same row or column, but on opposite edges of the map, may be paired.

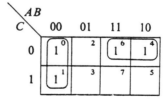

Figure 5.21 K-map for Equation 5.8.

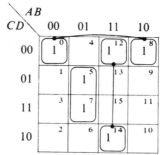

Figure 5.22 Map of Equation 5.9.

■ EXAMPLE 5.9

Determine a second-order network logically equivalent to the odd-parity check circuit given in Fig. 2.7.

SOLUTION The function $f(X_1, X_2, X_3, X_4)$, which is 1 when an odd number of the variables X_1, X_2, X_3, or X_4 are 1, is depicted in Fig. 5.23. None of the minterms in the classic checkerboard pattern of Fig. 5.23 can be combined to form products of fewer variables. Thus, the only second-order realization is the direct implementation of the expanded sum-of-products form. Realization of Fig. 5.23 requires eight four-input NAND gates and one eight-input NAND gate. Realization of Fig. 2.7 directly would require nine NAND gates, each with only two inputs. A second-order realization of the odd-parity check over eight variables will require 129 NAND gates. ■

Next consider the function

$$f(A, B, C, D) = \sum m(5, 7, 10, 13, 15)$$
$$= BD + A\bar{B}C\bar{D} \tag{5.10}$$

which is displayed on the map in Fig. 5.24. Here we see that a set of four adjacent minterms combines into a single term, with the elimination of two literals. In other words, if $B = 1$ and $D = 1$, then $f = 1$, regardless of the values A and C. Also note that m_{10} does *not* combine with any other minterm of the function. The adjacencies must be row or column adjacencies. For example, the diagonally adjacent m_{10} and m_{15} cannot be combined.

Other possible sets of four 1's, which form a single product, are shown in Fig. 5.25. Note the adjacency on the edges and corners.

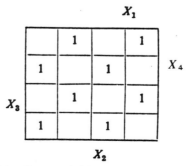

Figure 5.23 Four-variable odd-parity check function.

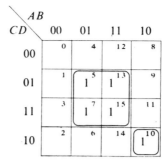

Figure 5.24   Map of Equation 6.10.

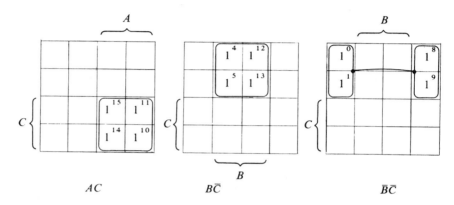

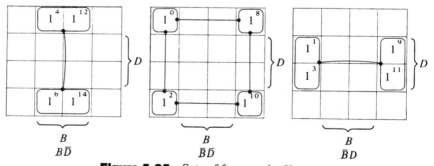

Figure 5.25   Sets of four on the K-map.

Since sets of two minterms combine to eliminate one variable, and sets of four combine to eliminate two, it is to be expected that sets of eight will combine to eliminate three variables. Figure 5.26 illustrates some of these sets.

The same general principles apply as we go to the five- and six-variable maps, but we also have logical adjacency between squares, or sets of squares, in the same position on different sections of the map.

Some combinable sets of 1's on a five-variable map are shown in Fig. 5.27. Some sets on a six-variable map are shown in Fig. 5.28. We note that here the "map-to-map" adjacency works both horizontally and vertically but not diagonally. For example, 32 combines vertically with 0 and horizontally with 48, but 0 and 48 do *not* combine, even though they are in the same position in their respective sections of the map. The reader should also keep in mind that the "end-to-end" adjacency applies only to the individual four-variable section. For example,

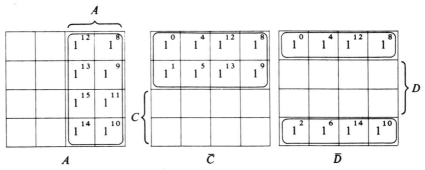

Figure 5.26   Sets of eight on the K-map.

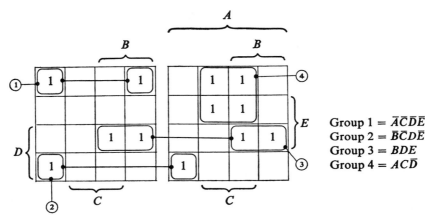

Figure 5.27   Sets on a five-variable map.

Group 1 = $\overline{A}\overline{C}\overline{D}E$
Group 2 = $\overline{B}\overline{C}D\overline{E}$
Group 3 = BDE
Group 4 = $AC\overline{D}$

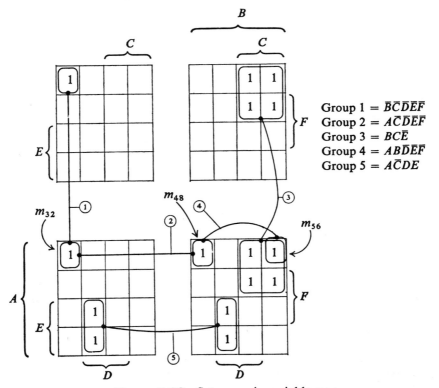

Group 1 = $\overline{B}\overline{C}\overline{D}EF$
Group 2 = $A\overline{C}\overline{D}EF$
Group 3 = $BC\overline{E}$
Group 4 = $AB\overline{D}EF$
Group 5 = $A\overline{C}DE$

Figure 5.28   Sets on a six-variable map.

48 and 56, at opposite ends of a row in a single section, combine, whereas 32 and 56, at opposite ends of separate sections, do not combine.

We have seen that a set of two *logically* adjacent minterms eliminates one variable, that a set of four eliminates two variables, that a set of eight eliminates three variables, etc. The way to test whether a set is, in fact, logically adjacent is to determine whether sufficient variables remain constant over the entire set. On an n-variable map, a pair of minterms is adjacent if $n - 1$ variables remain constant over the pair; a set of four minterms is adjacent if $n - 2$ variables remain constant; a set of eight minterms is adjacent if $n - 3$ variables remain constant over the set, etc. For example, in Fig. 5.28, the fact that 0 and 48 do not combine into a pair is consistent with the fact that they correspond to input combinations differing in the values of two variables, A and B.

The process of simplifying a function on a K-map consists of nothing more than determining the smallest set of adjacencies that covers (contains) all the minterms of the function. Let us illustrate the process by a few examples.

■ **EXAMPLE 5.10**

Simplify $f(A, B, C, D) = \sum m(0, 2, 10, 11, 12, 14)$.

SOLUTION

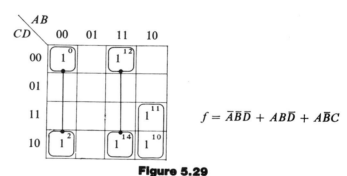

$$f = \bar{A}\bar{B}\bar{D} + AB\bar{D} + A\bar{B}C$$

Figure 5.29

■

Here there are choices. We could also combine m_2 and m_{10}, or m_{10} and m_{14}. However, there is no reason to use these combinations since m_2, m_{10}, and m_{14} are already covered by (contained in) the necessary pairings with m_0, m_{11}, m_{12}, respectively. We generalize this observation by stating the following rule. *All prime products that cover a minterm covered by no other product will be part of every minimal sum-of-products expression.* A product is prime if and only if it is not covered by a product of fewer variables. It follows that all prime products which are *essential* (a formal definition will follow in Chapter 6) products should be identified and included in the sum-of-products expression prior to the inclusion of any other products. Once this is done, completing a minimal sum-of-products expression will follow immediately for many functions of a small number of variables.

■ **EXAMPLE 5.11**

Simplify $f(A, B, C, D) = \sum m(0, 2, 8, 12, 13)$.

SOLUTION

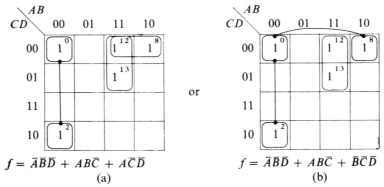

or

$$f = \overline{A}\overline{B}\overline{D} + AB\overline{C} + A\overline{C}\overline{D}$$
(a)

$$f = \overline{A}\overline{B}\overline{D} + AB\overline{C} + \overline{B}\overline{C}\overline{D}$$
(b)

Figure 5.30 Alternate maps for Example 5.11.

In this case, there are two equally valid choices. ∎

Note that, in Fig. 5.30a, m_{12} is covered by the terms $AB\overline{C}$ and $A\overline{C}\overline{D}$; in Fig. 5.30b, m_0 is covered by $\overline{A}\,\overline{B}\overline{D}$ and $\overline{B}\overline{C}\overline{D}$. Covering a minterm more than once causes no trouble. In terms of AND-OR realization, it simply means that when the variable values are such that the particular minterm is 1, the output of more than one AND gate will take on the value 1.

∎ EXAMPLE 5.12

Simplify $f(A, B, C, D) = \sum m(1, 5, 6, 7, 11, 12, 13, 15)$.

SOLUTION

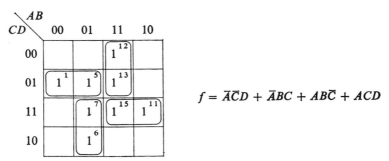

$$f = \overline{A}\overline{C}D + \overline{A}BC + AB\overline{C} + ACD$$

Figure 5.31

∎

Example 5.12 illustrates a possible pitfall in using K-maps. The temptation is great to use the set of four in the center, but when we include the essential pairings with the other four minterms, we find that the four in the center have been covered. This emphasizes the importance of determining the essential products first, that is, the products containing at least one minterm that can be combined in no other way.

We will now present a few more examples, without further comment. It is suggested that the reader study the first two carefully and then try to work the others before looking at the answers. Once mastered, Karnaugh maps will seem almost second nature, but mastery requires practice.

■ EXAMPLE 5.13

$$f(A, B, C, D, E) = \sum m(0, 1, 4, 5, 6, 11, 12, 14, 16, 20, 22, 28, 30, 31)$$

SOLUTION

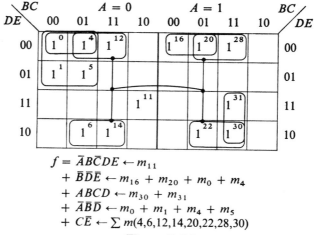

$$
\begin{aligned}
f = {} & \bar{A}B\bar{C}DE \leftarrow m_{11} \\
 & + \bar{B}D\bar{E} \leftarrow m_{16} + m_{20} + m_0 + m_4 \\
 & + ABCD \leftarrow m_{30} + m_{31} \\
 & + \bar{A}\bar{B}\bar{D} \leftarrow m_0 + m_1 + m_4 + m_5 \\
 & + C\bar{E} \leftarrow \sum m(4,6,12,14,20,22,28,30)
\end{aligned}
$$

Figure 5.32

■ EXAMPLE 5.14

$$f(A, B, C, D, E, F) = \sum m(2, 3, 6, 7, 10, 14, 18, 19, 22, 23, 27, 37, 42, 43, 45, 46)$$

SOLUTION

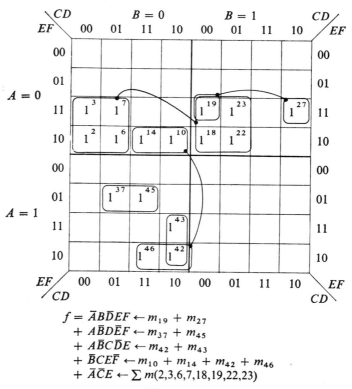

$$
\begin{aligned}
f = {} & \bar{A}B\bar{D}EF \leftarrow m_{19} + m_{27} \\
 & + A\bar{B}D\bar{E}F \leftarrow m_{37} + m_{45} \\
 & + A\bar{B}C\bar{D}E \leftarrow m_{42} + m_{43} \\
 & + \bar{B}CE\bar{F} \leftarrow m_{10} + m_{14} + m_{42} + m_{46} \\
 & + \bar{A}\bar{C}E \leftarrow \sum m(2,3,6,7,18,19,22,23)
\end{aligned}
$$

Figure 5.33

■ **EXAMPLE 5.15**

$$f(A, B, C, D) = \prod M(7, 9, 13)$$

SOLUTION

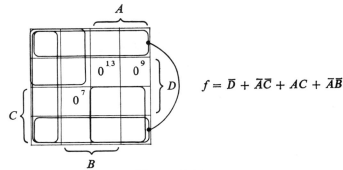

$$f = \bar{D} + \bar{A}\bar{C} + AC + \bar{A}\bar{B}$$

Figure 5.34

■ **EXAMPLE 5.16**

$$f(A, B, C, D) = \sum m(0, 1, 2, 4, 5, 8, 10)$$

SOLUTION

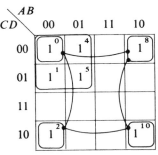

$$f = \bar{B}\bar{D} + \bar{A}\bar{C}$$

Figure 5.35

■ **EXAMPLE 5.17**

$$f(A, B, C, D, E) = \sum m(0, 1, 3, 4, 5, 7, 8, 9, 10, 12, 13, 21, 24, 25, 26, 28, 29)$$

SOLUTION

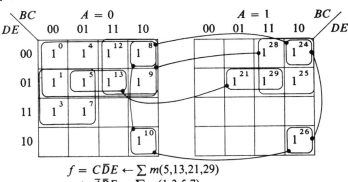

$$
\begin{aligned}
f = \; & C\bar{D}E \leftarrow \sum m(5,13,21,29) \\
& + \bar{A}\bar{B}E \leftarrow \sum m(1,3,5,7) \\
& + \bar{A}\bar{D} \leftarrow \sum m(0,1,4,5,8,9,12,13) \\
& + B\bar{C}\bar{E} \leftarrow \sum m(8,10,24,26) \\
& + B\bar{D} \leftarrow \sum m(8,9,12,13,24,25,28,29)
\end{aligned}
$$

Figure 5.36

5.7 MAP MINIMIZATIONS OF PRODUCT-OF-SUMS EXPRESSIONS

With only minor changes, the procedure described in Section 5.5 can be adapted to product-of-sums design. We have seen that each 1 of a function is produced by a single minterm and that the process of simplification consists of combining minterms into products that have fewer literals and produce more than a single 1. We have also seen that each 0 of a function is produced by a single maxterm. It would seem reasonable, then, to expect that maxterms might combine in a similar fashion.

Consider the function

$$f(A, B, C) = \prod M(3, 6, 7)$$
$$= (A + \bar{B} + \bar{C})(\bar{A} + \bar{B} + C)(\bar{A} + \bar{B} + \bar{C}) \tag{5.11}$$

which is mapped in Fig. 5.37. As was done for the first example of a sum-of-products realization, we use the associative and distributive laws (the duals of those used previously) to develop an expression equivalent to that in Equation 5.11, but including only two products corresponding to the two groups marked in Fig. 5.37.

$$f(A, B, C) = (A + \bar{B} + \bar{C})(\bar{A} + \bar{B} + C)(\bar{A} + \bar{B} + \bar{C})$$
$$= (A + \bar{B} + \bar{C})(\bar{A} + \bar{B} + \bar{C})(\bar{A} + \bar{B} + C)(\bar{A} + \bar{B} + \bar{C})$$
$$= (A\bar{A} + \bar{B} + \bar{C})(\bar{A} + \bar{B} + C\bar{C})$$
$$= (\bar{B} + \bar{C})(\bar{A} + \bar{B})$$

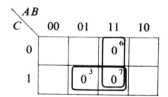

Figure 5.37 K-map of Equation 5.11.

The procedure for determining a minimal product-of-sums form is the dual of the sum-of-products procedure. Groups of 0's conforming to the same patterns defined in Figs. 5.25 through 5.28 are identified. A minimal realization is formed from a minimal number of such groups covering all 0's in the function. Each group is realized as a sum term, with the literals entered as the *complements* of those that would be used in a product-of-sums realization. Let us consider some examples.

■ **EXAMPLE 5.18**

Obtain a minimal product-of-sums realization of $f(A, B, C, D) = \sum m(0, 2, 10, 11, 12, 14) = \prod M(1, 3, 4, 5, 6, 7, 8, 9, 13, 15)$.

SOLUTION

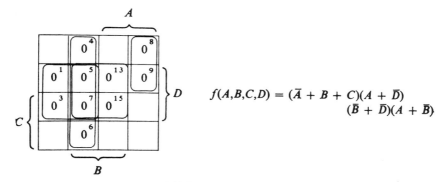

$$f(A,B,C,D) = (\bar{A} + B + C)(A + \bar{D})$$
$$(\bar{B} + \bar{D})(A + \bar{B})$$

Figure 5.38

■

■ EXAMPLE 5.19

$$f(A, B, C, D) = \sum m(0, 2, 8, 12, 13) = \prod M(1, 3, 4, 5, 6, 7, 9, 10, 11, 14, 15)$$

SOLUTION

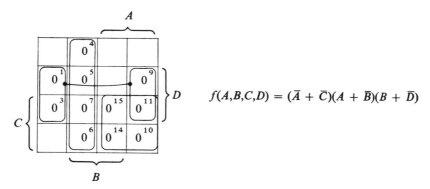

$$f(A,B,C,D) = (\bar{A} + \bar{C})(A + \bar{B})(B + \bar{D})$$

Figure 5.39

■

If we compare Examples 5.18 and 5.10, we see that the sum-of-products form is preferable to the product-of-sums form for this function, since there are three products in the sum-of-products form and four sums in the product-of-sums form. Comparing Examples 5.19 and 5.11, we see that the product-of-sums form is preferable. The number of products and sums is the same, but there are fewer literals in the sums. It would be useful to have some method of determining in advance which form (sum-of-products or product-of-sums) will be best for a particular function. Unfortunately, no such method exists. Assuming there is no hardware preference, the designer should try both forms.

5.8 INCOMPLETELY SPECIFIED FUNCTIONS

Recall that the basic specification of a switching function is the truth table, that is, a listing of the values of the function for the 2^n possible combinations of n variables. Our design process basically consists of translating a (generally) verbal description of a logical job to be done into a truth table and then finding a specific function that realizes this truth table and satisfies some criterion of minimum cost. Thus far, we have assumed that the truth values were strictly specified for all the 2^n possible input combinations. This is not always the case.

Sometimes the circuit we are designing is a part of a larger system in which certain inputs occur only under circumstances such that the output of the circuit will not influence the overall system. Whenever the output has no effect, we obviously *don't care* whether the output is a 0 or 1. Another possibility is that certain input combinations never occur due to various external constraints. Note that this does not mean that the circuit would not develop some output if this forbidden input occurred. Any switching circuit will respond in some way to any input. However, since the input will never occur, we don't care whether the final circuit responds with a 0 or 1 output to this forbidden-input combination.

When such situations occur, we say that the output is *unspecified*. This is indicated on the truth table by entering an **X** as the functional value instead of 0 or 1.* Such conditions are commonly referred to as *don't-cares*, and functions including don't-cares are said to be *incompletely specified*. A realization of an incompletely specified function is any circuit that produces the same outputs for *all input combinations for which output is specified*.

■ **EXAMPLE 5.20**

For some reason, of no interest to us, it is necessary to design a combinational logic circuit that we shall call a "divisible-by-3 detector," which will be incorporated within a digital computer designed to operate in the binary-coded-decimal (BCD) mode. The circuit will have four input lines, A, B, C, D, representing a BCD digit, and a single output Z that will be 1 if and only if this digit is divisible by 3. The most significant bit of the BCD number is A, B is the next significant, etc. The circuit is illustrated in Fig. 5.40.

This problem is interesting because there are only 10 BCD digits, 0000 to 1001, that can occur on lines A, B, C, D. The bit combinations 1010 through 1111 will never appear on these lines. Thus, we do not care what output the circuit would have for these input combinations.

SOLUTION   The functional values for Z may be entered directly on the Karnaugh map, since the BCD digits are the same as the minterm numbers. Thus, 1's are entered in Fig. 5.41 in the squares corresponding to m_3, m_6, and m_9. Since m_{10} through m_{15} represent input combinations that will never occur, **X**'s are entered in these squares.

Since the output is optional for the don't-cares, we assign them to be 0's or 1's in whatever manner will result in the simplest realization. On the K-map, this

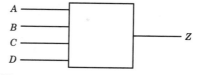

Figure 5.40   Divisible-by-3 detector.

*The symbol for the unspecified output condition is unfortunately not standard. Other symbols used include *d* and −.

means that we group the X's with the 1's whenever this results in a larger group and ignore them if there is no advantage to be gained from their use. In this case, we use the X's in squares 11, 13, and 15 together with the 1 in square 9 to form the product AD, use the X in square 14 with the 1 in square 6 to form the product $BC\bar{D}$, and reuse square 11 with the 1 in square 3 to form $\bar{B}CD$. The X's in squares 10 and 12 are not used because they cannot be grouped with any of the 1's to produce a larger group.

$$Z = AD + BC\bar{D} + \bar{B}CD$$

This expression will evaluate to 1 for each combination of values for which $Z = 1$ in Fig. 5.41 and to 0 wherever $Z = 0$ in this map. Should any of the don't-care conditions occur, the expression would evaluate to 1 for those that are included in groupings, to 0 for those that are not. ■

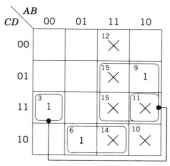

Figure 5.41 K-map for divisible-by-3 detector.

It would be convenient to have some compact algebraic form for incompletely specified functions. Because each row in the truth table corresponds to an input combination, we simply add a list of the rows for which the output is unspecified, enclosed by parentheses and preceded by a "d" to signify "don't care." Thus, a minterm list corresponding to Fig. 5.41 would be

$$f(A,B,C,D) = \sum m(3,6,9) + d(10,11,12,13,14,15)$$

■ **EXAMPLE 5.21**
___

Obtain a minimal sum-of-products representation for

$$f(w, x, y, z) = \sum m(0, 7, 8, 10, 12) + d(2, 6, 11)$$

SOLUTION

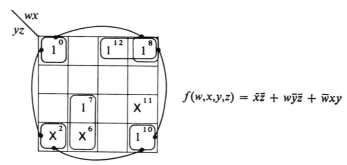

$$f(w,x,y,z) = \bar{x}\bar{z} + w\bar{y}\bar{z} + \bar{w}xy$$

Figure 5.42

■

We use the don't-cares in squares 2 and 6 to obtain larger sets than would otherwise be possible, but we ignore the **X** in 11, since the only possible combination is with 10, which is already covered.

■ **EXAMPLE 5.22**

Obtain a minimal product-of-sums realization for the function of Example 5.21.

SOLUTION

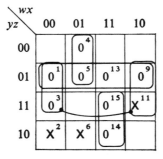

$$f(w,x,y,z) = (y + \bar{z})(x + \bar{z})(w + \bar{x} + y)$$
$$(\bar{w} + \bar{x} + \bar{y})$$

Figure 5.43

■

Since we want the product-of-sums form in Example 5.22, we design in the 0's, again using the don't-cares if they improve the combination, ignoring them otherwise. Again, we start with necessary sets. For example, 4 combines only with 5, and 14 only with 15. We could combine 4 with the don't-care in 6 (or 14 with 6), but this would cover nothing not covered by the combination of 4 and 5. The don't-care at 11 is used because it does place 3 in a larger set $(x + \bar{z})$ than could otherwise be obtained.

■ **EXAMPLE 5.23**

Obtain a minimal sum-of-products representation for

$$f(A, B, C, D, E) = \sum m(1, 4, 6, 10, 20, 22, 24, 26) + d(0, 11, 16, 27).$$

SOLUTION

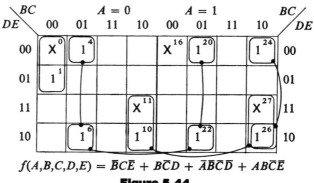

$$f(A,B,C,D,E) = \bar{B}C\bar{E} + B\bar{C}D + \bar{A}\bar{B}\bar{C}\bar{D} + AB\bar{C}\bar{E}$$

Figure 5.44

■

Here the don't-care in 16 could be used in several ways, but it would not increase the number of squares covered by any product.

5.9 LOGIC HAZARDS

Under certain circumstances, a minimal gate realization of a logic function may not be a satisfactory solution of a design problem. Consider the following dialogue that might well have taken place prior to the design of this circuit.

Application Engineer: Can you design for me a circuit that can be controlled by two inputs, x_1 and x_2 to generate any of the four output waveforms given in Fig. 5.45b.

Design Engineer: No problem! I assume that a two-bit Gray code counter-based circuit will be satisfactory.

Application Engineer: I don't care what is inside of your circuit package, but the output must conform faithfully to Fig. 5.45 with no extraneous or transient output changes (glitches). In addition, the frequency (when $x_1 = x_2 = 1$) must be a very stable 5 mHz $\pm$ 0.0001%.

Design Engineer: Hmmm.

The design engineer doesn't let on that he is unsure whether a similar design which he had already completed for a less exacting application will satisfy the requirement of no transient output transitions. He is confident that he will find some way to eliminate extraneous output changes, if indeed any appear.

We will not examine the design of the 2-bit Gray code counter whose output waveforms are given in Fig. 5.45a. We will concentrate on the combinational

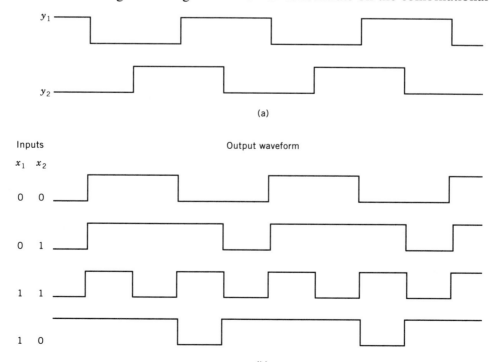

Figure 5.45 Waveform generation.

logic network that will use the counter output to generate the four desired waveforms as shown in Fig. 5.46a. The Karnaugh map for this logic network is given in Fig. 5.46b. Noting that the Gray code counter outputs are periodically $y_1 y_2 = 00 -> 01 -> 11 -> 10 -> 00$, the entries in each column of the map are the output values for each of the four periods of the count for the particular pair of control values heading that column. The minimal sum-of-products realization of this circuit is given in Fig. 5.46c.

In general, the existence of unanticipated output changes in Fig. 5.46c will be a function of the relative gate delays of the various gates in the network. To show that such a transition can occur, let us consider a particular combination of gate delays and a particular point in the output waveform. Although gates from the same logic family used to construct a circuit will usually have approximately the same gate delay, variations in the delay in individual gates can occur. In the circuit under consideration, a problem can appear if the delay in gate 4 is greater than the delay in the other AND gates. Consequently, we have postulated a delay of 20 nsec for gate 4 and 10 nsec for the other gates.

Let us assume that the desired output waveform is the one corresponding to $x_1 = 0$ and $x_2 = 1$. In this column, as in the other three columns, the states $y_1 y_2$ change in the order $00 -> 01 -> 11 -> 10 -> 00$. Let us focus our attention on the case in which y_2 remains constant at 1 and y_1 changes from 0 to 1. Under these circumstances, gates 1, 3, and 5 have a constant output of 0 and, therefore, do not affect the circuit output z. This is emphasized in Fig. 5.47a where the assumed input values are indicated.

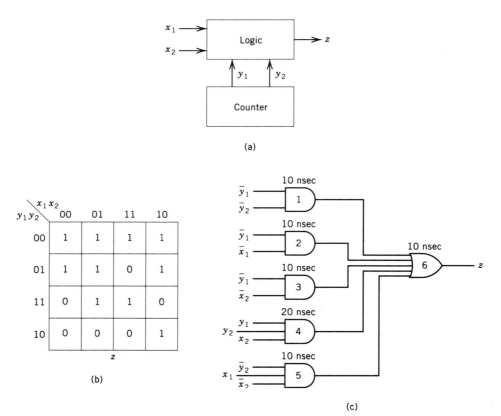

Figure 5.46   Realization of the waveform generator.

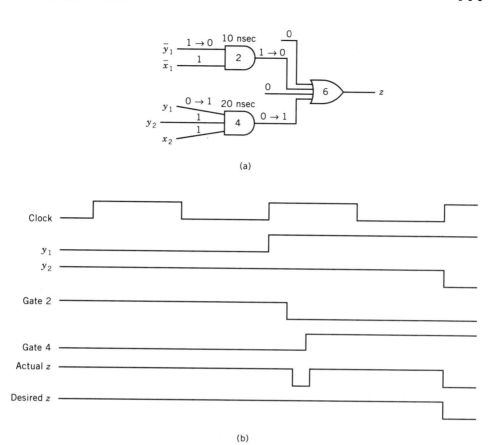

(a)

(b)

Figure 5.47 Malfunction in the waveform generator.

In Fig. 5.47b we depict the actual behavior of the waveform generator, as y_1 changes from 0 to 1 under the input and delay conditions stated above. Notice that the output of gate 2 goes to 0, 10 nsec before gate 4 goes to 1. Thus, there is a 10 nsec period during which the outputs of all five AND gates are 0. The output z will correspondingly go to 0 for a short period following the delay in the OR gate. (The actual output pulse will be less than 10 nsec wide due to rise time effects, but this is of no help.) The actual behavior on line z is depicted just above the desired output waveform in Fig. 5.47b.

The phenomenon that caused the malfunction described above is called a hazard. A hazard occurs in a two-level AND-OR network whenever the output is to remain 1 before and after a single input change, but no single AND gate has output 1 both before and after the change. Similar hazards occur in two-level OR-AND, NAND-NAND, NOR-NOR, and multilevel networks as well. It is possible to avoid hazards in the design of two-level networks subject to single input changes, as will be discussed in the next section.

The glitch resulting from a logic hazard is a problem here because of the rigid specification on the circuit output. When a subcircuit like that in Fig. 5.47a is buried within a large clock-mode (to be defined in Chapter 9) digital system and its output serves as input to another subcircuit synchronized by the same clock, hazards may usually be ignored. We shall see that transient circuit values, which occur in response to a clock transition, disappear and all gate outputs settle to their proper values before the next clock transition.

Do we conclude then that the applications engineer was merely an alarmist in demanding that no transient output transitions be allowed? No! The circuit that is to be driven by the waveform generator may not have the same clock or may not be clock-mode! Special circuits that must be connected to digital systems and will respond to every input transition, anticipated or not, are not uncommon.

5.10   ELIMINATION OF HAZARDS

Fortunately, the existence or nonexistence of hazards is a property of the combinational logic realization of a Boolean function. As we shall see, it is possible to eliminate these hazards by altering the realization. The Karnaugh map of the output network of the waveform generator is repeated in Fig. 5.48a. The realization of this map that led to the network of Fig. 5.46 is given as Equation 5.12.

$$z = \bar{y}_1\bar{y}_2 + \bar{y}_1\bar{x}_1 + \bar{y}_1\bar{x}_2 + y_1y_2x_2 + \bar{y}_2x_1\bar{x}_2 \tag{5.12}$$

The arrow in Fig. 5.48a points to the source of the problem discussed in the previous section. This point is the transition of y from 0 to 1 in the $x_1x_2 = 01$ column while $y_2 = 1$. Notice that not all the 1s in this column are contained in a single product grouping. The state of the circuit moves from the $\bar{y}_1\bar{x}_1$ product to the product $y_1y_2x_2$ when y_1 goes from 0 to 1. No product is 1 both before and after the transition.

In Fig. 5.48b we have added a redundant product to those selected in Fig. 5.48a. Note that this product will be 1 both before and after the transition of y_1 from 0 to 1 in the $x_1x_2 = 01$ column. This leads to a new expression for the output of the waveform generator given by Equation 5.13.

$$z = \bar{y}_1\bar{y}_2 + \bar{y}_1\bar{x}_1 + \bar{y}_1\bar{x}_2 + y_1y_2x_2 + \bar{y}_2x_1\bar{x}_2 + \bar{x}_1x_2y_2 \tag{5.13}$$

When the network of Fig. 5.46b is modified to incorporate the new product term and agree with Equation 5.13, the extraneous output pulse of Fig. 5.47b will disappear.

Will the output of the waveform generator satisfy the specification provided by the application engineer once the above modification is included? In arguing that the answer is yes, we note that changes in the circuit inputs are infrequent. In this case, only transient outputs arising from state changes are of concern. Therefore, the Karnaugh map of Fig. 5.48b need only be examined on a column-

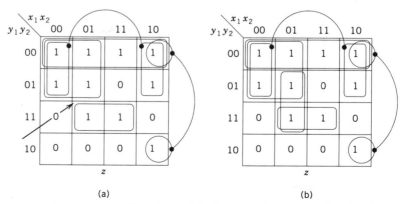

(a)   (b)

Figure 5.48   Elimination of the hazard in the waveform generator: (a) A. (b) Z with hazard eliminated.

by-column basis to determine if any additional hazards exist. Clearly, the first and last columns present no problems, since in each case, all the 1s are included in single product terms. The third column is not a problem, since y_1y_2 will never change directly from 00 to 11 or vice versa in a single clock period.

Our analysis of the hazard in the waveform generator leads us to the following theorem that we shall state without further proof.

THEOREM 5.3 Transient output transitions in sum-of-product networks caused by hazards may be eliminated by overlapping the hazard transitions with redundant product terms, provided that no two inputs are allowed to change simultaneously.

■ EXAMPLE 5.24

Obtain a hazard-free realization of the output network given by the Karnaugh map of Fig. 5.49a subject to any change in a single input.

SOLUTION   We note that a single input change can cause a transition between minterms m_1 and m_9, as well as between m_{11} and m_{15}. A single change in a state variable can cause a transition between m_5 and m_7. None of these pairs of minterms are included in the same cubes in the Karnaugh map of Fig. 5.49a. This situation is remedied by adding three redundant products, as shown in the map of Fig. 5.49b.

The sum-of-products network realization of the final Karnaugh map of Fig. 5.49b is given as Equation 5.14.

$$z = \bar{y}_2y_1\bar{x}_2 + y_2y_1x_1 + y_2y_1x_2 + \bar{y}_2y_1\bar{x}_1 + y_1\bar{x}_2x_1 + y_1x_2\bar{x}_1 + y_2x_1x_2 \qquad (5.14)$$

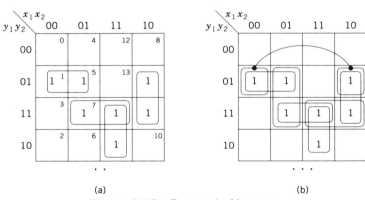

(a)   (b)

Figure 5.49   Removal of hazards.   ■

5.11   SHANNON'S EXPANSION THEOREM

This section focuses on proofs of two theorems that will be applied in upcoming chapters. The first almost self-evident theorem is attributed to Claud Shannon [14]. Applications of this theorem will appear in Section 7.6 and in Chapter 15. We adopt a notation that is most convenient for use in these applications. Given a Boolean function $f(x_1, x_2, ..., x_n)$, we define $f_1(1)$ as the function $f(1, x_2, ..., x_n)$ and $f_1(0)$ as $f(0, x_2, ..., x_n)$. Similar expressions are defined for an evaluation of f for other variables at 0 and 1.

THEOREM 5.4

$$f(x_1, x_2, ..., x_n) = x_1 f_1(1) + \bar{x}_1 f_1(0) \tag{5.15}$$

PROOF As guaranteed by Theorem 5.1, we express f as an expanded sum of products. Either x_1 or $\bar{x}_1$ appears in each product in this expression. Factoring x_1 from each product in which it appears and factoring $\bar{x}_1$ from the remaining products yield

$$f(x_1, x_2, ..., x_n) = x_1 g + \bar{x}_1 h \tag{5.16}$$

where g and h are functions of the remaining $n-1$ variables. Because Equation 5.16 is valid for all values of x_1, it must be valid for $x_1 = 0$ and $x_1 = 1$. Substituting these two values separately into Equation 5.16 yields

$$f_1(0) = f(0, x_2, ..., x_n) = h$$
$$f_1(1) = f(1, x_2, ..., x_n) = g$$

Replacing g and h in Equation 5.16 by $f_1(1)$ and $f_1(0)$, respectively, yields the original statement of the theorem given in Equation 5.15.

The second theorem that provides a mechanism for direct generation of the complement of a Boolean expression, will be applied in Chapters 6 and 7.

THEOREM 5.5
Let F be a Boolean sum-of-products expression of a particular Boolean function and F_D the dual of that expression. Then

$$\overline{F(x_1, x_2, ..., x_n)} = F_D(\bar{x}_1, \bar{x}_2, ..., \bar{x}_n) \tag{5.17}$$

PROOF We are given that F is a sum of products, so we may write $\bar{F} = \overline{(P_1 + P_2 + ... + P_r)}$, where each P_i is a product of complemented and uncomplemented literals from the set $\{x_1, x_2, ..., x_n\}$. By DeMorgan's theorem

$$\bar{F} = \overline{(P_1)} \cdot \overline{(P_2)} ... \overline{(P_r)}$$

DeMorgan's theorem is applied again, this time to each complemented product. For $\bar{P}_i$ the result is

$$\bar{P}_i = S_i = \bar{a}_1 + \bar{a}_2 + ... + \bar{a}_n$$

where a_j is the jth literal as it appears in P_i. Therefore,

$$\bar{F} = (S_1) \cdot (S_2) ... (S_r) = F_D(\bar{x}_1, \bar{x}_2, ..., \bar{x}_n)$$

PROBLEMS

5.1 Convert the following Boolean forms to minterm lists.
(a) $f(w, x, y, z) = wy + x(w + y\bar{z})$
(b) $f(U, V, W, X, Y) = \bar{V}(\bar{W} + \bar{U})(X + \bar{Y}) + \bar{U}\bar{W}\bar{Y}$
(c) $f(V, W, X, Y, Z) = (X + \bar{Z})(\overline{Z + WY}) + (VZ + W\bar{X})(\overline{Y + Z})$

5.2 Convert the forms of Problem 5.1 to maxterm lists.

5.3 By using the Karnaugh map, determine minimal sum-of-products realizations of the following functions.
(a) $f(A, B, C, D) = \sum m(0, 4, 6, 10, 11, 13)$
(b) $f(w, x, y, z) = \sum m(3, 4, 5, 7, 11, 12, 14, 15)$
(c) $f(a, b, c, d) = \prod M(3, 5, 7, 11, 13, 15)$
(d) $f(V, W, X, Y, Z) = \sum m(0, 2, 3, 4, 5, 11, 18, 19, 20, 23, 24, 28, 29, 31)$

5.4 A logic circuit is to be designed having four inputs $y_1, y_0, x_1,$ and x_0. The pairs of bits $y_1 y_0$ and $x_1 x_0$ represent two-bit binary numbers with y_1 and x_1 as the most significant bits. The only circuit output z is to be 1 if and only if the binary number $x_1 x_0$ is greater than or equal to the binary number $y_1 y_0$. Determine a minimal sum-of-products expression for z.

5.5 Determine a minimal sum-of-products expression equivalent to each of the following Boolean expressions.

(a) $f(A, B, C, D, E) = (\bar{C}\bar{E} + CE)(\bar{A} + B)D + (\overline{\bar{A} + B})D\bar{C}E$

(b) $f(w, x, y, z) = (\overline{\bar{w} + x}) + (\overline{\bar{x} + z}) + (\overline{\bar{y} + \bar{z}})$

5.6 By using Karnaugh maps, determine minimal product-of-sums realizations for the functions of Problems 5.3(b) and 5.3(c).

5.7 (a) Determine the minimal product-of-sums realization for the function of Example 5.15 and compare the cost to that of the minimal sum-of-products realization.

(b) Repeat (a) for Example 5.16.

5.8 A prime number is a number that is only divisible by itself and 1. Suppose the numbers between 0 and 31 are represented in binary in the form of the five bits

$$x_4 x_3 x_2 x_1 x_0$$

where x_4 is the most significant bit. Design a prime detector. That is, design a combinational logic circuit whose output Z will be 1 if and only if the five input bits represent a prime number. Do not count 0 as a prime. Base your design on obtaining a minimal two-level expression for Z.

5.9 A two-output, four-input logic circuit is to be designed that will carry out addition modulo-4. The addition table for modulo-4 addition is given in Fig. P5.9. For example, $(3 + 3)$ mod-4 = 2. Therefore, a 2 is entered in row 3, column 3 of the table, etc. The input numbers are to be coded in straight binary, with one input number given by $x_2 x_1$ and the other by $y_2 y_1$. The output is also to be coded as the binary number $z_2 z_1$. That is, $z_2 z_1 = 00$ if the sum is 0, 01 if the sum is 1, 10 if the sum is 2, and 11 if the sum is 3.

(a) Determine a two-level Boolean expression for z_1.

(b) Determine a two-level Boolean expression for z_2.

5.10 Determine minimal sum-of-products realizations for the following incompletely specified functions.

X Y	0	1	2	3
0	0	1	2	3
1	1	2	3	0
2	2	3	0	1
3	3	0	1	2

$Z = (X + Y)$ Mod-4

Figure P5.9

(a) $f(A, B, C, D) = \sum m(1, 3, 5, 8, 9, 11, 15) + d(2, 13)$

(b) $f(W, X, Y, Z) = \sum m(4, 5, 7, 12, 14, 15) + d(3, 8, 10)$

(c) $f(A, B, C, D, E) = \sum m(1, 2, 3, 4, 5, 11, 18, 19, 20, 21, 23, 28, 31)$
$+ d(0, 12, 15, 27, 30)$

(d) $f(a, b, c, d, e) = \sum m(7, 8, 9, 12, 13, 14, 19, 23, 24, 27, 29, 30)$
$+ d(1, 10, 17, 26, 28, 31)$

(e) $f(u, v, w, x, y, z) = \sum m(0, 2, 14, 18, 21, 27, 32, 41, 49, 53, 62)$
$+ d(6, 9, 25, 34, 55, 57, 61)$

5.11 Determine minimal realizations of the functions of Problem 5.10 in terms of wired-AND connections of NAND gates, and AOI gates.

5.12 The four lines into the combinational logic circuit depicted in Fig. P5.12 carry one binary-coded decimal digit. That is, the binary equivalents of the decimal digits 0–9 may appear on the lines $x_0 x_1 x_2 x_3$. The most significant bit is x_0. The combination of values corresponding to binary equivalents of the decimal numbers 10–15 will never appear on the lines. The single output Z of the circuit is to be 1 if and only if the inputs represent a number that is either 0 or a power of 2. Construct the logic block diagram of a minimal two-level realization of the circuit.

5.13 A shaft-position encoder provides a four-bit signal indicating the position of a shaft in steps of 30°, using a reflected (Gray) code as listed in Fig. P5.13. It may be assumed that the four possible combinations of four bits not used above will never occur. Design a minimal sum-of-products realization of a circuit to produce an output whenever the shaft is in the first quadrant (0–90°).

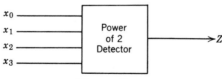

Figure P5.12

Shaft Position	Encoder Output $E_3 E_2 E_1 E_0$			
0–30°	0	0	1	1
30–60°	0	0	1	0
60–90°	0	1	1	0
90–120°	0	1	1	1
120–150°	0	1	0	1
150–180°	0	1	0	0
180–210°	1	1	0	0
210–240°	1	1	0	1
240–270°	1	1	1	1
270–300°	1	1	1	0
300–330°	1	0	1	0
330–360°	1	0	1	1

Figure P5.13

5.14 A circuit receives two 3-bit binary numbers: $A = A_2A_1A_0$ and $B = B_2B_1B_0$. Design a minimal sum-of-products circuit to produce an output whenever A is greater than B.

5.15 In digital computers, letters of the alphabet are coded in the form of unique combinations of five or more bits. One of the most common codes is the six-bit ASCII code. This code is given in Fig. P5.15 in terms of the octal equivalents. For example,

$$C + 03|_8 = \underbrace{000}_{0} \underbrace{011}_{3}$$

Design a minimal sum-of-products circuit that will receive this code as an input and produce an output whenever the letter is a vowel. The alphabet uses only 26 of the possible codes and the others are used for numerals and punctuation marks. However, you may assume that the data are pure alphabetic, so that only the codes listed will occur.

5.16 Shown in Fig. P5.16 are the ASCII codes for the 10 decimal numerals. Assume that a device receives alphanumeric data in this code, that is, either these codes for numerals or the codes for letters as listed in Problem 5.15. Design a minimal circuit that will develop a 1 out when the data are numeric, a 0 out when they are alphabetic. You may assume that any possible six-bit codes not used for either letters or numerals will not occur.

5.17 Referring again to the ASCII code listed in Problems 5.15 and 5.16, design an "error circuit," that is, a circuit that will put out a signal if a code other than one of the 36 "legal" alphanumeric codes is received.

A	01	N	16
B	02	O	17
C	03	P	20
D	04	Q	21
E	05	R	22
F	06	S	23
G	07	T	24
H	10	U	25
I	11	V	26
J	12	W	27
K	13	X	30
L	14	Y	31
M	15	Z	32

Figure P5.15

0	40
1	41
2	42
3	43
4	44
5	45
6	46
7	47
8	48
9	49

Figure P5.16

5.18 In a certain computer, three separate sections of the computer proceed independently through four phases of operation. For purposes of control, it is necessary to know when two of three sections are in the same phase at the same time. Each section puts on a two-bit signal (00, 01, 10, 11) in parallel on two lines. Design a circuit to put out a signal whenever it receives the same phase signal from any two or all three sections.

5.19 Determine a hazard-free sum-of-products realization of the Boolean function given in Problem 5.3(b).

5.20 Determine a hazard-free sum-of-products realization of the five-bit prime detector described in Problem 5.8.

5.21 Determine a hazard-free realization of the power of 2 detector described in Problem 5.12. Take advantage of "don't cares" whenever possible to minimize the number of products included in the hazard-free sum of products.

BIBLIOGRAPHY

1. Karnaugh, M. "The Map Method for Synthesis of Combinational Logic Circuits," *Trans. AIEE*, **72**, Pt. I:593–598 (1953).
2. Veitch, E. W. "A Chart Method for Simplifying Truth Functions," in *Proceedings of ACM*, Pittsburgh, Pa., May 2, 3, 1952, pp. 127–133.
3. Tanana, E. J. "The Map Method," in E. J. McCluskey and T. G. Bartee (eds.), *A Survey of Switching Circuit Theory*. McGraw-Hill, New York, 1962.
4. Caldwell, S. H. *Switching Circuits and Logical Design*. Wiley, New York, 1958.
5. McCluskey, E. J. *Logic Design Principles*. Prentice-Hall, Englewood Cliffs, N.J., 1986.
6. Muroga, S. *VLSI System Design*. Wiley, New York, 1982.
7. *The TTL Data Book*. Texas Instruments Inc., Dallas, Tex., 1984.
8. Heinbuch, D. V. *The CMOS3 Cell Library*. Addison-Wesley, Reading, Mass., 1988.
9. Hill, F. J. and G. R. Peterson. *Digital Logic and Microprocessors*. Wiley, New York, 1984.
10. Unger, S. H. *Asynchronous Sequential Switching Circuits*. Wiley-Interscience, New York, 1969.
11. Hayes, J. "Pseudo-Boolean Logic Circuits," *IEEE Trans. Computers*, 602–612 (July 1986).
12. Roth, C. H., Jr. *Fundamentals of Logic Design*, 2nd ed. West Publishing, St. Paul, Minn., 1979.
13. Mange, D. *Analyse et Synthese des Systemes Logiques*. Editions Georgi, St.-Saphorin, Switzerland, 1978.
14. Shannon, C. E. "Symbolic Analysis of Relay and Switching Circuits," *Trans. AIEE*, **57**:713–723 (1938).

CHAPTER 6

ALGORITHMS FOR OPTIMIZATION OF COMBINATIONAL LOGIC

6.1 IMPACT OF LOGIC SYNTHESIS

Much of the forthcoming treatment of the Quine–McCluskey algorithm [3] for the tabular minimization of two-level combinational logic was included in the first edition of *Introduction to Switching Theory and Logical Design*, published in 1968. It soon became clear that for manual design the tabular approach is tedious in comparison to the Karnaugh map, considered in Chapter 5. The Quine-McCluskey method provides the basis for a completely defined algorithm and consequent software implementation. Early implementations were typically of exponential order, with respect to the number of variables, and were therefore applicable to only slightly more complex problems than could be successfully addressed using a Karnaugh map. Engineering judgment could usually be used to divide a design into subproblems that could be addressed at the Karnaugh map level. When the pieces of the design were reassembled, the result was usually a more than two-level logic network, to which sum-of-products optimization would not have been applicable. For several years there was little incentive to treat carefully the contents of this chapter.

The advent of logic synthesis, as first introduced in Section 1.1, has dramatically altered this assessment of the importance of logic optimization algorithms. The processing of a complex system design by a hardware synthesis program will deprive (or relieve, depending on the point of view) the designer of many opportunities to insert engineering judgment into the design process. One alternative to implementation of this engineering judgment is logic optimization routines, both two-level and multilevel. Typically, such routines will be embedded within CAD packages with little advertisement of expected performance. It is hoped that consideration of the following sections will serve as a first step in the readers' acquisition of the background needed to make individual performance assessments. It is not our goal to bring the reader to the level of authors of logic optimization routines.

6.2 CUBICAL REPRESENTATION OF BOOLEAN FUNCTIONS

At this point, it is convenient to introduce a new representation for Boolean functions that will provide convenient terminology for further work. We are all familiar with the notion of a geometric representation of a continuous variable as

119

a distance along a straight line (Fig. 6.1*a*). In a similar fashion, a switching variable, which can take on only two values, can be represented by two points at the ends of a single line (Fig. 6.1*b*). We extend this to represent two variables by points in a plane (Fig. 6.2*a*). Similarly, the four possible values of two switching variables can be represented by the four vertices of a square (Fig. 6.2*b*). The extension to three variables, as shown in Fig. 6.3, should be obvious. The extension to more than three variables, requiring figures of more than three dimensions, is geometrically difficult but conceptually simple enough.

In general, we say that we represent the various possible combinations of n variables as points in n space, and the collection of all 2^n possible points will be said to form the *vertices of an n-cube*, or a *Boolean hypercube*.

To represent functions on the *n*-cube, we set up a one-to-one correspondence between the minterms of *n* variables and the vertices of the *n*-cube. Thus, in the 3-

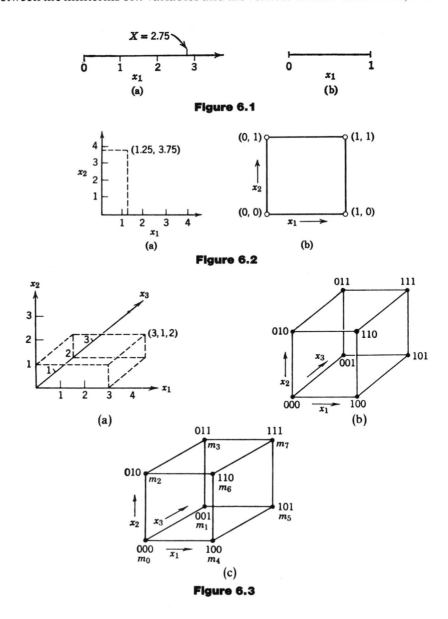

Figure 6.1

Figure 6.2

Figure 6.3

cube, the vertex (000) corresponds to m_0, the vertex (001) to m_1, etc., as shown in Fig. 6.3c. The cubical representation of a function of n variables then consists of the set of vertices of the n-cube corresponding to the minterms of the function. For example, the function

$$f(A, B, C) = \sum m(0, 2, 3, 7) \tag{6.1}$$

would be represented on the 3-cube as shown in Fig. 6.4a, where the vertices corresponding to $m_0, m_2, m_3,$ and m_7 are indicated by heavy back dots. These vertices, corresponding to the minterms, may also be referred to as the *0-cubes* of the function.*

Two 0-cubes of a function are said to form a 1-cube if they differ in only one coordinate. In the function of Fig. 6.4, we have three 1-cubes, consisting of the pairs of 0-cubes 000 and 010, 010 and 011, and 011 and 111. The 1-cubes may be denoted by placing an x in the coordinate having different values and darkening the line between the pair of 0-cubes (Fig. 6.4b). In a similar fashion, a set of four 0-cubes, whose coordinate values are the same in all but two variables, are said to form a 2-cube of the function. Pictorially, a 2-cube may be represented as a shaded plane. Figure 6.5 shows the cubic representation of a function exhibiting five 0-cubes, five 1-cubes, and one 2-cube.

When all the vertices (0-cubes) of a k-cube are in the set of vertices making up a larger k-cube, we will say that the smaller cube is *contained in*, or *covered by*, the larger cube. In Fig. 6.5, for example, the 0-cube 100 is contained in the 1-cubes x00, 10x, and 1x0 and in the 2-cube 1xx. Similarly, the 1-cubes 1x0, 10x, 11x, and 1x1 are all contained in the 2-cube 1xx.

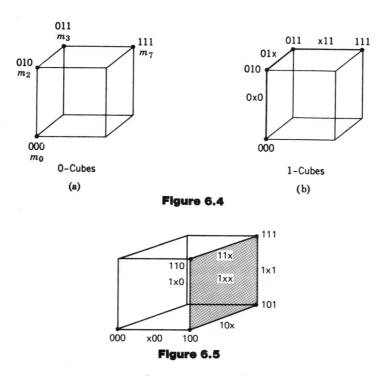

Figure 6.4

Figure 6.5

*The 0-cubes, 1-cubes, 2-cubes, etc., are more formally known as *k-dimensional product-subcubes* ($k = 0, 1, 2,$ etc.) but this lengthy expression is generally shortened to *k-cube*. See Reference [1], pp. 100–101.

The correspondences between the cubical representation and the Karnaugh map should be clear to the reader. The 0-cubes correspond to the squares on the K-map, the 1-cubes to pairs of adjacent squares, etc. The cubical *nomenclature* can be applied directly to the K-map without reference to the cubical *representation*, but the cubical representation makes the origin and significance of these terms much more obvious.

6.3   DETERMINATION OF PRIME IMPLICANTS

The most comprehensive algorithm for determining a minimal-cost two-level realization is a special case of the following:

Select the least cost subset *D* of a set of objects *A* so that some cover criterion is satisfied.

In the context of finding a minimal sum-of-products realization, set *A* is all possible Boolean cubes of any dimension that fall within a given Boolean function *f*. The set *D* will consist of all products in a minimal sum-of-products realization as given by Definition 5.1. This method is based principally on the work of W. V. Quine and E. J. McCluskey [2] and will be called the Quine–McCluskey algorithm. In this chapter we will develop the method in tabular form. Conversion to a computer program is straightforward.

The first step, represented by block I of Fig. 6.6, is to find a subset *B*, consisting of all cubes that are actual candidates for inclusion in the minimal realization. The members of *B* will be called *prime implicants*. The formal definition of prime implicant will be given later in this section, as we develop the process implementing block I.

The starting point for block I of Fig. 6.6 is the minterm list of the function. If the function is not in this form, it must be converted to this form by the methods discussed earlier. Let us assume the following function:

$$f(A, B, C, D) = \sum m(0, 2, 3, 6, 7, 8, 9, 10, 13) \qquad (6.2)$$

The cubical representation and K-map for this function are shown in Fig. 6.7.

The first step is to find all possible 1-cubes of the function. By definition, a 1-cube is found by combining two 0-cubes, which are identical except in one variable. First, convert the minterms to binary form and then count (and list) the number of 1s in the binary representation (Fig. 6.8). Now reorder the minterm list in accordance with the number of 1s in the binary representations (Fig. 6.9a). Separate the minterms into groups with the same number of 1's by horizontal lines. This grouping of minterms is made to reduce the number of comparisons that must be made in determining the 1-cubes. If two minterms are to combine, their binary representation must be identical except in one position, in which one minterm will have a 0, the other a 1. Thus, the latter has one more 1 than the former, so that the two minterms must be in adjacent groups in the list of Fig. 6.9a.

With the minterms grouped, the procedure for finding the 1-cubes is quite simple. Compare each minterm in the top group with each minterm in the next lower group. If two minterms are the same in every position but one, place a check ($\vee$) to the right of both minterms to show that they have been covered and enter the 1-cube in the next column (Fig. 6.9b). List the cubical form of the 1-cube with an x in the position that does not agree and also list the decimal numbers of the combining minterms.

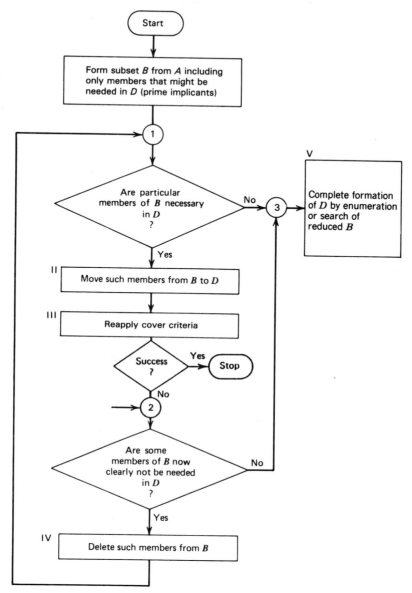

Figure 6.6 Generalized Quine–McCluskey algorithm.

In this case, m_0 (0000) combines with m_2 (0010) to form (00x0). The combination of two vertices into a 1-cube represents an application of the distributive law just as does combining two squares on a Karnaugh map. For example, this combination of m_0 and m_2 is equivalent to the algebraic operation

$$\bar{A}\bar{B}\bar{C}\bar{D} + \bar{A}\bar{B}C\bar{D} = \bar{A}\bar{B}\bar{D}(C + \bar{C}) = \bar{A}\bar{B}\bar{D} \qquad (6.3)$$

Minterm m_0 also combines with m_8 to form (x000). This completes the comparison between minterms in the first two groups, so we draw a line beneath the resultant 1-cubes. Next, the second and third groups of minterms are compared in the same manner. This comparison results in five more 1-cubes, formed from m_2 and m_3, m_8 and m_9, etc. A line is drawn beneath these five 1-cubes to indicate completion of comparisons between the second and third groups. Note that each minterm in a group must be compared with *every* minterm in the other group,

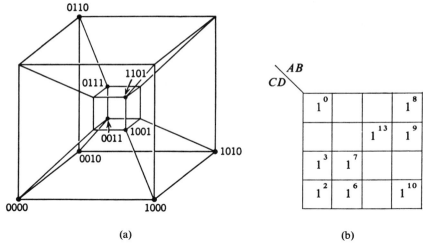

(a)   (b)

Figure 6.7   (*a*) Cubical representation. (*b*) K-map.

Minterm	Binary Form	No. of 1's
m_0	0000	0
m_2	0010	1
m_3	0011	2
m_6	0110	2
m_7	0111	3
m_8	1000	1
m_9	1001	2
m_{10}	1010	2
m_{13}	1101	3

Figure 6.8   Conversion of minterm list to binary form.

1's	Minterms		1-Cubes		2-Cubes	
0	m_0	0000 ✓	0,2	00x0 ✓	*0,2,8,10	x0x0
			0,8	x000 ✓	*2,3,6,7	0x1x
1	m_2	0010 ✓	2,3	001x ✓		
	m_8	1000 ✓	2,6	0x10 ✓		
			2,10	x010 ✓		
	m_3	0011 ✓	*8,9	100x		
	m_6	0110 ✓	8,10	10x0 ✓		
2	m_9	1001 ✓				
	m_{10}	1010 ✓	3,7	0x11 ✓		
			6,7	011x ✓		
3	m_7	0111 ✓	*9,13	1x01		
	m_{13}	1101 ✓				

(a)   (b)   (c)

Figure 6.9   Determination of prime implicants by Quine–McCluskey.

even if either or both has already been checked ($\checkmark$) as having formed a 1-cube. *Every* 1-cube must be found, although there is no need to check off a minterm more than once.

This comparison process is repeated between successive groups until the minterm list is exhausted. In this case, all the minterms have been checked, indicating that all combine at least into 1-cubes. Thus, none of the minterms will appear explicitly in the final sum-of-products form.

The next step is a search of Fig. 6.9*b* for possible combinations of pairs of 1-cubes into 2-cubes. Again, cubes in each group need to be compared only with cubes in the next group down. In addition, cubes need be compared only if they have the same digit replaced by an x. In this case, 1-cube 0,2 (00x0) need be compared only with 8,10 (10x0). They differ in only one of the specified (non-x) positions and therefore combine to form the 2-cube 0,2,8,10 (x0x0), which is entered in the next column (Fig. 6.9*c*). The two 1-cubes are also checked off to show that they have combined into a 2-cube. Next, 1-cube 0,8 (x000) is found to combine with 2,10 (x010) to form 0,8,2,10 (x0x0), which is the same 2-cube as that already formed. Therefore, the two 1-cubes are checked off, but no new entry is made in the 2-cube column.

To clarify the above process, consider the algebraic interpretation. First, we have

$$m_0 + m_2 + m_8 + m_{10} = \bar{A}\bar{B}\bar{C}\bar{D} + \bar{A}\bar{B}C\bar{D} + A\bar{B}\bar{C}\bar{D} + A\bar{B}C\bar{D}$$
$$= \bar{A}\bar{B}\bar{D}(C + \bar{C}) + A\bar{B}\bar{D}(C + \bar{C}) \tag{6.4}$$
$$= \bar{A}\bar{B}\bar{D} + A\bar{B}\bar{D}$$
$$= \bar{B}\bar{D}(A + \bar{A}) \tag{6.5}$$
$$= \bar{B}\bar{D}$$

Here we see the reason for requiring the x to be in the same position in both 1-cubes. The x represents the variable eliminated. If the same variable (C) had not been eliminated from both 1-cubes in Equation 6.4, it is obvious that the second elimination (of A) could not have been made in Equation 6.5. The second combination of 1-cubes into the same 2-cubes is seen in the following:

$$m_0 + m_8 + m_2 + m_{10} = \bar{A}\bar{B}\bar{C}\bar{D} + A\bar{B}\bar{C}\bar{D} + \bar{A}\bar{B}C\bar{D} + A\bar{B}C\bar{D}$$
$$= \bar{B}\bar{C}\bar{D}(A + \bar{A}) + \bar{B}C\bar{D}(A + \bar{A}) \tag{6.6}$$
$$= \bar{B}\bar{C}\bar{D} + \bar{B}C\bar{D}$$
$$= \bar{B}\bar{D}(C + \bar{C}) \tag{6.7}$$
$$= \bar{B}\bar{D}$$

Equations 6.6 and 6.7 represent the same elimination of variables as Equations 6.4 and 6.5, except in the reverse order.

No further combinations can be made between the first and second groups of Fig. 6.9*b*, so we draw a line beneath the 2-cube formed. The second and third groups are then compared in the same fashion, resulting in another 2-cube. This completes the comparison of 1-cubes, and any unchecked entries—8,9 and 9,13 in this case—are marked with an asterisk *, to indicate that they are *prime* implicants.

Definition 6.1 A *prime implicant* is any cube of a function that is not totally contained in some larger cube of the function.

If any minterms had failed to combine in the first step, they would also have been marked as prime implicants. The importance of prime implicants will become evident shortly.

Finally, the 2-cubes are checked for possible combination into 3-cubes. Since the x's are in different positions, these cubes do not combine and are thus prime implicants. For this example, the determination of prime implicants is now complete. In the general case, the same procedure continues as long as larger cubes can be formed.

It is also possible to accomplish prime implicant determination by comparison of minterm numbers in decimal form. Again, the first step is to list the minterms, grouped according to the number of 1's in the binary representation. Only the decimal minterm numbers are listed (Fig. 6.10a). The minterms that can be combined will differ by a power of 2. Consider the combination of m_{13} (1101) and m_9 (1001):

$$13 = 1 \times 2^3 + \quad 1 \times 2^2 \quad + 0 \times 2^1 + 1 \times 2^0$$
$$9 = 1 \times 2^3 + \quad 0 \times 2^2 \quad + 0 \times 2^1 + 1 \times 2^0$$

Each 1 in a binary number has the numeric weight of a power of 2. The two combining minterms have the same 0s and 1s, except in one position. Thus, the number of the minterm with the extra 1 must be larger than the number of the other minterm by a power of 2. The procedure, then, is as follows: Compare each minterm number with each *larger* minterm number in the next group down. If they differ by a power of 2, they combine to form a 1-cube.

To represent the 1-cube, list the two minterm numbers, followed by their numeric difference in parentheses. For the above example, the listing would be 9,13(4). The number in parentheses indicates the position in which the x would appear in the binary representation. Since the binary digit with the weight of 4 is the third from the right, the (4) in parentheses corresponds to the x in the third position in the binary notation 9,13 (1x01). Except for this change in notation, the process of comparing minterms and listing 1-cubes is the same as before. The resultant 1-cube column is shown in Fig. 6.10b.

In the binary-character method, it was seen that 1-cubes could combine only if the x's were in the same position. Since the number in parentheses indicates the

1's	Minterms
0	0 ✓
1	2 ✓
	8 ✓
2	3 ✓
	6 ✓
	9 ✓
	10 ✓
3	7 ✓
	13 ✓

(a)

1-Cubes
0,2(2) ✓
0,8(8) ✓
2,3(1) ✓
2,6(4) ✓
2,10(8) ✓
*8,9(1)
8,10(2) ✓
3,7(4) ✓
6,7(1) ✓
*9,13(4)

(b)

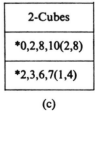

2-Cubes
*0,2,8,10(2,8)
*2,3,6,7(1,4)

(c)

Figure 6.10 Prime implicant determination, decimal notation.

position of the x, it follows that only 1-cubes with the same number in parentheses can combine. Therefore, compare each 1-cube with all 1-cubes in the next lower group that have the same number in parentheses. If the lowest minterm number of the 1-cube in the lower group is greater by a power of 2 than the corresponding number in the other cube, they combine. List the 2-cube by listing all four minterm numbers, followed by both powers of 2, in parentheses. For example, in Fig. 6.10b, 0,2(2) and 8,10(2) have the same number in parentheses and 8 is greater than 0 by 8. The resultant 2-cube is thus 0,2,8,10(2,8).

Figure 6.10c shows the complete 2-cube column in decimal notation. To compare 2-cubes, the same procedure applies. Just as both x's, in both 2-cubes, have to be in the same position for combination, so *both* numbers in parentheses must be the same for both 2-cubes. A comparison of Figs. 6.9 and 6.10 will show that the resultant charts are identical. Furthermore, the procedures are identical in principle, differing only in the substitution of mental subtraction of decimal numbers for visual comparison of binary numbers.

6.4 SELECTION OF OPTIMUM SET OF PRIME IMPLICANTS

In this section we turn our attention to obtaining the realization D from the set of prime implicants B. That this is possible is guaranteed by Theorem 6.1, first proved by Quine [2]. All the prime implicants determined for the example of the previous section are indicated on the 4-cube and the map of Fig. 6.11. Interpreting the k-cubes as sets of 1's on the K-map, we see that each k-cube will be represented in a sum-of-products realization by a product term with $n - k$ literals. Thus, any set of k-cubes containing all the 0-cubes (minterms) of the function will provide a sum-of-products realization.

THEOREM 6.1 Any sum-of-products realization that is minimal in the sense of Definition 5.1 must consist of a sum of prime implicant products.

PROOF Consider any set of k-cubes containing all the 0-cubes of the function. Any cube that is not a prime implicant is, by definition, contained in some prime implicant. Therefore, we can replace this cube in the set by any prime implicant containing it. Since the prime implicant is of higher dimension, its product representation will have fewer literals than that of the cube it replaces. The replacement will therefore lower the cost of the sum-of-products realization.

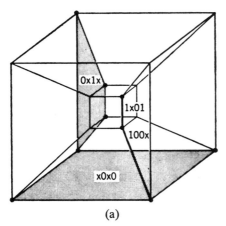
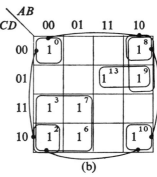

(a)   (b)

Figure 6.11 Prime implicants of $\sum m(0, 2, 3, 6, 7, 8, 9, 10, 13)$.

Theorem 6.1 tells us that we can restrict our attention to the prime implicants in selecting a sum-of-products realization. (In the future we will use the term *prime implicant* to mean either the cubes themselves or the corresponding product.) However, the theorem does not say that *any* sum of prime implicants covering all the minterms is minimal. To find a minimal sum, we construct a *prime implicant table* (Fig. 6.12).

Each column corresponds to a minterm. At the left of each row are listed the prime implicants of the function, arranged in groups according to cost (number of literals in the product). Also included are an extra column on the left and an extra row at the bottom, for reasons that will be explained shortly. In each row (except the bottom), checks are placed in the columns corresponding to the minterms contained in the prime implicant listed on that row. For example, the first prime implicant listed in Fig. 6.12 contains the 0-cubes 0,2,8, and 10, so checks are placed in the first row in columns 0, 2, 8, and 10.

The completed prime implicant table is then inspected for columns containing only a single check mark. In this example, we find only one check in the first column, on the first row. This indicates that the corresponding prime implicant is the only one containing m_0 and must be included in the sum-of-products realization. We therefore say that it is an *essential prime implicant* and mark it with an asterisk in the leftmost column. We also place checks in the bottom row in columns 0, 2, 8, and 10 to indicate that these are included in an already selected prime implicant (Fig. 6.13).

Continuing the search, we find a single check in column 3, indicating that the prime implicant 0x1x is the only one covering m_3. We mark this prime implicant

		0	2	3	6	7	8	9	10	13
	x0x0 (0,2,8,10)	✔	✔				✔		✔	
	0x1x (2,3,6,7)		✔	✔	✔	✔				
	100x (8,9)						✔	✔		
	1x01 (9,13)							✔		✔

Figure 6.12 Prime implicant table—first step.

		0	2	3	6	7	8	9	10	13
*	x0x0 (0,2,8,10)	✔	✔				✔		✔	
	0x1x (2,3,6,7)		✔	✔	✔	✔				
	100x (8,9)						✔	✔		
	1x01 (9,13)							✔		✔
		✔	✔				✔		✔	

Figure 6.13 Prime implicant table: second step.

as essential and check off the minterms it covers, which were not covered by the first one selected. Finally, we find that 1x01 is also essential (Fig. 6.14). The determination of essential prime implicants corresponds to block II of Fig. 6.6. When the search for essential prime implicants is complete, we inspect the bottom row to see if all columns have been checked. This constitutes block III of Fig. 6.6. If so, then all the 0-cubes are contained in the essential prime implicants and the sum of the essential prime implicants is the minimal sum-of-products realization. This is the case for our example. The final step is to determine the actual products. For the above example, we have

$$x0x0 \rightarrow \bar{B}\bar{D}$$

$$0x1x \rightarrow \bar{A}C$$

$$1x01 \rightarrow A\bar{C}D$$

Thus, the minimal sum of products is

$$f(A, B, C, D) = \sum m(0, 2, 3, 6, 7, 8, 9, 10, 13) = \bar{B}\bar{D} + \bar{A}C + A\bar{C}D \qquad (6.8)$$

This very simple example was chosen to clarify the exact significance of the various steps rather than as a practical illustration of the value of the technique.

In the just completed example, all the minterms of the function were covered by essential prime implicants (satisfaction of the cover criteria) so the algorithm terminated before reaching block IV of Fig. 6.6. For more complex functions, it will be possible to identify certain prime implicants that will never be needed in the realization, once essential prime implicants have been moved to set D. To make possible identification of unneeded prime implicants and completion of the Quine–McCluskey algorithm, we provide the following definitions.

Definition 6.2   Two rows **a** and **b** of a reduced prime implicant table, which cover the same minterms, that is, have checks in exactly the same columns, are said to be interchangeable.

Definition 6.3   Given two rows **a** and **b** in a reduced prime implicant table, row **a** is said to *dominate* row **b** if row **a** has checks in all the columns in which row **b** has checks and also has a check in at least one column in which row **b** does not have a check.

THEOREM 6.2   Let **a** and **b** be rows of a reduced prime implicant table, such that the cost of **a** is less than or equal to the cost of **b**. Then, if **a** dominates **b** or **a**

		0	2	3	6	7	8	9	10	13
*	x0x0 (0,2,8,10)	✔	✔				✔		✔	
*	0x1x (2,3,6,7)		✔	✔	✔	✔				
	100x (8,9)						✔	✔		
*	1x01 (9,13)							✔		✔
		✔	✔	✔	✔	✔	✔	✔	✔	✔

Figure 6.14   Prime implicant table: final.

and **b** are interchangeable, there exists a minimal sum of products that does not include **b**.

Theorem 6.2 which follows directly from definitions 6.2 and 6.3 is applied as necessary as part of the lower decision diamond or of block IV of the flowchart of Fig. 6.6.

■ EXAMPLE 6.1

Determine the minimal sum-of-products form for

$$f(A, B, C, D, E) = \sum m(1, 2, 3, 5, 9, 10, 11, 18, 19, 20, 21, 23, 25, 26, 27)$$

SOLUTION   Figure 6.15a shows the prime implicant table for the function. The essential prime implicants on this table do not cover all the minterms of the function. To make it easier to select the appropriate terms, a reduced prime implicant table is set up, listing only the 0-cubes not contained in the essential prime implicants as columns and those nonessential prime implicants containing any of the uncovered 0-cubes as rows (Fig. 6.15b). In the reduced table, we note that there could be no possible advantage to using **e**, since it covers only m_5, whereas **d**, which has the same cost, covers both m_5 and m_1. That is, **d** dominates **e**. Thus,

		1	2	3	5	9	10	11	18	19	20	21	23	25	26	27
xx01x	*a		✔	✔			✔	✔	✔	✔					✔	✔
x10x1	*b					✔		✔						✔		✔
0x0x1	c	✔		✔	✔	✔										
00x01	d	✔			✔											
x0101	e			✔								✔				
1010x	*f										✔	✔				
10x11	g									✔			✔			
101x1	h											✔	✔			
			✔	✔		✔	✔	✔	✔	✔	✔	✔	✔		✔	✔	✔

(a)

(b)

(c)

Figure 6.15   Application of dominance.

removing **e** from the table cannot prevent us from finding a minimal sum. We next note that **g** and **h** cover the same minterm at the same cost. That is, **g** and **h** are interchangeable. Thus, removing **h** from the table cannot prevent our finding a minimal sum. Since rows **e** and **h** are not needed in Fig. 6.15b, the answer to the lower diamond in the flowchart of Fig. 6.6 is yes. Thus, these rows are removed as called for by block IV of the flowchart to form a still further reduced table, as shown in Fig. 6.15c. On this table, it may be seen that **d** is the only prime implicant left to cover m_5 and **g** is the only prime implicant left to cover m_{23}. We therefore mark them with double asterisks ** to indicate that they are *secondary essential* prime implicants. Selection of secondary essential prime implicants is another application of block II of Fig. 6.6 on the second pass through the flowchart. We also note that these essential prime implicants cover all the remaining minterms, so the process terminates with success on this second pass. The minimal sum is given by

$$f(A, B, C, D, E) = \underbrace{\mathbf{a} + \mathbf{b} + \mathbf{f}}_{\text{Essential}} + \underbrace{\mathbf{d} + \mathbf{g}}_{\substack{\text{Secondary} \\ \text{Essential}}} \tag{6.9}$$

$$= \bar{C}D + B\bar{C}E + A\bar{B}C\bar{D} + \bar{A}\bar{B}\bar{D}E + A\bar{B}DE \tag{6.10}$$

∎

In Example 6.1, one application of row dominance was sufficient to complete the design. In some cases, the first selection of secondary essential prime implicants may leave a large group of minterms still uncovered. In this case, repeat the process, that is, reduce the table by removing secondary essential rows and apply row dominance again. This process may be repeated as many times as necessary. If, during some pass prior to finding a cover, no secondary essential prime implicants are identified or no prime implicants are dropped, the Quine–McCluskey algorithm is exited at point 3. Example 6.2 is provided to illustrate this case.

The Quine–McCluskey method can be easily modified to handle "don't care" terms. In determination of prime implicants, the "don't cares" are included in the minterm list. This assures that "don't care" vertices will be included, as needed, to make the prime implicant cubes as large as possible. Once the prime implicants are available, the "don't cares" are no longer of interest. Only the minterms for which the function must be 1 are listed across the top of the prime implicant table. It is only necessary that the selected minimal set of prime implicants cover these vertices whose functional value is 1.

■ EXAMPLE 6.2

Use the Quine–McCluskey method to find a minimal sum-of-products representation of

$$f(v, w, x, y, z) = \sum m(0, 4, 10, 13, 15, 16, 22, 23, 26) \\ + d(5, 11, 12, 14, 18, 21, 24)$$

SOLUTION Although the process was not shown, the 1-vertices and the "don't cares" were merged into the list (0, 4, 10, 13, 15, 16, 22, 23, 26, 5, 11, 12, 14, 18, 21, 24) for finding prime implicants. The resulting prime implicants for this function are listed in the table in Fig. 6.16 and identified by the boldface letters **a** through **j**. For

v w x y z		0	4	13	15	10	26	16	22	23
0 x 1 0 x	b		✔	✔						
0 1 1 x x	c			✔	✔					
0 1 x 1 x	d				✔	✔				
1 x 0 x 0	f						✔	✔		
0 0 x 0 0	a	✔	✔							
x 1 0 1 0	e					✔	✔			
x 0 0 0 0	g	✔						✔		
1 0 x 1 0	h								✔	
**1 0 1 1 x	i								✔	✔
1 0 1 x 1	j									✔

Figure 6.16 Prime implicant table for Example 6.2.

convenience, the minterms, for which $f = 1$, are listed out of numerical order in this prime implicant table.

No prime implicants in Fig. 6.16 are essential, but **i** dominates both **h** and **j**. Therefore, **i** is secondary essential, as marked in the table. Theorem 6.2 cannot be reapplied following the selection of **i**, and the Quine–McCluskey algorithm is exited at point 3.

All except the cost information of the reduced table in Fig. 6.16 may be expressed by Boolean Equation 6.11.

$$(a + g)(a + b)(b + c)(c + d)(d + e)(e + f)(f + g) = 1 \qquad (6.11)$$

Each of the sums corresponds to one of the uncovered minterm columns. That we must select either **a** or **g** is expressed by

$$(a + g) = 1$$

and so on for each of the columns. Because each of the individual sums must be 1, the AND combination of these seven sum terms must be 1. The task is completed by selecting a minimal-cost set of prime implicants satisfying Equation 6.11.

One approach to completing the solution (Petrick's method, [5]) calls for repeated application of the distributive law until Equation 6.11 is transformed into sum-of-products form. Each of these products of prime implicants may be used to complete a realization of the original function f. The minimal-cost realization may be identified after computing the cost of each product.

A method based on branching and reentry of the Quine–McCluskey algorithm will find a minimal-cost solution, usually without generating all possible solutions. For this example, the procedure begins by observing that either **a** or **g** must be selected and then reentering the Quine–McCluskey method at point 2 for each of these choices. The resulting activities of the algorithm are summarized in order in Fig. 6.17. The process finds two sum-of-products realizations of equally low cost. These realizations are

$$\mathbf{i} + \mathbf{a} + \mathbf{c} + \mathbf{f} + \mathbf{d} = v\bar{w}xy + \bar{v}\bar{w}\bar{y}\bar{z} + \bar{v}wx + v x \bar{z} + \bar{v}wy \qquad \text{and}$$

$$\mathbf{i} + \mathbf{g} + \mathbf{b} + \mathbf{d} + \mathbf{f} \qquad \blacksquare$$

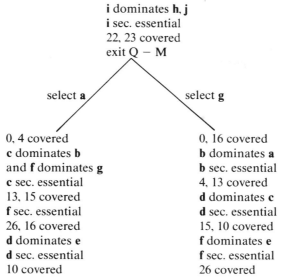

Figure 6.17 Solution tree for Example 6.2.

Whenever the Quine–McCluskey algorithm terminates without a realization, an as yet uncovered minterm m_i is selected. For each possible selection of a prime implicant covering m_i, the Quine–McCluskey algorithm, which incorporates Theorem 6.2, is reentered. This approach guarantees that the least-cost realization will be among the alternatives found by the process. Ultimate identification of the least-cost realization from these alternatives will require specific cost criteria.

Like the K-map, the Quine–McCluskey method can be adapted to product-of-sums design by designing on the 0's instead of the 1's of the function. Only two changes are necessary. In the determination of prime implicants, list the zeros (maxterms) of the function plus the don't-cares (if any). Then proceed exactly as before until the last step, the conversion of the prime implicants, at which point you convert to sums instead of products, using complemented variables, just as with the K-map method.

The real importance of the Quine–McCluskey method lies in the fact that it is a formal procedure, in no way dependent on human intuition. It is therefore suitable for computer mechanization, and a number of programs based on this method have been written [6]. A thorough understanding of the basic principles of this method is thus important to logic designers even though they may seldom use it personally.

6.5 MULTIPLE-OUTPUT CIRCUITS

So far we have considered only the implementation of a single function of a given set of variables. In the design of complete systems, we frequently wish to implement a number of different functions of the same set of variables. We can implement each function completely independently by the techniques already developed, but considerable savings can often be achieved by the sharing of hardware between various functions.

■ EXAMPLE 6.3

The circuit **BB** is to serve as an interface between the two computers of Fig. 6.18. The first four letters of the alphabet must intermittently be transmitted from computer 1 to computer 2. In computer 1, these letters are coded on three lines, x_1, x_2, x_3, as shown in Fig. 6.19a. In computer 2, they are coded on two lines, y_2 and y_1, as shown in Fig. 6.19b.

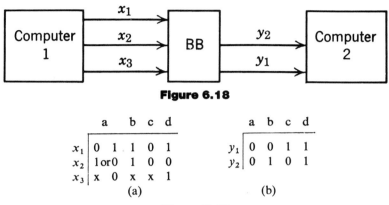

Figure 6.18

	a	b	c	d	
x_1	0	1	1	0	1
x_2	1 or 0	1	0	0	
x_3	x	0	x	x	1

(a)

	a	b	c	d
y_1	0	0	1	1
y_2	0	1	0	1

(b)

Figure 6.19

The translating functions $y_1(x_1, x_2, x_3)$ and $y_2(x_1, x_2, x_3)$ are easily compiled directly on Karnaugh maps, as in Fig. 6.20. To accomplish this, we determine from Fig. 6.19b which letters of the alphabet require 1 for either y_1 or y_2. We then determine from Fig. 6.19a which input combinations correspond to these letters. For example, y_1 is to be 1 for c or d (Fig. 6.19b). From 6.19a, we see that c or d is represented by $x_1 x_2 x_3 = 000,001,101$, so we enter 1 in the corresponding squares of Fig. 6.20a.

From these maps, the minimal realizations of y_1 and y_2, when considered individually, are easily seen to be those of Fig. 6.21. Alternatively, we note that y_1 and y_2 could be expressed as

$$y_1 = \bar{x}_1 \bar{x}_2 + x_1 \bar{x}_2 x_3 \quad \text{and} \quad y_2 = x_1 x_2 + x_1 \bar{x}_2 x_3 \tag{6.12}$$

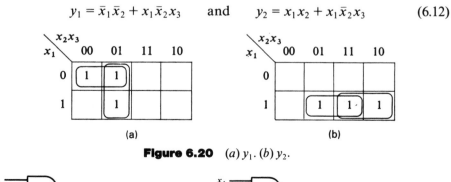

(a) (b)

Figure 6.20 (a) y_1. (b) y_2.

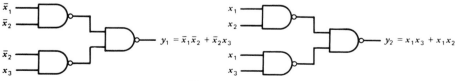

Figure 6.21

Taking advantage of the common term, $x_1 \bar{x}_2 x_3$, permits the implementation found in Fig. 6.22. Notice that this configuration requires only five NAND gates, as opposed to six in Fig. 6.21.

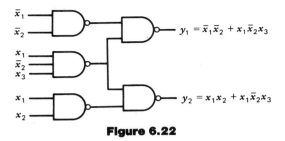

Figure 6.22

6.6 PROGRAMMED LOGIC ARRAY

In terms of chip area, the programmed logic array (PLA) is the most efficient structure for realizing multiple-output combinational logic functions on a VLSI chip. Logically, a PLA consists of a set of inputs, with an array of programmable connections to a set of AND gates, and an array of programmable connections from the AND gate outputs to a set of OR gates. Figure 6.23 depicts this logical organization of a PLA for an example with three input variables and three output lines. The three input variables with their complements provide the inputs to the AND array. There are four AND gates, each with six inputs to accommodate the three input variables and their complements, so that any AND function of the three variables can be realized. There is an OR gate to realize each of

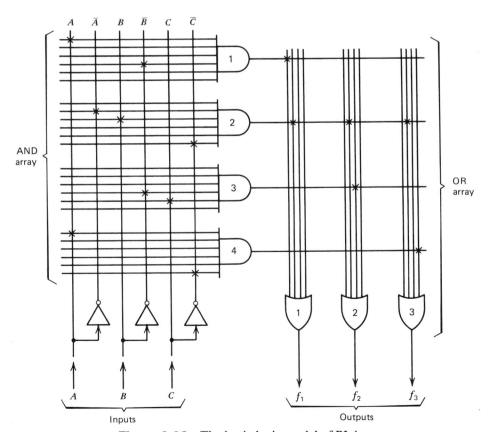

Figure 6.23 The basic logic model of PLA.

the three output functions. In each of the arrays, an X marks a point where a connection is made.

In this example, AND-gate 1 realizes $A\bar{B}$, AND-gate 2 realizes $\bar{A}B\bar{C}$, AND-gate 3 realizes $\bar{B}C$, and AND-gate 4 realizes $A\bar{C}$. OR-gate 1 is connected to AND-gates 1 and 2 and so realizes $A\bar{B} + \bar{A}B\bar{C} = f_1$. Similarly, the other two OR-gates realize $f_2 = \bar{A}B\bar{C} + \bar{B}C$ and $f_3 = \bar{A}B\bar{C} + A\bar{C}$.

Note that in both arrays, lines that are not connected have no effect. The AND and OR functions are generally realized directly in the connection matrices, rather than in physically distinct gates. Figure 6.23 does represent accurately the logical character of the PLA, but it is unwieldy, and the simplified form of Fig. 6.24 is commonly used. Here we have just one input line to each gate, representing all the input lines, with X connections at as many intersections as required to indicate the inputs of that gate. The diagram of Fig. 6.24 is thus exactly equivalent to that of Fig. 6.23.

Practical PLA's will be much larger than the simple example we have shown here. Typical PLA's have 10–20 inputs, 30–60 AND gates, and 10–20 OR gates, so that a single PLA can realize a considerable number of complex functions.

A NMOS realization of a sum-of-products PLA is depicted in Fig. 6.25. NOR gates are used to realize both the AND plane and OR plane levels of the realization. The AND gates are realized as

$$x_1 x_2 ,..., x_n = \overline{\bar{x}_1 + \bar{x}_2 + ... \bar{x}_n} \tag{6.13}$$

and the OR gates by complementing the output of NOR gates. In terms of chip area and speed, the double NOR gate alternative is the ideal. Only those devices that would remain after the programming activity is complete are shown in Fig. 6.25. Initially, a potential device would have existed at the intersection of each horizontal and vertical line.

6.7   MINIMIZATION OF MULTIPLE-OUTPUT FUNCTIONS

Just as for the single-output case, a multiple-output sum of products or product-of-sums minimization consists of two steps. The first step is finding all prime implicants of the set of functions. The second step is selection of an optimal subset, these prime implicants to be used in the realization of each function. A prime implicant, which covers only minterms, for which a given function is 1 or X, is said to be applicable to that function.

So far the "cost" of a prime implicant has been defined only to the extent that a product of fewer variables will always cost less to realize than a product involving a greater number of variables. This notion is usually sufficient for use in obtaining the best single-output realization. As we approach multiple-output functions, it is necessary to be more precise by defining the cost of a sum-of-products realization, as given in Equation 6.14.

$$Cost = \text{number of first-level gates } +$$

$$a \times (\text{total number of first-level gate inputs}) +$$

$$b \times (\text{total number of second-level gate inputs}) \tag{6.14}$$

The values of a and b will be less than 1, but will vary with the technology with which the sum-of-products expressions will be realized.

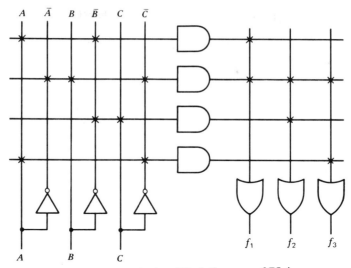

Figure 6.24 A simplified diagram of PLA.

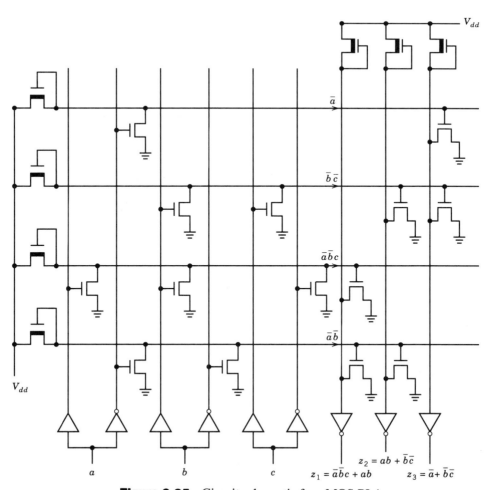

Figure 6.25 Circuit schematic for nMOS PLA.

An inspection of Fig. 6.25 suggests that because of the uniform structure of the PLA, the chip area occupied by a prime implicant would be independent of the number of variables in the prime implicant. **If the costs of all prime implicants are the same, then $a = 0$; any dominated prime implicant can be discarded without reference to cost.**

The cost will always increase with the number of prime implicants used in a multiple-output realization. Because of the uniform PLA structure, the cost of the PLA may not be impacted by connecting more than the minimum number of AND gate outputs to a particular second-level OR gate, that is, $b = 0$. Letting $b = 0$ implies that once a prime implicant is determined to be essential in one function, it can be immediately used without penalty in any function to which it is applicable.

Letting $a = b = 0$ in Equation 6.14 will be called the first-order PLA cost criterion. Using this criterion reduces the number of passes through the Quine–McCluskey algorithm and decreases the likelihood of needing Petrick's method or simplifies its application. Making $a = b = 0$ cannot be considered the universal PLA cost criterion, because extra connections may be disadvantageous, if PLA folding is used.

A tabular algorithm for identifying the prime implicants of a set of functions of common variables is given in Section 6.9. For functions of a small number of variables, Karnaugh maps may be used for this purpose.

Let us assume that we wish to find a minimal sum-of-products realization for the three functions

$$f_\alpha(A, B, C, D) = \sum m(2, 4, 10, 11, 12, 13)$$

$$f_\beta(A, B, C, D) = \sum m(4, 5, 10, 11, 13)$$

$$f_\gamma(A, B, C, D) = \sum m(1, 2, 3, 10, 11, 12)$$

We first draw the Karnaugh maps of all three functions (Fig. 6.26). We also draw maps of the *intersections*, or products, of the three functions, by pairs and all together. Recall that, in terms of K-maps, the intersection of two functions consists of their common squares.

We now locate the prime implicants on the maps, starting with the product of all the functions. The only prime implicant of $f_\alpha \cdot f_\beta \cdot f_\gamma$ is 101x. We now look for prime implicants of the function pairs, but we do not mark a grouping as a prime implicant if it is a prime implicant of a higher-order function product. For example, we do not mark 101x on $f_\beta \cdot f_\gamma$ since it is a prime implicant of $f_\alpha \cdot f_\beta \cdot f_\gamma$. On $f_\alpha \cdot f_\gamma$, we mark only x010 and 12, since the other group is also in $f_\alpha \cdot f_\beta \cdot f_\gamma$. Similarly, we mark only 4 and 13 on $f_\alpha \cdot f_\beta$. Now we go to the maps of the functions themselves, marking only those prime implicants that do not also appear in the function products.

Let us now use the first-order PLA cost criteria to select those prime implicants required in the realization of each function. A generalized prime implicant table is given in Fig. 6.27a. The columns correspond to the minterms in each function. The rows correspond to prime implicants ordered according to the number of literals in the product. This ordering will be ignored in the application of the PLA cost criteria. The prime implicants are identified by number in Fig. 6.27a. The corresponding notation for each numbered prime implicant is given immediately below in Fig. 6.27b. Also indicated are the functions to which each minterm is applicable.

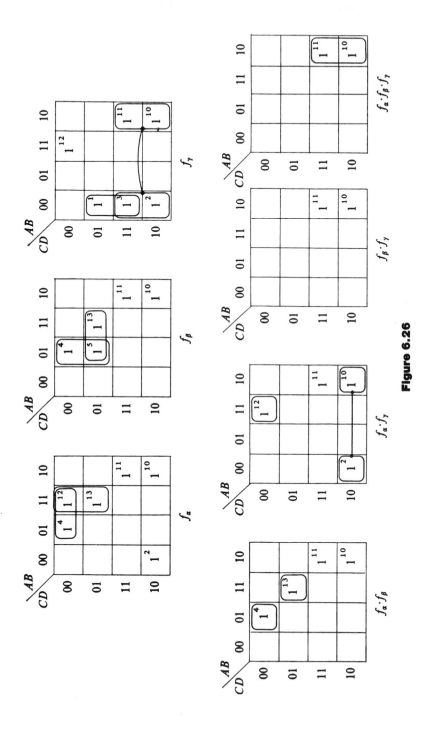

Figure 6.26

	Fcn.	PI	f_α						f_β					f_γ					
			2	4	10	11	12	13	4	5	10	11	13	1	2	3	10	11	12
	γ	**a**												✓	✓	✓	✓		
	α	**b**		✓			✓												
	α	**c**					✓	✓											
	β	**d**							✓	✓									
	β	**e**								✓			✓						
*	γ	**f**												✓		✓			
*	$\alpha\gamma$	**g**	✓		✓										✓		✓		
*	$\alpha\beta\gamma$	**h**				✓	✓				✓	✓						✓	✓
	$\alpha\beta$	**i**		✓					✓										
*	$\alpha\gamma$	**j**					✓												✓
	$\alpha\beta$	**k**						✓					✓						
			✓		✓	✓	✓				✓	✓		✓	✓	✓·	✓	✓	✓

$$\textbf{a} = \text{x01x}; \ \ \textbf{b} = \text{x100}; \ \ \textbf{c} = \text{110x}; \ \ \textbf{d} = \text{01x0}; \ \ \textbf{e} = \text{x101}; \ \ \textbf{f} = \text{00x1}$$
$$\textbf{g} = \text{x010}; \ \ \textbf{h} = \text{101x}; \ \ \textbf{i} = \text{0100}; \ \ \textbf{j} = \text{1100}; \ \ \textbf{k} = \text{1101}$$

(b)

Figure 6.27   Multiple-output prime implicant table.

On each row, a check is placed in a minterm column, only if the covering prime implicant is included in that function. In the first row, for example, columns 2, 3, 10, and 11 are checked only under f_γ, because the prime implicant **a** = x01x is included only in that function. Some of these minterms appear in other functions but are not checked on row **a**. For prime implicant **g** = x010, columns 2 and 10 are checked under both f_α and f_γ, because this 1-cube is included in both functions.

We next check for essential prime implicants, just as in the single-output Quine–McCluskey method, by looking for columns with only one check. In this case, the essential prime implicants do not cover all the minterms of all three functions, so a reduced prime implicant table is formed (Fig. 6.28).

The next step is to apply dominance to eliminate rows in exactly the same manner as for the single-output case. Note that the entire table must be considered in determining dominance, not just that part for a single function. For example, in the table of Fig. 6.28, row **d** would dominate row **i** if only function f_β were considered. However, if the whole table is considered, it is seen that **i** has a check in a column in which **d** does not. Therefore, **d** does not satisfy the conditions for the dominance of **i**.

Application of the PLA cost criteria assigns equal cost to prime implicants **b**, **c**, **i**, and **k**. Therefore, **b** may be discarded because it is dominated by **i**, and **c** may be discarded because it is dominated by **k**. Now **i** and **k** are secondary essential in f_α.

Fcn.	PI	f_α		f_β		
		4	13	4	5	13
α	**b**	✔				
α	**c**		✔			
β	**d**			✔	✔	
β	**e**				✔	✔
** $\alpha\beta$	**i**	✔		✔		
** $\alpha\beta$	**k**		✔			✔

Figure 6.28   Reduced Multiple Output
Prime Implicant Table

The $b = 0$ portion of the PLA cost criteria allows us to use these two prime implicants to cover minterms 4 and 13 in f_β as well. Now only minterm 5 in f_β remains. The prime implicants **d** and **e** cover this minterm and are interchangeable. First, **e** is discarded and then **d** becomes essential. Altogether, prime implicants **d**, **f**, **g**, **h**, **i**, **j**, and **k** are required in the realization of the functions, which may be expressed in terms of these prime implicants as follows:

$$f_\alpha = \mathbf{g} + \mathbf{h} + \mathbf{i} + \mathbf{j} + \mathbf{k}$$
$$f_\beta = \mathbf{d} + \mathbf{h} + \mathbf{i} + \mathbf{k}$$
$$f_\gamma = \mathbf{f} + \mathbf{g} + \mathbf{h} + \mathbf{j}$$

6.8   MINIMIZATION STRATEGY WITHOUT THE PLA COST CRITERIA

Occasionally, it is necessary to obtain a set of minimal multiple-output expressions that will be realized by some technology other than a PLA. In this case, the cost of extra input connections to first- and second-level gates will not be 0, that is, a and b are nonzero. Allowing for arbitrary values of a and b will significantly reduce opportunities to simplify the prime implicant table and force the Quine–McCluskey algorithm to default to enumeration (Petrick's method) early in the process. Instead, let us assume that although a and b are nonzero, $a \ll 1$ and $b \ll 1$. Under these conditions, the primary goal will still be to use as few first-level AND gates as possible. Because a is nonzero, we must disallow the discarding of a prime implicant dominated by a smaller cube. For reasons to be made clear, we choose to proceed at the outset as if $b = 0$, leading to the following definition of an optimal set of prime implicants.

Definition 6.4   An optimal set of prime implicants includes the smallest number of prime implicants that will cover the minterms of all functions and includes the smallest total number of first-level input connections subject to this constraint.

The following strategy, consistent with Definition 6.4, may in some cases lead to a suboptimal solution with respect to the total number of second-level input connections.

Step 1   Use the multifunction prime implicant table to determine an optimal set of prime implicants.

Step 2   Realize each individual function by selecting a minimal subset of the prime implicants determined in Step 1.

An implementation of Step 2 would entail followup tabular minimization of each individual function, beginning with the prime implicants selected in Step 1. For functions of a small number of variables, we may be less formal.

■ **EXAMPLE 6.4**

Rework the example of Section 6.7 using the process described above.

SOLUTION   Each simplification of Fig. 6.27 leading finally to Fig. 6.28 was consistent with Definition 6.4, so we continue Step 1 from Fig. 6.28. Thus far, prime implicants **f**, **g**, **h**, and **j** are included in the optimal set.

With $a \neq 0$, prime implicants **i** and **k** have a higher cost than the other five prime implicants in Fig. 6.28. Therefore, none of the prime implicants may be discarded. We, therefore, exit the Quine–McCluskey algorithm and form Equation 6.15 for use with Petrick's method.

$$(b + i)(c + k)(d + i)(d + e)(e + k) = 1 \tag{6.15}$$

Expanding Equation 6.16 to the sum-of-products form results in

$$cei + bcde + eik + dik + bdk = 1 \tag{6.16}$$

It is easy to see that either set **cei** or **bdk** should be added to those prime implicants already selected. Using **cei** completes an optimal set of prime implicants: {**c, e, f, g, h, i,** and **j**}. As expected, this set contains the same number of prime implicants as is found in the previous section, but the sets are not the same. To accomplish Step 2, we enter these prime implicants on the Karnaugh maps of the individual functions to form Fig. 6.29. The following expressions for the functions are easily determined from the maps. Notice that one less prime implicant is included in each of the expressions for f_α and f_β than in the previous section.

$$f_\alpha = c + g + h + i$$
$$f_\beta = e + h + i$$
$$f_\gamma = f + g + h + j$$

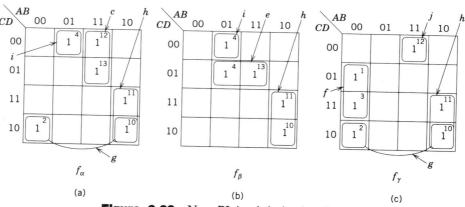

Figure. 6.29   Non–PLA minimization Step 2.

Either by locating the cubes on the maps or by the method given earlier, we convert these to algebraic notation:

$$f_\alpha = AB\bar{C} + \bar{B}CD + A\bar{B}C + \bar{A}B\bar{C}\bar{D} \tag{6.17}$$

$$f_\beta = A\bar{B}C + B\bar{C}D + \bar{A}B\bar{C}\bar{D} \tag{6.18}$$

$$f_\gamma = \bar{A}\bar{B}D + \bar{B}C\bar{D} + A\bar{B}C + AB\bar{C}\bar{D} \tag{6.19}$$

The realization of these equations is shown in Fig. 6.30. ■

6.9 TABULAR DETERMINATION OF PRIME IMPLICANTS

A tabular method of prime implicant determination for multiple-output functions can form the basis of a computer implementation. For a demonstration of the tabular method, we consider the same functions for which a multiple-output realization was determined in the previous section.

$$f_\alpha(A, B, C, D) = \sum m(2, 4, 10, 11, 12, 13)$$

$$f_\beta(A, B, C, D) = \sum m(4, 5, 10, 11, 13)$$

$$f_\gamma(A, B, C, D) = \sum m(1, 2, 3, 10, 11, 12)$$

The minterms from all three functions are tabulated together in Fig. 6.31a. The column provided under each function contains a 1 on the row corresponding to each minterm in that function. Two 0-cubes may be combined into a 1-cube, if they differ in only one variable value and they are common to at least one function. Entries are placed in the common function columns on the row of the resulting 1-cube in Fig. 6.31b. Note that 0001 and 0101 are not combined into a 1-cube, because their entries in the function columns are disjoint.

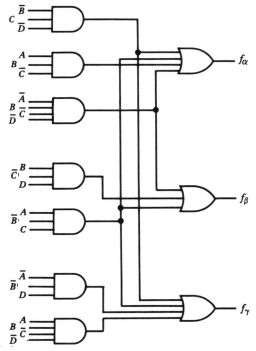

Figure 6.30 Minimal multiple-output realization.

0-cubes

				f_α	f_β	f_γ	PI
0	0	0	1			1	✔
0	0	1	0	1		1	✔
0	1	0	0	1	1		i
0	0	1	1			1	✔
0	1	0	1		1		✔
1	0	1	0	1	1	1	✔
1	1	0	0	1		1	j
1	0	1	1	1	1	1	✔
1	1	0	1	1	1		k

(a)

1-cubes

				f_α	f_β	f_γ	PI
0	0	–	1			1	f
0	0	1	–			1	✔
–	0	1	0	1		1	g
0	1	0	–		1		d
–	1	0	0	1			b
–	0	1	1			1	✔
–	1	0	1	1			e
1	0	1	–	1	1	1	h
1	1	0	–	1			c

(b)

2-cubes

				f_α	f_β	f_γ	PI
–	0	1	–			1	a

(c)

Figure 6.31 Tabulation of prime implicants.

A check is entered in the prime implicants (PI) column for a minterm, only if a 1-cube has been formed that is applicable to all functions covering that minterm. Minterm 0100 is not checked, although it is covered by $010-$ in f_β and -100 in f_α. It would be checked only if the same 1-cube were applicable to both these functions. The unchecked 0-cubes are prime implicants, identified in the PI column by the same letter used in the previous section.

As before, 1-cubes common to a function may be combined to form a 2-cube only if they differ in one variable value, and the blanks match. The process of determining the entries in the 2-cube function columns and the 1-cube PI column is the same as just described. Note that -010 is not checked, because prime implicant **a** is applicable only to f_γ. As expected, the prime implicants found above are the same as found on the Karnaugh maps in the previous section.

6.10 ESPRESSO*

The Quine–McCluskey algorithm, together with Petrick's method or a branching technique, will always lead to an optimal two-level realization. The downside is that computation time and memory space, of exponential order with respect to the number of inputs, are required by any method that lists all minterms' or prime implicants. For a 20-input function, the total number of possible minterms is 2^{20}. The actual number of minterms in a function will be less, but 2^{19} might be typical for a 20-input function. The number of possible prime implicants will also depend on the function, but might be only slightly less than the number of minterms. However the list of prime implicants might be found, the computational complexity of any procedure using the list would increase exponentially with the number of inputs. Determination of these prime implicants by a tabular method would be even worse. Any method that would guarantee exactly optimal realizations will be of exponential order.

The first step of a hardware synthesis program is much more likely to produce an initial description of a combinational logic function as a Boolean algebraic expression, rather than a minterm list. *Espresso* [12] is a technique that, when applied to an arbitrary sum of products, will produce an equivalent sum-of-

*This section is not essential to continuity.

products expression and then, when applied repetitively, will produce a succession of expressions, each with fewer products than its predecessor. Espresso is generally regarded as the most effective of the minimization techniques that avoid minterm lists. Although it cannot guarantee an absolute minimum realization, Espresso has been demonstrated to produce the same results as the Quine-McCluskey method on almost all members of a set of complex functions submitted to both programs. At its core it employs a heuristic strategy based on approximation.

Before attempting a summary of the Espresso algorithm, let us introduce a supporting computational technique and two related definitions. Some form of *tautology* computation appears at almost every step of Espresso. As defined in Chapter 1, a tautology is a statement that is always true or always logical 1. A tautology computation determines whether a Boolean function evaluates to 1, subject to the constraint that some other Boolean expression is 1. The notation $F|_{g=1}$ will be used to express a tautology computation on F, subject to the constraint that $g = 1$. The notation $(F - c^i)$ will be used to represent a sum-of-products expression F, from which a particular product or cube, c^i, has been removed.

Definition 6.5 A cube in a sum-of-products expression is *relatively essential* to that expression if and only if

$$(F - c^i)|_{c^i} \neq 1$$

■ EXAMPLE 6.5

Show that the cube bd is not relatively essential to the Boolean expression

$$F = \bar{a}\bar{c}d + \bar{a}bc + acd + ab\bar{c} + bd \tag{6.20}$$

SOLUTION   We use a tautology computation to determine whether bd is essential.

$$\begin{aligned}
(F - bd)|_{bd=1} &= (\bar{a}\bar{c}d + \bar{a}bc + acd + ab\bar{c})|_{bd=1} \\
&= \bar{a}\bar{c} + \bar{a}c + ac + a\bar{c} \\
&= \bar{a} \cdot 1 + \bar{a} \cdot 1 = 1
\end{aligned}$$

Therefore, bd is not relatively essential in expression F. The Karnaugh map of the corresponding function given in Fig. 5.31 indicates that bd was not an essential prime implicant and is not included in the minimal sum-of-products realization.   ■

In general, an essential prime implicant of a Boolean function must be relatively essential in every sum-of-prime-implicants representation of a function. However, a product or prime implicant that is relatively essential to some expression for a function is not necessarily an essential prime implicant of that function.

Two Boolean cubes are said to have a nonzero *consensus*, if and only if they intersect or would intersect except for the value of one variable.

Definition 6.6 Let the consensus of two Boolean cubes a and b be represented by $a \odot b$. If $a \wedge b \neq 0$, then

$$a \odot b = a \wedge b$$

If $a \wedge b = 0$, but for some variable x_i, $a = x_i c$ and $b = \bar{x}_i d$ and $c \wedge d \neq 0$, then

$$a \odot b = c \wedge d$$

Otherwise, $a \odot b = 0$.

THEOREM 6.3   If a and b are cubes in a sum-of-products representation of a Boolean function f, and their consensus is not 0, then part of each of the cubes is covered by some other prime implicant of the function.

PROOF   If the cubes intersect, they are partially covered by each other. If the consensus is $c \wedge d$, we may express f in sum-of-products form as

$$
\begin{aligned}
f &= g + a + b \\
&= g + x_i c + \bar{x}_i d \\
&= g + x_i c + x_i cd + \bar{x}_i d + \bar{x}_i cd \\
&= g + x_i c + \bar{x}_i d + cd
\end{aligned}
\tag{6.21}
$$

Therefore, $c \wedge d$ is the prime implicant cited in the theorem or is covered by that prime implicant.

In effect, there is a nonzero consensus of two cubes if they intersect or one variable prevents the intersection. The Karnaugh map in Fig. 6.32a depicts the consensus of $\bar{a}b\bar{c}$ and abd, whereas Fig. 6.32b depicts the consensus of $a\bar{c}\bar{d}$ and bd.

A somewhat simplified version of the Espresso algorithm itself is depicted in Fig. 6.33. As a vehicle for explaining the various procedures within Espresso, we choose a single-output function of only four inputs with no "don't cares." The same example is a trivial Karnaugh map problem. (*Although a Karnaugh map within the pages of this section might undermine confidence in the algebraic evaluations, the reader is bound by no such reservation. If reference to a map will enhance understanding at any stage, use one!*) It would, likewise, require much less effort, if approached via Quine–McCluskey rather than by the process we are about to illustrate. A problem of a complexity for which Espresso is suited cannot be addressed within the pages of a book.

We enter block I of the algorithm, given a Boolean function defined by the sum of products presented in Equation 6.22.

$$F = \bar{a}\bar{b}\bar{c}\bar{d} + \bar{a}b\bar{c} + abd + acd \tag{6.22}$$

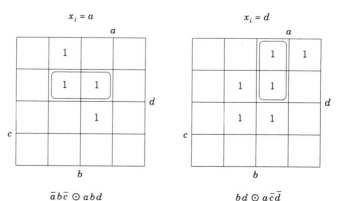

$\bar{a}b\bar{c} \odot abd$ $bd \odot a\bar{c}\bar{d}$

Figure 6.32   Consensus examples.

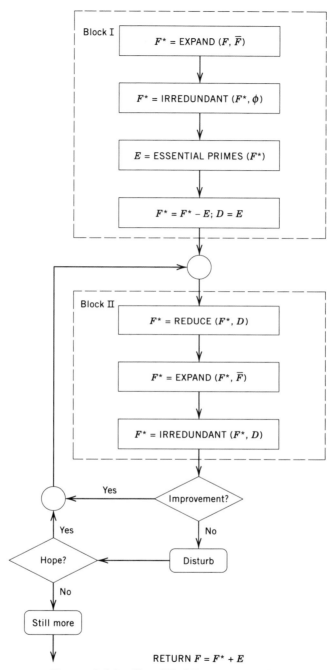

Figure 6.33 Simplified Espresso algorithm.

The first procedure applied to the function is EXPAND. If possible, EXPAND replaces each cube (e.g., a k-cube) in F by the set of all distinct covering $k+1$ cubes that do not intersect $\bar{F}$. Duplicate cubes are deleted. The resulting cubes are similarly expanded until no cubes can be added. A sum-of-products representation of $\bar{F}$ is required to accomplish the expansion. Fortunately, any sum-of-products expression, not necessarily a minimal expression, will do. Such an expression may be obtained by the application of Theorem 5.5 to the available expression of F, followed by the distributive law, $x \cdot \bar{x}$, and the elimination of included products. To save computation time, Espresso makes use of a recursive process involving unate functions. (See Section 7.2 for a definition of unate.) The application of Theorem 5.5 to F as given by Equation 6.22 and proceeding as discussed yields $\bar{F}$ as given by Equation 6.23.

$$
\begin{aligned}
\bar{F} &= F_D(\bar{a}, \bar{b}, \bar{c}, \bar{d}) \\
&= (a + b + c + d)(a + c + \bar{b})(\bar{a} + \bar{b} + \bar{d})(\bar{a} + \bar{c} + \bar{d}) \\
&= (a + c + \bar{b}d)(\bar{a} + \bar{d} + \bar{b}\bar{c}) \\
\bar{F} &= a\bar{d} + a\bar{b}\bar{c} + \bar{a}c + c\bar{d} + \bar{a}\bar{b}d + \bar{b}\bar{c}d
\end{aligned}
\tag{6.23}
$$

The expansion of $\bar{a}\bar{b}\bar{c}\bar{d}$ is depicted in Fig. 6.34. As each cube is generated, EXPAND must check for an intersection with $\bar{F}$.

We check the leftmost cube $\bar{b}\bar{c}\bar{d}$ as follows:

$$
\begin{aligned}
\bar{b}\bar{c}\bar{d} \wedge \bar{F} &= (\bar{b}\bar{c}\bar{d}) \wedge (c\bar{d} + \bar{a}c + a\bar{d} + \bar{b}\bar{c}d + a\bar{b}\bar{c} + \bar{a}\bar{b}d) \\
&= 0 + 0 + a\bar{b}\bar{c}\bar{d} + 0 + a\bar{b}\bar{c}\bar{d} + 0
\end{aligned}
\tag{6.24}
$$

Because Equation 6.24 is not 0, we conclude that $\bar{b}\bar{c}\bar{d}$ intersects $\bar{F}$. A similar computation was required for each of the cubes generated in Fig. 6.34. An attempt was made to expand the other three cubes in F. Each resulting 2-cube intersected $\bar{F}$. The F^* resulting after the application of EXPAND is given in Equation 6.25.

$$
F^* = \bar{a}\bar{c}\bar{d} + \bar{a}\bar{b}\bar{c} + abd + acd
\tag{6.25}
$$

In general, F^* at this point will consist of a large number of prime implicants of the function but probably not all prime implicants. The second procedure IRREDUNDANT of block I actually consists of the following three consecutively applied subprocedures. Each of the subprocedure names characterizes the function F^* remaining after the application of that subprocedure.

$$
\text{IRREDUNDANT} = \begin{cases} \text{REDUNDANT} \\ \text{PARTIALLY\_REDUNDANT} \\ \text{MINIMAL\_COVER} \end{cases}
$$

The procedure REDUNDANT separates all the cubes that might be redundant from those which are *relatively essential*. This computation was illustrated in Example 6.5. REDUNDANT identifies all the cubes in Equation 6.25 as relatively essential. Therefore, the next two subprocedures leave F^* unchanged. These will be discussed in more detail when applied in block II.

Only the relatively essential prime implicants (in this case, all) need be checked by ESSENTIALPRIMES. In the absence of a minterm list, a prime implicant will be declared essential, if it can be determined that it is not totally covered by any set of other prime implicants of the function F, whether or not

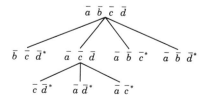

* = intersects $\bar{F}$.

Figure 6.34   Application of EXPAND.

these prime implicants are currently included in F^*. If c^i can be covered by other prime implicants, such a cover can be found from the consensus of c^i and each of the other cubes in F^*. That is, prime implicants not explicit in F^* but that overlap c^i are found by consensus. ESSENTIALPRIMES has proved to be a time-consuming procedure in Espresso implementations, but is executed only once and results in considerable time savings in the other procedures.

For each of the cubes in Equation 6.25, this consensus computation is tabulated in Fig. 6.35. The consensus of each of the column-heading cubes c^i with each of the other cubes c^j is computed in column i. The sum of these consensus cubes is the fifth entry in each column. If this sum covers the cube c^i, that cube is not essential. The sum of essential prime implicants E from Fig. 6.35 is given by Equation 6.26.

$$E = \bar{a}\bar{c}\bar{d} + acd \tag{6.26}$$

Now

$$F^* = F^* - E = \bar{a}b\bar{c} + abd$$

The essential primes are regarded as "don't cares" through the rest of the process and will be added to F^* when the final result is returned.

It is possible that F^* obtained by block I will be a minimal sum of products. If not, we assert that F^* is merely a local minimum, dependent on the form of the original sum-of-products expression F. To have hope of moving away from this local minimum, it is necessary to back up and reexecute EXPAND and IRRE-DUNDANT for a different initial expression for F^*. Block II which is executed repetitively as long as it continues to reduce the number of cubes in F^* begins with the process REDUCE. In obvious contrast to EXPAND, *the goal of REDUCE is make each cube of F^* as small as possible* while continuing to cover the original F^*. For each cube c^i, the expression

$$F^*(i) = (F^* - c^i) + D \tag{6.27}$$

covers all of F^* covered by cubes other than c^i or by don't cares D, given in this case by Equation 6.26. Therefore, $c^i \wedge \bar{F}^*(i)$ is the part of c^i covered only by c^i, and REDUCE must replace c^i by the smallest cube covering $c^i \wedge \bar{F}^*(i)$.

Relatively Essential Cubes

	$\bar{a}\bar{c}\bar{d}$	$\bar{a}bc\bar{b}$	abd	acd
$\bar{a}\bar{c}\bar{d}$	✔	$\bar{a}b\bar{c}\bar{d}$	Φ	Φ
$\bar{a}b\bar{c}$	$\bar{a}b\bar{c}d$	✔	$b\bar{c}d$	Φ
abd	Φ	$b\bar{c}d$	✔	$abcd$
acd	Φ	Φ	$abcd$	✔
$\sum_{j} C^j \odot C^i$	$\bar{a}b\bar{c}d$	$b\bar{c}(\bar{a} + d)$	$bd(a + \bar{c})$	$abcd$
$C^i \wedge (\sum) = C^i$	No	Yes	Yes	No
Essential	Yes	No	No	Yes

Figure 6.35 Identifying essential cubes by consensus.

This time we leave the computation of the complement using Theorem 5.5 to the reader.

$$F^*(\bar{a}b\bar{c}) = (\bar{a}b\bar{c} + abd - \bar{a}b\bar{c}) + D$$
$$= abd + \bar{a}\bar{c}\bar{d} + acd$$
$$\bar{F}^*(\bar{a}b\bar{c}) = c\bar{d} + c\bar{a} + \bar{a}d + a\bar{d} + \bar{b}\bar{c}d + ab\bar{c}$$
$$(\bar{a}b\bar{c}) \wedge \bar{F}^* = 0 + 0 + \bar{a}b\bar{c}d + 0 + 0 + 0 = \bar{a}b\bar{c}d$$

Now

$$F^* = \bar{a}b\bar{c}d + abd$$

Continuing the application of REDUCE to abd yields Equation 6.28 after a similar computation.

$$F^* = \bar{a}b\bar{c}d + ab\bar{c}d \qquad (6.28)$$

In this example, the order in which REDUCE was applied to the cubes of F^* did not matter. In general, the results will vary with the order, and Espresso implements a heuristic technique for selecting the sequence of cubes for REDUCE.

The application of EXPAND to F^* uses the same complement of F given in Equation 6.23, thus allowing the cubes to expand to cover "don't cares." The expansion of the 0-cubes in Equation 6.28 is depicted in Fig. 6.36. The repeated cube is deleted by EXPAND, yielding

$$F^* = b\bar{c}d + \bar{a}b\bar{c} + abd \qquad (6.29)$$

The first two subprocedures of IRREDUNDANT use the don't care function D as a parameter when applied in block II. In REDUNDANT a cube c^i of F^* is relatively essential only if it is not covered by $(F^* + D - c^i)$. We apply the tautology computation to cube $b\bar{c}d$.

$$(D + F^* - (b\bar{c}d))|_{b\bar{c}d=1} = (\bar{a}\bar{c}\bar{d} + acd + \bar{a}b\bar{c} + abd)|$$
$$= \bar{a} + a = 1$$

The result is the same for the other three cubes of F^*, so none are relatively essential!

We now apply PARTIALLY_REDUNDANT which will cause cubes covered by $D + RE$ to be declared totally redundant and discarded. RE is the sum of relatively essential cubes. Because $RE = 0$ for the example under consideration, the tautology calculation will necessarily reveal that none of the cubes are totally redundant. The cubes of F^* that are neither relatively essential nor totally redundant are called *partially redundant* and are given by the expression R_p. For the continuing example, R_p is given by

$$R_p = \bar{a}b\bar{c} + abd + b\bar{c}d \qquad (6.30)$$

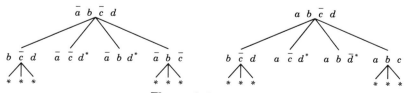

Figure 6.36

The object of the third subprocedure, MINIMAL, is to determine a minimal set of cubes from R_p that cover all of this expression not covered by RE and the don't care set D. MINIMAL utilizes an array **PI**, with a column corresponding to each of the prime implicant cubes making up R_p. The determination of the rows of **PI** is the topic of the next few paragraphs.

For each cube c^i in R_p, there will be a set S_i of one or more sum-of-products expressions, each a sum of cubes G_{nc} such that

$$cubecheck = (R_p - G_{nc}) + D + RE \qquad (6.31)$$

will cover c^i, if and only if one cube is deleted from G_{nc}. For the continuing example, the following tautology checks reveal that Equation 6.31 always covers cube $b\bar{c}d$, if G_{nc} is a single cube:

$$(R_p - b\bar{c}d + D)|_{b\bar{c}d} = (\bar{a}b\bar{c} + abd + \bar{a}\bar{c}\bar{d} + acd)|_{b\bar{c}d}$$
$$= \bar{a} + a + 0 + 0 = 1$$

$$(R_p - abd + D)|_{b\bar{c}d} = (\bar{a}b\bar{c} + b\bar{c}d + \bar{a}\bar{c}\bar{d} + acd)|_{b\bar{c}d}$$
$$= \bar{a} + 1 + 0 + 0 = 1$$

$$(R_p - \bar{a}b\bar{c} + D)|_{b\bar{c}d} = (b\bar{c}d + abd + \bar{a}\bar{c}\bar{d} + acd)|_{b\bar{c}d}$$
$$= 1 + a + 0 + 0 = 1$$

From the first computation, we see that if either of the cubes abd or $\bar{a}b\bar{c}$ were included with $b\bar{c}d$ in G_{nc}, the resulting instance of *cubecheck* would not cover $b\bar{c}d$. From either of the last two computations, we see that *cubecheck* would still cover $b\bar{c}d$, if $G_{nc} = abd + \bar{a}b\bar{c}$. Therefore,

$$S_{b\bar{c}d} = \{b\bar{c}d + \bar{a}b\bar{c}; b\bar{c}d + abd\}$$

Similar computations will reveal that

$$S_{abd} = \{abd + b\bar{c}d\} \qquad \text{and} \qquad S_{\bar{a}b\bar{c}} = \{b\bar{c}d + \bar{a}b\bar{c}\}$$

A row may now be entered in **PI** corresponding to each member function in each set S_i. The row corresponding to a particular function G_{nc} will have a 1 in each column representing a cube in G_{nc}. The four functions in the three sets just determined result in the four-row array given in Fig. 6.37.

THEOREM 6.4 A set of cubes whose columns will include at least one 1 in each of the **PI** rows will be a cover of R_p.

PROOF The functions G_{nc} were chosen so that selecting a cube in each function in set S_i will insure coverage of cube c^i.

Figure 6.37 has very much the appearance of a Quine–McCluskey table without a minterm list and with rows and columns interchanged. It is apparent from Fig. 6.37 that $b\bar{c}d$ is a cover of R_p and that this cube together with the

$$\mathbf{PI} = \begin{bmatrix} & \bar{a}\,b\,\bar{c} & & a\,b\,d & & b\,\bar{c}\,d & \\ & & & 1 & & 1 & \\ & 1 & & & & 1 & \\ & & & 1 & & 1 & \\ & 1 & & & & 1 & \end{bmatrix} \begin{array}{l} \left.\vphantom{\begin{matrix}1\\1\end{matrix}}\right\} S_{b\bar{c}d} \\[4pt] \\ S_{abd} \\ S_{\bar{a}b\bar{c}} \end{array}$$

Figure 6.37 Prime implicant array.

relatively essential prime implicants (none for the example) are a cover of F^*. For larger examples a heuristic strategy is applied that finds only an approximately minimal cover of the rows of **PI**.

Referring to Fig. 6.33, we see that there is by default improvement in F^* after the first pass through block II. On the second pass through block II, REDUCE will leave $F^* = b\bar{c}d$ unchanged, as will EXPAND and IRREDUNDANT. No improvement will be indicated after the second pass through block II and Espresso will eventually return

$$F = b\bar{c}d + \bar{a}\bar{c}\bar{d} + acd \qquad (6.32)$$

For larger problems the expression F^* will be reduced, then expanded and minimized again and again by block II. At some point, block II will be able to make no further reduction in F^*. When this occurs, there is no guarantee that the number of cubes in F^* is an overall minimum, in contrast to a mere local minimum. The implementation of Espresso includes additional heuristic strategies that attempt to dislodge the process from a possible local minimum and return it to block II. For this discussion we refer the reader to Reference [12].

The subprocedure MINIMAL is the most complex and time-consuming of the repetitively executed activities within Espresso. The implementation of the several interdependent tautology calculations once again makes use of recursion on unate functions. One might try to argue that the heavy dependence on MINIMAL leaves Espresso a facade for an implementation of Quine–McCluskey without minterm lists. Actually, MINIMAL is executed only on small subsets of the set of all prime implicants. If MINIMAL included a mechanism for finding the set of all prime implicants, it could be considered an alternative implementation of Quine–McCluskey. To emphasize the value of the other features of Espresso, MINIMAL was provided with all prime implicants and executed by itself on a set of 32 circuits [12]. Data for the three largest (most inputs) of these examples are given in Fig. 6.38. Clearly, the time consumed by Espresso is much less than that required by the modified Quine–McCluskey approach.

It is hoped that the reader is convinced that a faster although approximate alternative to Quine–McCluskey does exist. To provide a manageable example, we have neglected multiple outputs and an initial "don't care" set, both of which are key features of the implementation. The examples in Fig. 6.38 are, of course, multiple output. As might be expected, the actual algorithm is not algebraic. The cubes in a Boolean expression are treated as rows in an array. The occasional would-be implementer of Espresso is referred to [12].

6.11 FIELD PROGRAMMABLE COMBINATIONAL LOGIC ARRAYS

In our discussion of PLA's, we assumed that programming would be accomplished by making minor modifications in the masks (at connection points in two layers) prior to the integrated circuit fabrication process. This is convenient, if the PLA is part of a larger system fabricated on a single chip. If the logic array, used in the realization of multiple-output logic functions, is to be packaged separately, it would be preferable to be able to program the part after it is manufactured and packaged. Such a device is called field programmable.

It is much easier to manufacture a field programmable logic array, if only one array (layer) of connections is programmable. In that case, devices are fabricated to establish connections at every possible connection point. In parts that can be

Ckt.	Inputs	Outputs	Total PI's	Cubes in Cover		Time in Seconds	
				Expr.	MINIMAL	Expr.	MINIMAL[a]
1	29	7	671	141	140	13.0	591.3
2	32	7	302	32	32	5.3	24.2
3	25	8	1188	110	110	3.8	598.2

[a]Includes the time to find all prime implicants using EXPAND.

Figure 6.38 Examples of two-level minimization.

programmed only once (permanently), a high-current vulnerable fuse is included in each device path. Unwanted connections are destroyed in the programming process. Logic arrays are also available in which one array of connections can be reprogrammed after erasing previous connections by shining ultraviolet light through a window in the package.

A two-level logic array in which only the first-level AND array is programmable is usually called a PAL (programmed array logic). Just the opposite is the field programmable ROM or read-only memory that includes a fixed AND array and a programmable OR array. Because the logic minimization techniques of this chapter do not apply to the ROM, we will defer discussion of that part until Chapter 8.

Figure 6.39 shows a simplified diagram of a PAL network. Included are 12 input variables and 16 AND gates, with a connection array permitting any gate to realize any AND function of any of the 12 variables. The AND gates are connected to six OR gates in a fixed pattern with no overlap, that is, no AND gate can be used in more than one function. The rationale behind the PAL is that there are few situations in which it is necessary to realize a large number of functions of exactly the same large set of variables. The more common situation is one in which a large number of functions of a large number of variables must be realized, but there is little, if any, overlap between the functions. The PAL of Fig. 6.39 could be used in situations where there are six functions to be realized involving a total of 12 variables, whether or not any of the variables occur in more than one function. In situations with little overlap, it is unlikely that any product terms can be used in more than one function, so that the flexibility of the OR array in allowing us to connect any AND gate to any OR gate is wasted.

■ EXAMPLE 6.6

Realize the following functions using the PAL of Fig. 6.39:

$$f_1(A, B, C, D, E, F) = AB\bar{D}F + C\bar{D}E\bar{F} + AC\bar{E}F + A\bar{B}CD\bar{F}$$

$$f_2(D, E, F, G, H, K) = D\bar{E}K + E\bar{F}G\bar{H}K + D\bar{G}\bar{H}K + \bar{D}FGH\bar{K}$$

$$f_3(A, B, C, F, G, H, K, M, N) = A\bar{B}CG\bar{H}\bar{K}N + \bar{B}CG\bar{H}\bar{K}MN$$

$$f_4(G, H, K, M, N) = G\bar{H}KM\bar{N} + GH\bar{K}\bar{M}N$$

$$f_5(A, B, G, H, J, K, M, N) = A\bar{B}G\bar{H}JKM\bar{N} + \bar{A}BG\bar{H}JKMN$$

$$f_6(A, C, D, E, F, G, K, M, N) = \bar{A}C\bar{D}E\bar{F}GK\bar{M}N + A\bar{C}DEFGKMN$$

Although there is overlap, none of the functions involve the same set of variables, and two functions, f_1 and f_4, are completely disjoint. Nevertheless, all six functions can be realized in a single PAL, as shown in Fig. 6.40. By comparison, a realization in discrete NAND's would require 20 chips.   ■

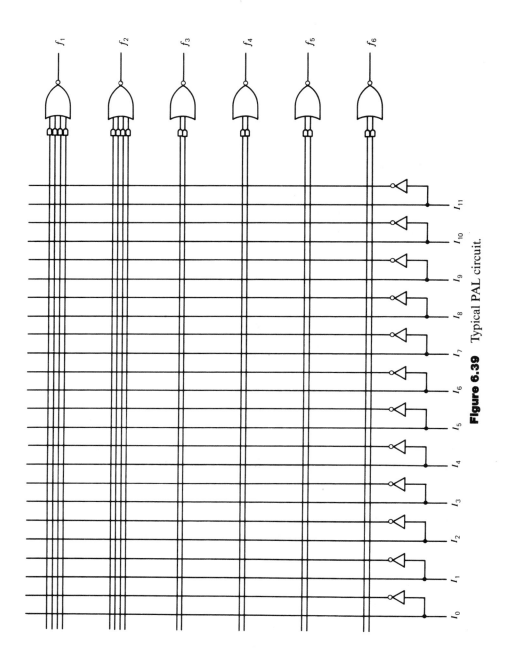

Figure 6.39 Typical PAL circuit.

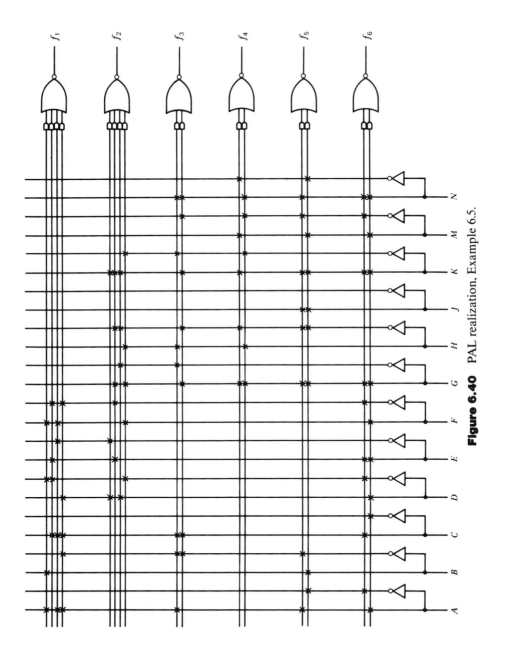

Figure 6.40 PAL realization, Example 6.5.

PAL circuits generally have from 10–16 inputs and come in a variety of configurations in terms of the number of OR gates and AND gates driving each. The relative economy of this approach will depend on how closely an available PAL matches the requirements of a particular application.

Because the AND gate outputs are uniquely and irrevocably connected to OR gate inputs, multiple-output logic minimization is not applicable to PAL design. For the separate minimization of each individual function, the AND input cost coefficient a will be 0. The coefficient b is not applicable. An optimal PLA will always require less chip area than a ROM or PAL realization of the same set of functions. The PLA will be the choice when the realization is to be embedded within a VLSI chip. In field programmable applications the PAL is particularly attractive when the total number of variables involved is large, but each function is a function of only a small subset of the variables.

PAL's [15] would not have achieved their significant and lasting popularity were it not for the existence of parts including memory elements, as well as combinational logic PAL's. This permits the realization of finite-state machines within individual PAL packages. A recent competitor [16] to the PAL, called a GAL, adds routing multiplexers as well as memory elements to the two-level logic of the PAL. Recently available field programmable gate array packages distribute memory among elements that can be connected to multilevel logic networks. As the capacity to fabricate complex VLSI circuits continues to increase, it may be expected that ever more complex approaches to user programming of parts will appear. We shall return to this subject in Chapter 12.

PROBLEMS

6.1 Use the method characterized by Fig. 6.9 to determine the set of all prime implicants (for a potential sum-of-products realization) for the Boolean functions listed below.

(a) $f(W, V, X, Y, Z) = \sum m(0, 1, 3, 8, 9, 11, 15, 16, 17, 19, 24, 25, 29, 30, 31)$

(b) $f(A, B, C, D) = \sum m(0, 1, 4, 5, 6, 7, 8, 9, 10, 12, 14)$

(c) $f(a, b, c, d) = \sum m(0, 2, 3, 7, 8, 10, 11, 12, 14)$

(d) $f(A, B, C, D, E) = \sum m(0, 1, 2, 3, 4, 6, 9, 10, 15, 16, 17, 18, 19, 20, 23, 25, 26, 31)$

(e) $f(a, b, c, d, e, f) = \sum m(0, 2, 4, 5, 7, 8, 16, 18, 24, 32, 36, 40, 48, 56)$

6.2 Determine the set of prime implicants for the product-of-sums realization of the Boolean function given in Problem 6.1(c).

6.3 The process of the Quine–McCluskey minimization of the Boolean function

$$f(A, B, C, D, E) = \sum m(0, 1, 5, 6, 11, 14, 15, 18, 19, 20, 21, 24, 26, 27, 28, 29)$$

has already been partially accomplished with the determination of prime implicants as tabulated below.

$$\text{prime implicants} = A\bar{C}D, AC\bar{D}, \bar{A}\bar{B}\bar{C}\bar{D}, \bar{A}\bar{B}\bar{D}E, BC\bar{D}E, AB\bar{D}\bar{E},$$
$$AB\bar{C}\bar{E}, B\bar{C}DE, \bar{A}BDE, \bar{A}BCD, \bar{A}CD\bar{E}$$

Construct a Quine–McCluskey table and follow the algorithm precisely to determine a minimal realization.

(a) Step 1: List all essential prime implicants.
Step 2: Then list all prime implicants that may now be deleted because they are dominated or interchangeable with another PI.

(b) Repeat Step 1, listing all secondary essential prime implicants. Then repeat Step 2, listing all prime implicants that may now be deleted because they are dominated or interchangeable with another PI.

(c) Repeat part (b), if necessary.

(d) Repeat part (b), if necessary, etc.

6.4 The prime implicants for the Boolean functions listed in Problem 6.1 were determined in that problem.

(a–e) Follow the process outlined in Problem 6.3 to find a minimal sum-of-products realization of the Boolean functions given in parts (a–e) of Problem 6.1.

6.5 Using the results of Problem 6.2, find a minimal product-of-sums realization of the Boolean function listed in Problem 6.1(c).

6.6 The process of Quine–McCluskey minimization of the Boolean function

$$f_1(A, B, C, D, E) = \sum m(0, 1, 5, 14, 15, 21, 24, 26, 27, 28)$$

where $d(6, 11, 18, 19, 20, 29)$ are "don't care" minterms has already been partially accomplished with the determination of prime implicants as tabulated below. Don't cares make up some of the prime implicants, but only the minterms for which $f_1 = 1$ are listed as covered.

Prime Implicant	Covered Minterms
$A\bar{C}D$	26, 27
$AC\bar{D}$	21, 28
$\bar{A}\bar{B}\bar{C}\bar{D}$	0, 1
$\bar{A}\bar{B}\bar{D}E$	1, 5
$\bar{B}C\bar{D}E$	5, 21
$AB\bar{D}\bar{E}$	24, 28
$AB\bar{C}\bar{E}$	24, 26
$B\bar{C}DE$	27
$\bar{A}BDE$	15
$\bar{A}BCD$	14, 15
$\bar{A}CD\bar{E}$	14

Construct a Quine–McCluskey table and follow the procedure of Problem 6.3 to determine a minimal sum-of-products realization.

6.7 The process of Quine–McCluskey minimization of the Boolean function

$$f_2(A, B, C, D, E) = \sum m(1, 4, 6, 8, 13, 14, 20, 21, 23)$$

where $d(0, 3, 5, 9, 10, 24, 26, 29, 30)$ are "don't care" minterms has already been partially accomplished with the determination of prime implicants as tabulated below. Don't cares make up some of the prime implicants, but only the minterms for which $f_2 = 1$ are listed as covered.

Prime Implicant	Covered Minterms	Prime Implicant	Covered Minterms
$\bar{A}\bar{B}D$	1, 4	$\bar{A}\bar{B}CE$	1
$\bar{A}C\bar{D}$	1, 8	$\bar{A}BC\bar{E}$	4, 6
$\bar{A}DE$	1, 13	$ABCE$	21, 23
$\bar{B}C\bar{D}$	4, 20, 21	$\bar{A}CD\bar{E}$	6, 14
$B\bar{C}\bar{E}$	8		
$C\bar{D}E$	13, 21		
$BD\bar{E}$	14		

Follow and document the process of Problem 6.1 in finding a minimal sum-of-products realization of the Boolean function f_2.

6.8   Use the Quine–McCluskey method to determine a minimal sum-of-products realization of each of the Boolean functions, with "don't care" lists given below.

(a) $f(A, B, C, D, E) = \sum m(4, 5, 10, 11, 15, 18, 20, 24, 26, 30, 31)$
    $+ d(9, 12, 14, 16, 19, 21, 25)$

(b) $f(A, B, C, D, E) = \sum m(0, 2, 3, 6, 9, 15, 16, 18, 20, 23, 26)$
    $+ d(1, 4, 10, 17, 19, 25, 31)$

(c) $f(a, b, c, d, e, f) = \sum m(0, 2, 4, 7, 8, 16, 24, 32, 36, 40, 48)$
    $+ d(5, 18, 22, 23, 54, 56)$

(d) $f(A, B, C, D) = \sum m(0, 1, 4, 6, 8, 9, 10, 12) + d(5, 7, 14)$

(e) $f(a, b, c, d) = \sum m(2, 4, 8, 11, 15) + d(1, 10, 12, 13)$

(f) $f(W, X, Y, Z) = \sum m(1, 4, 8, 9, 13, 14, 15) + d(2, 3, 11, 12)$

6.9   A circuit receives two two-bit binary numbers $Y = y_1 y_0$ and $X = x_1 x_0$. The two-bit output $Z = z_1 z_0$ should equal 11 if $Y = X$, 10 if $Y > X$, and 01 if $Y < X$. Design a minimal sum-of-products realization.

6.10   Five people judge a certain competition. The vote of each is indicated by 1 (pass) or 0 (fail) on a signal line. The five signal lines form the input to a logic circuit. The rules of the competition allow only 1 dissenting vote. If the vote is 2–3 or 3–2, the competition must continue. The logic circuit is to have two outputs, xy. If the vote is 4–1 or 5–0 to pass, $xy = 11$. If the vote is 4–1 or 5–0 to fail, $xy = 00$. If the vote is 3–2 or 2–3, $xy = 10$. Design a minimal sum-of-products circuit.

6.11   A five-bit binary number $N = x_4 x_3 x_2 x_1 x_0$ appears at the inputs to a combinational logic circuit. The circuit has two outputs:

z_1 indicates the number is evenly divisible by 6
z_2 indicates the number is evenly divisible by 9

Design a minimal sum-of-products realization.

6.12   A logic circuit has five inputs: x_4, x_3, x_2, x_1, and x_0. Output z_0 is to be 1 when a majority of the inputs are 1. Output z_1 is to be 1 when fewer than four of the inputs are 1, provided that at least one input is 1. Output z_2 is to be 1 when two, three, or four of the inputs are 1. Design a minimal sum-of-products circuit.

6.13   Using Karnaugh maps, determine all prime implicants that might be used in a multiple-output minimal sum-of-products realization of the three sets of three Boolean functions given below.

(a) $f_1(a, b, c) = \sum m(0, 1, 3, 5)$
   $f_2(a, b, c) = \sum m(2, 3, 5, 6)$
   $f_3(a, b, c) = \sum m(0, 1, 6)$

(b) $f_1(A, B, C, D) = \sum m(0, 1, 2, 3, 6, 7)$
   $f_2(A, B, C, D) = \sum m(0, 1, 6, 7, 14, 15)$
   $f_3(A, B, C, D) = \sum m(0, 1, 2, 3, 8, 9)$

(c) $f_1(A, B, C, D) = \sum m(4, 5, 10, 11, 12)$
   $f_2(A, B, C, D) = \sum m(0, 1, 3, 4, 8, 11)$
   $f_3(A, B, C, D) = \sum m(0, 4, 10, 12, 14)$

6.14 The following is a tabulation of the values of three Boolean functions f_1, f_2, and f_3. All prime implicants of interest for a minimal multiple-output sum-of-products realization of these functions have already been determined and are also listed below. The minterms actually covered by these prime implicants can be identified in the process of constructing the multifunction prime implicant table.

PI	In Functions
$\bar{A}\,\bar{C}$	f_1
$\bar{A}\,\bar{B}$	f_1
$\bar{B}\,C$	f_1, f_3
$B\bar{C}$	f_2
$\bar{A}\,B$	f_2
$\bar{A}\,C$	f_3
$A\bar{C}$	f_3
$A\bar{B}$	f_3
$\bar{A}\,B\bar{C}$	f_1, f_2
$\bar{A}\,BC$	f_2, f_3
$AB\bar{C}$	f_2, f_3
$A\bar{B}C$	f_1, f_2, f_3

A	B	C	f_1	f_2	f_3
0	0	0	1	0	0
0	0	1	1	0	1
0	1	0	1	1	0
0	1	1	0	1	1
1	0	0	0	0	1
1	0	1	1	1	1
1	1	0	0	1	1
1	1	1	0	0	0

Although no PLA need be constructed, the goal is to find minimal multiple-output sum-of-products expressions suitable for an optimal PLA realization, with nonzero cost assigned only to gates and 0 additional cost assigned to gate inputs.

Use a Quine–McCluskey prime implicant table to answer the following questions.

(a) List all essential prime implicants and indicate the functions to which they apply.

(b) List all dominated or interchangeable prime implicants that may be dropped from the process following part (a).

(c) List all secondary essential prime implicants and indicate the functions to which they apply.

(d) Write the minimal multiple-output sum-of-products expressions for f_1, f_2, and f_3.

6.15 The following are minterm lists of two Boolean functions: f_1 and f_2. All prime implicants of interest for a minimal multiple-output sum-of-products realization of these functions have already been determined and are also listed below in the Quine–McCluskey table with the covered minterms.

$$f_1 = \sum m(3, 4, 5, 7, 9, 13, 15) + d(11, 14)$$
$$f_2 = \sum m(3, 4, 7, 9, 13, 14) + d(0, 1, 5, 15)$$

	3	4	5	7	9	13	15	3	4	7	9	13	14
BD		1	1		1		1		1			1	
ABC							1						1
$\bar{A}CD$	1			1				1		1			
$\bar{A}B\bar{C}$		1	1								1		
$A\bar{C}D$					1	1					1	1	
CD	1			1		1							
AD				1	1	1							
$\bar{C}D$											1	1	
$\bar{A}C$									1				
$\bar{A}D$								1	1				

Repeat the process of Problem 6.14 in the determination of an optimal PLA realization (0 cost assigned to gate inputs).

6.16 Determine all prime implicants relevant to a multiple-output minimal (PLA) sum-of-products realization of the following two Boolean functions, f_1 and f_2, and then follow the process of Problem 6.14 to find that realization.

$$f_1(A, B, C, D) = \sum m(0, 2, 6, 7, 15) + d(8, 10, 14)$$
$$f_2(A, B, C, D) = \sum m(0, 1, 3, 7, 15) + d(8, 10, 14)$$

6.17 Find minimal multiple-output sum-of-products expressions suitable for a PLA realization for each of the sets of Boolean functions listed in Problem 6.13.

6.18 Find minimal multiple-output sum-of-products expressions suitable for a PLA realization for each of the sets of Boolean functions listed below.

(a) $f_1(A, B, C, D, E) = \sum m(0, 1, 2, 3, 6, 7, 20, 21, 26, 27, 28)$
  $f_2(A, B, C, D, E) = \sum m(0, 1, 6, 7, 14, 15, 16, 17, 19, 20, 24, 27)$
  $f_3(A, B, C, D, E) = \sum m(0, 1, 2, 3, 8, 9, 16, 20, 26, 28, 30)$

(b) $f_1(A, B, C, D, E) = \sum m(0, 1, 2, 8, 9, 10, 13, 16, 17, 18, 19, 24, 25)$
  $f_2(A, B, C, D, E) = \sum m(0, 1, 3, 5, 7, 9, 13, 16, 17, 22, 23, 30, 31)$
  $f_3(A, B, C, D, E) = \sum m(2, 3, 8, 9, 10, 11, 13, 15, 16, 17, 18, 19, 22, 23)$

6.19 In the yard tower of a railroad yard, a controller must select the route of freight cars entering a section of the yard from point A as shown on the control panel, Fig. P6.19. Depending on the positions of the switches, a car can arrive at any one of four destinations. Other cars may enter from points B or C. Design a circuit that will receive as inputs signals S_1 to S_5, indicating the positions of the corresponding switches, and will light a lamp, D_0 to D_4, showing which destination the car from A will reach. For the cases when cars can enter from B or C (S_2 or S_3 in the 0 position), all output lamps should light, indicating that a car from A cannot reach its destination safely.

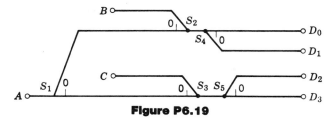

Figure P6.19

6.20 Find minimal multiple-output sum-of-products expressions suitable for a PLA realization for each of the sets of Boolean functions with "don't care" lists given below.

(a) $f_1(a, b, c, d) = \sum m(0, 2, 9, 10) + d(1, 8, 13)$
$f_2(a, b, c, d) = \sum m(1, 3, 5, 13) + d(0, 7, 9)$
$f_3(a, b, c, d) = \sum m(2, 8, 10, 11, 13) + d(3, 9, 15)$

(b) $f_1(A, B, C, D) = \sum m(2, 3, 6, 10) + d(8)$
$f_2(A, B, C, D) = \sum m(2, 10, 12, 14) + d(6, 8)$
$f_3(A, B, C, D) = \sum m(2, 8, 10, 12) + d(0, 14)$

(c) $f_a(W, X, Y, Z) = \sum m(0, 5, 7, 14, 15) + d(1, 6, 9)$
$f_b(W, X, Y, Z) = \sum m(13, 14, 15) + d(1, 6, 9)$
$f_c(W, X, Y, Z) = \sum m(0, 1, 5, 7) + d(9, 13, 14)$

(d) $f_1(A, B, C, D, E) = \sum m(2, 8, 10, 12, 18, 26, 28, 30)$
$+ d(0, 14, 22, 24)$
$f_2(A, B, C, D, E) = \sum m(2, 3, 6, 10, 18, 24, 26, 27, 29)$
$+ d(8, 19, 25, 31)$
$f_3(A, B, C, D, E) = \sum m(1, 3, 5, 13, 16, 18, 25, 26)$
$+ d(0, 7, 9, 17, 24, 29)$

6.21 Determine the consensus of each of the following pairs of cubes.

(a) $abcd \odot cd\bar{g}\bar{h}$

(b) $ab\bar{c}\bar{d} \odot cdgh$

(c) $abc\bar{d} \odot cdgh$

6.22 Which of the cubes in the following Boolean expression are not relatively essential?

$$\bar{a}ce + \bar{a}\bar{b}\bar{c}\bar{e} + \bar{a}\bar{b}d\bar{e} + \bar{b}cd\bar{e} + \bar{a}b\bar{d}\bar{e}$$

6.23 Determine the Boolean expression obtained after a single application of the Espresso routine EXPAND is applied to

$$abd + \bar{b}cd + \bar{a}b\bar{c} + abc\bar{d}$$

After an application of EXPAND, which cubes in the resulting expression are relatively essential? (Same as homework.)

6.24 Manually execute the Espresso algorithm on parts (c) and (d) of Problem 6.1. Carefully indicate the resulting expression after each successive step of the algorithm.

(a) Obtain any convenient sum-of-products expression from the minterm list by application of the distributive law to the expanded sum of products.

(b) After an application of EXPAND in block I of Fig. 6.33, list all relatively essential prime implicants. Apply ESSENTIALPRIMES to find all essential prime implicants.

(c) Apply and reapply block II of Fig. 6.33 until no further improvement results.

6.25 Repeat Problem 6.21 for the functions, defined in parts (c) and (d) of Problem 6.8, that include "don't cares."

6.26 Use the PAL of Fig. 6.39 to realize the following six functions of 10 variables. Construct a PAL diagram similar to that of Fig. 6.40.

$$ab \oplus cd; \quad cd \oplus ef; \quad f \oplus g; \quad g \oplus h; \quad h \oplus k; \quad k \oplus m$$

6.27 Determine the number of AND gates and the number of inputs to each AND gate necessary in a PLA realization of the set of functions given in Problem 6.26. Assume the minimum number of primary inputs to the PLA. Compare the total number of AND gate inputs in the PLA to the total number of AND gate inputs in Fig. 6.39. What conclusions might be drawn regarding required chip area, if one were considering realizing the two configurations in the same VLSI technology.

6.28 Add one output OR gate to the PAL in Fig. 6.39. Use this PAL to realize the BCD to biquinary code converter defined by Fig. P6.28. It may be assumed that the unused input combinations are "don't cares."

6.29 Without resorting to a formal multiple-output realization, construct a PLA diagram realizing the code converter of Problem 6.28. (**Hint: B_5 and B_0 can be expressed in terms of the products used in the other five outputs.**) Compare the chip area required for this PLA with a "same technology" VLSI realization of the PAL designed in Problem 6.28.

6.30 If a PLA version of an Espresso implementation is available, apply the program to the set of functions defined in part (d) of Problem 6.20.

6.31 If available, use an implementation of Espresso to design a realization of a 2-out-of-5 (refer to Chapter 2) to BCD code converter. The 22 unused input combinations may be considered "don't cares." First, disallow common AND gates (PAL format) in the realization of the four functions. Reapply Espresso to obtain a minimal multiple-output (PLA) realization. Compare the results.

	BCD				Bi-Quinary						
Digit	D_8	D_4	D_2	D_1	B_5	B_0	Q_4	Q_3	Q_2	Q_1	Q_0
0	0	0	0	0	0	1	0	0	0	0	1
1	0	0	0	1	0	1	0	0	0	1	0
2	0	0	1	0	0	1	0	0	1	0	0
3	0	0	1	1	0	1	0	1	0	0	0
4	0	1	0	0	0	1	1	0	0	0	0
5	0	1	0	1	1	0	0	0	0	0	1
6	0	1	1	0	1	0	0	0	0	1	0
7	0	1	1	1	1	0	0	0	1	0	0
8	1	0	0	0	1	0	0	1	0	0	0
9	1	0	0	1	1	0	1	0	0	0	0

(a)

D_8 —○ B_5
D_8 —
D_4 — Code —○ B_0
Converter —○ Q_4
D_2 — —○ Q_3
—○ Q_2
D_1 — —○ Q_1
—○ Q_0

(b)

Figure P6.28

BIBLIOGRAPHY

1. Miller, R. E. *Switching Theory*, Vol. I. Wiley, New York, 1965.
2. Quine, W. V. "The Problem of Simplifying Truth Functions," *Am. Math. Monthly*, **59**:8, 521–531 (Oct. 1952).
3. McCluskey, E. J. "Minimization of Boolean Functions," *Bell System Tech. J.*, **35**:5, 1417–1444 (Nov. 1956).
4. McCluskey, E. J. *Logic Design Principles*. Prentice-Hall, Englewood Cliffs, N.J., 1986.
5. Petrick, S. R. "On the Minimization of Boolean Functions," in *Proceedings of Symposium on Switching Theory*, ICIP, Paris, France, June 1959.
6. Bartee, T. C. "Automatic Design of Logical Network," in *Proceedings of Western Joint Computer Conference*, 1959, pp. 103–107.
7. Muroga, S. *VLSI System Design*. Wiley, New York, 1982.
8. Lawler, E. W. and D. E. Wood. "Branch and Bound Methods—A Survey," Oper. Res., **14**:669–719 (1966).
9. Goel, P. "An Implicit Enumeration Algorithm to Generate Tests for Combinational Logic Circuits," *IEEE Trans. Computers*, **C-30**:3, 215–222 (March 1981).
10. Hachtel, G. D., et al. "An Algorithm for Optimal PLA Folding," *IEEE Trans. CAD*, **CAD-1**:2, 63–77 (1982).
11. *PAL Programmable Array Logic Handbook*, 3rd ed. Monolithic Memories, Sunnyvale, Calif., 1983.
12. Brayton, R. K., G. D. Hachtel, C. T. McMullen, and A. L. Sangiovanni-Vincentelli. *Logic Minimization Algorithms for VLSI Synthesis*. Kluwer, Boston, 1984.
13. Wakerly, J. V. *Digital Design Principles and Practices*. Prentice-Hall, Englewood Cliffs, N.J., 1990.
14. Sandidge, R. S. *Modern Digital Design*. McGraw-Hill, New York, 1990.
15. Alford, R. C. *Programmable Logic Designers Guide*. Howard Sams and Co., Indianapolis, Ind., 1989.
16. Wilson, R. "New PLD Architectures Deliver Needed Flexibility," *Computer Design*, **22-6** (June 1988).

CHAPTER 7

VLSI REALIZATION OF COMBINATIONAL LOGIC

7.1 INTRODUCTION

The topic of this volume is computer-aided logic design. This would seem to exclude design approaches that draw the designer into the details of circuit design rather than allowing him or her to rely on the interconnection of already designed building blocks or cells. We will depart slightly from that restriction in the first few sections of this chapter as we introduce some circuit-dependent techniques applicable to NMOS design. As we move to CMOS in the later sections, we will return to using predesigned cells.

In Chapter 4 we looked briefly at the impact of transistor dimensions on the output levels and drive capability of the MOS transistor. In this chapter we will draw on that background, as necessary, without developing it in more detail. For the most part, it will be possible to limit references to transistor size to the discussion, while graphically depicting only circuit topology. A mechanism called a stick diagram was introduced for this purpose by Mead and Conway [1]. In the stick diagram polysilicon, metal p- and n-diffused areas are merely represented by distinctly styled lines. The line representation that we shall use for each of these regions is given in Fig. 7.1. In books that include color plates, the stick or line representations are assigned the same colors as the respective areas in a complete layout diagram.

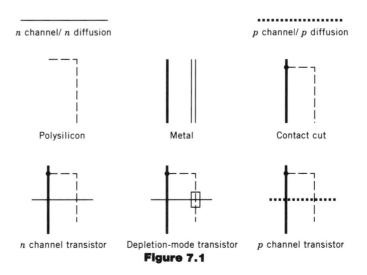

n channel/ *n* diffusion   *p* channel/ *p* diffusion

Polysilicon   Metal   Contact cut

n channel transistor   Depletion-mode transistor   *p* channel transistor

Figure 7.1

A contact cut connecting two types of areas is represented by a black rectangle. Also shown in Fig. 7.1 are n- and p-channel transistors. These devices are represented by a polysilicon line crossing n- and p-channel lines, respectively. The crossing is sufficient to show the existence of a transistor. If dimensions are required, these will be represented numerically or included in the discussion. A depletion region is represented by a small rectangular outline. Figure 7.1 depicts a depletion-mode transistor as a depletion area enclosing a polysilicon n-channel crossing.

Figure 7.2 uses interconnections of these stick building blocks to construct stick diagrams of two-input NAND gates in NMOS and CMOS. The topology of the NMOS gate is consistent with the layout diagram in Fig. 4.27. A stick diagram does not provide an accurate measure of the chip area required by a particular circuit. However, the topologies reflected in two stick diagrams will often provide a meaningful comparison of the chip areas required by two NMOS realizations of the same logic function. When devices within a gate, but not topology, are of interest, we will make use of device diagrams of the form of Fig. 4.23b.

7.2 PASS TRANSISTOR NETWORK REALIZATIONS

Two-level NAND-NAND and two-level NOR-NOR representations can be realized using NMOS and CMOS VLSI. Particularly for NMOS, the NOR-NOR realization is preferred over the NAND-NAND realization of the same function. The reason for this is the adverse impact on the β ratio of increasing the number of inputs and, therefore, the number of devices in series in the pull-down network. To compensate, it is necessary that the width of the diffusion area realizing the pull-down network increase in proportion to the increasing channel length. This was illustrated in Fig. 4.27. Thus, the overall chip area increases with the square of the number of inputs. The inputs to a NOR gate control parallel channels and thus do not impact the β ratio.

It is possible to realize logic functions with networks of pass transistors in much the same way as was classically done in relay contact networks. In many cases, a lower total device count (depletion-mode + enhancement-mode devices) can be achieved by not restricting realizations to networks of NOR gates. The likelihood that chip area will be reduced with the relaxation of this restriction is even greater.

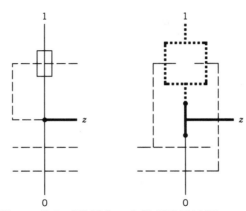

Figure 7.2 NMOS and CMOS NAND gates.

A generalization of one approach to the realization of combinational logic based on networks of pass transistors is depicted in Fig. 7.3. We shall refer to this approach as $f/\bar{f}$ design. If the realization is NMOS, the output inverter is realized by the circuit of Fig. 4.23. If the realization is CMOS, the output inverter symbol represents the circuit of Fig. 4.24. For convenience we label the output of the inverter $\bar{f}$, where f is the function to be realized by the pass network. Also listed are the two almost self-evident rules governing the design of the pass network. **Rule 1** tells us that connecting point p to V_{dd} whenever $f = 1$ is insufficient. If point p is allowed to float, the output of the inverter will be indeterminate. Point p must always be connected to something. It must be connected to 0, if $f = 0$. **Rule 2** is necessary to assure that chips are not destroyed by short circuits between power and ground.

How are we to treat a combination of input values for which f is a "don't care"? First, we must be a bit more careful in our definition of a "don't care."

Definition 7.1   A particular combination of input values that will not occur during normal operation of a circuit but might appear during testing is an *application don't care*.

Definition 7.2   A particular combination of input values that will not occur during normal operation or testing is an *absolute don't care*.

If one is certain that a combination of values is an absolute don't care, then the realization may connect point p to both 0 and 1, to 0, to 1, or to neither terminal for that combination of input values. If a combination of values is merely an application don't care, it must cause point p to be connected to either 0 or 1 but not both. An otherwise good chip may be destroyed if V_{dd} is connected to ground during a test. Application don't cares that will cause point p to float unconnected can complicate simulation for the purposes of test generation, if the combinational network is part of a larger sequential system (see Chapter 10). If p is unconnected during a particular clock period of the simulation, the charge on the gate capacitor of the inverter will retain its value from the previous clock period. This feature must be modeled by the simulation.

Treating all "don't cares" as application don't cares impacts on the minimization process for f and $\bar{f}$ and, therefore, on the chip area required by the pass transistor network. Consider the example given by the Karnaugh map of Fig. 7.4a. In the minimal sum-of-products representation of Fig. 7.4a, both design rules 1 and

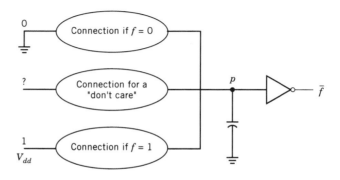

Rule 1: P must always be connected to 0 or 1.
Rule 2: P must never be simultaneously connected to 0 and 1.

Figure 7.3 $f/\bar{f}$ design.

2 are violated, if the "don't cares" are application don't cares. Minterm m_2 will cause point p to be connected to both 1 and 0, whereas m_{13} will leave point p unconnected. Both rules are satisfied after the modification of f, which requires an additional product, as shown in Fig. 7.4b.

The *stick diagram* of the NMOS realization of Fig. 7.4b is given in Fig. 7.5a. It will be our practice to call this form a stick diagram, although the inverters are represented by logic symbols. The somewhat simpler diagram of Fig. 7.5b, in which the existence of an enhancement device is represented by an X, retains all the information of the stick diagram for the pass transistor network. We shall call this representation a *pass network connection diagram* and use it frequently throughout the chapter. We shall call this particular connection diagram *uniform*, because connections to a given literal always appear in a single column and each variable and its complement occupy adjacent columns.

It may be observed that the stick diagram of the pass transistor network suggests quite clearly the actual mask layout of the network. If the rectangle occupied by the stick diagram is reduced, then the chip area required by the actual layout will be reduced. Notice that a device is saved by factoring A in the expression for $\bar{f}$, but it is unlikely that any saving in chip area will result. Some benefit in the form of a slightly reduced probability of failure does result from the elimination of each individual device. The factoring of B cannot be reflected in the realization because the metal columns representing B and $\bar{B}$ do not lie at either end of the network.

In general, the chip area required by the pass transistor network will be proportional to the number of diffusion paths to 0 and V_{dd} or simply to the number of prime implicants in f and $\bar{f}$. The cost function is, therefore, the same as used for the PLA in Chapter 6. The Quine–McCluskey algorithm may be adapted to the $f/\bar{f}$ design. Both f and $\bar{f}$ may be included as separate functions, with a third pseudofunction included to force the covering of all "don't care" minterms. The prime implicants used in the prime implicant table will be the union of those found using the "don't cares" and those found without them. A mechanism must be added to delete all other remaining prime implicants covering a "don't care" once one prime implicant covering that "don't care" is found to be essential.

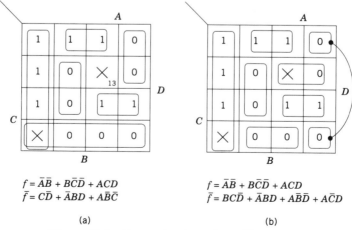

$$f = \bar{A}\bar{B} + B\bar{C}\bar{D} + ACD$$
$$\bar{f} = C\bar{D} + \bar{A}BD + A\bar{B}\bar{C}$$

(a)

$$f = \bar{A}\bar{B} + B\bar{C}\bar{D} + ACD$$
$$\bar{f} = BC\bar{D} + \bar{A}BD + A\bar{B}\bar{D} + A\bar{C}D$$

(b)

Figure 7.4 Determination of an $f/\bar{f}$ realization.

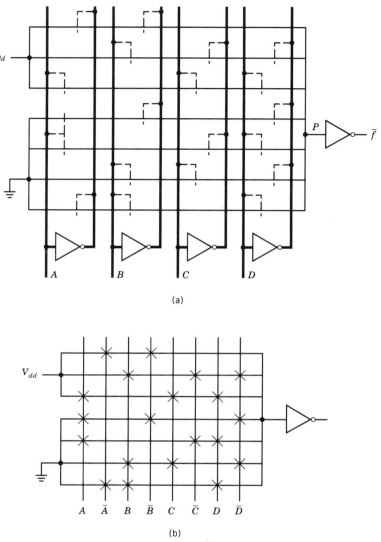

Figure 7.5   NMOS realization of Fig. 7.4b (a) Stick diagram. (b) Uniform pass network connection diagram.

Members of a certain class of Boolean functions called *unate functions* are particularly convenient to realize with $f/\bar{f}$ contact networks.

Definition 7.3   A Boolean expression is *positive* (negative) in x_i if and only if the complemented x_i (uncomplemented literal x_i) does not appear in the expression.

Definition 7.4   A Boolean function $f(x_1, x_2, ..., x_n)$ is *unate* if and only if there exists an expression for f that is either positive or negative in every variable x_i.

Definition 7.5   A Boolean function $f(x_1, x_2, ..., x_n)$ is *positive* if and only if there exists an expression for f that is positive in every variable x_i.

It can be proved that the minimal sum-of-products expression of a unate function is unique and either positive or negative in every variable. The minimal sum-of-products expression for a positive function is, therefore, positive in every variable, and it may be shown that every prime implicant in this expression covers the vertex for which the value of every variable is 1. (See Problem 7.2.).

■ EXAMPLE 7.1

Use Theorem 5.5 to realize the simple unate function

$$g_1 = \bar{A}B + CD$$

SOLUTION   Applying Theorem 5.5 that states

$$\overline{f(x_1, x_2, ..., x_n)} = f_d(\bar{x}_1, \bar{x}_2, ..., \bar{x}_n)$$

yields directly an expression for $\bar{g}_1$

$$\bar{g}_1 = (A + \bar{B})(\bar{C} + \bar{D})$$

Connecting to ground through a pass network implementing g_1 and to V_{dd} through the g_1 network results in the realization given in Fig. 7.6, with $\bar{g}_1$ represented by the output of the inverter.   ■

CMOS Perspective   The application of Theorem 5.5 is even more helpful in developing CMOS realizations. In NMOS the pass transistor network must necessarily be connected to the input of an inverter, because the pull-up is a single depletion-mode device. *CMOS circuit properties do not require an output inverter.* The network connected to V_{dd} will serve as the pull-up network, whereas the network connected to ground is the pull-down network. In the realization of a unate function, *p*-channel devices may be used exclusively in the pull-up network and *n*-channel devices in the dual pull-down network. If the gate inputs of all PMOS devices in a path from V_{dd} to the output point are 0, the voltage at every point in the path, including the output, will be V_{dd}. Likewise, an excited NMOS path to ground will cause the output to be exactly 0 V. An output inverter is not required to restore the logic levels. An output inverter might be included to drive a large capacitive load or permit a realization in the absence of a complete double-rail set of inputs.

For CMOS we make the following three more specific observations. (1) If the function to be realized is positive and the output inverter is accepted, only single-rail inputs are required. (2) If the set of complemented literals is available in place of the uncomplemented, any positive function may be realized without an output inverter. (3) If double-rail inputs are available, any unate function may be realized without an output inverter.

■ EXAMPLE 7.2

Realize the positive function $g_2 = AB + CD$ first assuming double-rail inputs and then with only single-rail inputs.

SOLUTION   Given double-rail inputs, the complemented literals are connected to the gate inputs of a PMOS pass network defined by

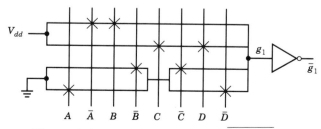

Figure 7.6   NMOS realization of $\overline{\bar{A}B} + CD$.

$$g_2 = AB + CD$$

between the gate output and V_{dd}. Note, for example, that the PMOS device controlled by $\bar{A}$ is on when $\bar{A} = 0$ or $A = 1$. Applying Theorem 5.5 yields

$$\bar{g}_2 = (\bar{A} + \bar{B})(\bar{C} + \bar{D})$$

This expression defines the NMOS pass network that connects the output point to ground. A device diagram of the resulting logic circuit is given in Fig. 7.7a. The stick diagram of the same circuit is shown in Fig. 7.7b. Because the function is unate, there is only one vertical metal line through the pass network per variable, and the required chip area is reduced accordingly. For completeness, the CMOS device diagram for the same function is also given in Fig. 7.7b.

If only single-rail inputs are available, we must use $g_2 = AB + CD$ to define the NMOS pull-down network and $\bar{g}_2 = (\bar{A} + \bar{B})(\bar{C} + \bar{D})$ to define the dual PMOS pull-up network. The inputs are now the uncomplemented variables, and an output inverter is required to produce g_2. This single-rail realization is illustrated in Fig. 7.7c.   ∎

Considerable difficulty is encountered in adapting the NMOS techniques to be discussed in succeeding sections to CMOS. For this reason, CMOS networks are often interconnections of standard building blocks or cells formed by realizing simple unate or positive functions.

7.3   STEERING OF 0, 1, x_i, AND $\bar{x}_i$ TO THE OUTPUT

In the previous sections Boolean functions were realized by providing connections through pass transistor networks to 0 or 1. In this section we shall generalize this approach somewhat by allowing pass network paths to a variable and its complement as well as to 0 and 1. As was done with $f/\bar{f}$ design in the previous section, we shall focus on the function f corresponding to the point at which the pass network is connected to the input of an inverter, although it is $\bar{f}$ that is actually

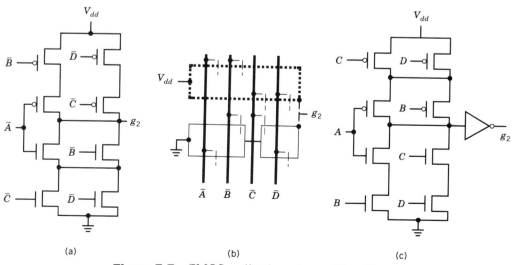

(a)   (b)   (c)

Figure 7.7   CMOS realization of $g_2 = AB + CD$.

realized at the inverter output. We omit the step of complementing the function, which we are asked to realize, at the beginning of the process.

If x_i and $\bar{x}_i$ are allowed as terminal points in the network, we call x_i the terminal variable, and the remaining variables are termed control variables. As illustrated in Fig. 7.8, a separate function of the control variables must be determined defining the connection to each of the terminals. These functions must be disjoint to insure satisfaction of Rule 2 of Fig. 7.3 and must satisfy Equation 7.1 as required by Rule 1.

$$f_0 + f_1 + f_x + f_{\bar{x}} = 1 \tag{7.1}$$

Clearly, the $f/\bar{f}$ approach is a special case of the steering method to be discussed in this section. One would expect that certain functions could be realized in less chip area by taking advantage of the increased generality. Also reducing the maximum number of pass transistors in series by 1 will result in a not insignificant reduction in the overall delay of the network, as discussed in Section 4.9.

■ **EXAMPLE 7.3**

Realize the complement of the function $f = \bar{A}\bar{B} + B\bar{C}\bar{D} + ACD$ (see Fig. 7.4b) using a pass network steering to 0, 1, D, and $\bar{D}$.

SOLUTION   The terminal variable is listed as the row variable in the Karnaugh map of Fig. 7.9a. The four control variable functions are clearly marked on the map and expressed in sum-of-products form as follows:

$$f_0 = A\bar{B}\bar{C} + \bar{A}BC, \qquad f_1 = \bar{A}\bar{B}$$
$$f_D = AC, \qquad\qquad f_{\bar{D}} = B\bar{C}$$

The NMOS network realization is given in Fig. 7.9b. This realization requires only 12 pass transistors, in contrast to the 19 required by the $f/\bar{f}$ realization of the same function given in Fig. 7.5b. Savings in chip area is apparent as well.   ■

CMOS Perspective   The connection to terminals D and $\bar{D}$ in Fig. 7.9b will present a problem in a CMOS version of this network. If the device in the path to D is

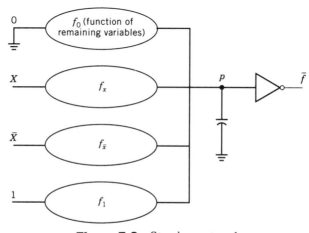

Figure 7.8   Steering network.

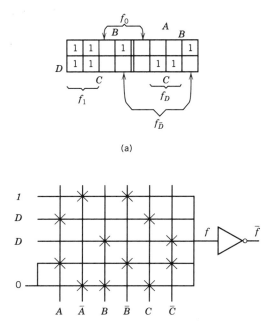

(a)

(b)

Figure 7.9 Steering realization of $\overline{A}\,\overline{B} + B\overline{C}\overline{D} + ACD$.

an n-channel device and $f_D = D = 1$, the voltage at the input to the inverter will be degraded as discussed in Section 4.7. If the device is p-channel, a similar degradation will occur if $f_{\overline{D}} = 1$ and $D = 0$. Once again, these degraded voltage levels are not allowed in CMOS. The result would be the simultaneous steady-state turn-on of both the pull-up and pull-down devices in the inverter. The problem can be avoided by letting C control a CMOS transmission gate (an n-channel device in parallel with a p-channel device).

In Example 7.3, D was designated as the terminal variable without providing evidence that this was the best choice. Selecting the optimal terminal variable for the pass network can only be done by considering each of the possibilities. If a function of a small number of variables is given as a list of minterms, this can be accomplished by mapping the minterms onto n separate Karnaugh maps, where n is the number of variables. Each map will have two rows and 2^{n-1} columns. The n maps are a subset of what will be identified in Chapter 16 as decomposition charts for n-variable functions. The variable whose value will identify a row will be unique in each of the n charts. To make evaluating the charts as fast as possible, the corresponding minterm number may be listed in each square of each of the n standard maps. Note that the significance of the variables in computing minterm numbers is defined by their position in the argument list for the function. This order is independent of the arrangement of variables on the Karnaugh map. The sets of standard two-row decomposition charts for four variables are given in Fig. 7.10. The reader is referred to Section 16.4 for the five-variable charts.

The function $f(A, B, C, D) = \sum m(4, 5, 6, 7, 8, 13, 14, 15)$ is entered on all four maps in Fig. 7.11. Marked in Fig. 7.11a are the functions of B, C, and D controlling the four pass networks to be connected to each of 0, 1, A, and $\overline{A}$. To satisfy Rules 1 and 2 of the previous section, it is important that each column on the map activate a logical path to one of the four terminals.

x_2	0				1			
x_3x_4	00	01	11	10	00	01	11	10
x_1 0	0	1	3	2	4	5	7	6
1	8	9	11	10	12	13	15	14

x_1	0				1			
x_3x_4	00	01	11	10	00	01	11	10
x_2 0	0	1	3	2	8	9	11	10
1	4	5	7	6	12	13	15	14

x_1	0				1			
x_2x_4	00	01	11	10	00	01	11	10
x_3 0	0	1	5	4	8	9	13	12
1	2	3	7	6	10	11	15	14

x_1	0				1			
x_2x_3	00	01	11	10	00	01	11	10
x_4 0	0	2	6	4	8	10	14	12
1	1	3	7	5	9	11	15	13

Figure 7.10 Four Variable (2 row) Decomposition Charts

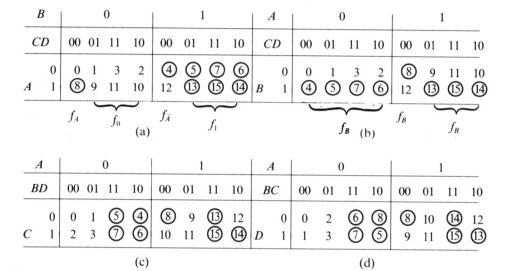

Figure 7.11 Mapping Example 7.3.

■ EXAMPLE 7.4

Identify the variable that when used as a terminal variable will provide for the minimal device count realization of

$$f(A, B, C, D) = \sum m(4, 5, 6, 7, 8, 13, 14, 15)$$

and obtain an abbreviated network diagram of this realization.

SOLUTION The maps of Figs. 7.11a, 7.11c, and 7.11d indicate that networks to each of the four possible terminals will be required if A, C, or D is chosen as the terminal variable. Figure 7.11b implies connections only to B and to $\bar{B}$. The resulting network is given in Fig. 7.12.   ■

7.4 TREE NETWORKS

In this section we shall explore still another approach to pass transistor realizations applicable primarily to NMOS technology. The networks will take the form

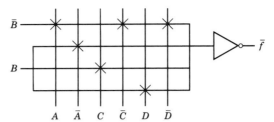

Figure 7.12   Realization of Fig. 7.11*b*.

of trees branching back from the output. The process of generating a tree network realization is essentially one of successively factoring a sum-of-products expression. When the minimal sum-of-products expression is not unique or "don't care" conditions are possible, the final tree network will depend on the starting sum-of-products expression, as well as the selected order of factoring.

Freeform Tree Procedure   Select a variable x_i to factor, then factor x_i from all products in which it appears, and finally factor x_i from the remaining products. If a product P is independent of these literals, first substitute $P = x_i P + \bar{x}_i P$. If possible, select a variable for which this manipulation can be avoided. For each expression ANDed with a factored literal, select a new variable to factor and repeat the process until all variables have been factored or a network realization is obtained. If a pass transistor controlled by either x_i or $\bar{x}_i$ is connected to a branch point, a device controlled by the other literal must always be connected to the same branch point. A branch may be terminated by a connection to 0, 1, or a single literal.

A network constructed according to the freeform tree procedure will satisfy Rules 1 and 2. A path through the tree to one and only one terminal point of the tree will be connected for each combination of values of the input variables. Each terminal will be fixed at 0 or 1 or will assume the 0 or 1 value of the terminal variable.

To begin the discussion of cases, let us consider a unique minimal sum-of-products expression

$$f_1(A, B, C, D) = \bar{A}\bar{B}C + \bar{A}\bar{C}D + AB\bar{C} + AC\bar{D} \tag{7.2}$$

on which we know in advance that factoring will be effective. A or $\bar{A}$ is found in each product so this variable is factored first.

$$f_1(A, B, C, D) = \bar{A}(C\bar{B} + \bar{C}D) + A(\bar{C}B + C\bar{D}) \tag{7.3}$$

The expressions $C\bar{B} + \bar{C}D$ and $\bar{C}B + C\bar{D}$ are already in factored form, if C is the variable selected to be factored. The network of Fig. 7.13 is a tree network repre-

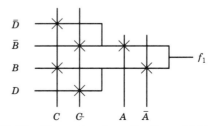

Figure 7.13   Tree network realization of $f_1(A, B, C, D)$.

senting the factoring of Equation 7.3. It is easily verified that the output of this network will be 1 whenever $f_1 = 1$ and 0 whenever $f_1 = 0$.

Figure 7.13 is called a *uniform* tree because the same variable C is factored from the two expressions that are ANDed with A and $\bar{A}$ after the first factoring. A very slight modification of the function defined by Equation 7.2 results in the function given by Equation 7.4, for which a uniform tree is not the minimal device count realization.

$$f_2(A, B, C, D) = \bar{A}\bar{B}C + \bar{A}BD + AB\bar{C} + AC\bar{D} \tag{7.4}$$

We begin by factoring $\bar{A}$ as before

$$f_2(A, B, C, D) = \bar{A}(\bar{B}C + BD) + A(\bar{C}B + C\bar{D}) \tag{7.5}$$

Now we see that one of the products in the expression ANDed with $\bar{A}$ does not include either C or $\bar{C}$. Instead, B and $\bar{B}$ may be factored in that expression, whereas C and $\bar{C}$ are factored, as before, in $(\bar{C}B + C\bar{D})$ to form the tree network of Fig. 7.14a. A tree network, formed without requiring a consistent order of factoring, will be called a *freeform* tree. A uniform tree is a special case of a freeform tree so the freeform tree will never require more devices than a uniform tree. The uniform tree may result in a minimal chip area realization, because for the freeform tree a single pair of metal lines may not be sufficient for some pairs of input literals. Additional area may be required to simply make connections in the freedom tree.

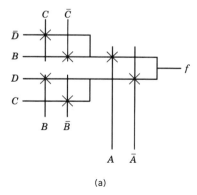

(a)

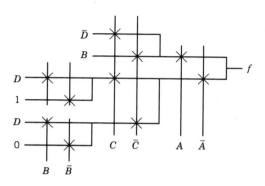

(b)

Figure 7.14 Realization of $f_2(A, B, C, D)$. (*a*) Freeform tree. (*b*) Uniform tree.

■ **EXAMPLE 7.5**

Develop a uniform tree realization logically equivalent to Fig. 7.14*a*.

SOLUTION   We select C and $\bar{C}$ as the second literal to be factored that requires further Boolean manipulation from Equation 7.4.

$$f_2(A,\,B,\,C,\,D) = \bar{A}\,(C\bar{B} + BD) + A(\bar{C}B + C\bar{D})$$
$$= \bar{A}\,[C\bar{B} + BD(C + \bar{C})] + A(\bar{C}B + C\bar{D})$$
$$= \bar{A}\,[C(\bar{B} + BD) + \bar{C}BD)] + A(\bar{C}B + C\bar{D})$$

Now selecting B and $\bar{B}$ as the third pair of literals for factoring results in the network of Fig. 7.14*b*. Notice that the term $\bar{C}BD$ cannot be implemented by

When $\bar{A} = \bar{C} = 1$ and $B = 0$, there would be no connection to the output. The network must be connected according to the completed factoring of $C(BD + \bar{B}0)$, so that there is a closed path for each combination of the values of factored variables.   ■

If a Boolean function is not already available in minimal sum-of-products form, the factoring may be accomplished by repeated application of Shannon's expansion theorem (see Section 5.11). The application of the theorem using K maps will permit an optimal assignment of "don't cares" for a tree network, which need not be the same as the assignment in a minimal sum-of-products expression. Should any combination of values represent a "don't care" condition, the corresponding functional value will necessarily be assigned as 1 or 0 in the process of generating the tree network. The application of Shannon's expansion theorem must continue until each coefficient function is 0, 1, or a single literal. Let us consider one more example.

■ **EXAMPLE 7.6**

Develop a uniform tree realization of the complement of the Boolean function given in Fig. 7.15*a*.

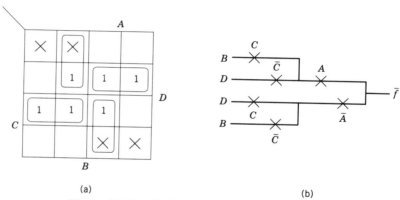

(a)

(b)

Figure 7.15   Uniform tree from Example 7.6.

SOLUTION   The map is similar to that of Example 5.12. Now, however, the 2-cube in the center of the map is included in the minimal sum-of-products expression

$$f = \bar{A}CD + BD + A\bar{C}D \tag{7.6}$$

No order of factoring the expression leads to a simple or elegant tree network. Note, however, that using two of the "don't cares" in the function results in four equal-size cubes (1-cubes) in four quadrants of the map. The enclosed "don't cares" are assigned the value 1 and the other two "don't cares" the value 0. A first application of Shannon's expansion theorem leads to

$$f = Af_A + \bar{A}f_{\bar{A}}$$

where

$$f_A = CB + \bar{C}D \qquad \text{and} \qquad f_{\bar{A}} = CD + \bar{C}B$$

Selecting the variable C and applying Shannon's expansion theorem to f_A and $f_{\bar{A}}$ do not change these expressions, leading to the six-device uniform tree in Fig. 7.15b.   ∎

The tree network of Fig. 7.15b is actually a multiplexer. The control variables are A and C. The technique to be introduced in Section 8.5 for the realization of functions using multiplexer cells is also applicable to uniform tree realizations.

It may appear that functions of almost any complexity can be realized using tree networks. Actually, delay considerations limit the number of pass transistors in a path to two or three. The effective resistance (dependent on β in Equation 4.2) of a pass transistor is several orders of magnitude greater than a metal connecting line or even a polysilicon connection. For purposes of delay calculation, the pass network is modeled by an R-C transmission line network, as shown in Fig. 7.16. The delay through this network increases approximately as the **square** of the number of pairs of R-C elements in series. A mathematical argument supporting this assertion may be found in [1]. Therefore, the delay associated with a long pass transistor network may well be more than that of several gates in series connected by metal lines.

7.5   NEGATIVE GATE REALIZATIONS

In this section we shall explore a method of realization of Boolean functions by replacing the single enhancement device in the pull-down path of a NMOS inverter with a pass transistor network. NOR and NAND gates are special cases. The method to be presented and its extensions, called the negative gate method first introduced by Muroga [2], may also be used in conjunction with $f/\bar{f}$ realizations or in CMOS, if double-rail inputs are made available.

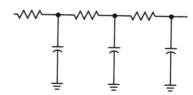

Figure 7.16   Delay model of pass transistor network.

Let Z as given by Equation 7.7 be a realization of a Boolean function f.

$$Z = \overline{f_1 g_1 + f_2 g_2 + ... + f_n g_n} \tag{7.7}$$

Now consider the special case in which each function and $f_2, ..., f_n$ and $\bar{g}_1, ..., \bar{g}_n$ are unate, and in addition, each function is positive and negative in the same set of variables. The vertex in the Boolean hypercube obtained by allowing the variables positive in each f_i to be 1 and the variables negative in each f_i to be 0 will be called the focal vertex. When one of the individual product terms $f_1 g_1, f_2 g_2, ..., f_n g_n$ is 1, the output Z will be 0.

Procedure   Each function f_i will be used to cover 0's while each corresponding $\bar{g}_i$ will be chosen to exclude any 1's that are covered by f_i from the product $f_i g_i$. Cover 0's nearest in the hypercube to the focal vertex first, covering 0's at increasingly greater distance from the focal vertex as i increases. Cover as many 1's as possible in each unate function $\bar{g}_i$, treating 0-vertices already covered by f_j for $j < i$ as "don't cares." Cover as many 0's as possible in each unate function f_i, treating 1-vertices covered by $\bar{g}_i$ as "don't cares." If the focal vertex is a 0-vertex, $\bar{g}_i = 0$ and $g_i = 1$.

It is also possible to realize any Boolean function in the form of Equation 7.6 using single-rail inputs by constraining each of the functions f_i and $\bar{g}_i$ as **positive functions**. Thus, the focal vertex will be the vertex of all 1's. We shall accept this constraint in all the following examples. To identify positive functions, we use once again the result (Problem 7.2) that **all prime implicants in minimal sum-of-products expression for a positive function cover the focal vertex.**

■ EXAMPLE 7.7

Use the negative gate format of Equation 7.6 to obtain a realization of the Boolean function given in Example 5.12.

SOLUTION   The function is repeated in the Karnaugh map of Fig 7.17a. Because the functional value of the focal vertex is 1, we first obtain $\bar{g}_1$ as a positive function covering as many 1's as possible. In this case, $\bar{g}_1 = BD + ABC$, as given in Fig. 7.17a. Next we select f_1 as a positive function covering as many 0's as possible. The 1's covered by $\bar{g}_1$ are now shown as "don't cares" in Fig. 7.17b. These "don't cares" may fall within f_1, but this function may not cover any of the remaining 1's in Fig. 7.17b. The four 0's that can be covered by this function are marked as f_1 in Fig. 7.17b. From the map we obtain the realization $f_1 = AB + BC + AC$.

Now all 0's already covered by $f_1 g_1$ are marked as "don't cares," as we search for a positive $\bar{g}_2$ in Fig. 7.17c to cover as many 1's as possible. We see that $\bar{g}_2 = B + CD + AD$ covers all the 1's. We now let $f_2 = 1$ to complete the realization. The switch-level device diagram of a complete NMOS realization is given in Fig. 7.17d.   ■

In the above example, the functional value of the focal vertex was 1, so that it was necessary to determine a $\bar{g}_1$ before finding f_1. In the following five-variable example, the focal vertex is a 0-vertex.

■ EXAMPLE 7.8

Determine a negative gate realization of the Boolean function given in Fig. 7.18a.

SOLUTION   In Fig. 7.18a we note that $\bar{g}_1 = 0$, and in order that f_1 be positive, it can cover only the focal vertex (i.e., $f_1 = ABCDE$). We next let that vertex be a "don't

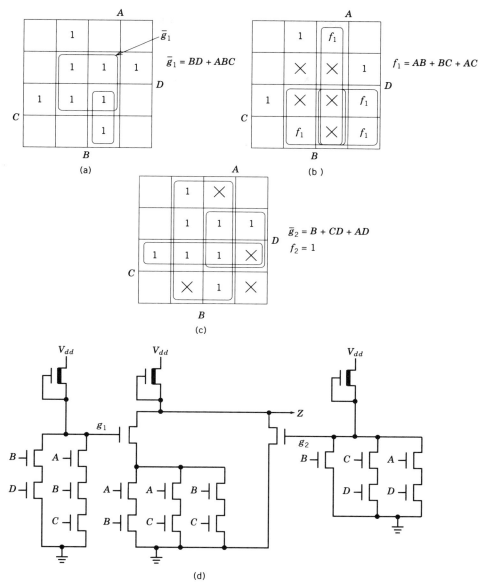

Figure 7.17 Realization of Example 7.7.

care" and determine $\bar{g}_2 = AC + CE + ABE$ in Fig. 7.18b. Letting the vertices covered by $\bar{g}_2$ be "don't cares" in Fig. 7.18c, we see that $f_2 = A + B + DE$ covers all the remaining 0's. ∎

■ EXAMPLE 7.9

In order to provide a reasonable β ratio, the negative gate realization of the function given in Example 7.8 must never have more than three transistors in series in a pull-down network. Modify the realization determined in that example to satisfy this constraint.

SOLUTION We see from Fig. 7.18b that g_2 will have no more than three transistors in a path. Similarly, the product $g_2 f_2$ will generate a pull-down path with no

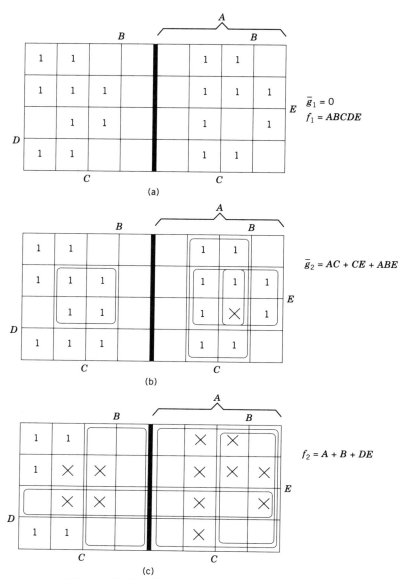

Figure 7.18 Karnaugh maps for Example 7.8.

more than three devices in series. The path corresponding to $f_1 g_1 = ABCDE$ would have five devices in series. We may satisfy the constraint by realizing f_1 as

$$f_1 = \overline{A} + \overline{B} + \overline{C} + \overline{D} + \overline{E} \tag{7.8}$$

The disadvantage of realizing f_1 as given in Equation 7.8 is that double-rail inputs must be made available. ∎

7.6 LOGIC DESIGN WITH CMOS STANDARD CELLS

The NMOS pass network techniques discussed in the previous sections are used most effectively, if the logic designer and layout designer are one and the same. The following two observations form the rationale in a CMOS design philosophy, which differs somewhat from that of NMOS. (1) To apply steering to 0, 1, x_i, and $\bar{x}_i$ or tree networks to CMOS design requires replacing each pass transistor with a two-device (n and p channels in parallel) transmission gate. This handicap has proved sufficient to limit single-inversion CMOS gates to the $f/\bar{f}$ networks, as in Fig. 7.7b. (2) A glance at Fig. 4.26 will convince the reader that the chip layout of a CMOS gate is significantly more complicated than a NMOS layout. In contrast to NMOS, we conclude for these reasons that **CMOS logic design usually consists of the interconnection of standard cell building blocks**, which have been laid out internally by specialists. We begin in this chapter with the standard cell realization of combinational logic. Problems associated with the design of complete systems using standard cells will be considered in Chapter 12.

Figure 7.19 illustrates the standard cell methodology in terms of a particular library [5] of 3-μ ($\lambda = 1.5$ μ) standard cells. As indicated in Fig. 7.19b, the height of most cells in the library is 150 μ (one exception is the 4020 multiplexer that will span two rows). The cell width will vary with the complexity of the circuit implemented in the cell. The cells are aligned in rows although the space between rows is not constant. There is no column alignment of cells, and the space between cells in a row is variable. The spaces between cells accommodate the cell-to-cell interconnection lines deposited in one of two metal layers during the fabrication process. The location of cells and the routing of connections between cells may be accomplished manually or by automatic place and route programs [6], a subject of Chapter 12.

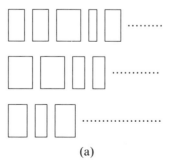

(a)

		Height, width (in μ)
1100	dual inverter	150×48
1130	3-input NOR	150×48
1140	4-input NOR	150×60
1220	2-input NAND	150×36
1230	3-input NAND	150×48
1740	4-input OR	150×72
1910	four × 2-input AND/OR	150×144
1930	two × 3-input AND/OR	150×84
1970	two × 2-input AND/OR	150×84
2310	exclusive-OR	150×72
4010	4 to 1 multiplexer	150×192
4020	8 to 1 multiplexer	363×384

(b)

Figure 7.19 Logic design with CMOS standard cells.

Figure 7.19b is a tabulation of a small subset of the combinational logic cells to be found in the typical library. These will be sufficient for the brief set of examples to follow. No cells with greater fan-in are available. Note that the 1910, 1930, and 1970 are unate functions, which may be realized as illustrated in Example 7.2. If only 150-μ high cells are used, the chip area (exclusive of interconnections) required by a particular CMOS realization of a combinational logic function is proportional to the sum of the widths of the cells used in the realization. The width of each cell is given in Fig. 7.19b.

■ EXAMPLE 7.10

Determine CMOS standard cell realizations of the Boolean function given in Example 5.12 in (a) NAND/NAND form, (b) NOR/NOR form, and (c) based on a 4010, four-input multiplexer cell. Compare the resulting total cell width for the three designs. Assume that dual-rail inputs are available. For convenience the minimal sum-of-products realization is given in Equation 7.9.

$$f = \bar{A}B\bar{C} + \bar{A}CD + A\bar{C}D + ABC \qquad (7.9)$$

SOLUTION   (a) A direct NAND/NAND realization of Equation 7.9 is impossible, because no four-input NAND gate is available in the library to implement the OR operation. (b) From the Karnaugh map of Fig. 7.20, we determine the following product-of-sums realization for Equation 7.9:

$$f = (A + B + C)(A + \bar{C} + D)(\bar{A} + C + D)(\bar{A} + B + \bar{C}) \qquad (7.10)$$

A four-input NOR gate is available. Using this together with four three-input NOR gates results in the following computation for total width:

$$\text{cell width} = 4 \times 48 + 60 = 252 \text{ μ} \qquad (7.11)$$

(c) The 4010 multiplexer is logically equivalent to one half the SN74153 dual multiplexer to be introduced in Section 8.5. With the techniques to be introduced in that section not yet available, we apply Shannon's expansion theorem A and then C, just as was done in Example 7.6. The result is the same expression that led to the NMOS multiplexer circuit of Fig. 7.16b.

$$f = \bar{A}\bar{C}(B) + \bar{A}C(D) + A\bar{C}(D) + AC(B) \qquad (7.12)$$

A and C will be connected to the control inputs of the 4010, with B and D each connected to two of the four multiplexer data inputs. Thus, the multiplexer realization, with a total cell width of 192 μ, uses less chip area than the NOR/NOR realization.   ■

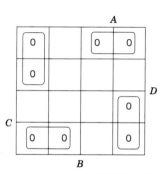

Figure 7.20   Karnaugh map for NOR/NOR realization: Equation 7.10.

■ EXAMPLE 7.11 (**Unnecessary for continuity)

By identifying a decomposition, using a procedure to be developed in Section 16.4, it can be shown that

$$f(x_1, x_2, x_3, x_4, x_5) = \sum m(1, 2, 7, 9, 10, 18, 19, 25, 31) \\ + d(0, 15, 20, 23, 26)$$

may be expressed as

$$f(x_1, x_2, x_3, x_4, x_5) = \bar{x}_3\bar{x}_4\Phi + \bar{x}_3 x_4\bar{\Phi} + x_3\bar{x}_4(0) + x_3 x_4\Phi \qquad (7.13)$$

where

$$\Phi(x_1, x_2, x_5) = x_2 x_5 + \bar{x}_1 x_5$$

Determine the total cell width of a CMOS standard cell realization of this function based on a 4010 multiplexer. Assume dual-rail inputs are available.

SOLUTION   The control inputs to the multiplexer are x_3 and x_4. The function Φ may be realized by using a 1970 two $\times$ two-input AND/OR cell (width 84). This function must be inverted by half an 1100 cell (width 24) to form $\bar{\Phi}$, which is connected to two of the four data inputs of the multiplexer. The remaining data input to the 4010 multiplexer (width = 192) is 0. Therefore, the total width of the three-cell realization is 300 μ. It will be left as a problem for the reader to find the best possible standard cell realization of f without using a multiplexer. It will almost certainly require more chip area than 150×300 μ² excluding interconnection space. A more general treatment of multiplexer design will be found in Section 8.5.   ■

7.7 GATE ARRAYS AND STANDARD CELLS

In the last section we introduced the concept of realization of logic functions utilizing building blocks, more complex than simple gates. We will continue to pursue this topic in Chapter 8 and again in Chapter 12, with the scope of the discussion broadened to include memory elements. This section will explore the distinctions between two technologies available for this purpose: gate arrays and standard cells. As is shown in Fig. 4.23, both provide for the interconnection of complex building blocks by way of two layers of metal interconnections in routing channels between rows of these circuit primitives. For this reason, the CALD tools used in designing with gate array and standard cell building blocks are very similar.

Notice that the standard cells of Fig. 7.21a are of differing widths, as discussed in the previous section. The gate array blocks of Fig. 7.21b are all identical in size and shape. In fact, after the first several steps in manufacturing, they are identical in every respect. Before the final interconnection step, each block in Fig. 7.21b is an array of gates or devices. The figure provides only a suggestion of the resources that might be available within a block. As part of the final step in the fabrication process, these blocks are individually customized by interconnecting the gates into complex logic blocks or sequential memory elements. It is possible to stockpile partially manufactured, or uncustomized, gate array wafers. Because fewer manufacturing steps are unique to a particular design, gate arrays enjoy certain costs advantages for small production quantities of a particular design.

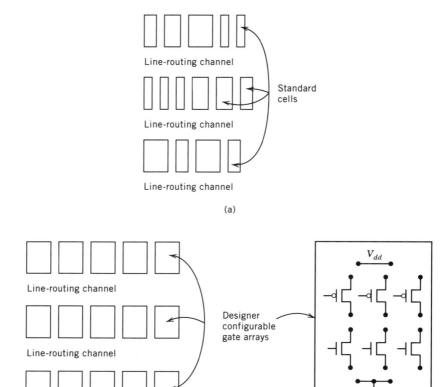

Figure 7.21 Placement of (*a*) cells and (*b*) gate arrays.

Some manufacturers have added a feature to the gate array concept that makes it possible to complete the manufacturing process before customizing a part. One approach includes fusible links between devices in the individual blocks and similar links between horizontal and vertical metal line segments in the routing channels. As is generally true for the field programmable PAL, discussed in Section 6.11, the fuses corresponding to unwanted connections are blown as a programming process. In a second approach, additional gates and memory elements are added to the gate array chips to make them *programmable* by writing connection data on the chip each time the system is initialized. Programmable gate arrays are attractive when only one or very few copies of a design are required. (See section 12.9.)

As suggested in Fig. 7.21*a*, individual standard cell types may be arranged in rows in arbitrary order. What will occupy each tiny increment of chip area in a standard cell design cannot be predicted until the design including place and route is complete. Therefore, all steps of the standard cell manufacturing process must be accomplished following design.

7.8 PRECHARGED CLOCKING OF CMOS COMBINATIONAL LOGIC

We shall learn in the next chapter that most sequential circuits, that is, logic circuits with the capacity to store information, depend on a periodic input signal called a clock to synchronize activity within the circuit. The simplest form of a clock is the single-phase clock shown in Fig. 7.22a. Each positive pulse triggers some type of activity in the sequential circuit. For reasons that will become clear in Chapter 9, the single-phase clock is inadequate for most NMOS and CMOS sequential circuits. In CMOS circuits complementary single-phase clocking, as illustrated in Fig. 7.22b, is the natural result of applying the single-phase clock to both n-channel and p-channel devices in the same circuit. The n-channel devices are on when the clock is 1; the p-channel devices are on when the clock is 0.

Most complete systems are sequential, so a clock is always present. Can this universally available clock be used to advantage in NMOS or CMOS sequential circuits? At the price of almost double the number of devices, CMOS has the advantage over NMOS of significantly lower power consumption. Could the clock be used in NMOS to somehow eliminate intervals where power is consumed, because both pull-up and pull-down networks conduct simultaneously? Could the clock be used in CMOS to avoid the penalty of almost doubling the number of devices (over NMOS) in each gate network? The answer to both questions is yes. In this section we shall focus on clocked CMOS logic.

Figure 7.23a depicts a generalized CMOS gate network realized according to Theorem 7.1. In Fig. 7.23b we replace the dual p-channel pull-up network with a clocked p-channel gate and add a clocked n-channel gate in series with the pull-down network. As will be shown in the next section, the clocked gate must be connected directly to ground to avoid charge-sharing. While the clock in Fig. 7.23b is 0, the p-channel gate will be closed, connecting the output to V_{dd}. The capacitor connected to the gate output is *precharged* to V_{dd}, and the output remains at this value, unless the output is connected to ground while *clock* = 1. If the input values are such that the output z is 0, a connected path through the pull-down network and, therefore, a path to ground will exist. (To avoid charge-sharing, a phenomenon to be discussed in Section 9.9, the clock must control the device in the pull-down network nearest the ground point.)

Figure 7.24a reduces the clocked CMOS gate network to an inverter, and Fig. 7.24b depicts the corresponding input/output waveforms. The half-cycles in which the output is precharged and the half-cycles where the output is valid are denoted P and V on the output waveform. The fact that the outputs of precharged CMOS domino logic gates are valid only during the last half of each clock cycle tells us that special care must be used in interconnecting these gates to form combinational logic networks and connecting network outputs to sequential circuits.

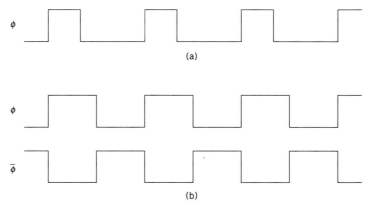

Figure 7.22 Clocking MOS. (a) Single-phase clock. (b) Complementary single-phase clock.

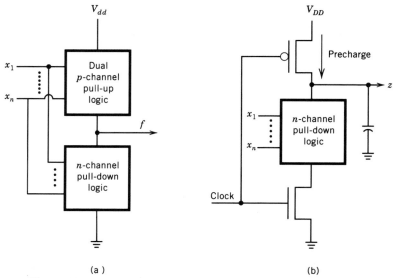

(a) (b)

Figure 7.23 Translation to precharged CMOS domino logic.

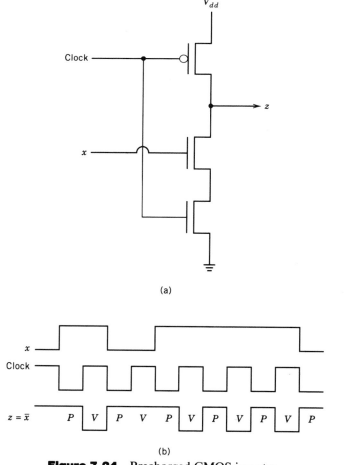

(a)

(b)

Figure 7.24 Precharged CMOS inverter.

We label networks of precharged CMOS gates as *domino networks,* because during each clock cycle all gate outputs are first charged to logic 1 like a row of dominoes. Then as the clock goes to 1, gates fall to 0 in order from input to output.

To illustrate a problem inherent to this type of network, consider the CMOS domino logic realization of $f = \bar{A}\bar{B}C + ABC$ given in Fig. 7.25. Consider the case where $A = B = C = 0$, for which the output of the first gate g will be pulled to 0 during the second half of the clock cycle. The inputs to the three devices in parallel in the rightmost gate network will also be 1. During the precharge cycle, the first gate output g will be 1. Thus, "device g" in the second gate network will have its input charged to V_{dd} volts. The input of device g will begin to discharge when the clock goes to 1 for the second half-cycle. There will be an interval at the beginning of the half-clock cycle during which a closed path will exist from output point f to ground. As shown in Fig. 7.25b, f will also begin to discharge to 0, although its final logical value should be 1. The value of f at the end of the clock cycle will depend on analog circuit parameters. As pointed out in Fig. 7.25b, f may be considered indeterminate.

The potential for the problem illustrated above will exist whenever the output of one clocked CMOS controls a device in the network of a succeeding CMOS gate. The precharge logic 1 value from the first gate will turn on any device it controls and hold it on momentarily after the clock goes to 1, although the output of that first gate will eventually discharge to 0.

The remedy is to insist that the output of each clocked gate be buffered by an unclocked inverter, as illustrated in Fig. 7.26. Now any device controlled by the

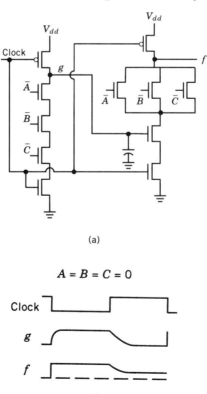

(a)

$$A = B = C = 0$$

(b)

Figure 7.25   Flawed precharged logic realization of $\bar{A}\bar{B}C + ABC$.

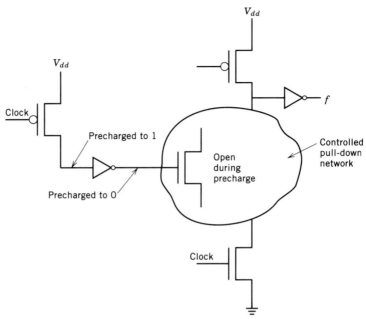

Figure 7.26 Using interstage inverters to eliminate domino logic indeterminate value.

output of this inverter y will not be turned on until y has reached its final value. A closed path will not exist in the controlled pull-down network until controlling logic variables have reached their final values. Thus, an unwanted discharge of line $\bar{f}$ will not happen. Using clocked domino logic with an unclocked inverter at the output of each gate requires that the logic expression to be realized be manipulated accordingly and essentially excludes negative gate realizations.

■ **EXAMPLE 7.12**

Because of the fan-in limitation, more than one gate will be required to realize the five-variable AND operation *ABCDE*. Obtain a CMOS domino logic realization of this function.

SOLUTION The resulting network is given in Fig. 7.27. Note that the required inverter opens an opportunity to AND two more variables into the product. ■

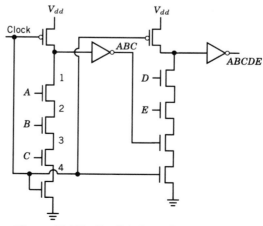

Figure 7.27 Realization of Example 7.12.

7.9   PRECHARGED CMOS PLA

The NMOS implementation of a PLA given in Fig. 6.25 used NOR gates to obtain both the AND and OR plane levels of the realization. The AND gates are realized as

$$x_1, x_2, ..., x_n = \overline{\overline{x}_1 + \overline{x}_2 + ... \overline{x}_n}$$

and the OR gates by complementing the output of NOR gates. In terms of chip area and speed, the double NOR gate alternative is the ideal.

In the previous sections we discussed CMOS alternatives to the simple NMOS pull-up to eliminate steady-state power consumption. Pull-up alternatives for implementing a PLA on a CMOS chip include (1) dual pull-up and pull-down networks, (2) precharged domino logic, and (3) replacing the depletion mode pull-up with a p-channel gate that is always turned on. The last approach, termed pseudo-NMOS, lacks the advantage of no steady-state power consumption. The dual network approach destroys the layout efficiency of the PLA, in addition to imposing the speed penalty of a series path of devices controlled by the appropriate literals in each prime implicant.

Implementing an NMOS-like NOR-NOR realization in precharged CMOS would result in the unacceptable behavior illustrated in Fig. 7.25. A CMOS domino logic realization of a PLA is given in Fig. 7.28. Each first-level AND gate is realized by a NAND, followed by the glitch-eliminating inverter. It is impossible to avoid devices in series in the NAND pull-down network, the speed disadvantage and the lack of field programmability not withstanding.

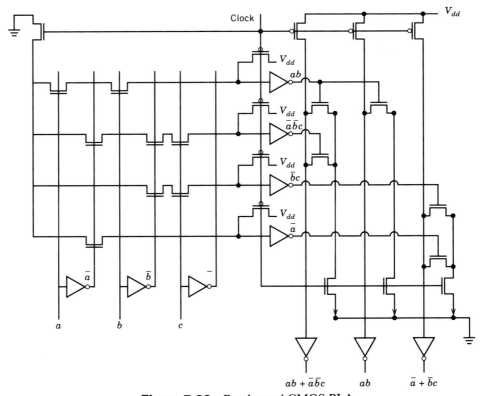

Figure 7.28   Precharged CMOS PLA.

PROBLEMS

7.1 Which of the following Boolean functions are positive? Which are unate?

(a) $\bar{A}\bar{B} + \bar{A}CD + BCD$

(b) $\bar{A}\bar{B} + \bar{A}CD + ABCD$

(c) $a \oplus b$

(d) $CD + BD + AB\bar{D}$

7.2 Prove that a minimal sum-of-products expression for a positive function is positive in all variables and that all prime implicants in the expression cover the vertex for which the values of all variables are 1.

7.3 Construct a modified stick diagram of a NMOS $f/\bar{f}$ realization of the function f given in the Karnaugh map of Fig. 7.4b. For f, use the sum-of-products expression $f = \bar{A}\bar{B} + BC\bar{D} + ACD$. For $\bar{f}$, use the expression for $\bar{f}$ determined from the dual of f as provided by Theorem 5.5. Note that metal lines may cross diffusion or polysilicon, but that polysilicon cannot cross diffusion (an unwanted transistor would result). Contacts (symbolized by a black dot) may be made between any of the three material types as needed. Contacts require space. Waste as little area as possible within the rectangle enclosing your design.

7.4 Obtain a pass network connection diagram of a NMOS realization of the function defined by Problem 5.12 using $f/\bar{f}$ design. Compare the total number of devices (depletion and enhancement) with that of a two-level NOR-NOR realization. Cite all advantages and disadvantages of the physical realizations of these two designs.

7.5 Realize using NMOS the function of Problem 7.4 by steering to $x, \bar{x}, 0$, and/or 1 and compare the total number of devices with the method $f/\bar{f}$ realization. Construct only the pass network connection diagram. Assume absolute "don't cares."

7.6 (a) Construct a stick diagram of the CMOS version of the network given in Fig. 7.6. Use no output inverter. Are dual-rail inputs required?

(b) Using transmission gates, construct the stick diagram of a CMOS version of the network given in Fig. 7.9b. Indicate any factors that might complicate the physical realization of this circuit.

7.7 Realize the Boolean function given in Example 5.12 by (a) steering to $x, \bar{x}$, 0, and/or 1, (b) a uniform tree, and (c) a freeform tree. Count the total number of devices used by each method. Is there any reason why the method resulting in the fewest devices might not result in the best physical realization? If so, explain.

7.8 Construct the pass network connection diagram of a freeform tree network realization of the Boolean function given below. The tree network will connect to the input of an inverter. The inverter input will be f. The inverter output will be $\bar{f}$.

$$f(a, b, c, d) = \sum m(0, 5, 7, 9, 11, 12) + d(1, 2, 4, 8, 13, 14)$$

Use as few devices in the tree network as possible.

7.9 (a) Obtain a negative gate realization of the function

$$f(a, b, c, d) = m_0 + m_{11} + m_{13} + m_{14} + m_{15}$$

Obtain only expressions for each unate function used in the realization.

(b) Obtain a pass network connection diagram for a steering to $x, \bar{x}, 0,$ and 1 realizations of this function and compare the total numbers of devices with the results of part (a). Weight a depletion transistor as equal to two enhancement devices.

7.10 Realize the function of Problem 7.3 above as a freeform tree. Compare the total number of devices with the realization considered in Problem 7.3.

7.11 Obtain a switch-level diagram of an optimal negative gate NMOS realization of the Boolean function defined by Fig. 7.4*b*. Use the two "don't cares" to the best advantage.

7.12 Construct a minimal device NMOS pass network connection diagram of a realization in which $0, 1, d,$ and $\bar{d}$ are steered to the output through paths controlled by the variables $a, b,$ and $c,$ for the Boolean function defined by the minterm list given below. Assume that the pass network is the input to an inverter and that $\bar{f}$ is to appear at the output of the inverter. The "don't cares" are application don't cares.

$$f(a, b, c, d) = \sum m(1, 2, 4, 5, 10, 11) + d(0, 9, 14)$$

7.13 The negative gate realization defined below must be modified to satisfy a constraint that no more than three devices may be connected in series in a pull-down network. Construct a switch-level diagram of a negative gate network satisfying this constraint, while adding as few devices as possible to the original realization. Assume double rail inputs are available.

$$z = \overline{g_1 (ABC + ABD + ACD + BCD) + g_2}$$
$$\bar{g}_1 = ABCD, \qquad \bar{g}_2 = A + B + C + D$$

7.14 Obtain a switch diagram of a CMOS realization of the negative gate expressions given in Problem 7.13. There is no constraint on devices in series. Double-rail inputs are available.

7.15 The simple NMOS pass network into a CMOS inverter given in Fig. P7.15 might seem to pass for a suitable CMOS realization of the function $AC + \bar{A}\bar{B}$. Would this circuit satisfy the CMOS assertion that no power is consumed in the steady state? Why or why not? Will the output voltage be degraded from the nominal values?

7.16 Construct a stick diagram of a CMOS $f/\bar{f}$ realization (f is to appear at the output of an inverter) of

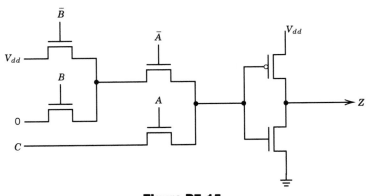

Figure P7.15

$$f = A + BCD$$

Assume single-rail inputs. Use the conventions of Figs. 7.1 and 7.2.

7.17 Use any of the cells tabulated in Fig. 7.20*b* to obtain a CMOS standard cell realization of the negative gate expression determined in Example 7.7. Compute the total cell width for your realization.

7.18 Construct a switch-level device diagram of the realization of the 4 to 1 multiplexer internal to the 4010 CMOS standard cell.

7.19 Determine the best CMOS standard cell realization of Equation 7.13 that you can design without using a 4010 multiplexer. Compute the combined cell width and the chip area required, excluding interconnection space.

7.20 Assuming single-rail inputs and basing the design on the focal vertex (1111), determine each of the Boolean expressions g_i and f_i needed to form a multilevel negative gate NMOS realization of the Boolean function

$$f(a, b, c, d) = \sum m(0, 3, 5, 6, 9, 10, 12, 15)$$

Combine these expressions into the single negative gate expression for f.

BIBLIOGRAPHY

1. Mead, C. and L. Conway. *Introduction to VLSI Systems*. Addison-Wesley, Reading, Mass., 1980.
2. Muroga, S. *VLSI System Design*. Wiley, New York, 1982.
3. Mukherjee, A. *Introduction to nMOS & CMOS VLSI Systems Design*. Prentice-Hall, Englewood Cliffs, N.J., 1986.
4. Dillinger, T. E. *VLSI Engineering*. Prentice-Hall, Englewood Cliffs, N.J., 1988.
5. Heinbuch, D. V. *The CMOS3 Cell Library*. Addison-Wesley, Reading, Mass., 1988.
6. Pucknell, D. A. and K. Eshraghian. *Basic VLSI Design*. Prentice-Hall, Englewood Cliffs, N.J., 1988.

CHAPTER 8

MULTILEVEL LOGIC USING COMPLEX (MSI) PARTS AND CELLS

8.1 THE PLACE FOR COMPLEX PARTS AND CELLS

In Chapters 5 and 6 and for the most part in Chapter 7, we concentrated on generating optimal two-level (sum-of-products or product-of-sums) realization of Boolean functions. There are many important combinational logic functions for which multilevel representations arise naturally and for sum-of-products or product-of-sums realizations would be prohibitively inefficient. One example is the parity check circuit of Section 2.6. In the three editions of *Introduction to Switching Theory and Logical Design*, we were content to introduce the most important instances of multilevel realizations and assume that similar designs would simply be accomplished by "engineering extrapolation." The notion of optimization (with respect to cost and delay) of multilevel logic was ignored. In this chapter we will introduce the standard structures, while not ignoring the more general logic optimization problem. We will not attempt the impossible, that is, to provide a comprehensive treatment of multilevel network manipulation.

Once we venture away from the restriction of sum-of-products and product-of-sums expressions, the possibility of connecting primitive elements, more complex than AND, OR, NAND, and NOR gates, into combinational logic networks arises immediately. In 1980 these more complex primitive elements were MSI parts to be mounted and interconnected with packages of gates on a printed circuit board. Today these elements include standard cells and gate array (including fusible) primitives to be interconnected by two layers of metallization on a VLSI chip. Design approaches for these two cases are similar but not identical.

8.2 DECODERS

One of the simplest and most useful multilevel and multiple-output logic structures is the decoder. Let us begin our discussion with a three-input eight-output example.

Assume we have a small memory array of eight locations. The address of the location to be accessed is a three-bit binary number. To access a location, we need to generate a signal of one of eight lines, corresponding to the addressed location. A 3-to-8 line decoder consists simply of eight AND gates realizing the eight minterms of three variables (Fig. 8.1a). For each possible combination of the inputs, there will be a 1 on just one of the output lines, and we assume that the 1 on the output line will make it possible to access the desired word. Circuits of this type are known as *n-to-2^n line decoders*, or more commonly, simply *decoders*.

Although decoders can be constructed of discrete gates, they are generally realized in the form of MSI circuits. There are a number of different types of decoders available in MSI form, including 3-to-8 line and 4-to-16 line. There are also decoders in which the number of output lines is not 2^n, such as the SN7442A BCD-to-decimal decoder, shown in Fig. 8.1*b*. Again, the circuit simply realizes minterms, but now only 10 of them, corresponding to the BCD codes for the decimal digits. As is usually the case in MSI decoders, NAND gates are used instead of AND gates, so that the polarity of the outputs is reversed, all outputs being at the 1 level except the selected output, which is at the 0 level.

Also note another peculiarity characteristic of MSI decoders—that the inputs are labeled A, B, C, D, but are taken in the opposite order to that used in this book, A being least significant and D most significant in determining the BCD code. Finally, note the two levels of inversion. The extra inversion provides power amplification to drive the NAND gates, so that a circuit driving the decoder sees the inputs as single-gate loads.

In theory, the networks of Fig. 8.1 could be extended directly to any number n inputs using 2^n gates with n inputs each. In practice, a fan-in limitation will eventually be reached in any technology. Thus, some type of multilevel network must be employed in large decoders. It is impossible to make an unqualified statement as to the most economical decoder design, because cost factors vary with different technologies. If we assume that decoders of the form of Fig. 8.1*a* are available, the

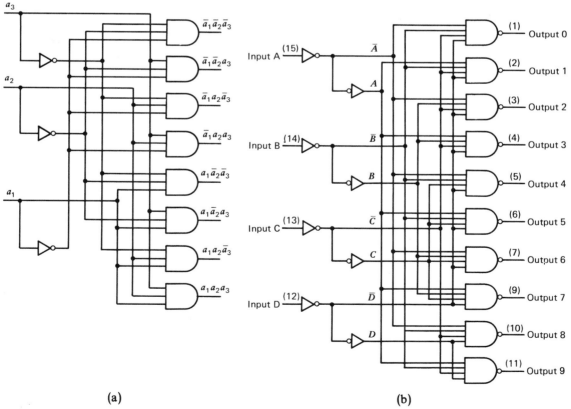

(a) (b)

Figure 8.1 Typical decoders.

following procedure will generally lead to the most economical n-input decoder. Given m, the number of inputs of the largest integrated circuit decoder available, determine the largest integer r satisfying Equation 8.1.

$$r \le \frac{n}{k} \le m, \qquad k = \text{a power of 2} \tag{8.1}$$

If $r = n/k$, we use k r-input decoders to form the minterms of disjoint subsets of r variables. These minterms will then be combined in as many levels of two-input AND gates as required. For example, suppose $n = 12$ and $m = 5$; then the largest r satisfying Equation 8.1 is given by

$$r = 3 = \tfrac{12}{4} \le 5 \tag{8.2}$$

So we use four three-input decoders, as shown in Fig. 8.2.

To keep the diagram readable, only a few connections are actually shown. The three-bit decoders may be considered copies of the circuit of Fig. 8.1a in integrated circuit form. There are 64 pairs of output lines, one from each of the upper two three-bit decoders. These pairs form the inputs to the upper 64 second-level AND gates. The outputs of these gates are the 64 possible minterms of the variables $a_1, a_2, a_3, a_4, a_5, a_6$. The lower 64 second-level gate outputs are the minterms of $a_7, a_8, a_9, a_{10}, a_{11}$, and a_{12}. The 2^{12} 12-bit minterms are formed by using all possible pairs of outputs of the second-level gates (one from the upper 64 and one from the lower 64) to form inputs to the final 2^{12} AND gates.

Assuming the fan-in level permits, we might decide that it would be better to combine the outputs of all four decoders in four-input AND gates, thus eliminating the second level of gates. Let us assume the cost is proportional to the number

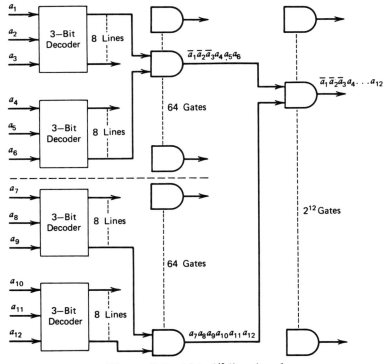

Figure 8.2 12-to-2^{12} line decoders.

of gate inputs. The number of gate inputs, exclusive of the three-bit decoders, is given by Equation 8.3. Clearly, the dominant cost is the output gates. No matter what the form of the decoder, the cost will always be dominated by the 2^n output gates, so these should have the minimum number of inputs: two.

$$\text{number of inputs} = \underbrace{2 \cdot 2 \cdot 64}_{\text{2nd Level}} + \underbrace{2 \cdot 2^{12}}_{\text{Output Gates}}$$

$$= 2^8 + 2^{13} \approx 2^{13} \tag{8.3}$$

Although fan-in problems limit the size of decoders of the form of Fig. 8.1, fan-out limits can also be a problem in very large decoders. In Fig. 8.2, for example, each of the 128 second-level gates provides an input to 64 of the output gates. Fan-out problems can be relieved by the use of a tree decoder, in which new input variables are introduced at each level of the tree. Figure 8.3 shows a tree decoder for three variables. There are still fan-out problems in that the variables introduced at the later levels must drive many gates. However, it is easier to provide heavy drive capability for a few input variables than for $2 \times 2^{n/2}$ gates at the second level in a circuit of the form of Fig. 8.2. Tree decoders, having more levels than the form of Fig. 8.2, are slower and contain almost twice as many gates, but for very large decoders they may be the only choice. Obviously, various combinations of the two basic forms could be devised.

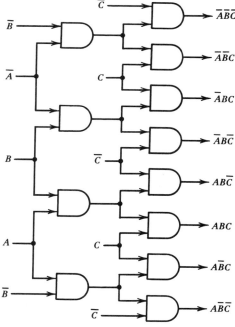

Figure 8.3   Tree decoder.

8.3   READ-ONLY MEMORY AS A LOGIC ELEMENT

Sometimes decoders are packaged independently, but usually they are incorporated into memory packages or memory arrays, as address decoders. A memory whose contents are fixed, that is, a memory for which the **write** operation does not apply, is called a *read-only memory* (ROM). ROM's may be incorporated into computers as control memory or included in the computers' primary memory address space. A ROM may also be used to realize combinational logic functions. In this application, each of the 2^n outputs of a decoder may be regarded as the realization of a minterm or 0-cube of the n input variables.

■ EXAMPLE 8.1

Consider the problem of converting from a BCD code to a seven-segment code, as discussed in Chapter 2. The corresponding codes are tabulated in Fig. 8.4. We require a circuit simultaneously producing seven functions of four inputs. We could apply the formal multiple-output techniques of Chapter 6, but with seven functions to be realized, it is obvious that this would be rather tedious. (In all 127 maps of functions and function products would be required.)

An easier design approach is found by noting that a full decoder produces all minterms of the input variables and that any function can be obtained by an ORing of minterms. In this case, since we are using BCD code, the decimal digits in Fig. 8.4 are the minterm numbers. So we can construct our code converter from a partial decoder, producing the 10 required minterms as shown in Fig. 8.5. To complete the design, we note that the table of Fig. 8.4 effectively lists the minterms making up each output function. For example,

$$B_6 = \sum m(2, 3, 4, 5, 6, 8, 9) \tag{8.4}$$

and

$$B_0 = \sum m(0, 2, 6, 8) \tag{8.5}$$

so we OR these minterms together as shown. The OR gates developing the other output functions would be similarly connected, although we have not shown the connections in the interest of clarity. ■

It can be seen that we have here the basis of a completely general design technique, not just for code converters but for logic circuits of any type. The resulting designs will seldom be even close to minimal in the classic sense of Chapter 6, but the simplicity of the design technique makes it quite attractive.

Dec.	BCD Input				Seven-Segment Output						
Digit	A_3	A_2	A_1	A_0	B_6	B_5	B_4	B_3	B_2	B_1	B_0
0	0	0	0	0	0	1	1	1	1	1	1
1	0	0	0	1	0	0	1	1	0	0	0
2	0	0	1	0	1	1	0	1	1	0	1
3	0	0	1	1	1	1	1	1	1	0	0
4	0	1	0	0	1	0	1	1	0	1	0
5	0	1	0	1	1	1	1	0	1	1	0
6	0	1	1	0	1	1	1	0	0	1	1
7	0	1	1	1	0	0	1	1	1	0	0
8	1	0	0	0	1	1	1	1	1	1	1
9	1	0	0	1	1	0	1	1	1	1	0

Figure 8.4 BCD to seven-segment code.

The larger the number of output functions, the greater the savings in design effort and the closer to minimal the circuits will be.

Figure 8.5 shows only the minterms that are connected to some output function. In practice, all minterms are realized by the decoder. The fact that the remaining six minterms are don't cares in this case would offer no advantage in an actual ROM realization.

There are many types of ROM's, but the type best suited to applications requiring logic-level outputs as a function of level inputs is the transistor-coupled, or semiconductor, ROM. A transistor-coupled ROM is shown in Fig. 8.6. The lines making up the interconnection matrix are known as *word lines* and *bit lines*. The word lines are driven by an n-to-2^n decoder, and the outputs are taken from the bit lines. Wherever a connection is to be made between a word line and bit line, a transistor is connected, emitter to the bit line and base to the word line. All collectors are tied to a common supply voltage.

In the absence of any input, all word lines are held at a sufficiently negative level to cut off the transistors so that the bit lines are at 0 V. When an input appears, the corresponding word line is raised to a sufficiently positive level to turn on the transistors, thus raising the connected bit lines to a positive level. For example, in Fig. 8.6, if the input were 100 (m_4), word line 4 would be turned on and

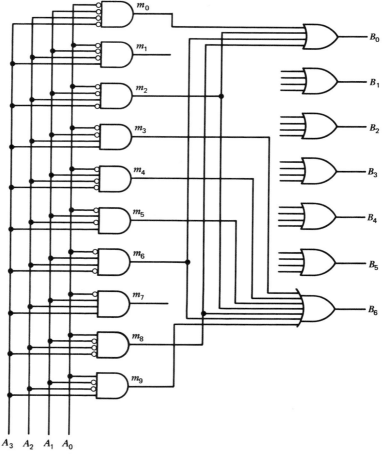

Figure 8.5   BCD to seven-segment converter.

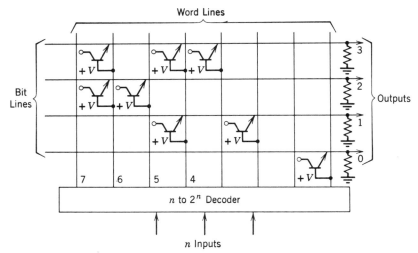

Figure 8.6 Transistor-coupled ROM.

the output would be $b_3 b_2 b_1 b_0 = 1010$. Comparing this to Fig. 8.5, we see that the decoder is the array of AND gates, and the bit lines with their associated transistors make up the array of OR gates. (Please note that we have described here only the basic principles involved; the precise details of the technology vary widely.)

ROM's come in a variety of sizes and in two basic types: *mask programmable* and *field programmable*. For the mask programmable type, the customer indicates, on a standard form provided by the manufacturer, the desired connection matrix. The manufacturer uses this data to create a mask used in the metallization phase of fabrication to connect transistors only where required. As with the PAL, the field programmable type, the ROM is provided to the customer with connections at every intersection and the customer programs the ROM, using current pulses literally to "blow out" the undesired connections. Because of this process, the connections are generally referred to as *links* or *fuses*. The choice between this field programmable ROM and the erasable ROM, or EPROM, is mostly a matter of cost. In terms of direct chip cost, the field programmable ROM will be cheaper than the mask programmable ROM since the manufacturer has to provide only one standard type, but the user must add the cost of programming each unit. For the mask programmable type, the initial cost of the mask is significant, but this is a one-time cost that will be cheaper than the cost of individual programming if spread over enough units. EPROM's are expensive and are generally used only in development labs where they are convenient for experimenting with new designs.

Recalling Sections 6.6 and 6.11, we now have three separate structures suitable for the realization of multiple-output combinational logic functions: ROM's, PLA's, and PAL's. Which is the proper choice for a particular application? There is less overlap in the application of these three options than might be expected. In the ROM the second level of logic in the sum-of-products form is programmable and the first level fixed. In the PAL the first level is programmable and the second level fixed. Because connections to both the AND and OR gate in the PLA depend on the functions being realized, this structure is seldom considered field programmable.

When realized on a custom VLSI chip, the PAL is a special case of a PLA. A PAL realization of a set of functions will never require less chip area than a

PLA realization of the same functions. Therefore, when a multiple-output design is to be incorporated on a chip, the choice is between a ROM and a PLA. **The designer will resort to the ROM when the number of functions is large and the individual functions are complex, thereby overwhelming available logic optimization software needed for PLA realization.**

For a separately packaged combinational logic realization, **a PAL will be chosen over a ROM, if the total number of variables is very large but the individual outputs are simple functions of small subsets of these variables.** For small-volume or experimental systems, the ROM always offers the advantage of the least design time.

8.4   BINARY ADDER

Chapters 5 and 6 dealt primarily with two-level realizations of Boolean functions. Such realizations will always be the fastest possible circuits and, for typical functions of a few variables, will have the minimal cost or a cost very close to minimal. There are, however, some very important functions that do not lend themselves to two-level realizations.

Let us consider the design of a circuit to accomplish the addition of two binary numbers. The addition of two n-bit binary numbers is depicted in Fig. 8.7. A two-level realization of this process would require a circuit with $n + 1$ outputs and $2n$ inputs. In a typical computer, n might be 32. Thus, an attempt at two-level design would seem a staggering problem, even for a computer implementation of Espresso.

Fortunately, a more natural approach to the problem is possible. Each pair of input digits X_i and Y_i may be treated alike. In each case, the carry C_i from the previous position is added to X_i and Y_i to form the sum bit S_i and the carry C_{i+1} to the next higher-order position. If exactly one or three of the bits X_i, Y_i, and C_i is 1, then $S_i = 1$. If not, then $S_i = 0$. The carry C_{i+1} will be 1 if two or more of X_i, Y_i, and C_i are 1. The carry C_i into the first digit position is available as an input at the same time X and Y become available. The addition process may thus be implemented one digit at a time, starting with the least significant digit. As soon as X, Y, and C_1 are available, C_2 may be generated, in turn making possible the generating of C_3, etc.

A circuit that accepts one bit of each operand and an input carry and produces a sum bit and an output carry is known as a *full adder*. From the above discussion, we can set down the equations and K-maps for the full adder as shown in Fig. 8.8. Two-level realizations of these two functions require a total of nine gates with 25 inputs, plus one inverter to generate $\bar{C}_i$. A NAND circuit for this form is shown in Fig. 8.9.

Since the full adder is of basic importance in digital computers, a great deal of effort has gone into the problem of producing the most economical realization.

$$S_{n+1} = C_{n+1} \quad \begin{array}{ccccc} C_n & & C_3 & C_2 & \\ X_n & & X_3 & X_2 & X_1 \\ Y_n & \cdots & Y_3 & Y_2 & Y_1 \\ S_n & S_{n-1} & S_3 & S_2 & S_1 \end{array}$$

Figure 8.7

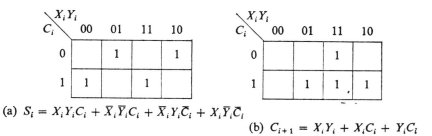

(a) $S_i = X_i Y_i C_i + \bar{X}_i \bar{Y}_i C_i + \bar{X}_i Y_i \bar{C}_i + X_i \bar{Y}_i \bar{C}_i$

(b) $C_{i+1} = X_i Y_i + X_i C_i + Y_i C_i$

Figure 8.8 Sum and carry functions for the binary full adder.

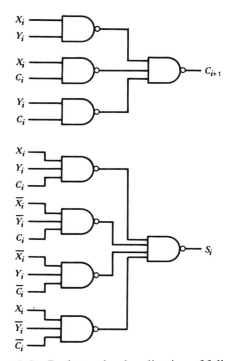

Figure 8.9 Basic two-level realization of full adder.

The form leading to best economy is a function of the technology used. Today complete full adders are normally realized as integrated circuits, so the forms of interest are those most suitable for realization in MSI form. One interesting form can be seen on the K-maps as shown in Fig. 8.10. We generate C_{i+1} as before and use it in the generation of S_i. Note that the map of $X_i + Y_i + C_i$ in Fig. 8.10a agrees with S_i in five of the eight squares. In Fig. 8.10b, we see the intersection of this map and a map of $\bar{C}_{i+1}$ (complement of Fig. 8.8b). It is only necessary to include one more term, $X_i Y_i C_i$, to form the sum

$$S_i = \bar{C}_{i+1}(X_i + Y_i + C_i) + X_i Y_i C_i \qquad (8.6)$$

as indicated in Fig. 8.10c. A realization of Equation 8.5 is shown in Fig. 8.11a, requiring eight gates with 19 inputs.

Another popular form of full adder utilizes two *half-adders*. A half-adder is a circuit for adding two binary bits. The K-maps and circuit for a half-adder are shown in Figs. 8.12a and 8.12b. To construct a full adder, we note that the equations for a full adder (Fig. 8.8) can be written in the alternate forms

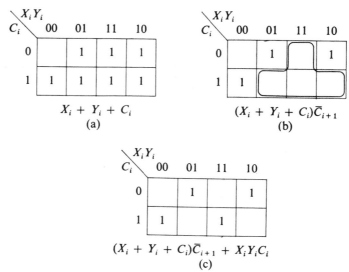

Figure 8.10   K maps for alternate realization of sum-bit in full adder.

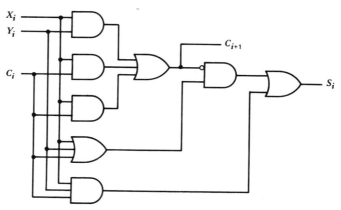

Figure 8.11   Circuit for realization of Fig. 8.10.

$$S_i = X_i \oplus Y_i \oplus C_i \qquad (8.7a)$$
$$C_{i+1} = (X_i \oplus Y_i)C_i + X_iY_i \qquad (8.7b)$$

These equations can be implemented with two half-adders plus an OR gate, as shown in Fig. 8.12c. Note one advantage of this circuit: It does not require any inverted inputs. A CMOS realization of this same adder, utilizing a transmission gate, is shown in Fig. 8.12d. It is likely that the transmission gate version would have less delay from C_i to C_{i+1} than any other CMOS realization.

The delay in the carry logic of a full adder is more critical than the delay associated with generating the sum bit. In the worst case a carry generated in the least significant bit might propagate through the carry logic in each successively more significant bit until it determines the final carry-out from the adder. This may be visualized in the context of the n-bit adder network depicted in Fig. 8.13. The adder network of Fig. 8.13 is an *iterative network*. Iterative networks exhibit properties analogous to those of sequential circuits. A complete discussion of iterative networks will be deferred to Section 16.6 to allow us to take advantage of this analogy.

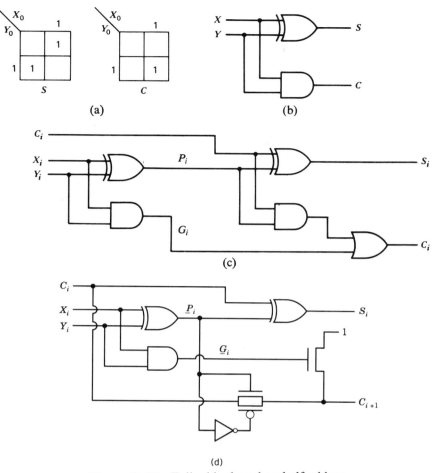

(a)

(b)

(c)

(d)

Figure 8.12 Full adder based on half-adders.

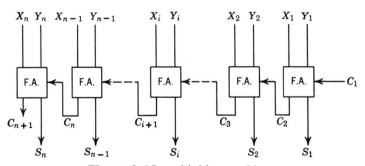

Figure 8.13 n-bit binary adder.

None of the full adder stages considered above ask the carry to propagate through more than two levels of logic. Increasing adder speed beyond that obtainable in the ripple carry adder of Fig. 8.13 will hinge on further reduction of the number of logic levels through which carry bits must propagate. The notion that a carry either propagates through an adder stage or is generated within that stage is clear in Figs. 8.12c and 8.12d. Two lines in each network are marked P_i and G_i. If the propagate line

$$P_i = X_i \oplus Y_i = 1$$

the input carry will merely propagate through stage i. Adder stage i will internally generate a carry, if $G_i = X_i Y_i = 1$. Two approaches have been used to reduce the number of stages through which a carry must propagate: *carry lookahead* [1,6] and *carry skip*. The most basic two-bit carry skip network in Fig. 8.14 ANDs the propagate lines from two successive stages. The line P_i that will identify a full adder stage as a propagate stage can be treated as an output of that stage. If $P_i = P_{i+1} = 1$, the carry may simply propagate around rather than through the two stages. C_i is connected through the multiplexer to the two-stage carry output, if $P_i P_{i+1} = 1$. Otherwise, the normally generated carry C_{i+2} is connected to output C_{i+2}^*. If $P_i P_{i+1} = 1$, the carry propagates through only two (the multiplexer) rather than four levels of logic. A variety of lengths and combinations of carry skips are possible.

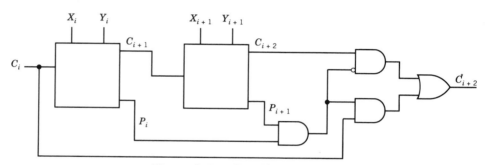

Figure 8.14   Two-bit carry skip.

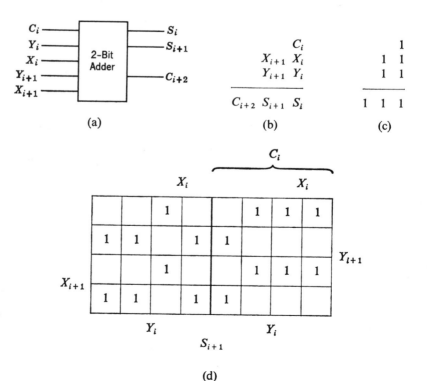

(a)

$$
\begin{array}{ccc}
 & C_i & \\
X_{i+1} & X_i & \\
Y_{i+1} & Y_i & \\
\hline
C_{i+2} & S_{i+1} & S_i
\end{array}
$$

(b)

$$
\begin{array}{ccc}
 & & 1 \\
 & 1 & 1 \\
 & 1 & 1 \\
\hline
1 & 1 & 1
\end{array}
$$

(c)

(d)

Figure 8.15   Two-binary-bit adder.

As a means of increasing speed, we note that it is not necessary to consider only one digit at a time. Consider the possibility of generating two sum-bits in a step, as illustrated in Fig. 8.15. The intermediate carry C_{i+1} is not actually generated but is implicit in the logic of the adder. The process is similar to the addition of three two-bit binary numbers, except that one of those numbers may not exceed 1. Thus, the sum will not exceed 7_{10}, as indicated in Fig. 8.15c. The function S_i may be generated as before, and C_{i+2} will be 1 whenever the sum is 4_{10} or more. Therefore,

$$C_{i+2} = X_{i+1}Y_{i+1} + (X_{i+1} + Y_{i+1})(C_i X_i + C_i Y_i + X_i Y_i) \tag{8.8}$$

The bit S_{i+1} will be 1 whenever the sum is 2, 3, 6, or 7_{10}, leading to the map in Fig. 8.15d. From this map, we determine the minimal second-order form for S_{i+1}, Equation 8.8. Thus, 13 gates are required for a two-level realization of S_{i+1}, compared to five gates for the two-level realization of S_i.

$$\begin{aligned}
S_{i+1} &= \bar{X}_{i+1}Y_{i+1}\bar{X}_i\bar{C}_i + X_{i+1}\bar{Y}_{i+1}\bar{X}_i\bar{C}_i + \bar{X}_{i+1}\bar{Y}_{i+1}X_iY_i \\
&+ X_{i+1}Y_{i+1}X_iY_i + \bar{X}_{i+1}Y_{i+1}\bar{Y}_i\bar{C}_i + X_{i+1}\bar{Y}_{i+1}\bar{Y}_i\bar{C}_i \\
&+ \bar{X}_{i+1}Y_{i+1}\bar{X}_i\bar{Y}_i + X_{i+1}\bar{Y}_{i+1}\bar{X}_i\bar{Y}_i + \bar{X}_{i+1}\bar{Y}_{i+1}X_iC_i \\
&+ X_{i+1}Y_{i+1}X_iC_i + \bar{X}_{i+1}\bar{Y}_{i+1}Y_iC_i + X_{i+1}Y_{i+1}Y_iC_i \tag{8.9}
\end{aligned}$$

For groups of three bits, a similar increase in functional complexity could be expected, etc. Thus, as conjectured earlier, the cost of a two-level realization of the entire n-bit addition would be prohibitive.

8.5 DESIGN WITH MULTIPLEXERS

A multiplexer is applicable when it is necessary to concentrate multiple sources of data on a single communications line. We shall see in subsequent chapters that each individual bit of a data bus may be realized as a multiplexer. Here we encounter the multiplexer as our first instance of a primitive, more complex than an individual gate, for use in realizing combinational logic. Figure 8.16 depicts the SN74153, a TTL dual four-input multiplexer. The 74153 is an MSI part, but the same configuration, or at least half of it, could appear as a CMOS standard cell or gate array primitive.

■ EXAMPLE 8.2

Implement the following function, using one-half of a 74153 multiplexer:

$$f(X, Y, Z) = \sum m(3, 4, 6, 7)$$

SOLUTION We start by drawing the Karnaugh map of the function (Fig. 8.17). We then connect X to the B address line and Y to the A address line on the multiplexer. With that connection, each column of the Karnaugh map corresponds to one AND gate in the multiplexer. For example, if $B, A = X, Y = 0, 0$, the top AND gate will be selected and the output will be determined by input 1C0. For this set of values for X and Y, the function is to be 0, so we connect logic 0 to input 1C0. For X, $Y = 0, 1$, the second gate down is selected and the output is determined by 1C1. For this set of values of X and Y, we see that the output is equal to Z, so we connect Z to input 1C1. In a similar manner, for $XY = 10$, the output is equal to $\bar{Z}$, so we connect $\bar{Z}$ to input 1C2, and we connect logic 1 to input IC2 to provide the outputs required for $XY = 11$. The complete circuit is shown in Fig. 8.18, with the strobe tied low so that the output will follow the address and data inputs. ■

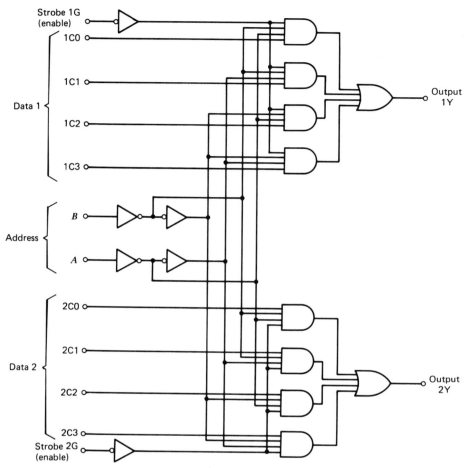

Figure 8.16 SN74153 dual four-input multiplexer.

Z \ X, Y	00	01	11	10
0	0	0	1	1
1	0	1	1	0

Figure 8.17 Map for Example 8.3.

It can be seen from the above example that any function of three variables can be realized with a four-input multiplexer. In the example we have all four possible combinations of 1 and 0 in the K-map columns, so this form can be used no matter where 1 and 0 appear on the map. In the same manner, any four-variable function can be realized with an eight-input multiplexer, and any five-variable function can be realized with a 16-input multiplexer. For an n-variable function realized with a multiplexer, $n - 1$ variables are connected to the address lines, and the data lines are driven by the remaining variable or its complement, or 0 or 1.

The above rules provide for the realization of *any* three-variable function with a four-input multiplexer, *any* four-variable function with an eight-input multiplexer, etc., but sometimes larger functions can be realized with smaller multiplexers and a small amount of external logic.

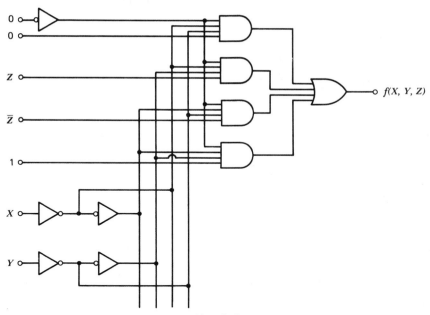

Figure 8.18 Circuit for Example 8.2.

■ EXAMPLE 8.3

Realize the function below, using a four-input multiplexer.

$$F(x_1, x_2, x_3, x_4) = \sum m(4, 5, 6, 7, 8, 13, 14, 15)$$

SOLUTION   The procedure described above would require an eight-input multiplexer, but let us see if we can do it with half of a SN74153. We connect x_3 to B and x_4 to A, and draw the map of the function with these two variables across the top, as shown in Fig. 8.19. Assume we are going to use the top half of the SN74153, as shown in Fig. 8.16. The first column of the map corresponds to $B, A = 0, 0$, for which the top AND gate is selected and the output is controlled by 1C0. From the map, we see that the output should be 1 in this column if $x_1, x_2 = 0, 1$, or if $x_1, x_2 = 1$, 0, so that these 1's will be realized by connecting 1C0 to $x_1 \oplus x_2$. The second column represents $B, A = 0, 1$, for which 1C1 controls the output. In this column the output is to be 1 if $x_2 = 1$, so we connect 1C1 to x_2. In a similar manner, we see that 1C2 and 1C3 should be connected to x_2. The complete circuit, utilizing one four-input multiplexer and one exclusive-OR gate, is shown in Fig. 8.20.   ■

$B\ A$ x_3x_4 x_1x_2	1C0 00	1C1 01	1C3 11	1C2 10
00				
01	1	1	1	1
11		1	1	1
10	1			

Figure 8.19 Map of example function.

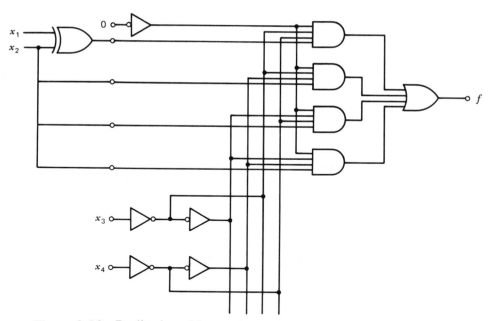

Figure 8.20 Realization of four-variable function with four-input multiplexer.

An obvious question in connection with the above design is why we chose to use x_3 and x_4 to drive the address lines. Why not x_1 and x_2, or some other combination? If we chose some other combination, would the resulting realization be more or less complex? As it turns out, the only way to find out which variables should be connected to the address lines to achieve the simplest circuit is to try all possibilities. This is another application of decomposition charts (see Chapter 16), already used in Section 7.3. Figure 8.21 shows the three two-variable by two-variable decomposition chart forms. Although the variables are rearranged on the three maps of Fig. 8.21, in every case the minterm numbers correspond to the ordering of the variables as x_1, x_2, x_3, x_4.

To use the charts, we circle the minterms of the function on all three charts, as shown in Fig. 8.22. We see that the first chart (Fig. 8.22a) corresponds to the K-map of Fig. 8.19. With B, A connected to x_3, x_4, the columns of this chart correspond to , 1C0, 1C1, 1C3, and 1C2, in that order from left to right. Each circled minterm corresponds to a 1 of the function and we read off the functions to be connected to the data lines just as we did from the map. If we connect B, A to $x_1 x_2$, the *rows* of this same chart correspond to 1C0, 1C1, 1C3, 1C2, in that order from top to bottom. From this we read off the following functions to be connected to the data lines if $x_1 x_2$ are connected to the address lines:

$$1C0 = 0, \qquad 1C1 = 1, \qquad 1C3 = x_3 + x_4, \qquad 1C2 = \bar{x}_3 \cdot \bar{x}_4$$

$x_3 x_4$	00	01	11	10
00	0	1	3	2
01	4	5	7	6
$x_1 x_2$ 11	12	13	15	14
10	8	9	11	10

$x_2 x_4$	00	01	11	10
00	0	1	5	4
01	2	3	7	6
$x_1 x_3$ 11	10	11	15	14
10	8	9	13	12

$x_2 x_3$	00	01	11	10
00	0	2	6	4
01	1	3	7	5
$x_1 x_4$ 11	9	11	15	13
10	8	10	14	12

Figure 8.21 Decomposition chart forms for four-variable functions.

$x_3 x_4$	00	01	11	10
00	0	1	3	2
$x_1 x_2$ 01	④	⑤	⑦	⑥
11	12	⑬	⑮	⑭
10	⑧	9	11	10

(a)

$x_2 x_4$	00	01	11	10
00	0	1	⑤	④
$x_1 x_3$ 01	2	3	⑦	⑥
11	10	11	⑮	⑭
10	⑧	9	⑬	12

(b)

$x_2 x_3$	00	01	11	10
00	0	2	⑥	④
$x_1 x_4$ 01	1	3	⑦	⑤
11	9	11	⑮	⑬
10	⑧	10	⑭	12

(c)

Figure 8.22 Decomposition charts for example function.

Moving to the second chart (Fig. 8.22b), if we connect B, A to x_2, x_4, the columns specify the data functions; if we connect B, A to $x_1 x_3$, the rows specify the data functions. For $B, A = x_2 x_4$, we read off the following data functions:

$$1C0 = x_1 \bar{x}_3, \qquad 1C1 = 0, \qquad 1C3 = 1, \qquad 1C2 = \bar{x}_1 + x_3$$

For $B, A = x_1 x_3$, we read off the following data functions:

$$1C0 = x_2, \qquad 1C1 = x_2, \qquad 1C3 = x_2, \qquad 1C2 = \overline{x_2 \oplus x_4}$$

On the third map (Fig. 8.22c), the pattern of circled minterms is the same as on the second, so the form of the data functions will be the same. We can see that our original choice of $B, A = x_3 x_4$ does lead to the simplest circuit.

8.6   MORE THAN TWO-LEVEL REALIZATIONS WITH BASIC PRIMITIVES

In this section we shall extend our point of view to consider arbitrary logic networks, but still use only AND, OR, NAND, NOR, and exclusive-OR gates as primitives. As a first example, consider the two-level expression given by Equation 8.10 and Fig. 8.23a.

$$f = A\bar{C}D + AC\bar{D} + B\bar{C}D + BC\bar{D} \tag{8.10}$$

A Boolean algebraic factoring operation (application of the distributive law) results in Equation 8.11. The corresponding realization is given in Fig. 8.23b. A second factoring to Equation 8.12 identifies an exclusive-OR network as given in the realization of Fig. 8.23c.

$$f = (A + B)\bar{C}D + (A + B)C\bar{D} \tag{8.11}$$
$$f = (A + B)(\bar{C}D + C\bar{D}) = (A + B)(C \oplus D) \tag{8.12}$$

Figure 8.23b includes one fewer gate and six fewer gate inputs (wiring or metallization connections in a standard cell or gate array realization) than Fig.

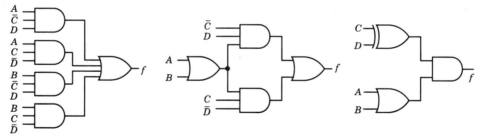

Figure 8.23 Improving logic realizations by factoring.

8.23a. If a standard cell or gate-array exclusive-OR primitive is available, as is likely, then Fig. 8.23c is much better by both criteria.

The use of exclusive-OR gates can often result in significant reductions in circuit cost. Sometimes, as in this case, exclusive-OR factors may be recognized in the course of algebraic manipulation. Sometimes the K-map of a function may help the designer to find such factors.

In some technologies certain of the basic logic primitives may be more economical to realize (for example less chip area). For TTL SSI, the NAND gate offered advantages over both AND and NOR. Because NAND is a universal primitive, it became advantageous to realize networks using only NAND gates. For two-level networks the method was presented in Chapter 5. In the following paragraphs we shall consider the process of replacing a multilevel logic network by an equivalent all-NAND realization. We will leave it to the reader to adapt the procedure, if some other primitive or pair of primitives is preferred.

The technique calls for partitioning the circuit into sections that have the AND-OR form, converting these to NAND-NAND form, and then reconnecting the sections, adding inversions where necessary. When the circuit does not divide conveniently into AND-OR sections, we use DeMorgan's law

$$\overline{A \cdot B} = \bar{A} + \bar{B}$$

which, in terms of gate realization, tells us that a NAND gate is equivalent to an OR gate with inversion at the inputs.

As a first example of AND-OR to NAND conversion, consider the circuit of Fig. 8.23b. In Fig. 8.24a we show this circuit divided into two sections: an AND-OR section and an isolated OR gate. Wherever we cut lines in dividing the circuit, we assign symbols to those lines to keep track of them for future interconnection. In Figs. 8.24b and 8.24c we convert the AND-OR section to NAND-NAND and replace the OR gate with a NAND with complemented inputs.

The procedure illustrated above can be extended to circuits of any complexity. Break the circuit up into sections small enough that the conversion to NAND

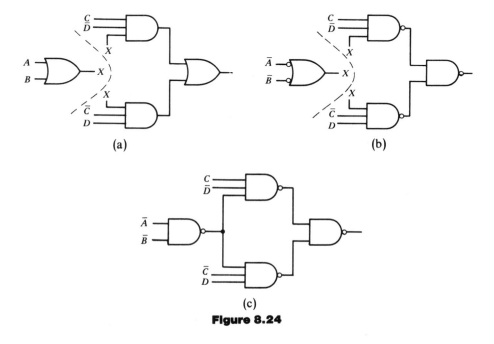

(a)

(b)

(c)

Figure 8.24

form is obvious, keep careful track of the identity of signals between sections, add inversions where required, and reconnect.

■ **EXAMPLE 8.4**

Convert the circuit shown in Fig. 8.25 to all-NAND form.

SOLUTION In Fig. 8.26a we divide the circuit into three sections: two AND-OR sections and an isolated AND. Note that there are more connections between sections than in previous examples and that every line that has been cut is identified. In Fig. 8.26b we show the initial conversion to NAND form. In this case, there are inputs to the OR gates in the AND-OR sections other than directly from the AND gates. We therefore initially convert these gates to the OR-with-inversion form, complementing the direct inputs to compensate. When we convert the upper-left AND to NAND, we delete the input inverter and complement the signal. The outputs of the AND's in the leftmost AND-OR group go to other parts of the circuit, so we identify these outputs as complements of the signals developed by the corresponding AND gates. At this point, we see the importance of keeping track of the identity of the intermediate signals. We know exactly what signals are required at the inputs of the various gates and we know what signals are available at the outputs of gates. For example, we need $\bar{X}$ at the input of the upper-right NAND gate and $\bar{X}$ is available from the upper-left NAND gate, so we can make a direct connection here. By contrast, we need Y at the input of the lower-right NAND gate but have $\bar{Y}$ from the lower-left NAND, so an inverter will be needed here. In a similar manner, we see that two other inverters are needed in connecting the final circuit, Fig. 8.26c. ■

The process of replacing an existing realization by an all-NAND network is not difficult to implement as a CAD tool once a network list data structure has been defined. We shall defer the consideration of alternative netlist formats until Section 12.2. This is not a disadvantage. Even if a netlist notation were available now, we would not wish to consider the details of an all-NAND algorithm at that level. **While all-NAND algorithms and implementations may differ slightly, one would expect only minor differences in the resulting networks in terms of complexity and delay.** The imposition of fan-in limits on individual NAND gates will complicate the procedure somewhat.

8.7 COMBINATIONAL MSI PARTS AND CELLS

There are certain complex functions that occur so frequently in design problems that it was found worthwhile to manufacture them as MSI parts and more recently to provide for their realization as complex standard cells. Such circuits

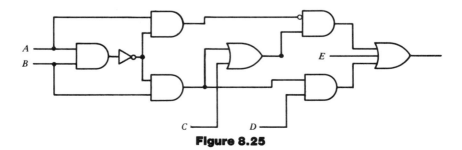

Figure 8.25

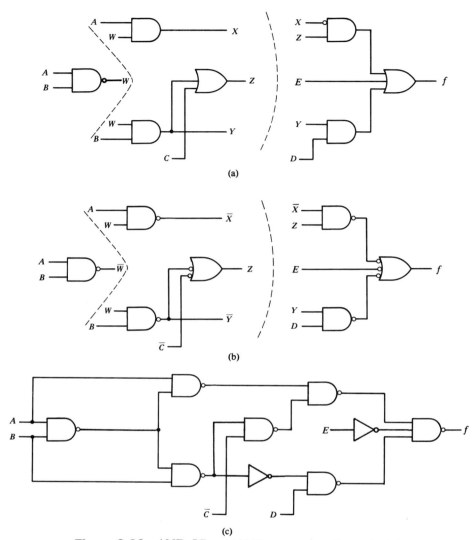

Figure 8.26 AND-OR to NAND conversion, Example 8.4.

are typically composed of 10–100 gates. In a logic diagram of a standard cell realization, it is often desirable to represent each distinct cell with a single symbol. If each symbol could unambiguously communicate the definition of the network it represents, then a more compact and manageable system diagram is obtained without loss of information. This suggests the value of a standard language of graphic symbols for combinational logic cells and parts. Unfortunately, we are not able to claim that all vendors and users of digital logic and all CAD tool developers have accepted a single language. The remainder of this section will be used to illustrate one block diagram language called dependency notation. Once envisioned as an IEEE standard for MSI parts, this language does have the advantage of the independence of a particular CAD tool, a MSI chip, or ASIC vendors. Individual vendor languages will share some of the features of dependency notation.

Decoders often include an enable line that allows the decoder to be turned on or off. Figure 8.27a shows a 2-to-4-line decoder with an active-low enable input E

and inverted outputs. Also illustrated in Fig. 8.27*b* is the dependency decoder symbol with input weight notation for matching each combination of input values to an active output line. The number associated with an input is the logarithm of the weights in base 2. Thus, the weight of B is 2^1, whereas the weight for A is 2^0. This notation is valid only when accompanied by the brackets and range of input sums in the form

$$\frac{0}{3}$$

Because of the inverter in the enable line, this is an active-low enable, that is, the input must go low (to 0) to enable the decoder. If $E = 1$, a logic 0 will be applied to all the AND gates, driving all the outputs to 0. If $E = 0$, the decoder will function in the normal manner, with one output low, all the others high. The notation EN indicates an enable input, the small triangle indicates that it is active-low ($E = 0$ enables the decoder). The outputs are also active-low, as denoted by the triangles at the output in Fig. 8.27*b*. Figure 8.27*c* is the symbol for a decoder with active-high outputs.

Before showing the symbols for multiplexers, we must first introduce a basic concept used in block diagram notation: AND dependency. A label G followed by a number (e.g., $G2$) indicates a gating input, one that is ANDed with any input or output labeled with the same number. Consider the symbol in Fig. 8.28. Whatever the function of this device may be, input $G1$ is ANDed with input 1 and input $G2$ is ANDed with output 2. Thus, input 1 will have no effect on the operation of the circuit unless $G1$ is active, and output 2 will be inactive unless $G2$ is active. By contrast, input 0 cannot be disabled since there is no $G0$ input. Gating

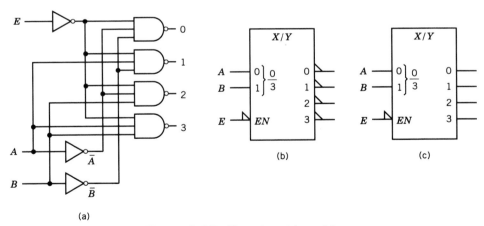

Figure 8.27 Decoder with enable.

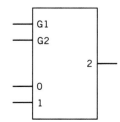

Figure 8.28 Gate dependency symbol.

inputs function in a manner similar to enable inputs, except that they affect only specified lines and can affect both inputs and outputs.

The block diagram symbol for a single multiplexer is shown in Fig. 8.29a. The label MUX identifies this device as a multiplexer. The symbol

$$G\,\frac{0}{3}$$

is shorthand for

$$G0, G1, G2, G3$$

The active gating signal is determined by the inputs enclosed in brackets, with weights equal to the powers of 2 indicated—in this case, 2^0 and 2^1. For example, if A and B are both active, the total weight is $2^1 + 2^0 = 2 + 1 = 3$, so $G3$ is active, in turn enabling data input 3. If only A is active, $2^0 = 1$, so $G1$ is active, enabling data input 1. The use of gating inputs does not preclude the use of enable inputs. Figure 8.29b shows the symbol for a multiplexer with an active-low enable.

We learned in Section 8.5 that more than one multiplexer may be packaged as a single MSI part. Figure 8.30 shows the dependency symbol for the 74153 dual 4-

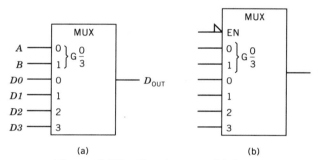

(a) (b)

Figure 8.29 Four-input multiplexer.

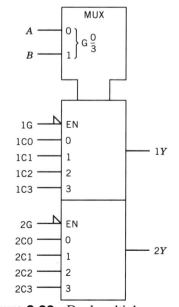

Figure 8.30 Dual multiplexer symbol.

to-1 multiplexer of Fig. 8.16. The control inputs A and B control both sections of the multiplexer, as indicated in the block diagram symbol by the "indented control box" at the top of the symbol. Inputs to a control block affect (or control) all lower sections uniformly.

Very often, multiplexers are used to realize system buses, in which case, the outputs may be vectors of four, eight, 16, or more bits. In Fig. 8.31a we see a detailed line-by-line symbol for a quad two-line to one-line multiplexer. This representation is tractable, but suppose there were eight or 16 outputs. In Fig. 8.31b we see a vector representation of Fig. 8.31a. This notation is defined to indicate that bits $A[0]$ and $B[0]$ are inputs to the first bit of the multiplexer with $C[0]$ as the output, etc., with $A[3]$ and $B[3]$ as inputs to the last bit. This type of notation will be useful whenever logic networks are merely replicated.

Occasionally, individual bits of vectors are treated repetitively by part of a logic network, but in distinct ways elsewhere in the same overall network. In these cases, it may be necessary to fork a single line representing a vector into two lines representing two subvectors, as illustrated in Fig. 8.32a. The notation for convergence of two vectors or a vector and scalar to form a single vector including all bits is illustrated in Fig. 8.32b. As shown, labels indicating the ordering and destination of individual bits must appear with each fork or convergence. Notice the distinctive curved lines of the fork in Fig. 8.32a, indicating division of the vector into separate components, as distinguished from Fig. 8.32c, which represents the fan-out of the same vector to different destinations.

The vector notation may be extended to symbols for other frequently encountered parts and cells. Figure 8.33 shows vector symbols for a four-bit magnitude comparator, a nine-bit odd/even parity checker, and a four-bit adder. Although

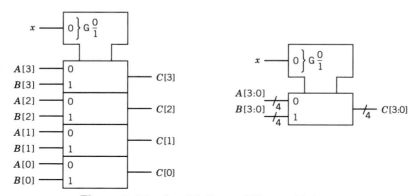

Figure 8.31 Quad 2-line to 1-line multiplexer.

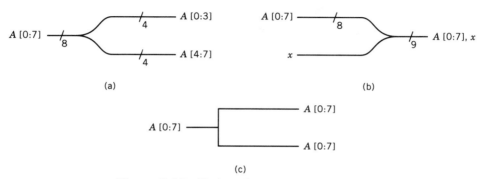

Figure 8.32 Fork and convergence notation.

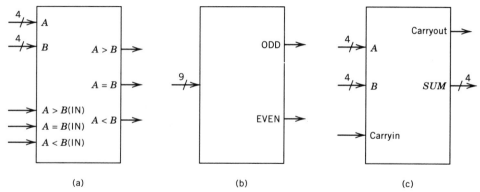

Figure 8.33   Vector symbols (*a*) Four-bit magnitude comparator. (*b*) Nine-bit odd/even parity generator/checker. (*c*) Four-bit binary adder.

vendors' data books will usually be much more complete, cell (or part) definitions like those given in Table 8.1 are usually sufficient to define the logical properties of such cells (or parts).

Table 8.1   Part or Cell Descriptions

I.   Four-Bit Magnitude Comparator: This device compares the magnitude of two four-bit binary numbers A_0, A_1, A_2, A_3 and B_0, B_1, B_2, B_3. If the first is greater, the output labeled $A > B$ will be 1. If the second is greater, the output labeled $A < B$ will be 1. If the two numbers are equal, the outputs will have the same values as the similarly labeled inputs. These inputs may reflect the comparison of less significant bits in the overall comparison of two many-bit vectors.

II.   Nine-Bit Odd/Even Parity Generator/Checker: This device has two outputs: labeled ODD and EVEN. ODD is 1 if there is odd parity over the nine input bits. EVEN is 1 if parity over these bits is even.

III. Four-Bit Binary Adder: This part handles four bits of a binary addition. A carry input is provided that passes the carry from the less significant bits. The outputs are four sum-bits and a carry to the next more significant bits. Some special logic circuitry not discussed in the previous chapter is included so that there are only three gate delays between the carry input and carry output. The most significant input bits are $A4$ and $B4$.

8.8   MULTILEVEL LOGIC MANIPULATION AND OPTIMIZATION

The issues relevant to computer-aided logic design with a standard cell are very similar to those that surface with gate arrays as the design medium. For reasons of style, we shall use standard cell terminology in the subsequent discussion. We leave it to the reader to extend the conclusions to gate arrays.

ASIC design may be divided into the three activities depicted in Fig. 8.34. Block I is engineering, blocks II and III represent the two steps of computer-aided logic design. Block II is the topic of this section. Once a final logic network has been determined, the individual cells must be positioned in rows in the standard cell array; interconnections must be established with two layers of metal in the space between the rows. This activity, called place and route in block III, is quite effectively implemented as a software routine.

The conclusion of Section 8.6, raised the issue of software implementation of the process of modifying an existing multilevel logic network to generate an

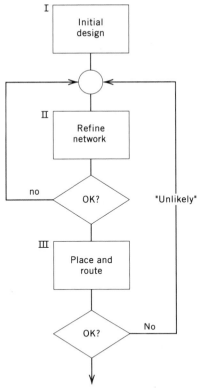

Figure 8.34 Interpretation of ASIC design.

equivalent all-NAND network. Finding the best all-NAND multilevel network realization of a combinational logic function is a special case of the overall multilevel logic network optimization problem, symbolized by block II above. Before a process to improve a multilevel logic network can be invoked, it must be armed with the following information:

A. Set of allowed primitives cells
B. Criterion for comparing the goodness of two realizations
C. Termination criterion

If the allowed primitives were NAND gates and inverters (one-input NAND's), the goodness criterion could be expressed simply as a gate input count. As expressed in Section 8.6, the process of NAND conversion accepted the existing network structure and directly converted sections of that network to NAND-NAND form. The economy of the final network would vary little with choices made during the conversion process. Therefore, the emergence of any all-NAND network would be a suitable termination criterion. If the set of allowed primitives is richer, criteria B and C are not so easily formulated.

A few relatively complex combinational logic primitives were introduced in Section 8.6. A sampling of simpler primitives is given in Fig. 8.35.

■ EXAMPLE 8.5
_____

The eight-bit odd-parity checker of Fig. 2.8 is to be realized as part of a standard cell ASIC. The three-bit parity checker of Fig. 8.35d is available as a standard cell, with cell width equal to 1.5 times that of an EXOR gate.

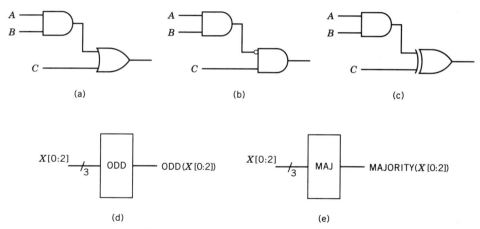

Figure 8.35 Sample cell primitives.

(a) Substitute as many three-bit odd-parity cells for series pairs of EXOR gates as possible.

(b) Determine the minimal cell area realization allowing EXOR gates and three-bit cells.

SOLUTION   An algorithm designed to search the network to insert three-bit odd-parity cells for EXOR gate pairs might identify those pairs of EXOR gates enclosed by dashed lines in Fig. 8.36a. Substitution of a three-bit cell for each of these pairs results in Fig. 8.36b, which requires a cell area equivalent to only six EXOR gates.

A smarter program, one aware of the general parity problem, or perhaps an engineer might devise the logically equivalent circuit of Fig. 8.36c. This network would occupy the cell area of 5.5 EXOR cells.   ■

The circuit with the least cell area in the above example also included one fewer internal connection than the next best alternative. This is an extra advantage when the place and route routine of block III of Fig. 8.34 is applied. There is always some chance that a single connection might make the difference between success and failure to route all the connections of a complex system in the available routing channels. We must conclude that the "goodness" criterion B reflects the number of internal connections, as well as total cell area.

The outer loop in the flowchart of Fig. 8.34 suggests place and route might be completed before the network refinement process terminates. The word "unlikely" has been marked on that loop, because place and route will normally be attempted only once, after the final network has been determined. One major reason for this approach is that a large digital system will usually be partitioned into disjoint sections for logic design and network refinement, whereas the place and route of an entire chip will be a single activity. Termination of network refinement must be based on the record of that process and prediction or estimation forthcoming of place and route.

The Espresso-algorithm for two level networks is also a special case of network refinement, block II. Recall that Espresso does not guarantee a globally optimal two-level network, although this is the usual result even for large circuits. The formulation of criteria for the termination of Espresso was a manageable problem. Generalizing to multilevel logic and allowing more complex primitives complicate that problem, perhaps by several orders of magnitude.

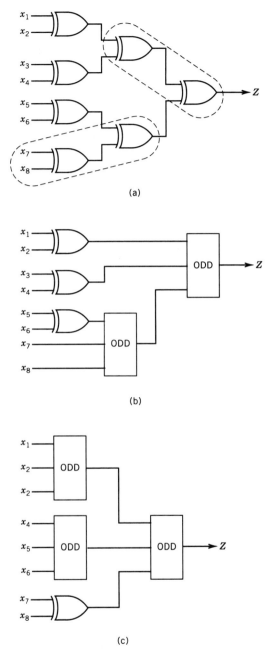

Figure 8.36 Reducing the cell area of a parity-check network.

■ EXAMPLE 8.6

Substitute as many as possible of the primitives of Fig. 8.35 for pairs of gates in the logic network of Fig. 8.25.

SOLUTION A network refinement program using a library of primitives including those of Fig. 8.35 would identify the gate pairs enclosed by dashed lines in Fig. 8.37a as potential candidates for replacement. Figure 8.37b shows the network after the substitution of two three-input primitives (Figs. 8.35a and 8.35b). The

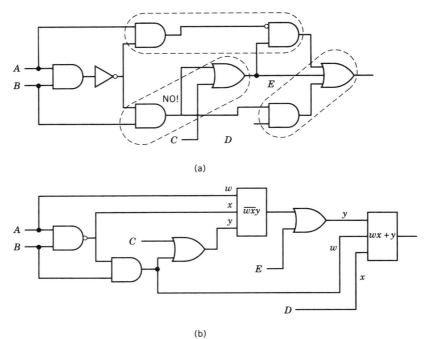

(a)

(b)

Figure 8.37   Substitution of three-input primitives.

three-input OR gate was first replaced by two two-input OR gates in series. The AND-OR pair marked "NO" would appear at first to be another example of the primitive in Fig. 8.35a. The fan-out from the AND gate causes the gate pair to have two outputs, a condition not provided for in the primitive.

In a standard cell environment, the substitution of two complex cells for three gates may not have been an improvement. In fact, the final network may require more cell area than the original. If the target technology is a gate array, the final network is probably an improvement. It is likely that a single gate or three-input primitive would each be obtained by a distinct personalization of one gate array module.

One option in the search for an effective multilevel logic manipulation program would be to allow Boolean expressions to be derived from a selected subsection of a logic network. A better subnetwork (perhaps two-level) could then be obtained from a manipulation of these expressions. It may be possible to identify a simple disjoint decomposition (see Chapter 16) once an appropriate network subsection has been identified.

■ **EXAMPLE 8.7**

Consider the alternatives available to a routine charged with decreasing the cell area required by the five-bit incrementer network of Fig. 8.38 (A_4 is the least significant bit).

SOLUTION   The AND-EXOR primitive of Fig. 8.35c appears applicable. Again, because of fan-out from two of the AND gates, only the rightmost gate pair can actually be replaced by the primitive.

A program, eager to replace subnetworks by Boolean expressions, might reduce the entire incrementer to five sum-of-products expressions in $A_4, A_3, A_2,$

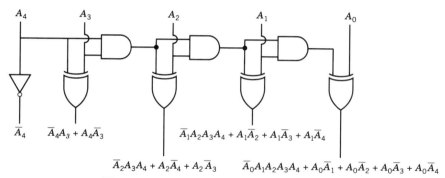

$\overline{A_4}$ $\qquad$ $\overline{A_4}A_3 + A_4\overline{A_3}$ $\qquad$ $\overline{A_1}A_2A_3A_4 + A_1\overline{A_2} + A_1\overline{A_3} + A_1\overline{A_4}$

$\overline{A_2}A_3A_4 + A_2\overline{A_4} + A_2\overline{A_3}$ $\qquad$ $\overline{A_0}A_1A_2A_3A_4 + A_0\overline{A_1} + A_0\overline{A_2} + A_0\overline{A_3} + A_0\overline{A_4}$

Figure 8.38 Five-bit incrementer network.

A_1, and A_0 and then apply Espresso. The result would be the implementation of the five expressions given in the figure. In terms of cell area, this would *not* be an improvement. ∎

Notice that the cells shown in Fig. 8.35 all have only three inputs and are necessarily less complex than, for example, the cells or parts of Fig. 8.33. All of the objects in Fig. 8.33 have been implemented as MSI parts. Because a complex MSI part utilized no more space on a printed circuit board than a much simpler part, there was little incentive to package a part as simple as those depicted in Fig. 8.35. Computer-aided logic design was in its infancy when state-of-the-art designs were realized with MSI parts. Complex parts were explicitly designed into systems by engineers. We have observed that *even the simple cells of Fig. 8.35 are a challenge to the computer-aided logic design systems available at the time of this writing* for use in standard cell and gate array ASIC design. The use of more complex cells continues to require engineering intervention.

Some design automation systems for gate arrays with fusible connections have incorporated libraries consisting of more than 40 of the 80 unique (subject to permutation of the input variables) three-input primitives. In a double-rail context or a technology where 0 cost could be assigned to an inverter, a library of 22 three-input primitives would be all-inclusive. An exhaustive library of four-input primitives would number 402, subject to complementation and permutation of the inputs, and 3984 with only permutation of the inputs allowed. Not surprisingly, vendors will be slow to include exhaustive libraries of four-input primitives in CAD systems. For a treatment of functional invariance over permutation and complementation of inputs, the reader is referred to an early work by Harrison [3].

We have barely scratched the surface of multilevel logic optimization, but it is hoped that the "scratch" has made the reader aware of many of the complicating factors involved. The reason we have introduced a problem we cannot solve is that *multilevel logic manipulation programs are already with us, good or bad*. It has become necessary to allow the user to input descriptions of combinational logic (recall Fig. 1.1) in formats other than the final Boolean expressions. In other cases, manipulation of logic expressions is necessary to adapt user input to hidden technology restrictions. Most readers will be users of, rather than developers of, CAD tools that include the manipulation of multilevel logic. They should be in a position to evaluate the effectiveness of such programs.

At this writing, the best known multilevel logic optimization program is MIS [10, 11]. Many of the developers of Espresso also participated in the MIS project.

MIS consists of two parts. First, the Boolean functions or logic network are reduced to two-level expressions. After two-level Espresso-like optimization, the program attempts to reconstruct an improved multilevel realization using a variety of techniques, including the identification of decompositions. The second part of the program tries to map sections of this improved multilevel form into complex cells available in the applicable technology library.

PROBLEMS

8.1 Use a ROM with 32 four-bit words to realize the 2-out-of-5 to BCD code converter introduced in Problem 6.28. Critical features of the logic block diagram should be shown, without indicating each individual connection. Assume the decoder is realized using 32 five-input AND gates. Compare the total number of gate inputs with the PLA and PAL realizations found in Problem 6.28. Assume that all 32 decoder outputs are realized consistent with a uniform ROM structure.

  (a) Discuss the factors that would be considered in choosing between an EPROM and a PAL realization of this code converter. Assume that either part will be integrated in the board-level design with MSI parts.

  (b) Compare the chip area occupied by ROM and PLA realizations of this code converter as part of a single-chip realization of a complex system.

8.2 Repeat Problem 8.1 for the BCD to biquinary code converter discussed in Problem 6.25. The availability of a four-input seven-output ROM may be assumed.

8.3 Develop an expression for the number of gate inputs required in a tree-type decoder with n input bits. Assume two-input gates and that one variable is introduced at each succeeding level of the tree.

8.4 Design a combinational circuit that will accomplish the multiplication of the two-bit binary number $X_1 X_0$ by the two-bit binary number $Y_1 Y_0$. Is a two-level circuit the most economical?

8.5 Verify that the circuit of Fig. 8.12c is a full adder.

8.6 The bits of two binary-coded decimal digits $X_3 X_2 X_1 X_0$ and $Y_3 Y_2 Y_1 Y_0$ serve as the input to a four-bit binary adder. The output bits of the adder are $Z_4 Z_3 Z_2 Z_1 Z_0$. Let the bits $S_3 S_2 S_1 S_0$ be the binary-coded decimal representation of the least significant bit of the decimal sum. Let $S_4 = 1$ if the most significant bit of the sum is 1, and let $S_4 = 0$ otherwise. Determine a minimal multiple-output realization of S_4, S_3, S_2, S_1, and S_0 as functions of Z_4, Z_3, Z_2, Z_1, and Z_0.

8.7 Given the function $f(A, B, C, D) = \sum m(1, 2, 4, 5, 6, 7, 8, 9, 10, 11, 14, 15)$

  (a) Realize this function using an eight-input multiplexer.

  (b) Realize this function using a four-input multiplexer plus external gates as required.

8.8 Realize the following two functions, using both halves of a SN74153 multiplexer and as little external gating as possible.

$$f_1(A, B, C, D) = \sum m(2, 4, 6, 10, 12, 15)$$
$$f_2(A, B, C, D) = \sum m(3, 5, 9, 11, 13, 14, 15)$$

8.9 The network of Fig. P8.9 has been designed for comparing the magnitudes of two two-bit binary numbers: a_2, a_1, and b_2, b_1. If $z = 1$ and $y = 0, a_2, a_1$ is larger. If $z = 0$ and $y = 1, b_2, b_1$ is larger. If $z = y = 0$, the two numbers are equal. Translate the network into a form that uses only NAND gates. Do not attempt to minimize the network.

8.10 Determine NAND equivalents of the circuits of Fig. P8.10.

8.11 Convert the circuit of Fig. P8.11 to NOR form.

8.12 Realize the following function in all-NAND form with a fan-in limit of 2. Assume double-rail inputs.

$$f(A, B, C) = \sum m(0, 3, 4, 6, 7)$$

8.13 Determine a minimal sum-of-products expression for the function realized by the all-NAND circuit given in Fig. P8.13.

8.14 Use two eight-bit adder cells of the form shown in Fig. 8.33c and any NAND, NOT, or exclusive-OR cells to obtain combinational logic network whose output will be the two's complement of an eight-bit binary number.

8.15 Construct the logic block diagram of a 16-bit adder composed of four of the four-bit adder cells defined in Fig. 8.33c.

8.16 Construct the logic block diagram of a network that will check parity over 10 bits. Use one cell of the form given in Fig. 8. 33b and any NAND, NOT, or exclusive-OR cells that may be required.

8.17 Modify the network obtained in Problem 8.14, so that it will compute the absolute value of an eight-bit two's complement number.

8.18 Use two of the comparator cells given in Fig. 8.33a and individual gate cells, as required, to construct the logic block diagram of a network with a single output Z, which will be 1 if and only if the eight-bit number X is greater than or equal to the eight-bit number Y.

8.19 Use odd-parity cells of the form given in Fig. 8.35d, together with exclusive-OR cells, to obtain a minimal cell area realization of an odd-parity checker over 10 input lines. Accept the cell area relation used in Example 8.5.

8.20 Let the relative cell area of the majority cell of Fig. 8.35e be 2 when compared to NOT cells and two-input AND or OR cells, each with relative cell area 1. Three input gates have relative cell area 1.5. Construct the logic

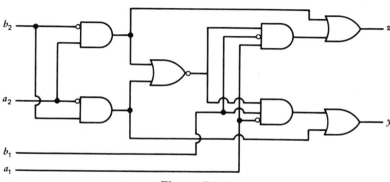

Figure P8.9

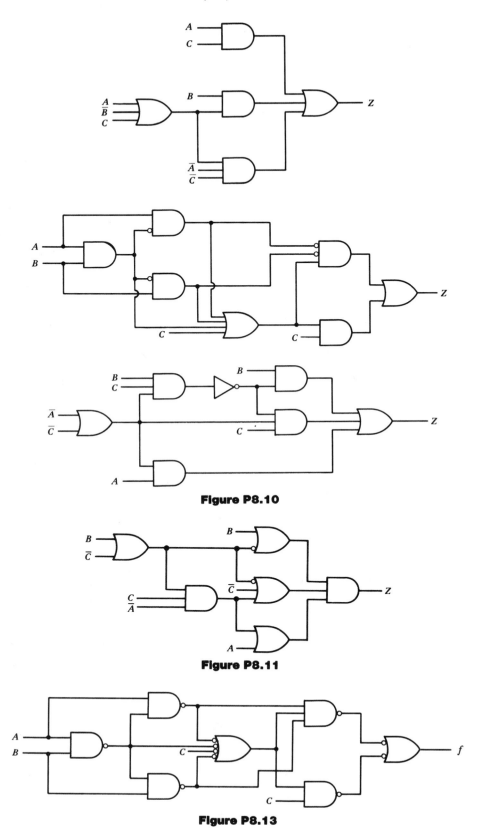

Figure P8.10

Figure P8.11

Figure P8.13

block diagram of a minimal cell area realization of a function that will be 1 if and only if three or more of the five-input variables are 1.

BIBLIOGRAPHY

1. Flores, I. *The Logic of Computer Arithmetic*. Prentice-Hall, Englewood Cliffs, N.J., 1963.
2. Blakeslee, T. R. *Digital Design with Standard MSI and LSI*, 2nd ed. Wiley-Interscience, New York, 1979.
3. Maley, G. R. and J. Earle. *The Logic Design of Transistor Digital Computers*. Prentice-Hall, Englewood Cliffs, N.J., 1963.
4. Ware, W. H. *Digital Computer Technology and Design*. Wiley, New York, 1963.
5. Harrison, M. A. *Introduction to Switching and Automata Theory*. McGraw-Hill, New York, 1965.
6. Hill, F. J. and G. R. Peterson. *Digital Systems: Hardware Organization and Design*, 3rd ed. Wiley, New York, 1987.
7. Barna, A. and D. Porat. *Integrated Circuits in Digital Electronics*. Wiley-Interscience, New York, 1973.
8. Sandige, R. S. *Digital Concepts Using Standard Integrated Circuits*. McGraw-Hill, New York, 1978.
9. Kraft, G. D. and W. N. Toy. *Mini/Microcomputer Hardware Design*. Prentice-Hall, Englewood Cliffs, N.J., 1979.
10. Brayton, R., R. Rudell, A. Sangiovanni-Vincentelli, and A. Wang. "MIS: A Multiple Level Logic Optimization System," *IEEE Trans. Computer-Aided Design Integrated Circuits and Syst.*, **CAD6**:6, 1062–1081 (November 1987).
11. Spickelmeier, R. (ed.). *OCT Release 3.0 Reference Manual*. University of California, Berkeley, Calif., 1990.

CHAPTER 9

COMPONENTS OF SEQUENTIAL SYSTEMS

9.1 SEQUENTIAL ACTIVITY

Our attention up to this point has been directed to the design of *combinational* logic circuits. Consider the general model of a switching system in Fig. 9.1. Here we have n input, or excitation, variables, $x_k(t)$, $k = 1, 2, ..., n$, and p output, or response, variables, $Z_i(t)$, $i = 1, 2, ..., p$, all assumed to be functions of time. If at any particular time the present value of the outputs is determined solely by the present value of the inputs, we say that the system is *combinational*. Such a system can be totally described by a set of equations of the form

$$Z_i = F_i(x_1, x_2, ..., x_n) \tag{9.1}$$

where the dependence on time need not be explicitly indicated, since it is understood that the values of all the variables will be those at some single time. If, on the other hand, the present value of the outputs is dependent not only on the present value of the inputs but also on the past history of the inputs, then we say that it is a *sequential system*.

As an example to clarify the distinction, consider two types of combination locks (Fig. 9.2). One type, the conventional combination padlock (Fig. 9.2a), has only one dial. The second type, sometimes used in luggage, has several dials (Fig. 9.2b). The inputs are the dial settings, the outputs the condition of the locks, open or closed. For the first type, the present condition of the lock (open or closed) depends not only on the *present* setting of the dial but also on the manner in which the dial has previously been manipulated. The second type will open whenever the three dials are set to the correct numbers; it does not matter what the previous settings were or in what order the dials were set. Obviously, then, the single-dial lock is a sequential device and should have been called a *sequential lock*. The multiple-dial lock is a mechanical combinational logic device.

A modern telephone handset and, indeed, the entire telephone network are sequential systems. Suppose you have dialed the first six digits of a telephone number and are now dialing the last digit. This seventh digit is the present input to the system, and the output is a signal causing you to be connected to the desired

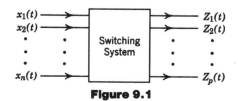

Figure 9.1

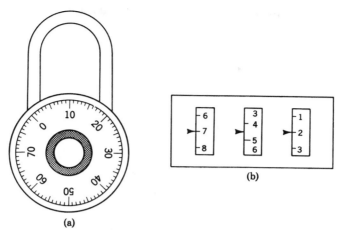

Figure 9.2 Two types of combination lock.

party. Obviously, the present input is not the only factor determining which number you reach, since the other six digits are equally important.

Digital computers are very important examples of sequential systems. There are many levels of sequential operation in a digital computer involving the sequential use of subsystems that are themselves sequential. Consider the addition of a number into the accumulator of a digital computer; the resultant sum depends both on the number and on what has been put into the accumulator previously.

9.2 MEMORY ELEMENTS

Clearly evident in the sequential devices and systems discussed above is the requirement to store information about previous events. In the combination lock, information about the previous dial settings is stored in the mechanical position of certain parts. In the original dial telephone system, each number was stored as it was dialed by electromechanical relays. In a computer, many different operations involve the storage of information.

There are many types of storage devices or memory elements. In electronic sequential circuits or systems, the most common memory device is the flip-flop.* Figure 9.3 shows the circuit for an RS flip-flop constructed from two NOR gates and the timing diagram for a typical operation sequence. We have also repeated the truth table for NOR for convenience in explaining the operation.

At the start, both inputs are at 0, the Y output is at 0, and the X output at 1. Since the outputs are fed back to the inputs of the gates, we must check to see that the assumed conditions are consistent. Gate 1 has inputs of $R = 0$ and $X = 1$, giving an output $Y = 0$, which checks. Similarly, at gate 2, we have $S = 0$ and $Y = 0$, giving $X = 1$. At time t_1, input S goes to 1. The inputs of gate 2 are thus changed from 00 to 01. After a delay (as discussed in Chapter 5), X changes from 1 to 0 at time t_2. This changes the inputs of gate 1 from 01 to 00, so Y changes from 0 to 1 at t_3. This changes the inputs of gate 2 from 01 to 11, but this has no effect on the outputs. Similarly, the change of S to 0 at t_4 has no effect. When R goes to 1, Y goes to 0, driving X to 1, thus "locking-in" Y so that the return of R to 0 has no further effect.

---

*The origin of the term *flip-flop* is uncertain, but the name has become standard.

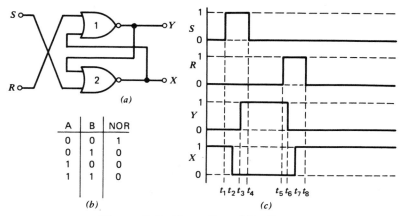

A	B	NOR
0	0	1
0	1	0
1	0	0
1	1	0

(b)

Figure 9.3 Operation of flip-flop.

A flip-flop operating in the above fashion is known as a *reset-set* or *R-S* flip-flop and has the block diagram symbol in Fig. 9.4*a*.

The S and R inputs correspond to those in Fig. 9.3, and the Q and $\bar{Q}$ outputs correspond to the Y and X outputs, respectively. A pulse on the S input will "set" the flip-flop—that is, drive the Q output to the 1 level and the $\bar{Q}$ output to the 0 level. A pulse on the R line will "reset" the flip-flop—that is, drive the Q output to the 0 level and the $\bar{Q}$ output to the 1 level. This behavior is summarized in tabular form in the *transition table*, shown in Fig. 9.4*b*. S^v, R^v, and Q^v indicate the values of the inputs and output at some arbitrary time t_v, and Q^{v+1} indicates the value to which the output will go as a result of an input at t_v. The reader should be satisfied that this table can be represented by the equation

$$Q^{v+1} = S^v + Q^v \bar{R}^v \qquad \text{(where } S^v \cdot R^v = 0\text{)} \qquad (9.2)$$

The don't care entries for Q^{v+1} in the last two rows reflect the fact that in normal operation both inputs should not be permitted to be 1 at the same time. There are

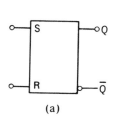

(a)

S^v	R^v	Q^v	Q^{v+1}
0	0	0	0
0	0	1	1
0	1	0	0
0	1	1	0
1	0	0	1
1	0	1	1
1	1	0	x
1	1	1	x

(b)

Figure 9.4 *R-S* flip-flop.

two reasons for this restriction. First, if both inputs are 1, both outputs will be driven to 0, which violates the basic definition of flip-flop operation, which requires that the outputs should always be the complements of each other. Second, if both inputs are 1 and then go to 0 at the same time, both NOR gates will see 00 at the inputs and both outputs will try to go to 1. However, because of the feedback, it is impossible for both outputs to be stable at 1 at the same time, with the result that the flip-flop will switch unpredictably and may even go into oscillation. With this restriction on the inputs, the R-S flip-flop functions reliably as a memory device in which the state of the outputs indicates which of the inputs was last at the 1 level.

In the discussion above we referred to *pulses* on the R and S lines, and it is important that the reader understand clearly what is meant by this term. Assume that $R = S = Q = 0$ and the flip-flop is to be set to 1. While R remains at 0, S must be raised to 1 for some minimum length of time, if the state change is to be accomplished reliably. Suppose S returns to 0 just before time t_3 in Fig. 9.3c. Instantaneously, both inputs to each of gates 1 and 2 would be simultaneously 0 while trying to drive the outputs X and Y to 1. In this case, the final state of the flip-flop would be unpredictable. If S remains 1 until Y is established at 1, then the flip-flop will stabilize at that value. We note that S must go back to 0 before the flip-flop can be reset again. It is most convenient for S to remain 1 just long enough to reliably set the flip-flop. Thus, we say that the flip-flop is set by a pulse, the duration of which is technology-dependent.

R-S flip-flops can be constructed from NAND gates as shown in Fig. 9.5a. In the timing diagram of Fig. 9.5c we see that the operation is similar to that of the NOR flip-flop of Fig. 9.3, except that the inputs are normally at the 1 level and negative pulses set or reset the flip-flop. The inversion circles at the inputs on the block diagram symbol (Fig. 9.5b) indicate that the state transitions occur following the negative-going transitions of the inputs, that is, the transitions from the more positive value of the input to the more negative value. By contrast, in the NOR flip-flop (Fig. 9.3) the absence of inversion circles at the inputs indicates that the state transitions occur following the positive-going transitions of the inputs.

As we will see in succeeding chapters, the design of systems containing flip-flops can be greatly simplified if we can synchronize state changes, that is, arrange that whenever any flip-flops are to change state, all will do so at the same time. Such operation can be achieved by using a common source of pulses to synchronize all state changes. Such a source of timing pulses is known as a *clock*, and clock pulses are usually periodic. The use of a separate timing signal implies that flip-flops must have separate inputs to control *what* happens and to control *when* it happens. Figure 9.6a shows how such separation of control can be achieved in the NOR R-S flip-flop, and Fig. 9.6b shows a typical timing sequence. Note that

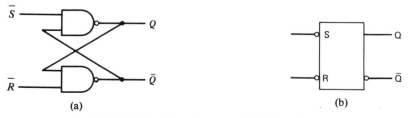

(a)   (b)

Figure 9.5 Negative-input R-S flip-flop.

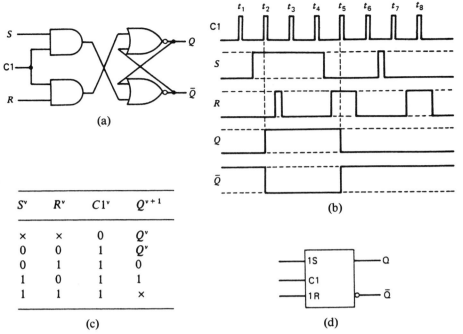

S^v	R^v	$C1^v$	Q^{v+1}
×	×	0	Q^v
0	0	1	Q^v
0	1	1	0
1	0	1	1
1	1	1	×

Figure 9.6 Clocked R-S flip-flop.

all we have done is to add AND gates so that S and R will have no effect on the flip-flop, except when a clock pulse is present. At t_1, both inputs are 0 so there is no change in the state of the flip-flop. At t_2, $S = 1$ and $R = 0$, so the clock pulse at t_2 sets the flip-flop. Between t_2 and t_3, R goes to 1 momentarily but this has no effect since the clock is 0 at this time. Between t_4 and t_5, S goes back to 0 and R goes to 1 so the pulse at t_5 resets the flip-flop.

Figures 9.6c and 9.6d show the transition table and block diagram symbol for this flip-flop, respectively. The first row of the transition table shows that the flip-flop will not change state as long as the clock is 0, regardless of S and R. When the clock is 1, the output will be determined by S and R in the same manner as in the circuit of Fig. 9.3. In normal clocked operation it is assumed, as in Fig. 9.6b, that S and R do not change during the clock pulse. The block diagram symbol (Fig. 9.6d) is an example of dependency notation. The input $C1$ is a *control input*, controlling the action of any other input with a label starting with the numeral 1. In this case, the effect of inputs $1S$ and $1R$ is *dependent* on the state of $C1$. When $C1$ is low, $1S$ and $1R$ have no effect on the flip-flop; when $C1$ is high, $1S$ and $1R$ control the operation in the normal manner associated with set and reset inputs to a flip-flop.

The clocked R-S flip-flop is rarely encountered in practice, having been supplanted by the clocked J-K and clocked D flip-flops. In the J-K flip-flop the restriction that both inputs may not be 1 at the same time, which may be inconvenient for the designer, is removed. The transition table and block diagram symbol for a J-K flip-flop are shown in Fig. 9.7. The behavior of the J-K is the same as the R-S (Fig. 9.6c), with J corresponding to S and K to R, except that when $J = K = 1$ at the time of a clock pulse, the flip-flop changes state. The characteristic equation for this flip-flop, determined from the last four rows of Fig. 9.7, is

$$Q^{v+1} = \bar{K}^v Q^v + J^v \bar{Q}^v \tag{9.3}$$

Note that J-K flip-flops are always clocked.

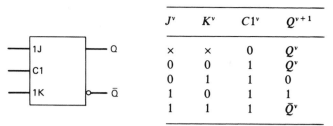

J^v	K^v	$C1^v$	Q^{v+1}
×	×	0	Q^v
0	0	1	Q^v
0	1	1	0
1	0	1	1
1	1	1	$\bar{Q}^v$

Figure 9.7 *J-K* flip-flop.

Another type of flip-flop is the D flip-flop, the block diagram symbol and transition table for which are shown in Figs. 9.8a and 9.8b. This flip-flop has only a single data input D and is the most direct realization of a memory device, in that it simply stores the current value of D whenever it is clocked. The equation for the D flip-flop is simply

$$Q^{v+1} = D^v \tag{9.4}$$

D flip-flops are available in IC form and can also be constructed by connecting a J-K as shown in Fig. 9.8c.

A fourth type of flip-flop is the T or trigger flip-flop (Fig. 9.9), which has only a single input T. Whenever T is pulsed, the flip-flop changes state, so that the equation is simply

$$Q^{v+1} = \overline{Q^v} \tag{9.5}$$

T flip-flops are realized by connecting J-K flip-flops as shown in Fig. 9.9b.

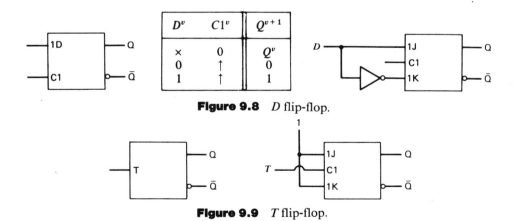

D^v	$C1^v$	Q^{v+1}
×	0	Q^v
0	↑	0
1	↑	1

Figure 9.8 *D* flip-flop.

Figure 9.9 *T* flip-flop.

9.3 WHY SEQUENTIAL CIRCUITS?

The reader is very familiar with the digital computer, the most important example of a sequential circuit. The computer is composed of many subsystems that are themselves sequential circuits. Often a hardware design or computer programming task can be formulated in either a combinational or sequential fashion. The thought process of the designer or the programmer may be similar as he or she endeavors to choose the best approach, sequential or combinational.

If the model of the tic-tac-toe game given in Fig. 9.10 were to be implemented as a special-purpose logic circuit, the game board might consist of switches by

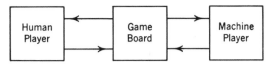

Figure 9.10 Model of tic-tac-toe game.

which the human player could indicate moves and lights to indicate the present state of the game. The machine player will consist of some sort of logic system to decide what moves should be made in response to the moves of the human player.

Clearly, the overall game progresses sequentially, but this does not mean that the machine player has to be a sequential device. It could be a combinational circuit, receiving from the game board information as to the state of the board at any given time and deciding on a move solely on the basis of that current information. There are nine squares on the board and three possible conditions (blank, O, X) for each square, so there are $3^9 = 19,683$ possible game situations. Thus, the combinational approach involves a circuit to implement the following logical requirement: Given any one of 3^9 possible input combinations, choose the proper one of nine possible responses. It is evident that the satisfaction of that requirement would involve a massive combinational logic circuit.

As a sequential approach, let us provide the machine player with some means of altering its structure as the game progresses to take advantage of the fact that not all the 3^9 situations can occur at any particular point in the game. Let us assume that the human player moves first. There are only nine possible moves, so the machine will start in a configuration designed to select the proper response to whichever move is made. Furthermore, when the machine selects its response, it then knows which seven responses are available to the human on the next move, and it can adjust to take this into account. A similar sequence of events is repeated at each successive move.

As another example, consider an eight-bit odd-parity checker. If all eight bits are available simultaneously, the desired function can be realized by a six-level combinational circuit described by the equation

$$Z = [(x_1 \oplus x_2) \oplus (x_3 \oplus x_4)] \oplus [(x_5 \oplus x_6) \oplus (x_7 \oplus x_8)] \quad (9.6)$$

requiring 21 gates in a NAND realization.

Next, let us assume that the eight-bit word is a unit of data that must be transmitted over a long distance. We could use eight lines, one for each bit, but this would be expensive. Instead, let us use a single line and send the eight bits in sequence, x_1 at time t_1, x_2 at t_2, etc. On a second line, we send *clock*, or synchronizing, pulses to mark these times t_1, t_2, etc. These two lines will provide the inputs to a circuit that is to put out a signal if the number of 1's in each group of eight bits is not even (Fig. 9.11a).

Each bit could be stored as it arrived until all eight were present. They could then be applied to the six-level combinational circuit, but this would be very expensive, requiring eight flip-flops for storage plus the 21-gate network. Instead, we note that the exclusive-OR is associative and rewrite Equation 9.6 in the form

$$z = \left(\left\{ \left[(\{[(x_1 \oplus x_2) \oplus x_3] \oplus x_4\} \oplus x_5) \oplus x_6 \right] \oplus x_7 \right\} \oplus x_8 \right) \quad (9.7)$$

Nothing would be saved by a combinational logic implementation of Equation 9.7. Instead, the first two bits are checked in an exclusive-OR network and the

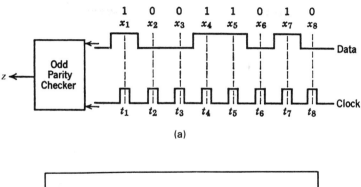

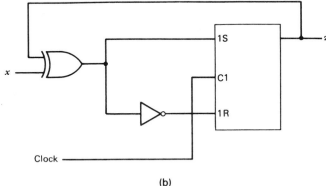

Figure 9.11 Sequential parity checker.

results stored until the third bit arrives. Then the third bit is compared to the stored results *in the same exclusive-OR* network. This result is stored in turn until the fourth bit arrives, etc. Figure 9.11b shows a simple circuit for realizing this method.

Initially, the flip-flop must be cleared to $z = 0$ before x_1 arrives so that the first operation is

$$x_1 \oplus z = x_1 \oplus 0 = x_1$$

This is stored in the flip-flop when a clock pulse arrives at time t_1. That is, if $x_1 \oplus z = 1$ when the clock pulse arrives, a pulse will be gated to the S input, setting the flip-flop to 1. Similarly, the flip-flop will be set to 0 if $x_1 \oplus z = 0$ when the clock pulse arrives. Then x_2 arrives, and the exclusive-OR forms

$$x_2 \oplus Z = x_2 \oplus x_1$$

which is stored in the flip-flop by the clock pulse at t_2. This procedure continues until, after the clock pulse at t_8, the output z is the desired parity check, that is, $z = 1$ only if there has been an odd number of 1's in the eight data bits. In fact, the output z provides an odd-parity check on whatever number of data bits have been received since the flip-flop was last cleared. Thus, Fig. 9.11b is a general n-bit odd-parity checker.

The tic-tac-toe machine and serial parity checker are examples of the type of situation in which the sequential approach is most often preferred. In neither case is there any speed advantage to using a combinational approach. In the tic-tac-toe machine, the speed of the game is dictated by the speed with which the human player makes his or her moves. The sequential parity checker is slower than the combinational circuit if we measure the speed of the combinational circuit *from the time that all eight bits are available*, but if the bits are going to arrive in sequence anyway, there is no difference in speed. The sequential circuit, however, is simpler, and therefore cheaper, than the combinational circuit.

9.4 A GENERAL MODEL FOR SEQUENTIAL CIRCUITS

The general model for sequential circuits, shown in Fig. 9.12, consists of two parts: a combinational logic section and a memory section. The memory stores information about past events required for proper functioning of the circuit. This information is represented in the form of r binary outputs, $y_1, y_2, ..., y_r$, known as the *state variables*. Each of the 2^r combinations of state variables defines a *state* of the memory; in general, each state corresponds to a particular combination of past events.

The combinational logic section receives as inputs the state variables and the *n* circuit inputs, $x_1, x_2, ..., x_n$. It generates *m* outputs, $z_1, z_2, ..., z_m$ and *p* excitation variables, $Y_1, Y_2, ..., Y_p$, which specify the next state to be assumed by the memory. The number of excitations is often equal to the number of state variables, but this depends on the type of memory. The combinational logic network may be defined by Boolean equations of the standard forms, that is,

$$z_i = f_i(x_1, x_2, ..., x_n, y_1, y_2, ..., y_r) \qquad (i = 1, 2, ..., m) \qquad (9.8)$$

$$Y_j = f_j(x_1, x_2, ..., x_n, y_1, y_2, ..., y_r) \qquad (j = 1, 2, ..., p) \qquad (9.9)$$

These are time-dependent equations, that is, they are valid at any time all the inputs and outputs are stable, as for any combinational circuit.

We have not yet specified the relationship between the inputs and outputs of the memory portion of Fig. 9.12. This relationship will depend on whether the memory is realized in terms of an array of memory elements or flip-flops or is merely a set of feedback loops with time delay. These two points of view will be considered in Sections 9.5 and 9.7.

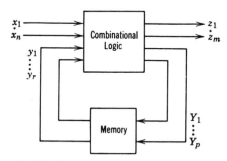

Figure 9.12 General model of sequential circuits.

9.5 CLOCK-MODE SEQUENTIAL CIRCUITS

A model for the *clock-mode*, or simply *clocked*, sequential circuit is shown in Fig. 9.13. The memory consists of r clocked flip-flops, which may be any type, although we show *J-K*. The excitations thus consist of r *J*-signals and r *K*-signals. We show the state variables coming from the center of the flip-flops, because either the true or complemented outputs, or both, may be used.

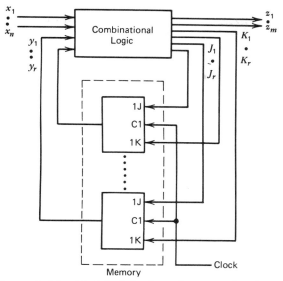

Figure 9.13 Model of clock-mode sequential circuit.

All inputs and outputs of the combinational logic section are *level* signals, that is, signals that may take on either the 1 or 0 values for arbitrary periods of time. As discussed above, the combinational logic is time-independent, so that the outputs and excitations may change at any time in response to changes in the circuit inputs. However, the state variables will change only when the flip-flops are triggered by a clock pulse. The J and K excitation variables may change at any time in response to changes in the x input variables, but no change in state will occur until a clock pulse arrives. The clock pulse contains no information in the sense of determining *what* state change will take place; it simply times, or synchronizes,* the state change. The values of the J and K signals at the time of the clock pulse determine *what* transition takes place. The times of arrival of the clock pulses are denoted by $t_v (v = 1, 2, 3, ...)$, and the values of the variables at these times are denoted by the superscript v. At time t_v, the excitation variables, $J_1, K_1, J_2, K_2, ..., J_r, K_r$, must be at stable levels determined by the current values of the inputs and state variables. The outputs are similarly stable at t_v and may be used as the inputs to other circuits synchronized by the same clock. The clock pulse arriving at t_v will cause a transition of the state variables from y_i^v to y_i^{v+1}, thus placing the circuit in a new state. The excitation and output variables will in turn adjust to this state change, as well as any input changes, reaching new stable levels before the next clock pulse, at t_{v+1}.

This circuit behavior may be described by the equations

$$z_i^v = f_{zi}(x_1^v \cdots x_n^v, y_1^v \cdots y_r^v) \qquad (i = 1, 2, ..., m) \qquad (9.10)$$

$$J_i^v = f_{Ji}(x_1^v \cdots x_n^v, y_1^v \cdots y_r^v) \qquad (i = 1, 2, ..., r) \qquad (9.11a)$$

$$K_j^v = f_{Kj}(x_1^v \cdots x_n^v, y_1^v \cdots y_r^v) \qquad (j = 1, 2, ..., r) \qquad (9.11b)$$

$$y_j^{v+1} = \bar{K}_j^v y_j^v + J_j^v \bar{y}_j^v \qquad (j = 1, 2, ..., r) \qquad (9.12)$$

Equation 9.12 characterizes the behavior of a *J-K* flip-flop, as discussed in Section 9.2. In that discussion, we used the notation Q^v and Q^{v+1} to refer to the

---

*Systems of this type are sometimes referred to as *synchronous* systems, but the word has been used so many different ways that the more precise term "clock-mode" is better.

"present" and "next" values of the flip-flop output. Here y^v refers to the value of y at the time of the clock time t_v, and y^{v+1} to the value at clock time t_{v+1}. This does not imply that y does not change to y^{v+1} until t_{v+1}; it generally changes immediately in the conclusion of the clock pulse t_v. However, this value (y^{v+1}) is the value it will have at t_{v+1}, and that is the value that is significant in terms of the next transition.

To illustrate the fact that the inputs, outputs, and state variables in a clock-mode circuit are only of interest at the time of a clock pulse, let us consider the circuit of Fig. 9.14a. This two-bit shift register circuit will have an output of 1 if and only if the present input value is *not* the same as the input two clock periods earlier, which is stored in flip-flop 2. Note first in Fig. 9.14b that y_1 is a near replica of the input x delayed by one clock period. At time t_1, for example, $x = 1$, so that the pulse at time t_1 causes y_1 to take on this value and remain 1 until after the clock pulse at time t_2. Succeeding values of x are similarly shifted into flip-flop 1 by succeeding clock pulses. Note next the brief pulse appearing on input line x at time t_b. This pulse does not overlap a clock pulse so it has no effect on y_1.

At time t_2, $y_1 = 1$, so that the corresponding clock pulse triggers this value into flip-flop 2. The value of y_2 remains 1 until after the pulse at time t_3. With each succeeding clock pulse, a new value is shifted from y_1 to y_2. Thus, y_2 is a replica of y_1 delayed one more clock period.

The output z may be found by continuously computing the exclusive-OR function of the values of y_2 and x. The resulting values of z are plotted as the last

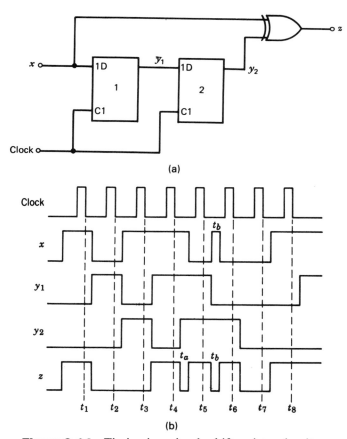

(a)

(b)

Figure 9.14 Timing in a simple shift register circuit.

line of Fig. 9.14*b*. Note the brief zero-going pulse at time t_a. This pulse occurs because both x and y_2 change values after time t_4, but y_2 changes first. Also note that the narrow input pulse at time t_b causes another zero-going output pulse at the same time. Neither of these output pulses overlaps a clock pulse. Suppose that this output serves as the input to another clock-mode sequential circuit. Just as the input pulse at time t_b did not affect the state of the circuit of Fig. 9.14*a*, the two output pulses will not affect the state of succeeding clock-mode circuits because they end before the occurrence of the next clock pulse.

9.6   SATISFYING THE CLOCK-MODE ASSUMPTION

The circuit of Fig. 9.14 will only function as described if the following clock-mode assumption is satisfied:

The states of the memory elements of a clock-mode sequential circuit may change only once each clock period and they must change simultaneously. Following a state change, every gate in the memory element input logic must reach a stable value before the next state change is triggered.

Usually, state changes will be triggered by a single system clock. It is the responsibility of the designer to manage the system clock and path delays so that the clock-mode assumption is satisfied.

The most elementary clocked *J-K* flip-flop, often called a *J-K latch*, is shown in Fig. 9.15. Note that the circuit is the same as for the *R-S* of Fig. 9.6*a*, except that the outputs are fed back to the input AND gates. Assume that the flip-flop is reset and $J = K = 1$ when a clock pulse arrives. The combination of $Q = 0$ and $\bar{Q} = 1$ will steer the clock pulse, via the upper AND gate to the lower NOR gate, to set the flip-flop. If the flip-flop is set, the clock pulse will be gated through the lower AND gate to the upper NOR gate, to reset the flip-flop.

This all seems reasonable at first glance, but let us consider the operation more carefully. If the circuit is reset, the 0 at Q will block the clock signal at the lower AND gate, preventing it from reaching the upper NOR gate, and the flip-flop will set. When it sets, Q will change to 1. If the clock is still high when this new value of Q feeds back to the lower AND, this signal to the upper NOR will go high, driving the flip-flop back to reset. Indeed, this circuit will oscillate between set and reset as long as the clock is held high with $J = K = 1$. The same sort of problem can occur when one flip-flop drives another that is controlled by the same clock. Consider the shift register of Fig. 9.14, which is repeated as Fig. 9.16 with a slightly different clock source. Note that clock pulse 5 is significantly wider than the other

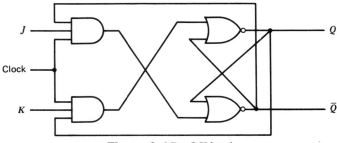

Figure 9.15   *J-K* latch.

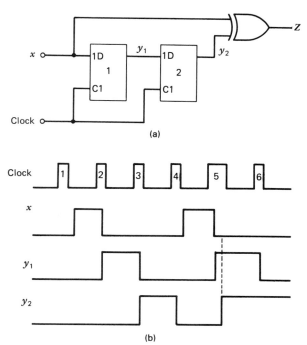

(a)

(b)

Figure 9.16 Double state-change problem.

pulses, which were tuned to the delays in the two D flip-flops shown. We see that $x = 1$ at the onset of clock pulse 5. This value is shifted into flip-flop 1 and appears as output y_1, while clock pulse 5 is still 1. This means that the input to flip-flop 2 is now 1, while clock pulse 5 is still on. Unfortunately, this causes y_2 to go to 1 also. This was *not* the desired behavior. Note that the value of x *was shifted two places in the shift register by 1 clock pulse*. Proper behavior calls for the shift of each bit in the register one place by each clock pulse.

The circuit of Fig. 9.16 violated the clock-mode assumption, because the wide clock pulse allowed the value of the J-K latch, y_2, to change twice. If latches are used as memory elements, satisfaction of the clock-mode assumption depends on the careful control of the width of all clock pulses to match the timing properties of the gates and latches used. This is an unattractive prospect in that speed and delay characteristics will vary with the individual gate and to a greater degree from technology to technology. Let us consider some approaches to insuring the satisfaction of the clock-mode assumption, regardless of the properties of individual components.

The master-slave R-S flip-flop of Fig. 9.17a is composed of two clocked R-S latches connected in series. The clock is connected directly to the first or *master*

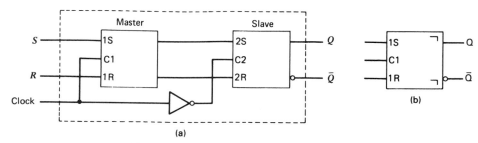

(a)

(b)

Figure 9.17 Master-slave R-S flip-flop.

latch, with the complement of the clock connected to the second or *slave* latch. When the clock is low, the master is disabled so that changes in *R* and *S* will have no affect on the flip-flop. The slave is enabled at this time but cannot change because it is driven by the master. When the clock goes high, the master is enabled and will behave as determined by the values of *R* and *S* at that time. At the same time, the inverted clock disables the slave, so it will not be affected by any changes at the output of the master. This requires that the inverted clock disable the slave before the outputs of the master can change, but this is easy to arrange by controlling the internal delays in the flip-flop. When the clock goes back down, the master is again disabled, isolating the flip-flop from changes in *R* and *S*. The inverted clock enables the slave, and the setting of the master, determined by *R-S* when the clock went high, is copied into the slave. Because the outputs of the complete flip-flop are taken from the slave, we are thus assured that the outputs cannot change until *after* the clock pulse has ended.

Although the master-slave flip-flop does consist of two pairs of cross-coupled gates, it is considered to be a single flip-flop for logical purposes, with the block diagram symbol (Fig. 9.17*b*) indicating only its signal characteristics, not its internal structure. The symbol (⌐) at the outputs indicates that a pulse on the *C*1 line is required to activate a transition in this flip-flop and that the outputs will change on the falling edge of the pulse.

The master-slave concept can be applied to any type of flip-flop and can be physically realized in a variety of circuits. Fig. 9.18 shows a typical TTL realization of a master-slave *J-K* flip-flop in all-NAND form. This circuit functions in the same manner as the elementary *J-K* shown in Fig. 9.15, except that it cannot change more than once in response to a single clock pulse, no matter how long. The outputs that are fed back to steer the clock to the proper side of the master are taken from the slave and thus cannot change until after the clock pulse has ended. In this form, there is no separate inverter for the clock, the necessary inversion being provided by the input NAND gates of the master, the outputs of which go directly to the slave as well as the master.

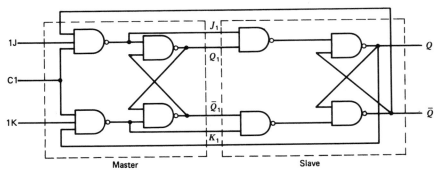

Figure 9.18 Master-slave *J-K* flip-flop.

9.7 THE STANDARD EDGE-TRIGGERED *D* FLIP-FLOP

A second type of flip-flop that can be used to assure that all state changes are synchronized at a single point in time is the edge-triggered *D* flip-flop, as shown in Fig. 9.19. The circuit is complex and most effectively described using the level-mode sequential circuit techniques to be developed in Chapter 15. For now consider Fig. 9.19*b*, which defines the input-to-output behavior of the circuit.

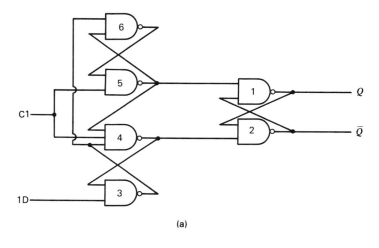

(a)

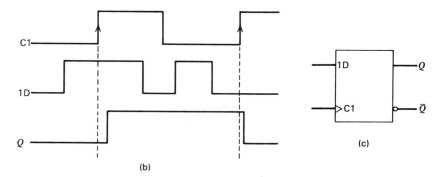

(b)

(c)

Figure 9.19 Positive edge-triggered D flip-flop.

Note that Q goes to 1 immediately following the first positive transition on the clock line, as marked by the arrow. The circuit is designed so that this clock transition triggers the state change, and state changes cannot take place, except at the time of a positive clock transition. We see, in fact, that D returns to 0 while the clock remains high, but no further change in Q occurs. Similarly, D is ignored while the clock is 0. A second change in Q does not take place until the second positive clock transition, also marked by an arrow. In order for the circuit to work reliably, the date input D must always be stable throughout a short interval around each positive clock transition. The "triangle" at the $C1$ input of the block diagram symbol (Fig. 9.19c) indicates that triggering takes place on the rising edge of the clock signal. A triangle preceded by an inversion bubble indicates triggering on the falling edge of the clock signal.

The transition period during which the input must be stable is a factor of both the circuit parameters and the rise time of the clock pulse. The value of D must be "set up," that is, established at a stable value a specified time before the clock transition starts, and must "hold" at that value until a specified time after the clock transition is complete. The manufacturer will provide specifications for t_{setup} and t_{hold}. The timing diagram of Fig. 9.20 will clarify the meaning of these terms. The total time during which the D input must be held stable is the sum of the setup time, the rise time, and the hold time.

Although the edge-triggered and master-slave flip-flops are similar in that both delay the switching of the outputs until a time when further changes at the inputs can have no effect, there are important differences. In a master-slave flip-

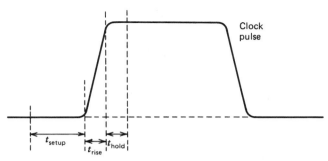

Figure 9.20   Transition time specifications.

flop, depending on the current state of the outputs, changes at the input *during* the clock pulse may or may not cause changes at the output. To ensure reliable operation, the data inputs must be held stable throughout the clock pulse. In the edge-triggered flip-flop, by contrast, the inputs need be held stable only at the time of the triggering edge. As a general rule, an increase in the time inputs must be held stable will require a slower clock, so that edge-triggered flip-flops are generally preferable. *D* flip-flops are almost invariably edge-triggered, but *J-K* flip-flops may be of either type, and the designer should check the data sheets carefully to be sure which type he or she is getting. For example, the 74107 *J-K* flip-flop is master-slave, but the 74LS107 is falling-edge-triggered.

Often memory elements include direct-reset and direct-set lines, by means of which the flip-flop can be set or reset independently of the values on the data or clock lines. These lines are generally used to establish initial values in flip-flops when systems are turned on or reset. For example, if we have a counter in a system, we might want to initialize the counter to zero when the system is started up. For this purpose, we could tie all the clear inputs together and apply the appropriate voltage to set all the flip-flops to 0. Figure 9.21 shows the *D* flip-flop of Fig. 9.19 with set and reset added. As shown in the schematic (Fig. 21*a*), the set and reset lines are applied directly to the output flip-flop. These lines are normally held high, in which case they have no effect on the operation of the flip-flop. If the set line goes low, it drives gate 1 high, setting the flip-flop; if the reset line goes low, it drives gate 2 high, resetting the flip-flop. The set and reset lines are also applied to gates 3, 5, and 6 to prevent the flip-flop from switching to the opposite state if the clock is high when the set and reset signal goes back to 1. Figure 9.21*b* illustrates the transition table including the set and reset lines, and Fig. 9.21*c* shows the block diagram symbol. The inversion bubbles on the *R* and *S* inputs indicate that these lines are *active-low*, that is, the low level causes the action associated with that input line. The first two lines of the transition table show that the reset and set lines override the other inputs. The next three lines indicate that the flip-flop acts as a normal *D* flip-flop when the reset and set are high. The last line indicates that the action is unpredictable if both go low at the same time. Most standard IC flip-flops include reset or set or both, and most, as in this case, are active-low.

9.8 CLOCK SKEW AND TWO-PHASE CLOCKING OF NMOS MEMORY ELEMENTS

In Section 9.6 it was argued that satisfaction of the clock-mode assumption required a clocking interval or clocking edge shorter in duration than the shortest logic propagation delay from output to input of two memory elements. This is a

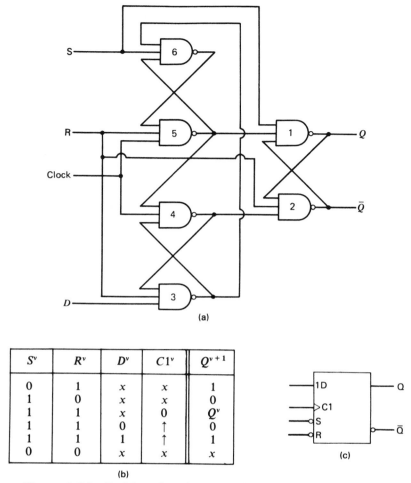

(a)

S^v	R^v	D^v	$C1^v$	Q^{v+1}
0	1	x	x	1
1	0	x	x	0
1	1	x	0	Q^v
1	1	0	↑	0
1	1	1	↑	1
0	0	x	x	x

(b)

(c)

Figure 9.21   Positive edge-triggered D flip-flop with reset and set.

necessary but not a sufficient condition for clock-mode design. A second issue that might invalidate the clock-mode assumption in a particular design is *clock skew*. It is important to note that an intolerably long clock-skew interval can occur in any technology. Clock skew is perhaps most critical in NMOS sequential circuits, so NMOS will be the context for defining the problem. One solution, which is necessary in NMOS but is usually not applied to other technologies, will be offered.

It was pointed out in Chapter 4 that the input gate capacitance of a NMOS inverter will store its past value for as much as a millisecond while the gate itself is unconnected. It is possible to take advantage of this gate capacitance as the storage mechanism in a NMOS sequential circuit. Such a sequential circuit, employing *complementary single-phase clocking*, is illustrated in Fig. 9.22a. The slash in the input output lines and feedback loops indicate that the single wires shown represent vectors of wires. Consider the two series inverters in the feedback loop to be a memory element representing similar elements for each of the state variables. This type of memory element has been termed a *dynamic memory element*.

At the beginning of each clock period T as cl goes to 1, the present state together with input values appear at the input to the combinational logic. Once the next

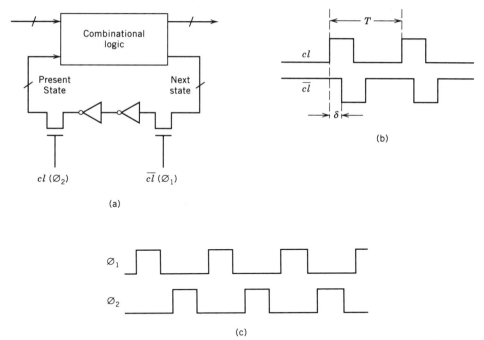

Figure 9.22   Eliminating the impact of clock skew with two-phase clocking.

state values have propagated to the output of the combinational logic, they will be retained by gate capacitances in the combinational logic even after cl returns to 0. When $\overline{cl}$ goes to 1, the next state values are gated into the dynamic memory elements (and retained by the inverter gate capacitances after $\overline{cl}$ returns to 0) to complete the activity of the clock period. It is intended that cl and $\overline{cl}$ will not be 1 simultaneously to ensure no more than one state change per clock period, satisfying the clock-mode assumption.

A problem may arise if for some reason the clock inputs are out of phase by some time interval δ as indicated in Fig. 9.22b. This is *clock skew*. Should δ be greater than the smallest of the propagation delays through combinational logic in series with the two inverters composing a dynamic memory element, then it is possible to have two consecutive state changes in the same clock period. Satisfaction of the clock-mode assumption may be assured by replacing the complementary clock inputs in Fig. 9.22c with connections to the nonoverlapping two-phase clock shown in Fig. 9.22c. The onset of Φ_1 may now be considered to represent the triggering edge in an analogy to sequential circuits employing the edge-triggered D flip-flop. The interval between the two clock phases will typically exceed the worst expected clock skew. The two-phase clock is much more common in NMOS than in other technologies, including CMOS.

The memory elements in Fig. 9.22a are called dynamic because the clock signals must cause the values stored by the inverters in the feedback loop to be refreshed each clock period. In contrast, a *static memory element* will store a value indefinitely in the absence of further input stimuli. We shall see in Chapter 11 that static memory elements are required in digital system data units where registers may be infrequently updated. Let us assume that the memory element is to be updated only during clock periods when a control input that we shall call LD is 1. As was the case for the dynamic element, the state variable signal path must pass through devices controlled by both phases of the nonoverlapping two-phase clock, to assure that the state of the memory element can change no more than once in a single clock period.

Two implementations of a static NMOS memory element, satisfying the above conditions, are given in Fig. 9.23. The element in Fig. 9.23b, in which the stored data are internally refreshed each period by clock phase Φ_2, will be the NMOS memory building block used in Chapter 11. Only during those periods in which LD is 1 will new data be transferred into the memory element by clock phase Φ_1.

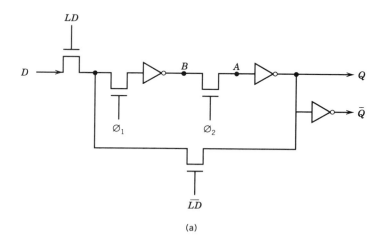

(a)

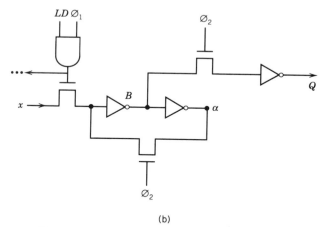

(b)

Figure 9.23   Static NMOS memory elements.

9.9   CONNECTING DATA TO NMOS BUSES

Busing of data is a common operation in any large digital system. When register transfer languages are introduced in Chapter 11, a special symbol will be defined to provide for connections to a bus. Implementation of bus connections varies with the technology. This section will focus on three separate NMOS implementations of bus connections. These include the (1) *active bus*, (2) *precharged bus*, and (3) *passive bus*. The value of an active bus will always be connected to 0 or 1. We

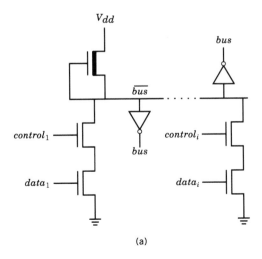

(a)

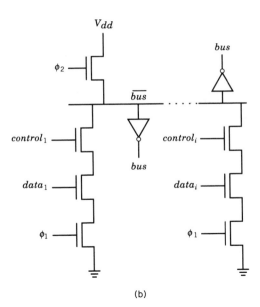

(b)

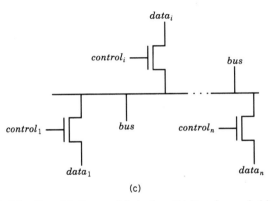

(c)

Figure 9.24 One-bit Buses. (*a*) Active. (*b*) Precharged. (*c*) Passive.

shall assume that the bus value will be 0 when no data is placed on the bus. The passive bus will be considered capable of assuming any one of three values 0, 1, or Z representing the high impedance or unconnected state.

One bit of an active bus is depicted in Fig. 9.24a. The wire interconnecting points where data may be placed on the bus are actually $\overline{bus}$. An inverter is added at each point, where data from the bus may be utilized, to generate the uncomplemented bus value. If none of the pull-down paths from $\overline{bus}$ to ground are closed, the depletion-mode pull-up device will hold $\overline{bus}$ at 1. If one of the control lines $control_i$ is 1, the value of $\overline{bus}$ will be the complement of the corresponding data input $data_i$. The designer must assure that no more than one of the control inputs to the bus is 1 at any time.

Replacing the depletion-mode pull-up device in Fig. 9.24a with an enhancement-mode device controlled by clock phase 2 (Φ_2) and including a device controlled by phase 1 (Φ_1) in series with each pull-down path results in the precharged bus shown in Fig. 9.24b. The advantage of the precharged bus, like the CMOS precharged domino logic of Section 7.8, is the absence of steady-state power consumption. Much like its CMOS logic counterpart, the line $\overline{bus}$ is charged to logic 1 during Φ_2 and assumes its correct value at the end of Φ_1. If no data is connected to the bus, $\overline{bus}$ remains at logic 1.

In NMOS, the passive bus is perhaps the simplest implementation of a bus. Shown in Fig. 9.24c is one bit of a passive bus, where the simple pass transistor is the mechanism by which data is connected to the bus. If none of the pass transistors are active, the bus is in the high-impedance state. When the device controlled by line $control_i$ is on, $data_i$ is connected to the bus. As always, the simultaneous connection of more than one data line to the bus must be avoided. Neither the precharged or passive buses are suitable for driving chip outputs. An active bus would be used for this purpose.

Very often the output of a bus is connected to the D input of a memory element, as illustrated for the precharged bus in Fig. 9.25. Consistent with the previous section, the contents of the bus are clocked into the memory element by clock phase Φ_1.

9.10   CHARGE-SHARING

Notice in the precharge bus configurations of Figs. 9.24b and 9.24c that clock phase Φ_1 controls the device nearest the ground point in every pull-down network. By design, the Φ_1-controlled devices are placed in this position to avoid a potential problem called charge-sharing.

A precharge bus network vulnerable to charge-sharing is depicted in Figs. 9.26a and 9.26b. In Fig. 9.26a $\Phi_2 = 1$, and $\overline{bus}$ is charged at the logic 1 value. The line $control$ is 1, whereas the data line labeled $data$ is 0. Thus, during the Φ_1 portion of the clock period, the middle device is turned on while the lower device remains off. Now, as shown in Fig. 10.25b, $\overline{bus}$ does not discharge to ground, but part of the charge at this point passes through the turned-on device to the node below. The amount of sharing of the original charge on $\overline{bus}$ will depend on the relative capacitance of the two nodes. The voltage at the $\overline{bus}$ point will fall below the logical 1 value. The logic value at any gate driven by $\overline{bus}$ must be considered indeterminate.

The potential for charge-sharing is eliminated by merely interchanging the positions in the pull-down network of the clocked device and the device driven by

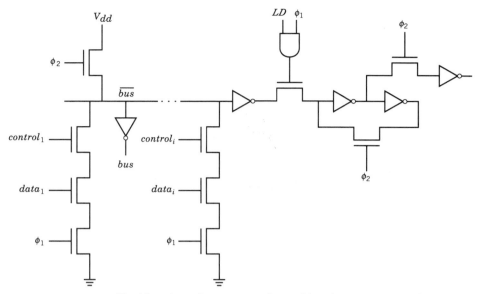

Figure 9.25 Clocking the value on a precharged bus into a memory element.

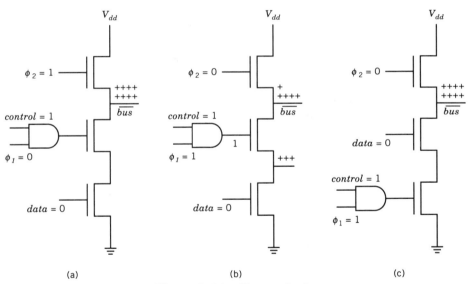

Figure 9.26 Charge-sharing.

the data bit. By clocking only the device nearest ground with Φ_1, all nodes in the pull-down path will either be charged to their final values during the precharged half-cycle or will be discharged through a path to ground during the Φ_1 half-cycle. The circuit of Fig. 9.26a is so modified to form Fig. 9.26c.

9.11 STATIC CMOS MEMORY ELEMENTS

Before discussing CMOS memory elements, it is necessary to consider one more CMOS combinational logic element that was not treated in Chapter 8. The device depicted in Fig. 9.27a is a *tri-state* inverter. If *enable* = x = 1, the output will be connected to 0. If *enable* = 1 and x = 0, the output will be connected to V_{dd}. If

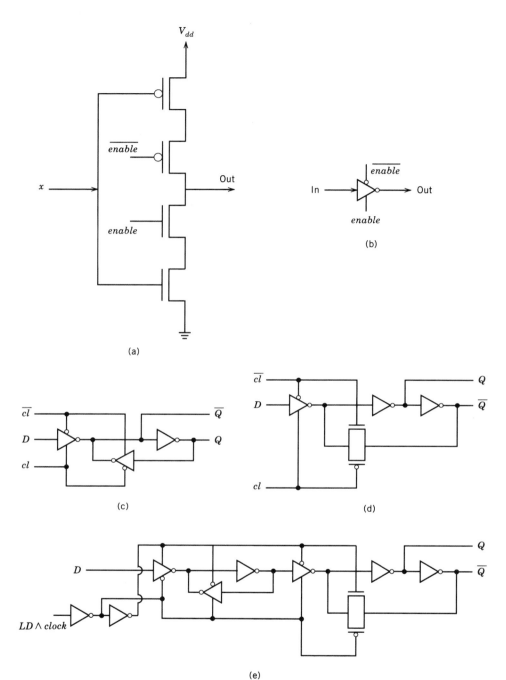

Figure 9.27 CMOS memory elements.

enable = 0, the output is unconnected and will retain its previous value for several clock periods. The logic symbol for the tri-state inverter is given in Fig. 9.27*b*.

The two static CMOS memory elements given in Figs. 9.27*c* and 9.27*d* both incorporate the tri-state inverter controlled by a complementary single-phase clock. In Fig. 9.27*c* the stored value is circulated internally when the clock is 0, and the stored value is updated from the *D* input when the clock is 1. The device in Fig. 9.27*d* works similarly, with one of the tri-state inverters replaced by a transmission gate. This latter device has the advantage of driving both the Q and $\bar{Q}$ outputs with low-impedance unclocked inverters.

As with NMOS, each state variable signal path must pass through devices controlled by both clock phases. Using either of the memory elements in Figs. 9.27c or 9.27d will necessitate the inclusion of a device controlled by $\bar{cl}$ elsewhere in each state variable path. In Section 8.6 it was pointed out that CMOS VLSI design is typically approached as the interconnection of standard cells. When static combinational logic is used, it is usually most convenient to apply the clock only to memory element cells. Connecting the two memory devices just discussed in series forms the master-slave memory element of Fig. 9.27e. Here the controlled clock signal, $LD \wedge clock$, must first take on the value 1 and then 0 in order to propagate D from input to output. Hence, the clock-mode assumption is satisfied. Clock-skew problems are avoided by generating the complement of the enabling signal within the cell. Usually, a single-phase clock will be used in CMOS in contrast to the NMOS two-phase clock.

The master-slave memory element of Fig. 9.27e has the same low-impedance driving advantage found in the device of Fig. 9.27d. Another incidental advantage is that this master-slave element may be adapted to designs in which enabling signals are either normally low or normally high. In one case, the stored data is recirculated in the master element. In the other case, it is refreshed in the slave.

9.12 A TRI-STATE CMOS BUS

The NMOS passive bus described in Section 9.9 may be called a *tri-state bus*, because the bus wire remains in the high-impedance state (distinct from 0 and 1) if nothing is actively connected to the bus. Most CMOS buses are tri-state. The CMOS-controlled inverter of Fig. 9.27a may serve as a CMOS bus driver. If *enable* = 1, the value of x is forced on the bus. The line labeled *enable* is the control line, whereas x is the data line. Any number of these devices may be connected to a bus wire. If the enable lines of all such devices are 0, the bus will remain in the high-impedance state.

9.13 CONCLUSION

In this chapter, we have attempted to give the reader an overview of the general characteristics of sequential circuits, of some of the factors governing their use, and of some of their special problems. We have treated the timing of clock-mode circuits in more detail to prepare the reader for Chapter 10. The reader should be able to consider the logical analysis and synthesis of such sequential circuits with confidence that physical realizations will work as postulated.

PROBLEMS

9.1 By means of a timing diagram similar to Fig. 9.3c, investigate the behavior of the R-S flip-flop when R and S pulses occur at the same time. Assume simultaneous 100-ηsec pulses on both R and S. Assume that the delays through the gates are 20 ηsec. Continue the timing diagram for a long enough period after the end of the pulses to determine what the flip-flop will do.

9.2 Construct a timing diagram similar to Fig. 9.6b showing the behavior of the flip-flop if the R and S inputs are both 1 during the clock pulses. Assume a 100-ηsec clock pulse and 20-ηsec delays through the gates. Continue the diagram at least 100 ηsec after the end of the clock pulse.

9.3 Figure 9.11 shows a sequential parity checker using an R-S flip-flop. Determine the design for a sequential parity checker using a J-K flip-flop. Verify its operation by means of a timing diagram showing a sequence of possible inputs and resulting outputs.

9.4 By means of a timing diagram, demonstrate how improper operation can occur in the J-K latch of Fig. 9.15 if the clock pulse is still high when the change at the outputs is seen at the input AND gates.

9.5 In the master-slave R-S flip-flop of Fig. 9.17, assume that both the master and the slave are clocked R-S flip-flops as shown in Fig. 9.6a. By means of a timing diagram, show a typical sequence of state changes in the flip-flop. This diagram should be similar to Fig. 9.6b, except that you should show S, R, Q_1, $\bar{Q}_1$, Q_2, $\bar{Q}_2$, CK_1, CK_2. Assume clock pulses are 100 ηsec in duration and the delays in all gates are 10 ηsec.

9.6 Repeat Problem 9.5 for the master-slave J-K flip-flop of Fig. 9.18.

9.7 Figure P9.7 shows the circuit diagram for the SN74109 J-K flip-flop, which is listed by the manufacturer as positive edge-triggered. By means of a timing diagram, determine when this flip-flop triggers and when it is enabled. Assume the delay through each gate is 10 ηsec.

9.8 Consider the circuit realization of an edge-triggered D flip-flop given in Fig. P9.8. Suppose that, prior to the occurrence of a clock pulse, the circuit output Q and circuit input D are stable at the values 0 and 1, respectively.

(a) Under the above conditions, what must be the logical values at points a, b, c, and d?

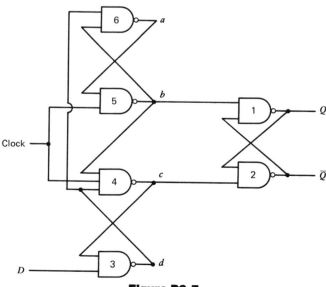

Figure P9.7

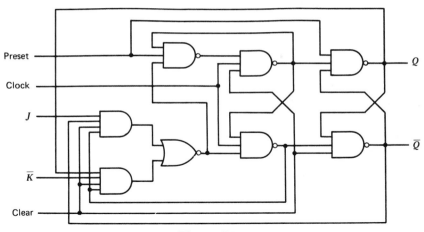

Figure P9.8

(b) Start with the initial conditions determined above and let the clock go to 1 and remain at that value for 200 ηsec. Construct as accurately as possible a timing diagram of D, a, b, c, d, and Q beginning 50 ηsec prior to the clock pulse and extending 100 ηsec after the trailing edge of the pulse. Assume that the delay associated with each gate is 20 ηsec. Be sure to verify that changes at the D input while the clock is at the 1 level do not affect the output.

9.9 Make the minimum modification of the NMOS D flip-flop of Fig. 9.23b necessary to add a normally reset line to circuit. Construct a logic block diagram from your results.

9.10 Make the minimal modification necessary to add normally high set and reset lines to the CMOS master-slave flip-flop of Fig. 9.24e. Construct a logic block diagram.

9.11 Develop a logic block diagram for a CMOS J-K memory element.

9.12 Consider the static NMOS memory element in Fig. 9.23b. Suppose the line leading through the pass transistor controlled by Φ_2 to output Q together with that pass transistor and the output inverter are deleted from the circuit, and the output is connected to point α. In one or two sentences describe the problems which might result in a digital system realized using this type of memory element.

9.13 Consider the precharged logic circuit of Fig. 7.27. Suppose that the circuit is left unchanged except that the clock line is connected to the device between points 1 and 2 in place of line A and that line A is connected to the gate of the device between point 4 and ground. Suppose the capacitance C_1 (point 1 to ground) is .3 pf and the capacitances C_2, C_3, and C_4 are each .1 pf. Suppose $A = 0$ and $B = C = 1$. Let $V_{dd} = 5$v. The node at point 1 will precharge to 5v when the *clock* $= 0$. Determine the steady state voltage at point 1 after *clock* goes to 1.

BIBLIOGRAPHY

1. Mukherjee, A. *Introduction to NMOS & CMOS VLSI Systems Design*. Prentice-Hall, Englewood Cliffs, N.J., 1986.
2. Mealy, G. H. "A Method for Synthesizing Sequential Circuits," *Bell System Tech J.,* **34:** 5, 1045–1080 (Sept. 1955).

3. Moore, E. F. *Sequential Machines: Selected Papers*. Addison-Wesley, Reading, Mass., 1964.

4. Peatman, J. B. *Digital Hardware Design*. McGraw-Hill, New York, 1980.

5. McCluskey, E. J. *Logic Design Principles*. Prentice-Hall, Englewood Cliffs, N.J., 1986.

6. *The TTL Data Book*. Texas Instruments Inc., Dallas, Texas, 1984.

7. Heinbuch, D. V. *The CMOS3 Cell Library*. Addison-Wesley, Reading, Mass., 1988.

CHAPTER 10

SYNTHESIS OF CLOCK-MODE SEQUENTIAL CIRCUITS

10.1 ANALYSIS OF A SEQUENTIAL CIRCUIT

In this chapter, we will develop a complete design procedure for clock-mode sequential circuits. In Chapter 13, we will develop similar procedures for fundamental-mode (pulseless) sequential circuits. The clock mode will be considered first because the clock-mode design process is the most straightforward throughout. A disadvantage of the clock mode as the initial topic might be the difficulty in constructing practical examples. Most real-life systems that include a clock contain a large number of states and are often better described by a system design language than a state diagram. Describing larger systems using a hardware description language will be the subject of Chapter 11.

As is the case with electrical networks, the analysis of a sequential circuit is easier than the design. To provide the reader with insight into the goal of the synthesis process, we will first analyze an already designed sequential circuit. Consider the circuit given in Fig. 10.1. Note that this is an example of the general clock-mode model presented in Chapter 9.

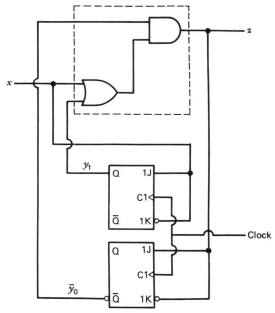

Figure 10.1 Sequential circuit example.

The *memory element input equations*, 10.1 and 10.2, can be determined by tracing connections in the simple combinational logic block.

$$J_{y1} = x, \qquad K_{y1} = \bar{x} \tag{10.1}$$

$$J_{y0} = z = (x + y_1)\bar{y}_0, \qquad K_{y0} = \overline{(x + y_1)\bar{y}_0} \tag{10.2}$$

From these expressions we can use the general next-state equation for the *J-K* flip-flop, $y^{v+1} = J \cdot \bar{y}^v + \bar{K}^v \cdot y^v$, to determine the *next-state equations* for the memory elements y_0 and y_1 as given by Equations 10.3a and 10.3b.

$$y_1^{v+1} = x^v \cdot \bar{y}_1^v + x^v \cdot y_1^v = x^v \tag{10.3a}$$

$$y_0^{v+1} = [(x^v + y_1^v)\bar{y}_0^v] \cdot \bar{y}_0^v + \overline{[(x^v + y_1^v)\bar{y}_0^v]}y_0^v$$
$$= (x^v + y_1^v)\bar{y}_0^v \tag{10.3b}$$

In both these expressions, $K = \bar{J}$, leading to a considerable simplification. In effect, the circuit could have been implemented equally well using D flip-flops insead of *J-K* flip-flops. Although it will usually not be possible to simplify the expressions so readily, the use of $J \cdot \bar{y}^v + \bar{K} \cdot y^v = y^{v+1}$ to determine next-state equations is a completely general procedure in which *J-K* flip-flops are used.

From Equations 10.3a and 10.3b, we can tabulate what we will define as the *transition table* for the sequential circuit. This is merely an evaluation of y_1^{v+1}, y_0^{v+1}, and z^v for all possible combinations of values of x^v, y_0^v, and y_1^v. The transition table for the circuit under discussion is given in Fig. 10.2. For example, if $x^v = y_0^v = 0$ and $y_1^v = 1$, then y_1^{v+1} and y_0^{v+1}, as given by Equations 10.3a and b, are 0 and 1, respectively. These values are entered in the square enclosed by the dashed line in Fig. 10.2. The reader can easily determine the remaining values in the table. The values for z^v are similarly determined. In this case, $z^v = y_0^{v+1}$.

Transition tables of the form of Fig. 10.2 will tend to become rather bulky as the number of variables increases. To simplify the tables, we will assign a shorthand notation, as follows. First, we represent the states by the letter q, with arbitrarily assigned decimal subscripts. For example, we will represent $y_1 y_0 = 00$ by $q_1, y_1 y_0 = 01$ by $q_2, y_1 y_0 = 11$ by q_3, and $y_1 y_0 = 10$ by q_4. In this example, there is only one input and one output. If there are multiple inputs and outputs, these values may be coded also, in which case we represent the set of input variables $(x_1, x_2, ..., x_n)$ by X and the set of output variables $(z_1, z_2, ..., z_q)$ by Z. Particular combinations of values will be represented by the decimal equivalents to the binary interpretations of the combinations. Thus, $x_2 x_1 = 00$ will be represented by $X = 0, x_2 x_1 = 01$ by $X = 1, z_2 z_1 = 11$ by $Z = 3$, etc. We also combine the output and next-state portions of the transition table. This notation results in the *state table*, as shown in Fig. 10.3a. Now the entry in the dashed square, for example, may be interpreted as

$y_1{}^v$	$y_0{}^v$		y_1^{v+1}, y_0^{v+1}			z^v	
		x^v	0		1	0	1
0	0		0 0	1	1	0	1
0	1		0 0	1	0	0	0
1	1		0 0	1	0	0	0
1	0		0 1	1	1	1	1

Figure 10.2 Transition table for circuit of Fig.10.1.

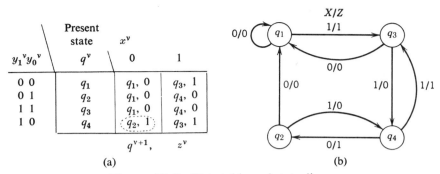

$y_1^v y_0^v$	Present state x^v q^v	0	1
0 0	q_1	q_1, 0	q_3, 1
0 1	q_2	q_1, 0	q_4, 0
1 1	q_3	q_1, 0	q_4, 0
1 0	q_4	q_2, 1	q_3, 1
		q^{v+1},	z^v

(a)

(b)

Figure 10.3 State table and state diagram.

follows: If the circuit is in state q_4 and receives input 0, the output at clock time will be 1 and the next state will be q_2.

The same information contained in the state table may be displayed more graphically by a *state diagram*. The state diagram corresponding to the state table of Fig. 10.3a is shown in Fig. 10.3b. There is a circle for each state and an arrow leaving each state for each input. The arrows terminate at the appropriate next states. Each is labeled with the number of the input that causes that particular transition. The inputs are indicated on the left of each slash and the output corresponding to that particular input present-state combination on the right, that is, X/Z. For example, the square enclosed by the dashed circle in Fig. 10.3a corresponds to $X = 0$, and the output entry is 1. Hence, the arrow is denoted 0/1. Although state tables and state diagrams contain the same information, both are useful at different steps in the synthesis process.

10.2 DESIGN PROCEDURE

The analysis process of the previous section followed a sequence of steps that successively transformed the circuit description as follows:

$$\text{Circuit} \rightarrow \begin{array}{c} \text{Memory} \\ \text{element} \\ \text{input} \\ \text{equations} \end{array} \rightarrow \begin{array}{c} \text{Next} \\ \text{state} \\ \text{equations} \end{array} \rightarrow \begin{array}{c} \text{Transition} \\ \text{table} \end{array} \rightarrow \begin{array}{c} \text{State} \\ \text{table} \end{array} \rightarrow \begin{array}{c} \text{State} \\ \text{diagram} \end{array}$$

The synthesis process resembles an analysis in reverse. The principal distinction is that synthesis begins prior to the existence of a state diagram. This diagram must be obtained from some other description of the problem. Usually, this is merely a hopefully unambiguous natural language description of the intended circuit function. The synthesis procedure is depicted in Fig. 10.4.

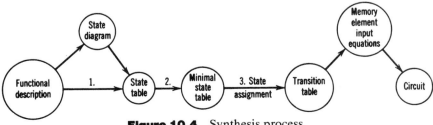

Figure 10.4 Synthesis process.

We will consider the complete process in this chapter. For the reader impatient to get back to the circuit, we first remark that considerable attention must be paid to Steps 1 and 2. Sections 10.3 and 10.4 are concerned with various approaches to obtaining a state table. The next two sections are devoted to finding an equivalent state table with a minimum number of states. A circuit can be obtained from the minimal state table without great difficulty. Obtaining the most economical circuit is not so easy. The complexity of the memory element input equations will vary with the assignment of combinations of flip-flop output values to states in the state table. The attempt to optimize this correspondence is called the *state assignment problem*.

The state diagram and state table can become cumbersome when a large number of input lines are present. If only a small subset of these inputs is sensed while the circuit is in any given state, a more convenient notation, called an *algorithmic state machine* (ASM) chart, can be used in place of the state diagram. The ASM chart [13] resembles a program flowchart, but can represent the same information as a state diagram. We shall feel free to make use of an ASM chart when convenient in a later example.

10.3   SYNTHESIS OF STATE DIAGRAMS

As depicted in Fig. 10.4, one approach to obtaining the state table involves the use of a state diagram. This method is particularly convenient for certain types of circuits that possess a readily identifiable state called a *reset state*. When such a state exists, there is a mechanism by which the circuit may be returned to the reset state from any other state in a single operation. This operation may be specified as a special input column in the state table, or it may function independently of the state table. For example, a special line may be connected to all flip-flops that will clear them to 0. This line may be connected to separate input points rather than those used in the implementation of the state table. Example 10.1 involves a simple circuit with a reset state.

■ EXAMPLE 10.1

The beginning of a message in a particular communications system is denoted by the occurrence of three consecutive 1's on a line x. Data on this line have been synchronized with a source of clock pulses. A clock-mode sequential circuit is to be designed that will have an output of 1 only at the clock time coinciding with the third of a sequence of three 1's on line x. The circuit will serve to warn the receiving system of the beginning of a message. It will be provided with a separate reset mechanism to place it in state q_0 following the end of a message. Typical behavior of the circuit is depicted in Fig. 10.5.

SOLUTION   The synthesis procedure begins by designating the reset state as q_0, as shown in Fig. 10.6a. Essentially, the circuit must keep track of the number of consecutive 1's it has received. It does this by moving to a different state with each input of 1. As shown in Fig. 10.6b, the circuit goes to state q_1 with the first 1 input and to state q_2 with the second. The outputs associated with both of these inputs are 0, as shown. The third consecutive input of 1 generates a 1 output and causes the circuit to go to q_3. Once in q_3, the circuit will remain in this state, emitting 0 outputs until it is externally reset to q_0. This reset mechanism is not shown in the state table.

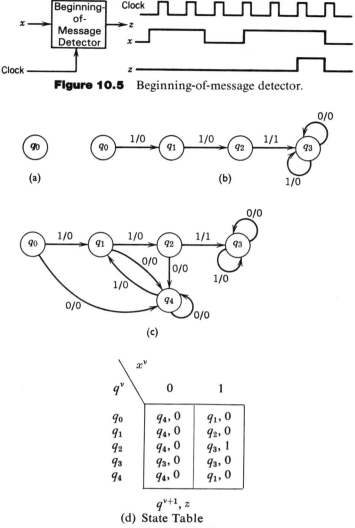

Figure 10.5 Beginning-of-message detector.

Figure 10.6 State diagram for beginning-of-message detector. (d) State table.

It is possible for an input of 0 to appear during any clock period, possibly interrupting a sequence of 1's. A state q_4 is provided for this eventuality. Every input of 0 (unless the circuit is already in q_3) generates a 0 output and causes the circuit to go to q_4, as shown in Fig. 10.6c. A second 0 merely causes the circuit to remain in state q_4. An input of 1 while the circuit is in q_4 may be the first of a sequence of three 1's. Thus, a 1 will cause the circuit to change state from q_4 to q_1.

All possible input sequences that can occur have been considered. The reader may verify this by noting that arrows for both 0 and 1 inputs have been defined, leaving each of the five states. Thus, Fig. 10.6c is the completed state diagram. This translates directly to the state table of Fig. 10.6d. ∎

The perceptive reader may have noticed that the circuit could be returned to q_0 for inputs of 0. This would eliminate the need for state q_4. As a general rule, already existing states should be used as next states whenever possible. At some point, the designer must stop defining new states. However, since state diagram construction is an intuitive procedure, there may be doubt as to whether an existing state will serve. *When in doubt, define a new state*, thus avoiding serious error.

■ EXAMPLE 10.2

Numbers between 0 and 7 expressed in binary form are transmitted serially on line x from a digital measuring device. The bits appear in descending order of significance, synchronized with a clock. Design a circuit that will have an output $z = 1$ at the clock time of the third bit of a binary number representing an extreme value (0 or 7) on line x. Otherwise, $z = 0$. The circuit will have two inputs, x and d, in addition to the clock. If $d = 1$ at a given clock time, then the most significant bit of a measurement will appear on line x at the next clock time.

Assume that $d = 1$ will cause the circuit to assume a reset state a. Construct a state table regarding line d as an external reset mechanism. Then revise the state table to include input d as part of the state table.

SOLUTION   Beginning with state a, we construct a tree of input arrows as shown in Fig. 10.7a. Each arrow leads to a new state from which two new arrows (in general, one for each possible input) emerge. In this problem, it is only necessary to generate two levels of the tree. The numbers 0 and 7 are detected at the time of the third input. The binary number is 0 if an input of 0 occurs when the circuit is in state d. An input of 1 for state g indicates 7. Outputs of 1 are shown for these cases.

The third input must take the circuit to a state for which subsequent outputs of 1 are impossible. Notice that states e and f satisfy this requirement. Once the circuit is in either of these states, it remains there permanently; the outputs are always 0. In fact, the circuit will behave in the same manner whether it is in state e or state f. Therefore, only one of these states is necessary, as emphasized by the dashed closed curve in Fig. 10.7a. In translating the state diagram to the state table in Fig. 10.7b, state f is omitted. When f would be a next state, it is replaced by e, as indicated by asterisks.

The state table with the effect of the reset input d included is given in Fig. 10.8b. The four combinations of the inputs x and d are coded as shown in Fig. 10.8a. The

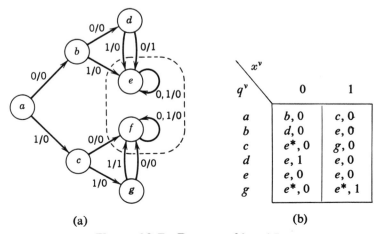

q^ν \ x^ν	0	1
a	$b, 0$	$c, 0$
b	$d, 0$	$e, 0$
c	$e^*, 0$	$g, 0$
d	$e, 1$	$e, 0$
e	$e, 0$	$e, 0$
g	$e^*, 0$	$e^*, 1$

(a)   (b)

Figure 10.7   Detector of 0 and 7.

two columns for $X = 0$ and $X = 1$ correspond to $d = 0$. These two columns are a direct copy of Fig. 10.7b. Anytime $d = 1$, the circuit is reset to state a, as shown. If the sequential circuit is implemented as given by Fig. 10.8b, no external reset mechanism is required. ∎

The response tree of Fig. 10.7 illustrates a general procedure that can be used in the absence of other insight. When no specific reset state exists, an arbitrary state may be used as a starting point. Preferably this state will have some past history to distinguish it clearly from other states. In the absence of a well-defined starting state, the tree may grow very large before the branches are terminated. In general, the designer should prune the tree steadily by routing arrows to existing states that lead to the same future behavior as might be defined for new states. If the designer is slow in applying insight, the size of a response tree can get out of hand rapidly.

10.4 FINITE MEMORY CIRCUITS

An equivalent version of the circuit of Fig. 10.1 is shown in Fig. 10.9a. The J-K flip-flops with inverters on the clear input are replaced with equivalent D flip-flops, with the clock input omitted for clarity.

Recalling that the output of a D flip-flop is equal to the value of the input at the immediately previous clock time, t^{v-1}, we see that in this case the present output is a function only of the present input and the immediately previous input and output, that is,

$$z^v = f(x^v, x^{v-1}, z^{v-1}) \tag{10.4}$$

This circuit is an example of a *finite memory* circuit, *finite* in the sense that it need "remember" only a finite number of past inputs and outputs in order to determine the appropriate present output and next state. Note that *finite* here has nothing to do with the memory capacity in bits, which is, of course, always finite.

A generalized version of a finite memory circuit is shown in Fig. 10.9b. Note that r previous inputs and s previous outputs are stored. The current output is a function of this stored information and the current input. The symbols X and Z are vectors that may represent several input and output lines. The blocks labeled D are not individual flip-flops but represent one-dimensional arrays of flip-flops, one for each input or output line.

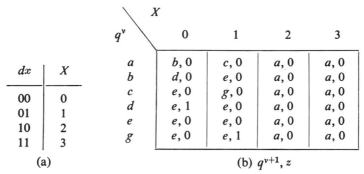

dx	X
00	0
01	1
10	2
11	3

(a)

q^v \ X	0	1	2	3
a	$b, 0$	$c, 0$	$a, 0$	$a, 0$
b	$d, 0$	$e, 0$	$a, 0$	$a, 0$
c	$e, 0$	$g, 0$	$a, 0$	$a, 0$
d	$e, 1$	$e, 0$	$a, 0$	$a, 0$
e	$e, 0$	$e, 0$	$a, 0$	$a, 0$
g	$e, 0$	$e, 1$	$a, 0$	$a, 0$

(b) q^{v+1}, z

Figure 10.8 Including input d in the state table.

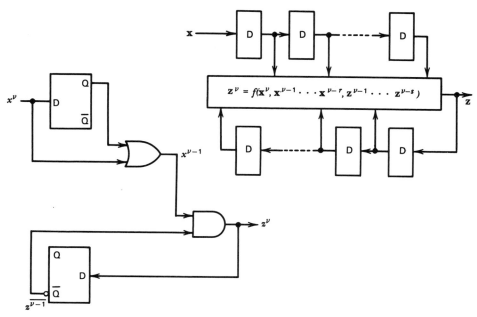

Figure 10.9 Finite memory circuits.

Not all sequential circuits or finite-state machines are finite memory. Consider a circuit whose output Z is initially 0. After some sequence of inputs S, $Z = 1$. The output remains 1 until a second identical sequence of inputs S is observed. At the end of this sequence, Z returns to 0 where it remains forever. Based on remembering only a finite number of past inputs, the circuit will be unable to determine whether an observed sequence S is the first or second such sequence. If only a finite number of consecutive output values are remembered, the machine will not distinguish between 0 or two input sequences S in its history, once that finite number of clock periods have elapsed following the last appearance of a sequence S. The sequential circuit just described is not finite memory. Example 10.5 will demonstrate that such an activity is nonetheless realizable as a finite-state machine.

It is often possible to recognize a finite memory circuit from the initial specifications, and this knowledge can often simplify the synthesis procedure. If the designer knows the circuit is finite memory, she or he can generally specify an upper bound on the memory size, that is, the maximum number of internal states. The designer can then set up either the transition table or state table directly, as shown in the following example.

■ **EXAMPLE 10.3**

Information bits are encoded on a single line x so as to be synchronized with a clock. Bits are encoded so that two or more consecutive 1's or four or more consecutive 0's should never appear on line x. An error-indicating sequential circuit is to be designed to indicate an error by generating a 1 on output line z coinciding with the fourth of every sequence of four 0's or the second of every sequence of two 1's. If, for example, three consecutive 1's appear, the output remains 1 for the last two clock periods.

SOLUTION   The circuit to be designed may be immediately recognized as finite memory. In this case, we may call it *finite input memory* in that only inputs need be remembered. In the worst case, the circuit will have to remember the value of three previous inputs. If these stored inputs are all 0 and the present input is 0, then the circuit output is 1. A circuit of the form shown in Fig. 10.10a can be made to do the job. As the input bits arrive, they are shifted through the string of flip-flops so that all times $y_2 = x^{v-1}, y_1 = x^{v-2}, y_0 = x^{v-3}$. The combinational logic then determines the output on the basis of these stored inputs and the present input.

The transition table for this circuit can be set up directly, as shown in Fig. 10.10b. First, list all eight combinations of state variables as the present-state entries. The next-state entries, $(y_2 y_1 y_0)^{v+1}$, follow directly from noting that the state variables shift to the right each clock time. For example, if the present values are 000 and the input is 1, the next values are 100. To complete the transition table, the appropriate outputs should be entered. The output is to be 1 if the three previous inputs and the present input are all 0 or if the immediately previous and present inputs are both 1. If the designer wishes to use the circuit form shown in Fig. 10.10a, the only step left is the design of the combinational logic to generate the output z. For this case,

$$z = xy_2 + \bar{x}\bar{y}_2\bar{y}_1\bar{y}_0 \qquad (10.5)$$

■

Although the circuit form of Fig. 10.10a will work, it may not be minimal. In only one case is it necessary for the circuit to "remember" three inputs back. If x^{v-1} = 1, it does not matter what x^{v-2} and x^{v-3} were. Thus, it appears that less than eight internal states may be sufficient. To check on this, the designer may convert the transition table to a state table simply by assigning an arbitrary state designation to each combination of state variables in the table of Fig. 10.10b. The resulting state table, Fig. 10.10c, may then be minimized by the formal techniques to be developed in later sections.

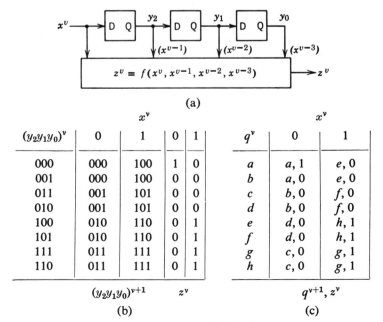

$(y_2 y_1 y_0)^v$	x^v 0	1	0	1
000	000	100	1	0
001	000	100	0	0
011	001	101	0	0
010	001	101	0	0
100	010	110	0	1
101	010	110	0	1
111	011	111	0	1
110	011	111	0	1
	$(y_2 y_1 y_0)^{v+1}$		z^v	
	(b)			

q^v	x^v 0	1
a	a, 1	e, 0
b	a, 0	e, 0
c	b, 0	f, 0
d	b, 0	f, 0
e	d, 0	h, 1
f	d, 0	h, 1
g	c, 0	g, 1
h	c, 0	g, 1
	q^{v+1}, z^v	
	(c)	

Figure 10.10 Synthesis of circuit with finite input memory of length 3.

In a finite input memory circuit of the general form of Fig. 10.10a, the memory essentially acts as a serial-to-parallel converter. The input information required to make an output decision is arriving in serial, that is, in time sequence, so the circuit simply stores the inputs until enough have been received to make the output decision and then applies them all at once, that is, in parallel, to the combinational logic implementing the output function. Any time it can be determined that the output of a sequential circuit depends on no more than some maximum number of previous inputs, then the approach discussed in the previous example can be used.

If the longest sequence has length n, we say that the finite input memory circuit has a *memory of length n*, and n memory elements will be required for each input line. Assume there is only one input line. Then the memory will have 2^n possible states, that is, 2^n distinct input messages can be stored. If the output is to be different for each of these 2^n messages, then the general form of Fig. 10.10a is the most economical, and the designer's task reduces to nothing more than realizing the output function.

If the outputs are the same for many of the input messages, then fewer memory elements may be sufficient. In the above example, we have a memory of length 3 so that eight distinct messages can be stored. As we have already seen, however, all messages with $x^{v-1} = 1$ produce the same output, so the memory need not distinguish among the four messages with this characteristic. Thus, by processing, or classifying, the messages as they arrive, we may be able to reduce the memory requirements.

If we reduce the memory in this manner, the circuit form will not be the same as Fig. 10.10a, in that the outputs of the memory elements will not be equal to previous inputs but will be the functions of them. Nevertheless, we may still use the state table derived from this basic circuit form as a starting point in the design process. The situation is somewhat analogous to first specifying a combinational function as a standard or expanded sum of products. This may not be minimal, but it allows us to express the function in Karnaugh map form, from which a minimal expression can be determined.

Once we have any state table describing the appropriate behavior, we can derive from it any number of equivalent tables, including a unique minimal table. For an assumed circuit form such as Fig. 10.10a, the transition table, with regard to next-state entries, is unique for a given memory length and number of inputs since the next-state entries are simply derived by shifting the inputs through the memory. We may thus set up standard transition tables for finite input memory circuits. These tables are shown in Fig. 10.11 for single-input circuits of length 2, 3, and 4. To form the complete transition tables for any specific circuit, the designer need only fill in the appropriate outputs. If the number of distinct outputs is more than half the number of states, the number of memory devices cannot be reduced, so the designer should use the standard circuit form and proceed directly to the design of the output logic. If the number of distinct outputs is less than half the number of states, the designer should generally minimize the corresponding state table before deciding on a circuit form.

Although *finite input memory* circuits are possibly the easiest to recognize and design, general finite memory circuits (of the form of Fig. 10.9b) are also important, and recognizing their finite memory character can often simplify the design process. This is illustrated in the next example.

$x^{v-1}x^{v-2}$	q^v	x^v 0	1
00	a	a	d
01	b	a	d
11	c	b	c
10	d	b	c

q^{v+1}

(a) $n = 2$

$x^{v-1}x^{v-2}x^{v-3}x^{v-4}$	q^v	x^v 0	1
0000	a	a	i
0001	b	a	i
0011	c	b	j
0010	d	b	j
0100	e	d	l
0101	f	d	l
0111	g	c	k
0110	h	c	k
1000	i	e	m
1001	j	e	m
1011	k	f	n
1010	l	f	n
1100	m	h	q
1101	n	h	q
1111	p	g	p
1110	q	g	p

q^{v+1}

(c) $n = 4$

$x^{v-1}x^{v-2}x^{v-3}$	q^v	x^v 0	1
000	a	a	e
001	b	a	e
011	c	b	f
010	d	b	f
100	e	d	h
101	f	d	h
111	g	c	g
110	h	c	g

q^{v+1}

(b) $n = 3$

Figure 10.11 Standard-state tables for single-input finite input memory circuits of length n.

■ EXAMPLE 10.4

Design a three-bit (modulo-8) up-down counter. The count is to appear as a binary number on three output lines, z_2, z_1, and z_0. The most significant bit is z_2. The count will change with each clock pulse. If the input $x = 1$, the count will increase. If $x = 0$, the count will decrease. Modulo-8 implies that increasing the count from 7 results in a count of 0, and decreasing it from 0 gives a count of 7.

SOLUTION   One can immediately recognize the up-down counter as a *finite output memory* machine of memory length 1. The value of the output z is a function of only the immediately previous output z^{v-1}, and the present input x. The output z, however, is a three-bit vector with components z_2, z_1, and z_0, so that three flip-flops are required to represent one past output.

A circuit configuration and the corresponding transition table are shown in Fig. 10.12. The transition table is constructed directly by noting that the desired present values of the outputs are the next values of the state variables. For example, for the circled entry in Fig. 10.13b, the previous count (present state) was 011, so if $x = 1$, the output (next state) should be 100.

Clearly, eight states are needed for a modulo-8 counter, so the table of Fig. 10.12b is minimal. If the circuit form of Fig. 10.12a is satisfactory, the designer can proceed directly to the design of the combinational logic circuits to realize the required outputs. However, even though the table is minimal, the form of Fig.

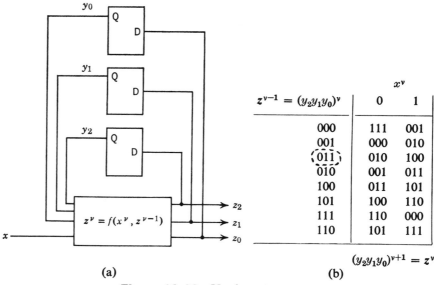

$z^{v-1} = (y_2y_1y_0)^v$	x^v	
	0	1
000	111	001
001	000	010
(011)	010	100
010	001	011
100	011	101
101	100	110
111	110	000
110	101	111

$$(y_2y_1y_0)^{v+1} = z^v$$

(a)   (b)

Figure 10.12   Up-down counter.

q^v	x^v	
	0	1
q_0	$q_7, 7$	$q_1, 1$
q_1	$q_0, 0$	$q_2, 2$
q_2	$q_1, 1$	$q_3, 3$
q_3	$q_2, 2$	$q_4, 4$
q_4	$q_3, 3$	$q_5, 5$
q_5	$q_4, 4$	$q_6, 6$
q_6	$q_5, 5$	$q_7, 7$
q_7	$q_6, 6$	$q_0, 0$

$$q^{v+1}, z$$

Figure 10.13   State table for up-down counter.

10.12 is not the only possible circuit form. Different types of flip-flops might be used, in which case the state variables need not be equal to the most recent outputs. For completeness, we assign state numbers to the transition table to obtain the state table of Fig. 10.13. The corresponding state diagram is given in Fig. 10.14. ■

Two distinct approaches to state table synthesis have been described, both of which should be mastered by the student. Sometimes the most difficult task for the novice designer is deciding which method to use in a specific case. If the problem can be identified as finite memory, the direct synthesis of the state table or transition table is generally preferable. The state diagram approach works best when a reset or starting state can be identified; as it happens, machines with reset states are often nonfinite memory machines. A nonfinite memory machine will usually be driven to some holding state or caused to cycle within a subset of its states after it has made the desired response to some input sequence. Examples 10.1 and 10.2 illustrate machines with holding states. If such a machine is to be reusable, there must be an input sequence that will return it to its initial state. A

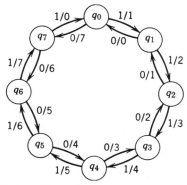

Figure 10.14 State diagram for up-down counter.

fixed sequence that always will return a sequential machine to its initial state is called a *synchronizing sequence*. An initial state that can be reentered in this manner will be called a *reset state*. Often the synchronizing sequence will be a single input on a special line called the reset line.

■ **EXAMPLE 10.5**

Repeat the design of the 0, 7 detector of Example 10.2, except that now the reset line d is to be omitted. Instead, a second 7 (three 1's in a row), coming any time after a three-bit group has been tested, will signal that a new character will start arriving at the next clock time.

SOLUTION The state diagram (Fig. 10.15a) through state g is the same as Fig. 10.7a with e and f combined, as discussed in Example 10.2. We now recognize e as the holding state characteristic of nonfinite memory circuits. The reset is now provided by a sequence of three 1's taking the circuit from e back to a via h and i. The resulting state table is shown in Fig. 10.15b. Since the sequence of three 1's must be applied after the circuit has reached state e, it is not, strictly speaking, a synchronizing sequence. ■

In some problems, it might not be possible to identify either a reset state or a bound on the memory. In such cases, one must arbitrarily choose some state whose past history can be specified as closely as possible to use as an initial state.

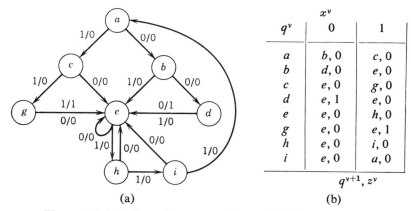

q^v	x^v	
	0	1
a	$b, 0$	$c, 0$
b	$d, 0$	$e, 0$
c	$e, 0$	$g, 0$
d	$e, 1$	$e, 0$
e	$e, 0$	$h, 0$
g	$e, 0$	$e, 1$
h	$e, 0$	$i, 0$
i	$e, 0$	$a, 0$
	q^{v+1}, z^v	

(a)　　　　　　　　(b)

Figure 10.15 State diagram and state table, Example 10.6.

From here, special care should be applied in the generation of a state table. Occasionally, the reader may encounter the dilemma of a finite memory machine with a reset state. In that case, either approach may be used.

10.5   EQUIVALENCE RELATIONS

The procedures discussed in Section 10.4 provide means of obtaining a state table for a given specification, but the resulting table is not unique and may not be optimum in terms of the ultimate realization. Since the number of flip-flops in a circuit increases as the number of states increases, we may say that a minimal state table is one with a minimum number of states. In some cases, different states in the original state table will be found to perform the same function. Or more specifically, it may not be possible, by means of output measurements, to distinguish between two or more states. In such cases, the states are said to be *equivalent* and may be replaced by a single state. In order to develop a more concrete definition of the equivalence of states, let us first pursue the concept of an *equivalence relation*.

An equivalence relation was defined in Section 3.6 as a relation satisfying the *reflexive*, *symmetric*, and *transitive* properties. An example of an equivalence relation might be "on the same team as." We recall that if xRx for every x in the set of interest, then the relation R is said to be *reflexive*. The relation "is greater than or equal to" is reflexive, whereas the relation "is greater than" is not reflexive.

If yRx whenever xRy, then the relation R is *symmetric*. On sets of usual interest, a symmetric relation will be reflexive as well. An exception is a relation such as "is a cousin of," which is symmetric. In the usual sense, one is not his or her own cousin, so this relation is not reflexive. Neither of the relations in the previous paragraph ($x \geq y$ or $x > y$) is symmetric. Finally, if xRy and yRz imply xRz, then R is a *transitive* relation.

An equivalence relation on a set will always partition the set into disjoint subsets, known as *equivalence classes*. For example, the set

$$\{A, a, B, b, C, c, D, d, E, e\}$$

is partitioned into the subsets

$$\{A, B, C, D, E\} \qquad \text{and} \qquad \{a, b, c, d, e\}$$

if xRy is interpreted to mean that x and y are the same case. A different group of disjoint subsets is obtained if xRy is taken to mean that x and y are the same letter.

Our intuitive concept of equality is an example of an equivalence relation since

$$x = x \qquad \qquad \text{(Reflexive)}$$

and

$$y = x \quad \text{if} \quad x = y \qquad \qquad \text{(Symmetric)}$$

$$\left. \begin{array}{l} x = y \quad \text{and} \quad y = z \\ \\ x = z \end{array} \right\} \qquad \qquad \text{(Transitive)}$$

then

10.6 EQUIVALENT STATES AND CIRCUITS

In order to define the notion of equivalent states and equivalent sequential circuits, we must introduce some new notation. State tables comprise tabular representations of two functions: the output function and the next-state function. If the outputs and next states are specified for every combination of inputs and present states, the circuits are called *completely specified* circuits. Let us denote the next-state function as δ an the output function as λ. For example, in the circuit of Fig. 10.16,

$$\lambda(q_2, 3) = 0 \qquad (10.6)$$

and

$$\delta(q_2, 3) = q_4 \qquad (10.7)$$

If a circuit is in some initial state, it will respond to a sequence of inputs with a specific sequence of outputs. Consider the case where the circuit of Fig. 10.16 is initially in the state q_1 and is subjected to the series of inputs 0 2 3 0 0 1. In order to determine the output sequence in response to this input sequence, it is also necessary to determine the next-state sequence

$$\lambda(q_1, 0) = 0 \qquad \text{and} \qquad \delta(q_1, 0) = q_3$$
$$\lambda(q_3, 2) = 2 \qquad \text{and} \qquad \delta(q_3, 2) = q_1$$
$$\lambda(q_1, 3) = 0 \qquad \text{and} \qquad \delta(q_1, 3) = q_2$$
$$\lambda(q_2, 0) = 0 \qquad \text{and} \qquad \delta(q_2, 0) = q_3$$
$$\lambda(q_3, 0) = 0 \qquad \text{and} \qquad \delta(q_3, 0) = q_3$$

and, finally,

$$\lambda(q_3, 1) = 1 \qquad \text{and} \qquad \delta(q_3, 1) = q_1$$

The output sequence and final state may now be summarized as a function of the initial state and input sequence as follows:

$$\lambda(q_1, 023001) = 020001$$
$$\delta(q_1, 023001) = q_1 \qquad (10.8)$$

Now let us take a different point of view. Suppose that the output sequence for a given initial state and input sequence is known. Is the sequence of internal states then of any real interest? The answer is "no." To the "outside world," only the output response of a circuit to a sequence of inputs is significant. In a sense, the function of a sequential circuit may be regarded as the translation of one sequence of signals to a second sequence.

Present State $\backslash x^v$	0	1	3	2
q_1	$q_3,0$	$q_1,0$	$q_2,0$	$q_2,0$
q_2	$q_3,0$	$q_3,0$	$q_4,0$	$q_4,0$
q_3	$q_3,0$	$q_1,1$	$q_1,3$	$q_1,2$
q_4	$q_4,0$	$q_4,0$	$q_2,0$	$q_2,0$

q^{v+1}, z

Figure 10.16 Typical state table.

We are thus led toward a meaningful definition of the equivalence of two sequential circuits. First, let us define the equivalence of two states within these circuits.

Definition 10.1 Let S and T be completely specified circuits, subject to the same possible input sequences. Let $(X_1, X_2, ..., X_n)$ represent a sequence of possible values of the input set X, of arbitrary length. States $p \in T$ and $q \in S$ are *indistinguishable* (equivalent), written $p \equiv q$, iff (if and only if) $\lambda_T(p, X_1, ..., X_n) = \lambda_s(q, X_1, X_2, ..., X_n)$ for every possible input sequence.

This definition may be applied equally well to different states of a single circuit. Less formally, p and q are equivalent if there is no way of distinguishing between them on the basis of any output sequences starting from these states. We use the terms *indistinguishable* and *equivalent* almost synonymously. *Indistinguishable* will be used when necessary to emphasize that indistinguishability is only one of many equivalence relations that might be defined on the states of a state table.

The equivalence of states in a single circuit is an equivalence relation. This can be easily verified as follows. Since the machine is completely specified, the output sequence resulting from the application of a given input sequence to a particular initial state q will always be the same. Therefore, $q \equiv q$, satisfying the reflexive law. Clearly, no distinction can be made between $p \equiv q$ and $q \equiv p$, so the symmetric law is satisfied. Furthermore, if p and q lead to identical output sequences for every input sequence and q and r lead to identical output sequences for every input sequence, certainly p and r must lead to identical output sequences for every input. That is, if $p \equiv q$ and $q \equiv r$, then $p \equiv r$, and the transitive law is satisfied.

Definition 10.2 The sequential circuits S and T are said to be equivalent, written $S \equiv T$ iff for each state p in T there is a state q in S such that $p \equiv q$, and, conversely, for each state q in S there is a state p in T such that $q \equiv p$.

Less formally, two circuits are equivalent if there is no way to distinguish between them simply by observing the response to input sequences.

As an example, consider the simple circuits described by the state tables of Fig. 10.17. We first examine circuit S. Suppose that this circuit is initially in state q_1. An input of 0 will take the circuit of state q_3 with an output of 0. As long as 0 inputs continue, the circuit will cycle back and forth between q_3 and q_1 with 0 outputs. Should a 1 input occur, the circuit will go to q_2, and the output will be 1. A 0 input would then send the circuit back to q_1.

Note that if the circuit were initially in state q_3, exactly the same output sequences as outlined above would occur. This is illustrated in Fig. 10.18 for two input sequences. These two input sequences are, in fact, sufficient to demonstrate that states q_1 and q_3 are equivalent. Readers who remain in doubt should test the circuit with input sequences of their own design until convinced.

S	$X = 0$	$X = 1$
q_1	$q_3, 0$	$q_2, 1$
q_2	$q_1, 1$	$q_2, 0$
q_3	$q_1, 0$	$q_2, 1$

T	$X = 0$	$X = 1$
p_1	$p_1, 0$	$p_2, 1$
p_2	$p_1, 1$	$p_2, 0$

Figure 10.17

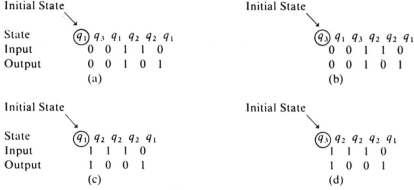

Figure 10.18

Note now that these or any other input sequences will also result in the same outputs for initial state p_1 in circuit T. Thus, p_1 is equivalent to both q_1 and q_3. Similarly, p_2 is equivalent to q_2. Thus, corresponding to every state in T, there is an equivalent state in S, and vice versa. Therefore, by Definition 10.2, S and T are equivalent.

Our basic problem is to find an optimum state table. In view of the above definitions, the problem can now be stated in more specific terms: Given a state table, find an equivalent state table with as few states as possible. A table with fewer states may require less memory, thus generally leading to a more economic realization. For example, circuit S of Fig. 10.17 has three states and would therefore require two flip-flops for realization. Circuit T, having only two states, would require only a single flip-flop.

10.7 DETERMINATION OF CLASSES OF INDISTINGUISHABLE STATES

Our basic approach to finding an optimum state table will be to partition the state table into the smallest possible number of *equivalence classes* of indistinguishable states. We will then show that an equivalent sequential circuit can be formed by defining one state corresponding to each class of indistinguishable states. There are many ways to partition the states of a sequential circuit into disjoint classes. Not all such partitions result in equivalence classes of indistinguishable states. The following theorem provides a test for this condition.

THEOREM 10.1 Let the states of a sequential circuit be partitioned into disjoint classes. $p \triangleq q$ denotes that states p and q fall into the same class in the partition. The partition is composed of equivalence classes of indistinguishable states (two indistinguishable states must be in the same class) if and only if the following two conditions are satisfied for every pair of states p and q in the same class ($p \triangleq q$) and every single input X.

$$(1)\ \lambda(p, X) = \lambda(q, X)$$
$$(2)\ \delta(p, X) \triangleq \delta(q, X)$$

PROOF A. (Necessity.) Assume that the partition consists of equivalence classes of indistinguishable states. Let p and q be any pair of states such that $p \triangleq q$. Therefore, $p \equiv q$. By Definition 10.1, we can write

$$\lambda(p, XX_1X_2 \cdots X_k) = \lambda(q, XX_1X_2 \cdots X_k) \tag{10.9}$$

for any sequence $X_1X_2 \cdots X_k$ and any single input X. Both sides of Equation 10.9 may equally well be written as the concatenation of a single output and the remaining sequence of outputs.

$$\lambda(p, X)\lambda[\delta(p, X), X_1X_2 \cdots X_k] = \lambda(q, X)\lambda[\delta(q, X), X_1X_2 \cdots X_k] \tag{10.10}$$

Therefore,

$$\lambda(p, X) = \lambda(q, X)$$

and

$$\lambda[\delta(p, X), X_1X_2 \cdots X_k] = \lambda[\delta(q, X), X_1X_2 \cdots X_k] \tag{10.11}$$

so that

$$\delta(p, X) \equiv \delta(q, X) \qquad \text{and} \qquad \delta(p, X) \triangleq \delta(q, X) \qquad \text{Q.E.D.}$$

B. (Sufficiency.) Assume that the states of a sequential circuit are partitioned into a set of equivalence classes such that two states fall into the same class if and only if conditions (1) and (2) are satisfied. Let p and q be any two states in the same equivalence class, that is, $p \triangleq q$. Suppose that $\delta(p, X_1X_2 \cdots X_i) \triangleq \delta(q, X_1X_2 \cdots X_i)$. Therefore, by condition (2),

$$\delta(p, X_1X_2 \cdots X_{i+1}) \triangleq \delta(q, X_1X_2 \cdots X_{i+1})$$

so that, by induction,

$$\delta(p, X_1X_2 \cdots X_R) \triangleq \delta(q, X_1X_2 \cdots X_R) \tag{10.12}$$

for all T. Since

$$\lambda(p, X_1 \cdots X_R) = \lambda(p, X_1)\lambda[\delta(p, X_1), X_2] \cdots \lambda[\delta(p, X_1 \cdots X_{R-1}), X_R]$$

and

$$\lambda(q, X_1 \cdots X_R) = \lambda(q, X_1)\lambda[\delta(q, X_1), X_2] \cdots \lambda[\delta(q, X_1 \cdots X_{R-1}), X_R] \tag{10.13}$$

and the sequences of symbols on each side of these expressions are identical, we have

$$\lambda(p, X_1 \cdots X_R) = \lambda(q, X_1 \cdots X_R)$$

for any k. Therefore, by Definition 10.1, $p \equiv q$, and the partition is composed of equivalence classes of indistinguishable states.   Q.E.D.

We are now ready to use Theorem 10.1 to partition example sequential circuits into equivalence classes of indistinguishable states. As a first example, we will use a successive partitioning technique to reduce the number of states in the error-detecting sequential circuit of Example 10.3. This technique, usually referred to as the Huffman–Mealy method, will be illustrated in Example 10.6.

■ EXAMPLE 10.6

Partition the states of the sequential circuit in Fig. 10.19 into equivalence classes of indistinguishable states.

SOLUTION   The method consists of two steps. The first step is to partition the states into the smallest possible number of equivalence classes so that states in

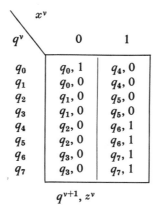

Figure 10.19 State table for error detector.

the same class have the same outputs, that is, satisfy condition (1) of Theorem 10.1. This is done by inspection of the state table. For the state table in Fig. 10.19, we see that three classes are required. Class **a** will include states whose output is 1 for input 0 and 0 for input 1. Class **b** will include states whose output is 0 for either input. Class **c** will include states whose output is 0 for input 0 and whose output is 1 for input 1.

$$\begin{array}{ccc} \mathbf{a} & \mathbf{b} & \mathbf{c} \\ (q_0) & (q_1, q_2, q_3) & (q_4, q_5, q_6, q_7) \end{array}$$

The classes of indistinguishable states will always be composed of subclasses of classes **a**, **b**, and **c**. These classes are the largest classes that will satisfy condition (1). Insisting on the satisfaction of condition (2) will only further subdivide the classes.

The partition satisfies the first condition of Theorem 10.1. If the next states for each state in class **b** (and for the states in class **c**) are in the same class for every possible single input, the second condition of Theorem 10.1 would be satisfied. In such a case, classes **a**, **b**, and **c** would be the desired equivalence classes. Let us determine for each state whether the next states are in classes **a**, **b**, and **c**. Below, the next-state class for a 0 input is written to the left of each state; the next-state class for a 1 input is written to the right.

Class	**a**	**b**			**c**			
States	0	1	2	3	4	5	6	7
Next Class	a c	a c	b c	b c	b c	b c	b c	b c

Clearly, class **b** contains pairs of states that do not satisfy condition (2) and are therefore not indistinguishable. We must separate class **b** into two classes **b** and **d**. The new partition may or may not consist of equivalence classes of indistinguishable states. To find out, the process of listing the classes containing the next states is repeated.

Class	**a**	**b**		**c**				**d**
States	0	2	3	4	5	6	7	1
Next Class	a c	d c	d c	b c	b c	b c	b c	a c

In this case, the four classes **a**, **b**, **c**, and **d** are equivalence classes of indistinguishable states. Note that all states in a class have next states in the same class for inputs or 0 and 1. Thus, condition (2) is finally satisfied.   ∎

The partitioning of states into classes whose outputs agree and whose next states are in the same class is graphically illustrated in Fig. 10.20 for the above example. It should be clear to the reader that the same inputs applied to two states in the same class will result in the same output and send the circuit to next states in the same class. Any input applied to both these states will yield the same output and again send the circuit to next states in the same class, and so on indefinitely. Therefore, the class containing the two initial states is an equivalence class. Similar arguments apply to the other classes.

Having arrived at a set of equivalence classes of states within a circuit S, as in the previous example, it remains to determine from them a minimal state circuit equivalent to S.

Equivalence Class Circuit   Corresponding to a completely specified sequential circuit S, form a circuit T, with one state p_j corresponding to each equivalence class C_j of the states of S. As each state in C_j has the same output $\lambda(C_j, X)$ for any input X, let $\lambda(p_j, X) = \lambda(C_j, X)$. Similarly, each state of C_j has, for any input X, a next state in the same class C_k. Therefore, we let $\delta(p_i, X) \equiv p_k$. Given this construction, it can be shown by an argument similar to the proof of Theorem 10.1 that for every class C_k, $p_k \equiv q_{ki}$ for every $q_{ki} \in C_k$. A circuit T, formed in this manner, is known as an equivalence class circuit.

THEOREM 10.2   The circuit T as defined above is equivalent to S. Furthermore, no other circuit equivalent to S has fewer states than T.

PROOF   Let q_j be any state in S. Since it falls in some equivalence class C_j, there is a corresponding state p_j in T such that, according to the above definition, $q_j = p_j$. Conversely, for any state p_j in T, there is a corresponding equivalence class C_j such that, by the above definition, p_j is equivalent to any state in the class. Thus, by Definition 10.2, the circuits are equivalent.

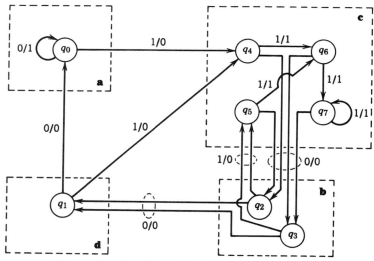

Figure 10.20   Illustration of equivalence classes.

No single state of any circuit can be equivalent to two states q_1 and q_2 of circuit S, which are in different equivalence classes. By definition, these states must have different output sequences for some input sequence. Thus, the minimal state equivalent of S can have no fewer states than T, and one state must be assigned to each class.

Q.E.D.

■ **EXAMPLE 10.6 (continued)**

Determine the minimal state equivalent of the sequential circuit in Fig. 10.19.

SOLUTION   Corresponding to each of the classes **a, b, c,** and **d,** we define states p_1, p_2, p_3, p_4 in a circuit T. The outputs and next states, defined according to Theorem 10.2, are listed in the state table of Fig. 10.21*a*. The corresponding state diagram is given in Fig. 10.21*b*. It is interesting to observe that an eight-state circuit, apparently requiring three flip-flops in the direct realization of Fig. 10.10, has been reduced to a four-state machine. Conceivably, one might have arrived at this four-state circuit directly by synthesizing the state diagram. Figure 10.21*b* has been deliberately arranged to illustrate this possibility. The pair of states p_2 and p_3 may have been set down as a starting point. These will be the states of the machine following a single 0 or 1, respectively. It is not always possible to be clever, and arriving at Fig. 10.21 after a minimization process would probably be the normal circumstance.

■

The state minimization process resulted in the saving of a flip-flop in the previous example and will do so again in Example 10.7. This will not always happen. It may be possible to reduce the number of states without reducing the number of flip-flops. Assume that the original number of states N_0 that satisfies Equation 10.15 is reduced to N_1. The number of states N_1 in the minimal circuit must be less

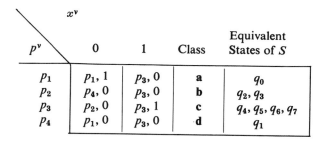

p^ν	x^ν		Class	Equivalent States of S
	0	1		
p_1	$p_1, 1$	$p_3, 0$	**a**	q_0
p_2	$p_4, 0$	$p_3, 0$	**b**	q_2, q_3
p_3	$p_2, 0$	$p_3, 1$	**c**	q_4, q_5, q_6, q_7
p_4	$p_1, 0$	$p_3, 0$	**d**	q_1

$p^{\nu+1}, z^\nu$

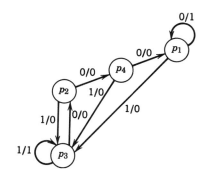

Figure 10.21   Minimal state equivalent of Fig. 10.19.

than or equal to 2^{r-1} if a saving in memory elements is to be achieved. A designer who is certain that sufficient states cannot be eliminated may decide to skip the state minimization process.

$$2^r \geq N_0 > 2^{r-1} \tag{10.15}$$

■ EXAMPLE 10.7

The state table for the beginning-of-message detector is repeated as Fig. 10.22a. Obtain a minimal state equivalent of this table.

SOLUTION   We first partition the states into classes satisfying condition (1) and then find the classes containing the next states. We note that state q_1 must be removed from class **a**.

Class	a				b
States	q_0	q_1	q_3	q_4	q_2
Next Class	a a	a b	a a	a a	a a

We thus define class **c** and repeat the process of classifying next states. This time we find that state q_3 must be removed from class **a**, so we define class **d**. After

Class	a			b	c
States	q_0	q_3	q_4	q_2	q_1
Next Class	a c	a a	a c	a a	a b

another pass, we find that states q_0 and q_4 are equivalent, and the process terminates. States $a, b, c,$ and d are defined corresponding to the respective classes.

Class	a		b	c	d
States	q_0	q_4	q_2	q_1	q_3
Next Class	a c	a c	a d	a b	d d

If we use the already tabulated information, the minimal state table of Fig. 10.22b is easily generated according to Theorem 10.2. Once again, one might have recognized at the outset that q_0 and q_4 are equivalent.   ■

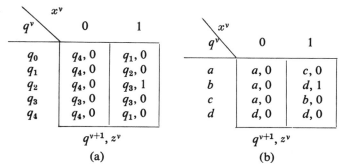

Figure 10.22   State tables for beginning-of-message detector.

10.8 SIMPLIFICATION BY IMPLICATION TABLES

An alternative method for determining the equivalence classes defined by Theorem 10.1 involves the use of the *implication table*, based on the work of M. C. Paull and S. H. Unger [2]. This method will probably seem more time-consuming than the successive partitioning method. However, unlike the successive partitioning method, it can be extended to the incompletely specified case, as discussed in Chapter 13.

Definition 10.3 A set of states $\mathbf{P}$ is *implied* by a set of states $\mathbf{R}$ if, for some specific input X_j, $\mathbf{P}$ is the set of all next states $\delta(r, X_j)$ for all present states r in $\mathbf{R}$.

Consider the next-state table shown in Fig. 10.23. Assume that $\mathbf{R}_1 = (q_1, q_2)$. Then the sets *implied* by $\mathbf{R}_1$ are $\mathbf{R}_2 = (q_3, q_4)$ and $\mathbf{R}_3 = (q_2, q_3)$, since $\delta(q_1, 0) = q_3$ and $\delta(q_2, 0) = q_4$, and $\delta(q_1, 2) = q_2$ and $\delta(q_2, 2) = q_3$. Similarly, set $\mathbf{R}_2 = (q_3, q_4)$ *implies* $\mathbf{R}_1 = (q_1, q_2)$ and $\mathbf{R}_4 = (q_1, q_3)$. Note that a given set, for example, $\mathbf{R}_1$, can both *imply* and *be implied*. By Theorem 10.1, the states in a set $\mathbf{R}$ are equivalent only if all the states in any set $\mathbf{P}$ implied by $\mathbf{R}$ are equivalent. This follows directly from the definition of implication and condition (2) of Theorem 10.1. Again, if we refer to Fig. 10.23, it is apparent that the states in $\mathbf{R}_1 = (q_1, q_2)$ cannot be equivalent unless the states in $\mathbf{R}_2 = (q_3, q_4)$ and $\mathbf{R}_3 = (q_2, q_3)$ are also equivalent. The implication table provides a systematic procedure for determining the equivalence classes of the states of the circuit.

Present State q^v	Input X^v			
	0	1	2	3
1	3	4	2	4
2	4	4	3	4
3	1	1	3	4
4	1	2	1	4
	q^{v+1}			

Figure 10.23 Next-state table.

■ EXAMPLE 10.8

Use an implication table to find a minimal state equivalent to the state table given in Fig. 10.24.

q^v	$x^v = 0$	$x^v = 1$	q^v	$x^v = 0$	$x^v = 1$
1	2, 0	3, 0	7	10, 0	12, 0
2	4, 0	5, 0	8	8, 0	1, 0
3	6, 0	7, 0	9	10, 1	1, 0
4	8, 0	9, 0	10	4, 0	1, 0
5	10, 0	11, 0	11	2, 0	1, 0
6	4, 0	12, 0	12	2, 0	1, 0
	q^{v+1}, z			q^{v+1}, z	

Figure 10.24 State table, Example 10.8.

SOLUTION   *For completely specified* state tables, it is sometimes possible to reduce the number of states informally before using a formal minimization technique. This is worthwhile since considerable labor is involved in both formal methods. Note in Fig. 10.24 that outputs and next states of states 11 and 12 are identical. Thus, these two states are indistinguishable, and state 12 may be eliminated from the table. Since state 12 is a next state of states 6 and 7, we replace these entries with 11. The state table thus modified is given in Fig. 10.25a. Note that only next-state information is provided in this table. State 9, which has different outputs than the other states, is listed separately at the bottom of the table.

In Fig. 10.25a states 5 and 7 are now seen to be indistinguishable. Eliminating state 7 and replacing all next-state entries of 7 with 5 yield the state table of Fig. 10.25b. No further reductions are apparent. This version may be further simplified using an implication table.*

The implication table is shown in Fig. 10.25. Vertically down the left side are listed all the states of the reduced state table except the first, and horizontally across the bottom are listed all but the last. The table thus contains a square for each pair of states. We start by placing an X in any square corresponding to a pair of states that have different outputs and therefore cannot be equivalent. In this case, q_9 has a different output than any other state, so we place an X in every square in row 9. The other states all have the same output, so no more squares are marked X.

We now enter in each square the pairs of states implied by the pair of states corresponding to that square. We start in the top square of the first column, corresponding to the pair of states (q_1, q_2). From the state table, we see that this pair implies the sets (q_2, q_4) and (q_3, q_5), so we enter 2–4 and 3–5 in this square. Proceeding to the next lower square, we see that (q_1, q_3) implies (q_2, q_6) and (q_3, q_5). We proceed in this manner to complete the first column. Note that there is only one entry in the square for (q_1, q_{11}). Since one of the implied sets is the single set (q_2), only one equivalence pair (q_3, q_1) is implied. We next proceed to the second column, the third, etc., until the table is complete (Fig. 10.25c).

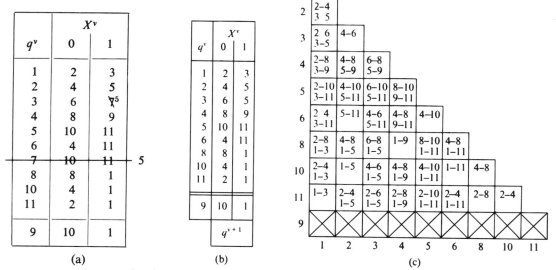

Figure 10.25   Simplification by inspection and implication.

*This reduced table could equally well form the starting point for the successive partitioning method.

The next step is to make a second "pass" through the table to see if any of the implied sets are ruled out as possible indistinguishable pairs by the squares marked X. In general, if there is an X in a square with "coordinates" j-k, then any square containing j-k as an implied set may be X'd out. In this case, q_9 cannot be equivalent to any other state, so any implied sets containing 9 are also ruled out. We place an X in every square containing a 9 (Fig. 10.25d). When this has been done, every square with coordinate 4 has been crossed out, indicating that state q_4 cannot be equivalent to any other state. We therefore proceed to cross out every square containing a 4 entry (Fig. 10.25d).

No other state is ruled out completely, so we now make a systematic check of all squares still not crossed out, moving from right to left. The first such square encountered is 8–11, which contains the entry 8–2. Square 8–2 has already been crossed out, so we must cross out square 8–11 (Fig. 10.25e). Next, we find entry 1–11 in square 6–10. Square 1–11 is not crossed out, so we leave square 6–10 as is. In a similar fashion, we find that squares 5–8, 3–8, and 1–8 must be crossed out (Fig. 10.25e). When we have completed this pass, we must make still another check of all remaining squares, since, for example, the elimination of square 1–8 might require the elimination of one of the remaining squares to the right. In general, this process must be repeated until a pass is completed without the elimination of any more squares. In this case, the final form is that shown in Fig. 10.25e. We have recopied the implication table twice for clarity in explaining the method, but in practice the complete procedure can be carried out on a single table.

In the completed implication table, each square not crossed out represents a pair of equivalent states. The equivalence classes may now be determined as follows. We first list the states corresponding to the columns of the implication table in reverse order (Fig. 10.26). We then check each column of the final implication table for any squares not crossed out, working from right to left. In Fig. 10.25e, there are no such squares in columns 11, 10, and 8, so dashes (—) are placed opposite 11, 10, and 8 in Fig. 10.26.

In column 6, we find square 6–10 not crossed out, so we enter the indistinguishable pair (6, 10) opposite 6 in Fig. 10.26. In column 5, we find the square 5–11 not crossed out, so we enter (5, 11) opposite 5 in Fig. 10.25 and recopy the previously determined pair (6, 10). In column 4, there are no squares not crossed out, so we simply recopy the already selected pairs opposite 4. In column 3, we find that q_3 is equivalent to both q_5 and q_{11}, and since q_5 and q_{11} are already equivalent, we add 3 to this set opposite 3 in Fig. 10.26. Similarly, column 2 adds q_2 to the set (q_6, q_{10})

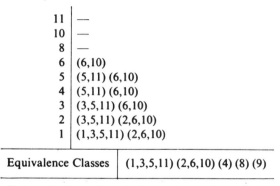

11	—
10	—
8	—
6	(6,10)
5	(5,11) (6,10)
4	(5,11) (6,10)
3	(3,5,11) (6,10)
2	(3,5,11) (2,6,10)
1	(1,3,5,11) (2,6,10)

| Equivalence Classes | (1,3,5,11) (2,6,10) (4) (8) (9) |

Figure 10.26 Determination of equivalence classes from the implication table.

and column 1 adds q_1 to the set (q_3, q_5, q_{11}). The resulting list opposite 1 in Fig. 10.26 comprises all equivalence classes with more than one member. Any states in the state table but not already on this list are not equivalent to any other state and must be included as single-state equivalence classes. These are added in the last line of Fig. 10.26.

Since there are five equivalence classes, the original state table of Fig. 10.24 reduces to the equivalent five-state table of Fig. 10.27. ∎

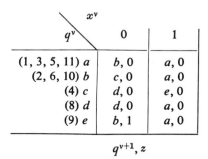

Figure 10.27 Minimal state table for Example 10.8.

■ **EXAMPLE 10.9**

Find the minimal table equivalent to the state table of Fig. 10.28.

SOLUTION First we recopy the next-state portion of the state table, partitioning the states into groups with like output, as shown in Fig. 10.29a.

We next check for states identical by inspection. States 3 and 11, which are in the same output group, have identical next-state entries, so 11 may be replaced by 3. State 10 has 11 as a next-state entry, so we replace it with 3, after which 10 is seen to be identical to state 2. We then replace the next-state entry of 10 with 2 at state 7, but no further identities are found, so we proceed to the implication table, Fig. 10.29b.

Present State q^v	Input x^v							
	0	1	2	3	0	1	2	3
1	6	2	1	1	0	0	0	0
2	6	3	1	1	0	0	0	0
3	6	9	4	1	0	0	1	0
4	5	6	7	8	1	0	1	0
5	5	9	7	1	1	0	1	0
6	6	6	1	1	0	0	0	0
7	5	10	7	1	1	0	1	0
8	6	2	1	8	0	0	0	0
9	9	9	1	1	0	0	0	0
10	6	11	1	1	0	0	0	0
11	6	9	4	1	0	0	1	0
	q^{v+1}				z^v			

Figure 10.28 State table, Example 10.9.

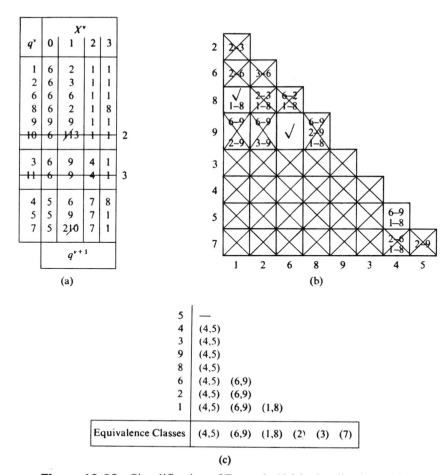

Figure 10.29 Simplification of Example 10.9 by implication table.

Note the ✔ in squares 1–8 and 6–9 of Fig. 10.29*b*. These indicate sets that imply only themselves and are therefore equivalent by inspection. Consider pair (q_1, q_8), which has identical next-state entries except for $X = 3$, for which the implied set is also (q_1, q_8). Thus, (q_1, q_8) does not imply any other sets; since they have the same output, they are equivalent. Squares marked with a check will never be crossed out in the processing of the implication table. If any pairs with identical next-state entries and outputs should be overlooked in the preliminary reduction by inspection, they should also be marked with a ✔ in the implication table. The minimal equivalent table is shown in Fig. 10.30. ∎

10.9 MEALY CIRCUITS AND MOORE CIRCUITS

In the state tables developed thus far, the output has been a function of both the input and the present state. This most general model of a sequential circuit is called a *Mealy model*. As an example, the state table of Fig. 10.13 is a Mealy model. This state table was a representation of an up-down counter deliberately configured to illustrate a finite-output memory circuit. Suppose the up-down counter circuit of Fig. 10.12 is adjusted slightly so that the outputs z_2, z_1, and z_0 are connected to the flip-flop outputs as shown in Fig. 10.31. This circuit remains an up-

q^v \ x^v		0	1	2	3
(4, 5)	a	$a, 1$	$b, 0$	$f, 1$	$c, 0$
(6, 9)	b	$b, 0$	$b, 0$	$c, 0$	$c, 0$
(1, 8)	c	$b, 0$	$d, 0$	$c, 0$	$c, 0$
(2)	d	$b, 0$	$e, 0$	$c, 0$	$c, 0$
(3)	e	$b, 0$	$b, 0$	$a, 1$	$c, 0$
(7)	f	$a, 1$	$d, 0$	$f, 1$	$c, 0$

q^{v+1}, z^v

Figure 10.30   Minimal state table, Example 10.9.

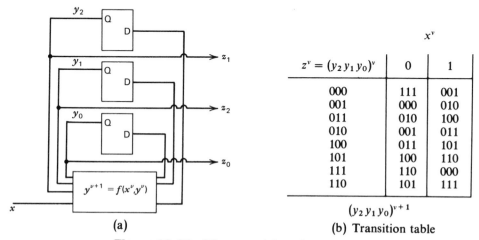

$z^v = (y_2 y_1 y_0)^v$ \ x^v	0	1
000	111	001
001	000	010
011	010	100
010	001	011
100	011	101
101	100	110
111	110	000
110	101	111

$(y_2 y_1 y_0)^{v+1}$

(a)   (b) Transition table

Figure 10.31   Moore model up-down counter.

down counter, with outputs delayed one clock period compared to those of Fig. 10.12.

The up-down counter transition table remains as shown in Fig. 10.12b, except that $(z_2, z_1, z_0) = (y_2, y_1, y_0)$. Using this relationship, the state table for Fig. 10.31 may be represented as shown in Fig. 10.32a. Note that the output values for each state are the same for $x = 1$ as for $x = 0$. That is, the output is a function of the present state only and not of the inputs. Thus, it is possible to express the state table in the more compact form of Fig. 10.32b on which the outputs are listed in a separate column, one entry for each state. A sequential circuit whose outputs are a function of present states only and can be represented by a state table of the form of Fig. 10.32b is a *Moore model*. The state diagram for a Moore model also differs from the Mealy model. As shown in Fig. 10.32c, output values are included within the circles containing the state identifiers. Only inputs are associated with the arrows interconnecting the states.

In the case of the two counters discussed above, we have Mealy and Moore circuits that perform the same function (except for the timing difference inherent in the Mealy and Moore models) and have the same number of states. However, Mealy and Moore circuits performing the same functions will not always have the same number of states. Let us translate the Mealy circuit for which the state diagram is shown in Fig. 10.33 to a Moore circuit with the same input/output function. Every output that is associated with a transition between states in the

q^v	x^v 0	1
q_0	$q_7, 0$	$q_1, 0$
q_1	$q_0, 1$	$q_2, 1$
q_2	$q_1, 2$	$q_3, 2$
q_3	$q_2, 3$	$q_4, 3$
q_4	$q_3, 4$	$q_5, 4$
q_5	$q_4, 5$	$q_6, 5$
q_6	$q_5, 6$	$q_7, 6$
q_7	$q_6, 7$	$q_0, 7$

q^{v+1}, z

(a)

q^v	x^v 0	1	z
q_0	q_7	q_1	0
q_1	q_0	q_2	1
q_2	q_1	q_3	2
q_3	q_2	q_4	3
q_4	q_3	q_5	4
q_5	q_4	q_6	5
q_6	q_5	q_7	6
q_7	q_6	q_0	7

q^{v+1}

(b)

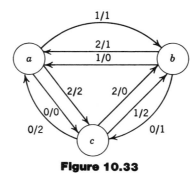

(c)

Figure 10.32 Moore model state tables and state diagrams.

Figure 10.33

machine of Fig. 10.33 must be associated with a state in the proposed Moore circuit. Since the inputs cannot be anticipated, the outputs corresponding to inputs in Fig. 10.33 must be associated with the next states following the same inputs in the Moore circuit.

Consider some particular state p of the Mealy circuit. Let $\{\lambda_i\}_p$ be the set of all distinct outputs λ_i of the Mealy circuit given by

$$\lambda_i = \lambda(q, I_i) \tag{10.16}$$

for any input I_i and any state q satisfying

$$\delta(q, I_i) = p \tag{10.17}$$

For example, in Fig. 10.33, we have three different transitions to state c. That is,

$$\delta(a, 0) = c \quad \text{where} \quad \lambda(a, 0) = 0$$
$$\delta(a, 2) = c \quad\quad\quad\quad \lambda(a, 2) = 2 \tag{10.18}$$
$$\delta(b, 0) = c \quad\quad\quad\quad \lambda(b, 0) = 1$$

Therefore, $\{\lambda_i\}_c = \{0, 1, 2\}_c$.

We may now define a set of states $\{p_{\lambda i}\}$ in the Moore circuit corresponding to the single state p of the Mealy circuit. That is, for every $\lambda_i \in \{\lambda_i\}_p$, let $p_{\lambda i}$ be a distinct state in the Moore circuit such that

$$\lambda(p_{\lambda i}) = \lambda_i \tag{10.19}$$

By definition, $\lambda(p_{\lambda i})$ is the output associated with $p_{\lambda i}$ in the Moore circuit.

In Fig. 10.33, for example, the set of states $\{C_0, C_1, C_2\}$ corresponds to the output set $\{0, 1, 2\}_c$. Similarly, for states a and b, the output sets are $\{0, 1, 2\}_a$ and $\{0, 1, 2\}_b$; the sets of states in the Moore circuit are $\{A_0, A_1, A_2\}$ and $\{B_0, B_1, B_2\}$. These states are listed in the state table of Fig. 10.34.

The next-state function for any input I in the Moore circuit may be defined as follows. Suppose

$$\delta(q, I) = p \tag{10.20}$$

for two states p and q in the Mealy circuit. Let $\{p_{\lambda j}\}$ and $\{p_{\lambda i}\}$ be the sets of states in the Moore circuit corresponding to p and q, respectively. Let $p_{\lambda j}$ be the member of $\{p_{\lambda i}\}$ such that

$$\lambda_j = \lambda(q, I) \tag{10.21}$$

Then we may define the next-state function in the Moore circuit for input I and any state $q_{\lambda i} \in \{q_{\lambda i}\}$ as

$$\delta(q_{\lambda i}, I) = p_{\lambda j} \tag{10.22}$$

Suppose the input to the Moore circuit of Fig. 10.34 is $X = 0$ when the circuit is in one of the set of states $\{A_0, A_1, A_2\}$. Recall from Fig. 10.33 that

	$X = 0$	$X = 1$	$X = 2$	λ
A_0	C_0	B_1	C_2	0
A_1	C_0	B_1	C_2	1
A_2	C_0	B_1	C_2	2
B_0	C_1	A_0	A_1	0
B_1	C_1	A_0	A_1	1
B_2	C_1	A_0	A_1	2
C_0	A_2	B_2	B_0	0
C_1	A_2	B_2	B_0	1
C_2	A_2	B_2	B_0	2

Figure 10.34 Moore circuit corresponding to the Mealy circuit of Fig. 10.33.

$$\delta(a, 0) = c \qquad \text{and} \qquad \lambda(a, 0) = 0 \qquad\qquad (10.23)$$

Therefore, we conclude that for present state a, $\lambda_j = 0$, as defined by Equation 10.21, and that from Equation 10.22

$$\delta(A_0, 0) = \delta(A_1, 0) = \delta(A_2, 0) = C_0 \qquad\qquad (10.24)$$

The remaining next-state entries in Fig. 10.34 may be determined in a similar manner.

Thus far we have only indicated that the circuits of Figs. 10.33 and 10.34 should have the same output sequence for any input sequence. The further refinement in notation required to complete a proof of this fact would take us too far afield. The reader who remains unconvinced can easily check the outputs of the circuit for a typical input sequence. In Fig. 10.35a, we see the sequence of outputs and sequence of states assumed by the Mealy circuit when subjected to the input sequence

$$1021002 \qquad\qquad (10.25)$$

when initially in state a. Similarly, we see in Fig. 10.35b the output and next-state sequences of the Moore circuit when subjected to the same input sequences. In the latter case, the initial state may be any member of A_0, A_1, A_2.

Note in this case that a nine-state Moore circuit was required to imitate a three-state Mealy circuit. This Mealy circuit was deliberately chosen to have distinct outputs for each transition to each of its three states. As we have seen, the situation is not this bad for many examples of practical interest.

The Moore circuit of Fig. 10.34 may be translated to an equivalent Mealy circuit by merely assigning an output to each input-present-state combination equal to the output associated with the next state for that combination in the Moore circuit. The result is Fig. 10.36.

Using the minimization procedure of the previous section, we find that the states of the circuit in Fig. 10.36 fall into the following equivalence classes:

$$(A_0 A_1 A_2)(B_0 B_1 B_2)(C_0 C_1 C_2) \qquad\qquad (10.26)$$

Initial State

Input	↓ 1 1 2 1 0 0 2	A_0 ⎱	1 1 2 1 0 0 2
New State	@ b a c b c a c	A_1 ⎰	B_1 A_0 C_2 B_2 C_1 A_2 C_2
Output	1 0 2 2 1 2 2	A_2 ⎰	1 0 2 2 1 2 2
	(a)		(b)

Figure 10.35 Input/output sequences.

A_0	C_0 0	B_1 1	C_2 2
A_1	C_0 0	B_1 1	C_2 2
A_2	C_0 0	B_1 1	C_2 2
B_0	C_1 1	A_0 0	A_1 1
B_1	C_1 1	A_0 0	A_1 1
B_2	C_1 1	A_0 0	A_1 1
C_0	A_2 2	B_2 2	B_0 0
C_1	A_2 2	B_2 2	B_0 0
C_2	A_2 2	B_2 2	B_0 0

Figure 10.36

The circuit formed from these equivalence classes is the original circuit of Fig. 10.33.

10.10   STATE ASSIGNMENT AND MEMORY ELEMENT INPUT EQUATIONS

Once a minimal state table has been obtained, it remains to design a circuit realizing that table. It must be recognized that the question of economics has not been exhausted with the minimization of the number of states. It is true that the number of memory elements required by such a state table is minimum, but we also require combinational logic to develop the output and memory element input equations. It may be that some state table other than a minimal table would require significantly less combinational logic and, therefore, less overall hardware than the latter. However, we will ignore this problem for now and concentrate our attention on realizing a given state table as economically as possible.

As shown in Fig. 10.4, the next step in the design process after obtaining a minimal state table is state assignment, the assignment of a specific combination of values of the state variables to each state, leading to the transition table. In Chapter 11 we observe that very large sequential systems are usually partitioned into separate control and data units. State assignment in the data unit is usually done at the beginning of the design process, and state minimization and, therefore, subsequent state assignment are not applicable. The state assignment problem is similarly trivial in cases where the finite memory model was applied with no subsequent reduction in the number of states, and in Moore model circuits in which each state corresponds to a distinct output (e.g., counters).

In a control sequential circuit or whenever state minimization techniques are applied, state assignment is necessary. Determination of the best possible state assignment is a difficult problem. We recall that each of the states will correspond to one of the 2^r combinations of the r state variables. The first problem in designing a circuit to realize a given state table is to decide which of the 2^r combinations shall be assigned to each state. If the number of states m satisfies

$$2^{r-1} < m \leq 2^r$$

then r state variables will be required. There will be

$$\frac{2^r!}{(2^r - m)!} \tag{10.27}$$

ways to assign the 2^r combinations of state variables to the m states.

The problem of assigning state-variable combinations to four states, say, $a, b, c,$ and $d,$ can be visualized as the assignment of the four states to the four vertices of the 2 cube. Three distinct assignments are shown in this fashion in Fig. 10.37. Any other possible assignments would amount to rotations or reversals of these three assignments and would thus correspond either to reversing the order of the variables or complementing one or both variables. Such changes will not affect the form of any Boolean function and are thus irrelevant with respect to cost. The three distinct assignments for three rows can be obtained in a similar manner. The distinct assignments for three and four rows are listed in Table 10.1. The number of distinct assignments for various numbers of states is tabulated in Table 10.2 [2].

To illustrate that the choice of state assignment can impact on the overall economy of a realization, we turn to the beginning-of-message detector developed

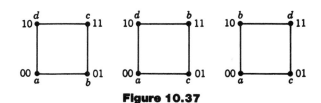

Figure 10.37

Table 10.1

States	Three-State Assignments			Four-State Assignments		
	1	2	3	1	2	3
a	00	00	01	00	00	00
b	01	11	11	01	11	10
c	11	01	00	11	01	01
d	—	—	—	10	10	11

Table 10.2

No. of States m	No. of State Variables r	No. of Distinct Assignments
2	1	1
3	2	3
4	2	3
5	3	140
6	3	420
7	3	840
8	3	840
9	4	10,810,800

in Examples 10.1 and 10.7. The minimal state table for the beginning-of-message detector is repeated in Fig. 10.38a. Figures 10.38b, 10.38c, and 10.38d show the resulting next-state portions of the transition tables for the three distinct state assignments suggested by Fig. 10.37. Because the output is 1 for only one combination of state-variable values, the assignment will not impact the realization of z, so the output portions of the transition tables have been omitted for clarity. Note that only the order of the states is changed; the state variables are always listed in K-map order, as required for transition tables. This is important, because the next step is to translate the transition tables into K-maps, known as the excitation maps, representing the memory element input equations.

In Chapter 9, we developed K-maps representing the next states as a function of the present state and inputs for the common types of flip-flops. As we synthesize, we will use this information in a slightly different form. In a J-K flip-flop, for example, we must have at our fingertips the values of J and K that will cause any particular state transition to take place. The values of J and K required to effect each of the four input transitions are given as a *transition list* in Fig. 10.39b. This table can be derived immediately from the K-map of Fig. 10.39a. Transition lists for the R-S and D flip-flops are given in Fig. 10.39c and d, respectively. For the D flip-flops the transition table is actually the excitation map for all state variables.

The left column in each diagram lists the four possible combinations of present value and desired next value for the output Q. The entries give the

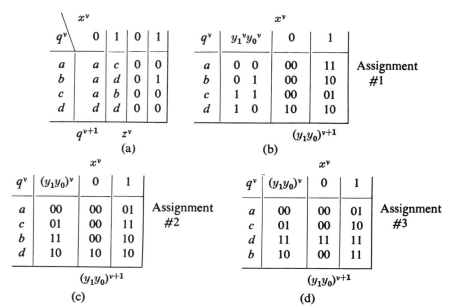

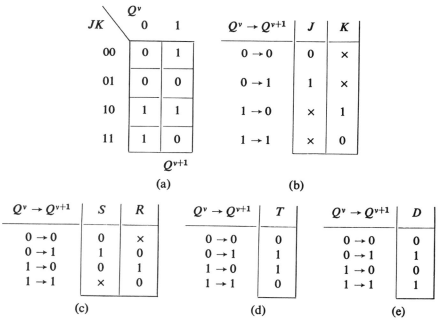

Figure 10.38 State table and possible transition table, beginning-of-message detector.

Figure 10.39 Transition lists.

necessary values at the inputs to achieve these combinations. Consider, for example, the diagram for the J-K flip-flop. If the present value is 0, that is, the flip-flop is cleared, and it is to remain that way, then J must equal 0 and K can be either 0 or 1 (don't care). If Q is now 0 and is to be 1, that is, the flip-flop is to be set, then $J = 1$ and $K = 0$ or 1. The other entries follow by similar reasoning.

To develop the excitation maps for assignment 1, we start with partial transition tables, each showing the transition in only a single-state variable (Figs. 10.40a and 10.40d). From these, we generate the excitation maps for the flip-flops

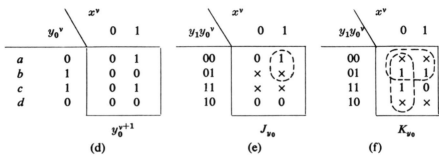

Figure 10.40 Translation of transition table to excitation maps.

controlling each state variable by reference to the transition list for the type of flip-flop to be used (JK, Fig. 10.39b).

Consider the entries on the second row. For $(y_1 y_0 x)^v = 010$, we see from Fig. 10.40a that $y_1^v = 0, y_1^{v+1} = 0$, so $J_{y1} = 0$ and $K_{y1} = \times$. For $(y_1 y_0 x)^v = 011, y_1^v y_1^{v+1} = 01$, so $J_{y1} = 1$ and $K_{y1} = \times$. Similarly, from Fig. 10.40d, for both $(y_1 y_0 x)^v = 000$ and $(y_1 y_0 x)^v = 001, y_0^v = 1 \rightarrow y_0^{v+1} = 0$, so $J_{y0} = \times$ and $K_{y0} = 1$. This process can be speeded up by filling in all the don't cares first. Note that if the present value of the state variable is 0, $K = \times$ regardless of the next value, while $J = \times$ if the present value is 1.

From these maps (Figs. 10.40b,c,e,f), we read off the minimal realizations, noting that the problem is multiple output, so products should be shared when possible. The minimal groupings are as shown. The equations for the assignment are

$$J_{y1} = x, \qquad J_{y0} = x\bar{y}_1, \qquad K_{y1} = y_0$$
$$K_{y0} = \bar{x} + \bar{y}_1, \qquad z = x\bar{y}_1 y_0 \tag{10.28}$$

Figure 10.41 shows the maps for assignments 2 and 3. The resulting equations are

Assignment 2

$$J_{y1} = xy_0, \qquad K_{y1} = \bar{x}y_0, \qquad z = xy_1 y_0$$
$$J_{y0} = x\bar{y}_1, \qquad K_{y0} = \bar{x} + y_1 \tag{10.29}$$

Assignment 3

$$J_{y1} = xy_0, \qquad K_{y1} = \bar{x}\bar{y}_0, \qquad z = xy_1\bar{y}_0$$
$$J_{y0} = x, \qquad K_{y0} = \bar{y}_1 \tag{10.30}$$

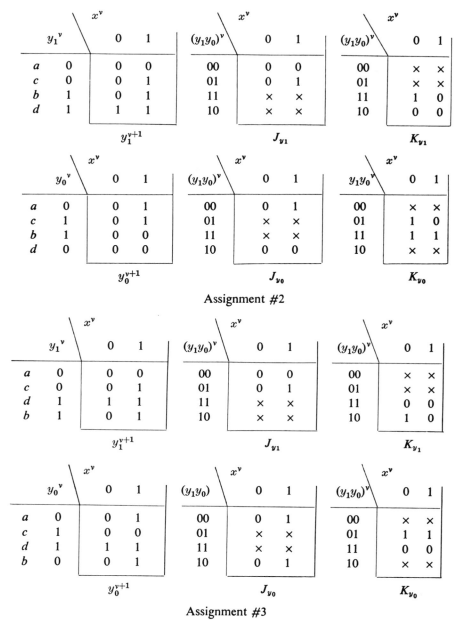

Figure 10.41 Excitation maps for alternate assignments.

Assignments 1 and 3 are seen to have the same basic gate cost, but 3 is preferable because it uses only AND gates while 1 requires one OR gate. The circuit for 3 is shown in Fig. 10.42.

■ EXAMPLE 10.10

Determine a circuit realization for the error detector circuit of Examples 10.3 and 10.6 using *S-R* flip-flops.

SOLUTION   The minimal state table for the circuit is shown in Fig. 10.43*a*, and the resulting transition table for assignment 2 is illustrated in Fig. 10.43*b*.

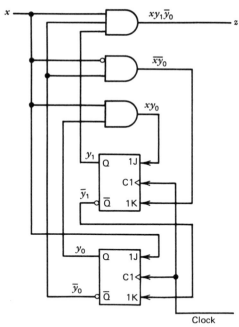

Figure 10.42   Final circuit, beginning-of-message detector.

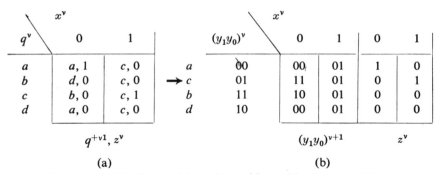

(a)

(b)

Figure 10.43   State table and transition table, Example 10.10.

The translation of the partial transition tables into the excitation maps for S-C flip-flops is shown in Fig. 10.44. Breaking up the transition tables into partial tables for each variable is not absolutely necessary. One could work directly from the complete transition table, but separating the variables reduces the likelihood of error. The equations for this assignment are the following:

Assignment 2

$$S_{y1} = \bar{x}y_0, \qquad R_{y1} = x + \bar{y}_0, \qquad S_{y0} = x, \qquad R_{y0} = \bar{x}y_1$$
$$z = \bar{x}\bar{y}_1\bar{y}_0 + x\bar{y}_1y_0 \tag{10.31}$$

We will leave it to the reader to verify that the equations for assignments 1 and 3 are as follows:

Assignment 1

$$S_{y1} = x + \bar{y}_1y_0, \qquad R_{y1} = \bar{x}y_1, \qquad S_{y0} = x, \qquad R_{y0} = \bar{x}\bar{y}_1$$
$$z = \bar{x}\bar{y}_1\bar{y}_0 + xy_1y_0 \tag{10.32}$$

y_1^v \ x^v	0	1
0	0	0
0	1	0
1	1	0
1	0	0

y_1^{v+1}

$y_1 y_0^v$ \ x^v	0	1
00	0	0
01	1	0
11	×	0
10	0	0

S_{y1}

$y_1 y_0^v$ \ x^v	0	1
00	×	×
01	0	×
11	0	1
10	1	1

R_{y1}

y_0^v \ x^v	0	1
0	0	1
1	1	1
1	0	1
0	0	1

y_0^{v+1}

$y_1 y_0^v$ \ x^v	0	1
00	0	1
01	×	×
11	0	×
10	0	1

S_{y0}

$y_1 y_0^v$ \ x^v	0	1
00	×	0
01	0	0
11	1	0
10	×	0

R_{y0}

Figure 10.44 Derivation of excitation maps, Example 10.10.

Assignment 3

$$S_{y1} = \bar{x}\bar{y}_1 y_0, \qquad R_{y1} = x + y_1 y_0, \qquad S_{y0} = x + y_1 \bar{y}_0, \qquad R_{y0} = \bar{x} y_0$$
$$z = \bar{x}\bar{y}_1\bar{y}_0 + x y_1 y_0 \tag{10.33}$$

Assignment 2 is seen to lead the simplest equations. We will leave it to the reader to draw the final circuit. ∎

As the number of states grows beyond three or four, consideration of all possible state assignments becomes prohibitive. A computer-aided logic design tool for implementing sequential circuits from state tables may very well include a provision for selecting a "usually better than arbitrary" state assignment. Such a routine would be an implementation of a set of rules for predicting a good state assignment based on experience with the realization of state tables with similar features. The program might feature an "expert system," perhaps the best-known application of artificial intelligence. The expert system would likely incorporate a few general rules and numerous lesser rules to reflect special cases. The set of rules or conjectures that might be offered to an expert system are far too numerous to discuss here. When a state assignment program is unavailable, the reader is best advised to look for state assignment clues in the original problem statement or to arbitrarily select some convenient assignment and tolerate the small number of extra gates that might result.

10.11 PARTITIONING AND STATE ASSIGNMENT*

For many state tables, the benefit obtained from finding an optimal state assignment does not justify the required effort. In a few cases, finding a state assignment consistent with the natural structure of the system under design will make a

---

*The material of this section is not essential to subsequent developments and may be omitted from a first reading.

significant difference in the economy of the realization. Often, this structure will be reflected in the original problem definition before the design is actually reduced to state table form. We shall use the state table to convincingly argue the benefit of identification of such a structure when it exists.

In Section 10.7, we encountered the concept of partitioning the states of a sequential circuit into equivalence classes such that two states p and q were in the same class ($p \overset{G}{=} q$) if and only if the following two conditions were satisfied for all inputs X:

$$(1) \ \lambda(p, X) = \lambda(q, X)$$

$$(2) \ \delta(p, X) \overset{G}{=} \delta(q, X)$$

Now let us consider a partition inducing equivalence classes that satisfy only condition (2), that is, the only requirement for states to be equivalent is for their next states to be equivalent.

Consider, for example, the state table of Fig. 10.45 that is a repeat of the state table for the up-down counter of Example 10.4. Note that the partitions $(0, 2, 4, 6)$ $(1, 3, 5, 7)$ and $(0, 4), (1, 5), (2, 6), (3, 7)$ meet condition (2). For states $(0, 4)$, the next states for $x = 0$ are $(3, 7)$ and for $x = 1$ are $(1, 5)$; for $(1, 5)$, the next states are $(0, 4)$ and $(2, 6)$ for $x = 0$ and $x = 1$, respectively, etc. For $(0, 2, 4, 6)$, the next-state group is $(1, 3, 5, 7)$ for both inputs, and vice versa. Thus, for all the states in any given block of a partition, the next states are in a single block of the same partition for any given input.

Partitions having the property described above are called *closed*, or *preserved*, partitions. To illustrate why such partitions are useful for state assignment, let us assign one of the state variables to the two-block partition discussed above, that is, let $y_1 = 0$ for states $0, 2, 4, 6$ and $y_1 = 1$ for states $1, 3, 5, 7$. Since present states $(0, 2, 4, 6)$ always lead to next states $(1, 3, 5, 7)$, and vice versa, if $y_1^v = 0$, then $y_1^{v+1} = 1$, and if $y_1^v = 1$, then $y_1^{v+1} = 0$. Thus, the next-state equation for y_1 is simply

$$y_1^{v+1} = \overline{y_1^v} \tag{10.34}$$

Assigning state variables always induces partitions of the states into groups corresponding to various combinations of values of the variables. If we can assign the state variables so as to induce closed partitions, we will, as in the above example, generally obtain simpler equations than would otherwise be the case. If a partition is closed, then the block containing the next state is determined by the block containing the present state and, possibly, the input. Since there are fewer

x^v

q^v	0	1
0	7	1
1	0	2
2	1	3
3	2	4
4	3	5
5	4	6
6	5	7
7	6	0

q^{v+1}

Figure 10.45 State table, up-down counter.

blocks than states, blocks can be identified by subsets of the state variables. If the partition is closed, the next values of the variables identifying that partition are thus determined by the present values of those variables and the inputs. The net result is to reduce the number of variables in the next-state equations, generally resulting in simpler equations.

Not only do closed partitions generally result in desirable assignments, they are also often closely related to the basic structure of the sequential circuit. In the above example, y_1 is a function only of itself, so that the circuitry developing y_1 will have no other inputs. In general, closed partitions result in the separation of the complete circuit into subcircuits, each responsible for computing some subset of the system variables and dependent only on some subset of the system variables. Figure 10.46 illustrates the manner in which an arbitrary circuit with four state variables might partition into such subcircuits. Note that each subcircuit has as inputs some, but not all, of the state-variable outputs from the other subcircuits. The flip-flop in circuit I, for example, has input equations that are functions of its own output y_0 and the overall circuit input x. Similarly, subcircuit II has as inputs only x, y_0, and its own inputs.

In general, a good partition of the states of a sequential circuit corresponds to a division of the circuit into subcircuits with as few interconnections as possible. Considerable research has been reported [6–8] on the problem of finding suitable partitions, most of it directed to formal search procedures for closed partitions and other types of partitions having similar desirable properties. Unfortunately, these procedures require a much more extensive theoretical development than suitable for a book at this level. In addition, their application is tedious in the extreme and they often lead nowhere after very considerable effort has been expended.

Rather than concentrating on formal partitions of the states, we feel it is generally more productive to be alert for partitions of the circuit that might be suggested by the original problem formulation. Often, the subcircuits will perform readily identifiable subfunctions, such as counting, shifting, keeping track of parity, etc. If the designer can recognize these subfunctions early in the process of analyzing the circuit, the state assignment problems may be considerably simplified. The four-bit sequential parity checker of the previous section is an interesting example.

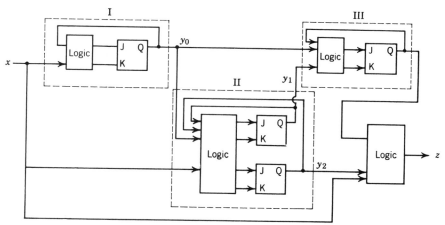

Figure 10.46   Typical partitioned sequential circuit.

■ EXAMPLE 10.11

Design a serial parity checker for four-bit binary characters. Starting from reset, the circuit receives four-bit characters serially on a single input line x. At the time of the fourth bit, the output is to be 1 only if the total number of 1's in the character is even. At all other times, the output is to be 0. Following receipt of the fourth bit, the circuit should return to reset to receive the next character.

SOLUTION   We approach the problem by trying to identify subfunctions within the parity checker. Quite obviously, one function that must be accomplished is keeping track of parity as the bits arrive. From our experience with the parity checker in Chapter 9, we observe that only one flip-flop or state variable is required for that purpose. Next we notice that the circuit must behave differently when the fourth bit of a character arrives than it does for the first three. It is thus necessary to keep track of the number of bits received. A simple four-state counter that will do this is shown in Fig. 10.47a. In Figs. 10.47b and 10.47c we see the state diagram for the parity check flip-flop. Figure 10.47b, in which the output z is always 0, is valid for the first three inputs. The clock pulse corresponding to the fourth input resets the checker to even parity regardless of x, as shown in Fig. 10.47c. The output is specified as 1 if and only if parity over the four-bit character has been even. Note that the values of y_1 and y_2 are unaffected by the present values of x and y_0. However, the values of y_0^{v+1} and z^v are functions of the present values of x, y_1, y_2, and y_0 as indicated by Fig. 10.47. In order to merge the two functions into a single sequential circuit, we tabulate the values of y_0^{v+1}, as shown in Fig. 10.48a. The two rows designated by asterisks correspond to Fig. 10.47c, which is valid for bit 4 or when $y_1^v = y_2^v = 1$. The remaining rows correspond to Fig. 10.47b. There are two don't cares in the table, since y_0 will always be 0 when $y_1 = y_2 = 0$. Figure 10.48b is a transition table for the two-bit binary counter. From these tables, the reader can verify the memory element input equations as given by Equation 10.35.

$$J_{y2} = K_{y2} = 1, \qquad J_{y1} = K_{y1} = y_2$$
$$J_{y0} = x(\bar{y}_1 + \bar{y}_2), \qquad K_{y0} = x + y_1 y_2 \qquad (10.35)$$
$$z = y_1 y_2 (x y_0 + \bar{x} \bar{y}_0)$$

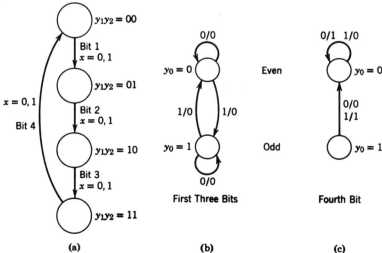

(a)   (b)   (c)

Figure 10.47   Counting and parity check subfunctions.

xy_1y_2	$y_0 = 0$	$y_0 = 1$
0 0 0	0, 0	×
0 0 1	0, 0	1, 0
0 1 0	0, 0	1, 0
*0 1 1	0, 1	0, 0
1 0 0	1, 0	×
1 0 1	1, 0	0, 0
1 1 0	1, 0	0, 0
*1 1 1	0, 0	0, 1

(a) y_0^{v+1}, z^v

y_1y_2	$x = 0$	$x = 1$
0 0	0 1	0 1
0 1	1 0	1 0
1 0	1 1	1 1
1 1	0 0	0 0

$(y_1y_2)^{v+1}$

(b)

Figure 10.48   Transition table for sequential parity checker.

With this selection of variables to perform distinct functions, the circuit breaks down into distinct subcircuits, as shown in Fig. 10.49.

Now that we have obtained a realization, let us examine the state table of the sequential parity checker. Figures 10.48*a* and 10.48*b* may be combined to form the transition table of Fig. 10.50*a* and the corresponding state table of Fig. 10.50*b*. Since y_0 is not an input to the two-bit counter, we would expect to find a closed partition of the states into four classes, one for each combination of values of $y_1^v y_2^v$. This partition is (04)(15)(26)(37). That this partition is closed may be verified from Fig. 10.50*c*, in which the classes of next states for each class are tabulated.   ∎

Obtaining a useful partition of the states of a sequential circuit depends on the natural structure of a machine and the designer's ability to recognize that structure. As with many aspects of engineering judgment, this ability grows with experience. It is suggested that the reader be alert to the possible existence of partitionable structure whenever a state diagram is developed.

Before returning to the up-down counter example, we add another small touch of formality to our approach. First, we adopt the symbol $\prod$ with appropriate subscripts to refer to a partition of the states of a state table.

Definition 10.4   Let $\prod_{ab}$ be the product $\prod_a \cdot \prod_b$ of two partitions $\prod_a$ and $\prod_b$. Two states q_1 and q_2 are in the same class in $\prod_{ab}$ if and only if q_1 and q_2 are in the same class in $\prod_a$ and in $\prod_b$.

Definition 10.5   The partition of a sequential circuit in which each state is in a separate class is denoted $\prod(0)$.

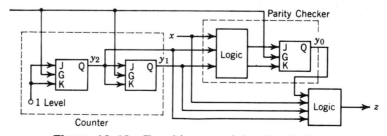

Figure 10.49   Four-bit sequential parity checker.

q	$y_0{}^v y_1{}^v y_2{}^v$	$x^v = 0$		$x^v = 1$	
q_0	0 0 0	0	01	1	01
q_1	0 0 1	0	10	1	10
q_2	0 1 0	0	11	1	11
q_3	0 1 1	0	00	0	00
q_4	1 0 0	—		—	
q_5	1 0 1	1	10	0	10
q_6	1 1 0	1	11	0	11
q_7	1 1 1	0	00	0	00

$(y_0 y_1 y_2)^{v+1}$

(a)

q^v	$x^v = 0$	$x^v = 1$
q_0	q_1, 0	q_5, 0
q_1	q_2, 0	q_6, 0
q_2	q_3, 0	q_7, 0
q_3	q_0, 1	q_0, 0
q_4	—	—
q_5	q_6, 0	q_2, 0
q_6	q_7, 0	q_3, 0
q_7	q_0, 0	q_0, 1

q^{v+1}, z^v

(b)

	Next-states	
Present states	$x^v = 0$	$x^v = 1$
(04)	(15)	(15)
(15)	(26)	(26)
(26)	(37)	(37)
(37)	(04)	(04)

(c)

Figure 10.50 State table for sequential parity checker.

■ EXAMPLE 10.12

Let us consider again the modulo-8 up-down counter of Example 10.4, for which the state table is repeated in Fig. 10.51a. We noted at the beginning of this section that there are two nontrivial closed partitions in this table, but let us assume that we are unaware of these and consider how our perception of the natural structure of this circuit might lead us to these partitions.

A natural partition for any counter is into blocks of states corresponding to odd counts and even counts. Such a partition will be closed if the number of states is even, since an odd count will always be followed by an even count, and vice versa. For this circuit, this corresponds to the partition (0, 2, 4, 6), (1, 3, 5, 7); as already suggested, we will assign y_1 to this partition. We can then set up a transition table for the subcircuit developing y_1, as shown in Fig. 10.51b. The function of this subcircuit is to identify which class the overall circuit is in, (0, 2, 4, 6) or (1, 3, 5, 7). As shown, if the present class is (0, 2, 4, 6), the next will be (1, 3, 5, 7), and vice versa. From this, we read off the equation

$$y_1^{v+1} = \overline{y_1^v} \tag{10.36}$$

as noted earlier. The corresponding circuit using a D flip-flop is shown in Fig. 10.51c.

Each state must be identified by a unique combination of state-variable values. In a complete state assignment, the product of all partitions induced by individual state variables must be $\prod(0)$. We have already developed a partial realization based on a closed partition $\prod_{y1}$ induced by state variable y_1. We must find $\prod_{y2}$ and $\prod_{y3}$ such that

$$\prod_{y1} \cdot \prod_{y2} \cdot \prod_{y3} = \prod(0) \tag{10.37}$$

x^v

q^v	0	1
0	7	1
1	0	2
2	1	3
3	2	4
4	3	5
5	4	6
6	5	7
7	6	0

q^{v+1}

x^v

Present class	y_1^v	0	1	0	1
(0, 2, 4, 6)	0	(1, 3, 5, 7)	(1, 3, 5, 7)	1	1
(1, 3, 5, 7)	1	(0, 2, 4, 6)	(0, 2, 4, 6)	0	0

Next class $\qquad\qquad y_1^{v+1}$

Figure 10.51 State table and design of y_1 subcircuit for up-down counter.

We have already verified that $(0, 4)\,(1, 5)\,(2, 6)\,(3, 7)$ is a closed partition. If we let $\prod_{y2} = (0, 1, 4, 5)\,(2, 3, 6, 7)$, then

$$\prod_{y1} \cdot \prod_{y2} = (0, 4)(1, 5)(2, 6)(3, 7)$$

and the next value of y_2 is only a function of y_1, y_2, and x. (We have already concluded that the next value of y_1 is only a function of y_1.) We may now develop a transition table for y_2, as given in Fig. 10.52a. From this table, we can read off the equation

$$y_2^{v+1} = x^v \cdot (y_1^v \oplus y_2^v) \ + \ \bar{x}^v \cdot \overline{(y_1^v \oplus y_2^v)} = \bar{x}^v \oplus (y_1^v \oplus y_2^v)$$

The resulting interconnection of the two subcircuits inducing the partition $(0, 4)$ $(1, 5)\,(2, 6)\,(3, 7)$ is shown in Fig. 10.52b.

All that remains is to select $\prod_{y3}$ to satisfy Equation 10.37. There are eight possible partitions. The one that will correspond naturally to the most significant bit of a counter is $\prod_{y3} = (0, 1, 2, 3)\,(4, 5, 6, 7)$. Since we have already designed subcircuits to develop y_1 and y_2, it is obvious that the final subcircuit will develop only y_3. The transition table for y_3 is given in Fig. 10.53. The complete counter circuit is shown in Fig. 10.54 with the already developed realizations of y_1 and y_2 represented by blocks I and II, respectively. It will be left to the reader to verify that the transition table of Fig. 10.53 leads to the equation

$$y_3^{v+1} = [(y_1^v \cdot y_2^v) \oplus y_3^v] \oplus \bar{x}^v \tag{10.38}$$

The reader may not feel that this procedure has produced a particularly economical design in view of the number of exclusive-OR gates, but this is the best realization using D flip-flops. Any other assignment will make y_1 and y_2 functions of more variables, leading to a more complex design. ∎

We hope that the above example, illustrating as it does some of the more interesting implications of state table partition theory, may motivate some readers to explore this fascinating subject in more depth (see References 8, 12).

The relationship between state assignment partitions and the natural structure of the circuit function raises questions as to the desirability of state table minimization. In Example 10.5, we arrived at an eight-state table, which would require three state variables. The reader who considers the basic structure will

y_2y_1	
00	$(0, 1, 4, 5) \wedge (0, 2, 4, 6) = (0, 4)$
01	$(0, 1, 4, 5) \wedge (1, 3, 5, 7) = (1, 5)$
11	$(2, 3, 6, 7) \wedge (1, 3, 5, 7) = (3, 7)$
10	$(2, 3, 6, 7) \wedge (0, 2, 4, 6) = (2, 6)$

Present class	y_2y_1	x^v 0	1	0	1
$(0, 4)$	00	$(3, 7)$	$(1, 5)$	1	0
$(1, 5)$	01	$(0, 4)$	$(2, 6)$	0	1
$(3, 7)$	11	$(2, 6)$	$(0, 4)$	1	0
$(2, 6)$	10	$(1, 5)$	$(3, 7)$	0	1
		Next class		y_2^{v+1}	

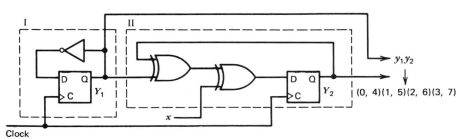

Figure 10.52 Development of y_2 subcircuit of up-down counter.

q^v	$(y_3y_2y_1)^v$	x^v 0	1	0	1
0	000	7	1	1	0
1	001	0	2	0	0
3	011	2	4	0	1
2	010	1	3	0	0
4	100	3	5	0	1
5	101	4	6	1	1
7	111	6	0	1	0
6	110	5	7	1	1
		q^{v+1}		y_3^{v+1}	

Figure 10.53 Development of y_3 subcircuit of up-down counter.

note two distinct three-state counting sequences. Perhaps a simpler and more natural structure would result with four state variables, two for each sequence. The extra state variables would require an extra flip-flop, but the simpler structure might simplify the excitation and output logic sufficiently to result in a net decrease in cost.

A special case in which the use of extra flip-flops is very likely to lead to lower cost is the finite memory circuit. Recall the error detector of Example 10.3, which was originally formulated as a finite memory circuit, as shown in Fig. 10.10. If we

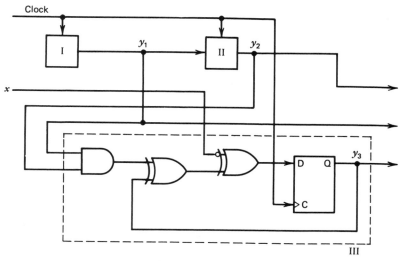

Figure 10.54

realize the circuit in this form, the only cost, in addition to the three D flip-flops, is for logic to realize the output equation, Equation 10.5. We minimized this same table for four states in Example 10.6 and made a state assignment in Example 10.10. The equations for the selected assignment are shown in Equation 10.31. The output equations for the two cases are comparable in cost, so the saving of the flip-flop is offset by the cost of three gates to realize the excitation equations of Equation 10.31. As a rough rule of thumb, one flip-flop is equal in cost to two to four gates, so the basic logic cost for the two forms would be comparable. On the other hand, the four-state version will have many more interconnections between logic cells and will thus require more area in the routing channels of a standard cell realization.

10.12 CONCLUSIONS

The first step in the design of sequential circuits, setting up the state diagram or state table, is the most important step. The initial specification of a circuit function is almost invariably vague and incomplete. The procedures developed in Sections 10.3 and 10.4 help clarify your understanding of the circuit function and force you to specify precisely the desired behavior in every possible circumstance. Possibly because it requires considerable time and effort to think the problem through carefully and thoroughly, there is a tendency for designers to skip this step. Instead, they try to apply the sort of intuition about structure that we discussed in connection with state assignment at the very beginning. They try to interconnect a set of hazily perceived subcircuits, such as counters, parity checkers, etc., in a trial-and-error fashion, usually producing a design that does many unexpected things. The time to apply intuition about circuit structure is after you thoroughly understand the circuit function.

If the circuit is finite memory and the transition table has been formulated accordingly, the only step left is to read the output equations directly from the transition table. As discussed above, when the finite memory form is applicable, it almost invariably produces the most economical designs and should be used.

If the circuit is not finite memory, the next step is to minimize the state table. As discussed, a minimal table does not necessarily result in the most economical design. However, it is best to start with a minimal table and consider adding more state variables only if there seems to be an obvious natural structure requiring more variables.

Finally, we hope that the reader will acquire some perspective about where best to apply efforts in this rather elaborate design process. The first step warrants the greatest attention because no amount of later effort can produce a good design from an initially incomplete or erroneous specification. The second step, state reduction, is rather tedious but has the virtue of producing guaranteed results (if we assume there are no clerical errors) and is usually time well spent. It is in the state assignment step that judgment must be applied. It is natural to seek simple, elegant designs and easy to spend more time on state assignment than on other parts of the design process. However, engineering time is expensive, and many hours spent eliminating a gate or two can be a very bad investment unless production of the circuit is to be enormous.

The above paragraph and indeed the bulk of this chapter are most pertinent to machines with relatively few states (≤ 16). As the total amount of information to be stored increases, so will the likelihood of uncovering a partitionable structure. For large circuits, some formal system description other than the state table would be very desirable. Such formal descriptions exist in the form of register transfer languages. Such a language will be a subject of Chapter 11,

PROBLEMS

10.1 Analyze the sequential circuit given in Fig. P10.1 in the manner discussed in Section 10.1. Obtain first the flip-flop input equations and then the state diagram representation of the circuit.

10.2 A clock-mode sequential circuit is to be designed featuring an external reset mechanism that will on occasion reset the circuit to state q_0. Determine a state diagram of the circuit so that it will generate an output of 1 for one clock period only coinciding with the second 0 input of a sequence consisting of exactly two 1's (no more than two) followed by two 0's. Once the output has been 1 for one clock period, the output will remain 0 until the circuit is externally reset to q_0.

10.3 Derive the state diagram and state table of a circuit featuring one input line X_1, in addition to the clock input. The only output line is to be 0 unless the input has been 1 for four consecutive clock pulses or 0 for four consecutive pulses. The output must be 1 at the time of the fourth consecutive identical output.

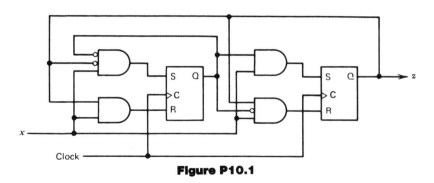

Figure P10.1

10.4 A sequential circuit is to be designed with two input lines x_1 and x_2 and a single output z. If a clock pulse arrives when $x_2 x_1 = 00$, the circuit is to assume a reset state that may be labeled q_0. Suppose the next six clock pulses following a resetting pulse coincide with the following sequence of input combinations: 01–10–11–01–10–11. The output z is to be 1 coinciding with the sixth of such a string of six clock pulses but 0 at all other times. The circuit cannot be reset to q_0 except by input 00. Define a special state q_1 to which the circuit will go once it becomes impossible for an output-producing sequence to occur. The circuit will thus wait in q_1 until it is reset. Determine a state diagram for this circuit.

10.5 A sequential circuit is to be designed in which the circuit output z^v is a function of only the current input x^v and the three previous inputs x^{v-1}, x^{v-2}, and x^{v-3}. A Boolean expression for z^v is given by

$$z^v = x^v \cdot x^{v-1} \cdot x^{v-2} \cdot x^{v-3} + \overline{x^v} \cdot \overline{x^{v-1}} \cdot \overline{x^{v-2}} \cdot \overline{x^{v-3}}$$

Treat the circuit as a finite-input memory circuit and determine an eight-state state table representation.

10.6 A variable-delay circuit is to provide a two- or three-clock period delay of an input signal x_i. That is, the circuit output x_d^v is to be equal to x_i^{v-3} if the control input $r = 1$ and is to be equal to x_i^{v-2} if $r = 0$. A simplified block diagram of the circuit is shown in Fig. P10.6. Treat it as a finite memory circuit and obtain a state diagram for the variable-delay sequential circuit.

10.7 A sequential circuit has two inputs, x_1 and x_2. Five-bit sequences representing decimal digits coded in the 2-out-of-5 code appear from time to time on line x_1, synchronized with a clock pulse on a third line. Each five consecutive bits on the x_1 line, which occur while $x_2 = 0$ but immediately following one or more inputs of $x_2 = 1$, may be considered a code word. The single output z is to be 0, except upon the fifth bit of an invalid code word. Determine a state table of the above sequential circuit.

10.8 A clock-mode sequential circuit has two input lines, a and b. During any clock period, $a^v a^{v-1} a^{v-2}$ and $b^v b^{v-1} b^{v-2}$ may be regarded as three-bit binary numbers. The output z is to be 1 if $a^v a^{v-1} a^{v-2} \geq b^v b^{v-1} b^{v-2}$. The inputs a^v and b^v are the most significant bits, as implied by the format shown. Obtain a state table for this sequential circuit.

10.9 Let aRb if and only if $a = b \pm 8X$, where X is any integer. (We say that a is congruent to b modulo-8). Show that R is an equivalence relation on the set of all integers.

10.10 Determine a minimal state table equivalent to the state table in Fig. P10.10. (**Hint: Five states are required.**)

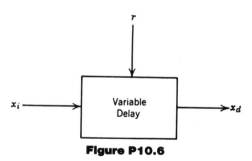

Figure P10.6

	q^{v+1}, Z^v	
q^v	$x = 0$	$x = 1$
1	1,0	1,0
2	1,1	6,1
3	4,0	5,0
4	1,1	7,0
5	2,0	3,0
6	4,0	5,0
7	2,0	3,0

Figure P10.10

10.11 Determine a minimal state table equivalent to the state table shown in Fig. P10.11.

10.12 Determine a circuit whose output is equivalent to the circuit shown in Fig. P10.12, but that has a minimum number of states.

10.13 The minimal state circuits satisfying the functions of previous problems are listed as follows. In each case where the circuit synthesized by the

Present State	Next State, Output			
	$x = 0$	$x = 1$	$x = 2$	$x = 3$
A	E,1	C,0	B,1	E,1
B	C,0	F,1	E,1	B,0
C	B,1	A,0	D,1	F,1
D	G,0	F,1	E,1	B,0
E	C,0	F,1	D,1	E,0
F	C,1	F,1	D,0	H,0
G	D,1	A,0	B,1	F,1
H	B,1	C,0	E,1	F,1

Figure P10.11

	q^{v+1}, Z	
q^v	$x = 0$	$x = 1$
q_1	$q_2,0$	$q_3,0$
q_2	$q_4,0$	$q_5,0$
q_3	$q_6,0$	$q_7,0$
q_4	$q_8,0$	$q_9,1$
q_5	$q_{10},0$	$q_{11},0$
q_6	$q_{12},1$	$q_{13},0$
q_7	$q_{14},0$	$q_{15},0$
q_8	$q_1,0$	$q_3,0$
q_9	$q_1,0$	$q_3,0$
q_{10}	$q_1,0$	$q_3,0$
q_{11}	$q_1,0$	$q_2,0$
q_{12}	$q_2,0$	$q_1,0$
q_{13}	$q_2,0$	$q_1,0$
q_{14}	$q_2,0$	$q_1,0$
q_{15}	$q_2,0$	$q_1,0$

Figure P10.12

reader was not minimal, obtain a minimal machine by the formal techniques of Sections 10.7 and 10.8.

10.2	6 states
10.5	6 states
10.7	14 states

10.14 Determine a minimal state equivalent of the state table found in Problem 10.8.

10.15 Determine a realization of a clocked J-K flip-flop in terms of a clocked S-C flip-flop and appropriate combinational logic.

10.16 Obtain a circuit realization of the state table found in Problem 10.5:
(a) Using S-R flip-flops.
(b) Using D flip-flops.
(c) Using J-K flip-flops.

10.17 Obtain a circuit realization of the state table given in Fig. P10.17 using J-K flip-flops.

10.18 Repeat Problem 10.17 using D flip-flops.

10.19 Obtain circuit realizations for the minimal state tables for
(a) Problem 10.2
(b) Problem 10.6
(c) Problem 10.14
using whichever type of flip-flop appears most suitable in each case.

10.20 There are three closed partitions of the eight states of the sequential circuit whose block diagram is given in Fig. P10.20. The state assignment is listed below. Indicate the three closed partitions, using the notation for closed partitions used in Section 10.11. (**Hint: The partitions can be identified without the laborious reconstruction of the state table**).

	y_1	y_2	y_3
a	0	0	0
b	0	0	1
c	0	1	0
d	0	1	1
e	1	0	0
f	1	0	1
g	1	1	0
h	1	1	1

	q^{v+1}, Z	
q^v	$x = 0$	$x = 1$
A	B,0	B,1
B	F,0	D,1
C	E,1	G,1
D	A,0	C,0
E	D,1	G,0
F	F,0	A,0
G	C,1	B,0

Figure P10.17

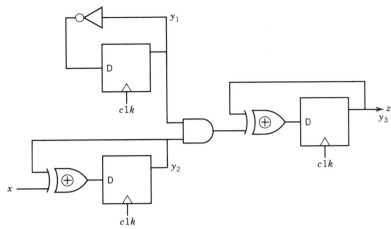

Figure P10.20

10.21 A circuit with an external reset mechanism, a single input x, and a single output z is to behave in the following way. At the first clock time following reset, the output is to be $z = 0$. At the second clock time, $z^v = x^v \cdot x^{v-1}$. At the third clock time, $z^v = x^v + x^{v-1}$, and at the fourth clock time, $z^v = x^v \oplus x^{v-1}$. This sequence of output functions is then repeated every four clock periods until the circuit is reset. The circuit can be readily partitioned into two subcircuits: a two-bit counter and a one-bit shift register. Take advantage of this fact to determine a state diagram and a realization of the circuit in terms of J-K flip-flops.

10.22 Obtain the state table of a serial translating circuit that performs the translations tabulated in Fig. P10.22. As indicated, there is only one input line X and one output line Z in the circuit. The inputs occur in characters of three consecutive bits. One character is followed immediately by another. A reset signal, which need not be included in the state table, will cause the translator to begin operation synchronized with the input. A translated output character must be delayed by only two clock periods with respect to an input character. (**Hint: A 12-state minimal state table can be obtained by partitioning the circuit into a three-state counter and a two-bit shift register.**)

10.23 There exists a closed partition of the 12 states of the sequential circuit determined in Problem 10.22. Find the partition which includes three classes of four states and verify that it is indeed closed.

X^v	X^{v-1}	X^{v-2}	Z^{v+2}	Z^{v+1}	Z^v
0	0	0	0	0	0
0	0	1	0	1	1
0	1	0	1	0	0
0	1	1	1	0	1
1	0	0	1	1	0
1	0	1	0	0	1
1	1	0	0	1	0
1	1	1	1	1	1

Figure P10.22

10.24 Obtain a realization of the state table derived in Problem 10.22 using *J-K* flip-flops.

10.25 Repeat Problem 10.7 using an approach that partitions the circuit into a modulo-5 counter and a two-bit counter for counting 1's appearing on line x_1. Determine a five flip-flop (*J-K*) realization of this circuit without minimizing the state table. Compare the overall complexity without minimizing the state table. Compare the overall complexity of this realization with a realization of a minimal equivalent of the state table determined in Problem 10.7.

BIBLIOGRAPHY

1. Huffman, D. A. "The Synthesis of Sequential Switching Circuits," *J. Franklin Inst.*, **257**: 3, 161–190; 4, 275–303 (March–April, 1954).
2. Paul, M. C. and S. H. Unger. "Minimizing the Number of States in Incompletely Specified Sequential Switching Functions," *IRE Trans. Electronic Computers*, EC-8: 3, 356, 357 (Sept. 1959).
3. Mealy, G. H. "A Method for Synthesizing Sequential Circuits," *Bell System Tech. J.*, **34**: 5, 1045–1080 (Sept. 1955).
4. Moore, E. F. "Gedanken Experiments on Sequential Machines," in C. E. Shannon and J. McCarthy (eds.), *Automata Studies*. Princeton Univ. Press, Princeton, N.J., 1956.
5. McCluskey, E. J. and S. H. Unger. "A Note on the Number of Internal Variable Assignments for Sequential Switching Circuits," *IRE Trans. Electronic Computers*, EC-8: 4, 439–440 (Dec. 1959).
6. Hartmanis, J. "On the State Assignment Problem for Sequential Machines, I," *IRE Trans. Electronic Computers*, EC-10: 2, 157–165 (June 1961).
7. Stearns, R. E. and J. Hartmanis. "On the State Assignment Problem for Sequential Machines, II," *IRE Trans. Electronic Computers*, EC-10: 4, 593–603 (Dec. 1961).
8. Hartmanis, J. and R. E. Stearns. *Algebraic Structure Theory of Sequential Machines*. Prentice-Hall, Englewood Cliffs, N.J., 1966.
9. Karp, R. M. "Some Techniques of State Assignment for Synchronous Sequential Machines," *IEEE Trans. Electronic Computers*, EC-13: 5, 507–518 (Oct. 1964).
10. Dolotta, T. A. and E. J. McCluskey. "The Coding of Internal States of Sequential Circuits," *IEEE Trans. Electronic Computers*, EC-13: 5, 549–562 (Oct. 1964).
11. Harrison, M. A. *Introduction to Switching and Automata Theory*. McGraw-Hill, New York, 1965.
12. Kohavi, Z. *Switching and Finite Automata Theory*, 2nd ed. McGraw-Hill, New York, 1978.

CHAPTER 11

VECTOR PROCESSES: DESCRIPTION AND REALIZATION

11.1 DESCRIPTION OF MORE COMPLEX SYSTEMS

In Chapter 10 we agreed that either a state diagram or a state table could be accepted as an unambiguous description of a design to be realized as a clock-mode sequential circuit. For many examples, with relatively few states and therefore few memory elements, a state table might also be the most efficient or convenient medium of description. This is particularly true when the next state is distinct for almost every combination of inputs. Under the broadest definition of a language, a state table could be called a hardware description language.

A digital hardware system may be considered a physical implementation of an algorithm. Algorithms that require large numbers of memory elements for execution rarely require the precision or descriptive efficiency offered by a state table. As design task complexity increases, it is quite natural to look to other language forms to bridge the gap between English and the hardware realization.

Historically, the earliest computers were developed with the availability of very few language tools, probably little more than English and Boolean algebra. For this reason, as well as the primitive state of available hardware, the development cycle for these machines was long and replete with error and redesign. When these machines were first used, the languages available for their programming were equally limited. The great potential of these machines was recognized, nonetheless, and much thought was devoted to ways to ease the burden of programming. Flowcharts came into use quickly and, over the years, a number of high-level programming languages have appeared. It was eventually realized that variations of these same tools could also be applied in the design process for digital hardware.

The flowchart has been an informal tool for digital design for many years. More recently, it was formalized as the ASM or "algorithmic state machine" chart [1, 2]. The use of high-level languages as a means of describing hardware is also a relatively recent innovation [3–5]. We shall introduce both techniques in this chapter. We will look first at the ASM chart as an alternative for the description of sequential circuits like those we considered in Chapter 10. Next we will extend the ASM chart to more easily represent systems that are naturally partitioned into separate control and data sections. This will provide a bridge to the introduction of the concept of a *hardware description language*. Hardware description languages have much in common with high-level programming languages, but their syntax is usually adjusted to reflect hardware structure. The flowchart is often used in the process of writing a program in a high-level language. Similarly, the ASM chart will prove helpful in deriving the representation of digital hardware in a high-level language. It will also continue to serve as a formal specification for a realization of the control portion of the larger digital system.

11.2 STANDARD SYMBOLS FOR THE ASM CHART

An ASM chart is a flowchart that describes the behavior of a sequential circuit. A precise set of rules has been formulated to govern the construction of ASM charts. Each individual state of the sequential circuit is represented by an ASM block. An ASM block is a small configuration of symbols that represent the name of the state, the flip-flop values for that state, the circuit outputs as a function of the state, any inputs that can occur while the circuit is in this state, and the next states as they depend on circuit inputs. When the ASM blocks representing all states of a sequential circuit are connected together, the resulting ASM chart is an alternative unambiguous description of that circuit.

There are only three symbols that may appear in an ASM block. Always included is the rectangular state box illustrated in Fig. 11.1. Within this box are listed the circuit outputs that can occur whenever the circuit is in the corresponding state, regardless of input values. At the left of this box is the name of the state. Immediately above the box is a list of the flip-flop values which define that state. In contrast to the state diagram, this approach expects that the assignment of variable values has been made before formulation of the ASM chart begins. As was concluded in Section 10.11, using information available in the initial problem definition to effect a state assignment may be preferable to using formal techniques to make that assignment later.

Two additional symbols may appear within a block, depending on the inputs that impact the circuit while in the corresponding state. Inputs without impact, while the circuit is in a particular state, are completely ignored in the state box. This feature is especially convenient if the circuit has many inputs, only a few of which can affect the circuit at any given time. Each input wire, or combination of input values, that can possibly be active when the circuit is in a particular state must be represented in a decision diamond to be to be associated with that state. A decision diamond is illustrated in Fig. 11.2a. Sometimes Boolean expressions of input wires may be entered in a decision diamond. Each value that may be

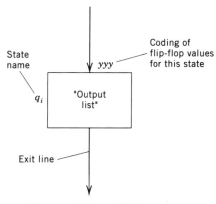

Figure 11.1 ASM state box.

assumed by an input wire or expression in a diamond is associated with an exit line from that diamond. These lines lead to the blocks corresponding to the next states of the circuit following the next clock pulse. If output values are also functions of these inputs, these exit lines pass through conditional output boxes of the form illustrated in Fig. 11.2b. The outputs that occur when the path to a conditional output box is satisfied are listed within the box. Outputs which are not listed in either the state box or a conditional output box in a particular ASM block are always inactive when the circuit is in that state.

An example of an ASM block is given in Fig. 11.3. The sequential circuit that this state partially represents includes three flip-flops, y_2, y_1, y_0. In this state, which is labeled q_1, these flip-flops assume the values 0, 1, 0. While the circuit is in this state, the circuit output Z_1 will always be 1. If $X_1 = 0$, the circuit output Z_2 will also be 1, and the next state will be q_2. If $X_1 = 0$, Z_2 will be 0 and the next state will depend on the input X_2. For $X_1 X_2 = 1, 0$ the next state will be q_3. For $X_1 X_2 = 1, 1$ the next state will be q_4.

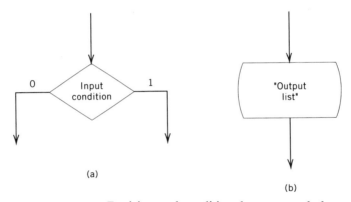

(a)

(b)

Figure 11.2 Decision and conditional output symbols.

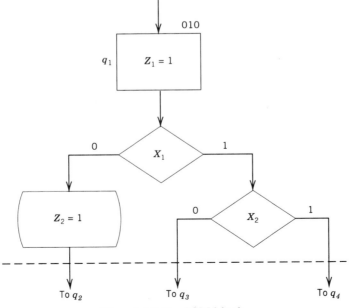

Figure 11.3 ASM block.

The ASM chart is primarily a medium in terms of which complex designs can be initially formulated prior to the development of a transition table that will lead to a hardware realization. For an example to illustrate a first complete ASM chart, we look back at the beginning-of-message detector, the reduced state table of which was given in Fig 10. 22b.

■ **EXAMPLE 11.1**

Construct an ASM chart of the beginning-of-message detector, whose state table is repeated in Fig. 11.4a.

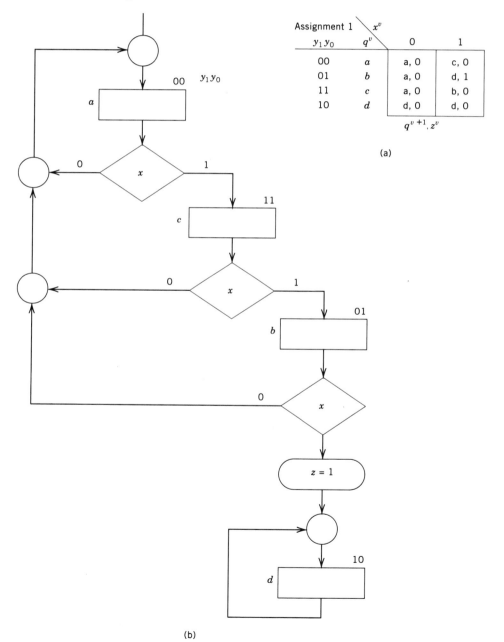

Assignment 1	x^v		
$y_1 y_0$	q^v	0	1
00	a	a, 0	c, 0
01	b	a, 0	d, 1
11	c	a, 0	b, 0
10	d	d, 0	d, 0
		q^{v+1}, z^v	

(a)

(b)

Figure 11.4 ASM chart for beginning-of-message detector.

SOLUTION   In this case, we simply begin by constructing an ASM block for state
a and then adding a block for each new state as it appears as a next state in the
state table. The result is Fig. 11.4*b*. To supply the memory element values, it was
necessary to anticipate a state assignment. We chose state assignment 1 of Section
10.10, one of the two most economical assignments. The output is only indicated
as $Z = 1$ in ASM block *b* for $x = 1$. Z is implicitly 0 elsewhere in the diagram.   ■

In practice, it is unlikely that an ASM chart would be of interest once a state
table is already available. The ASM chart might be the preferred alternative for
initial formulation of the "unambiguous problem description" in cases where
only a small subset of the large number of input lines will impact the circuit function at each present state. *The drawback of the ASM chart approach can only be
economy of realization*. The state assignment is fixed and Boolean expressions of
inputs are included in the charts, eliminating opportunity for further exploitation of "don't care" conditions. Both state assignment and "don't cares" are
immaterial, if output and next-state expressions are implemented using a ROM
and D flip-flops. The ASM chart of Fig. 11.4*a* may be translated directly to the D
flip-flop transition table given in Fig. 11.5*a*. A ROM implementation is depicted
in Fig. 11.5*b*.

y_1	y_0	x	y_1^{v+1}	y_0^{v+1}	z
0	0	0	0	0	0
0	0	1	1	1	0
0	1	0	0	0	0
0	1	1	1	0	1
1	0	0	1	0	0
1	0	1	1	0	0
1	1	0	0	0	0
1	1	1	0	1	0

(a)

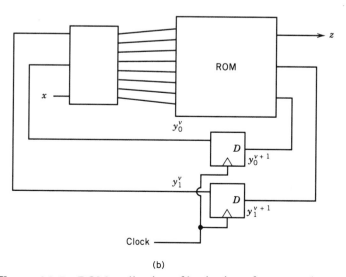

(b)

Figure 11.5 ROM realization of beginning-of-message detector.

11.3   CLOCK INPUT CONTROL

Until now, we have inferred that a sequential circuit only operates in the clock mode if the clock or inputs of all flip-flops on the circuit are tied *directly* to the system clock, as depicted in Fig. 11.6a. Under certain circumstances, particularly when D-type flip-flops are used, this approach to clocking can be unduly restrictive. A simple illustration of a sequential circuit with direct clocking is given in Fig. 11.6b. If control signal $CS_1 = 1, x_1$ is to be shifted into the flip-flop shown. If $CS_2 = 1, x_2$ is shifted in. Regardless of the control signals that might be present, it is always necessary that the value to be stored in the D flip-flop after a clock pulse be present at the flip-flop input before the clock pulse arrives. Therefore, if both control signals are 0, the current value of the flip-flop must be routed back to the D input. This is accomplished by the upper AND gate in Fig. 11.6b.

The network in Fig. 11.6b can be simplified by a departure from the direct clocking procedure. Suppose that a clock pulse is allowed to reach the clock input of a flip-flop only if the contents of the flip-flop are to be modified. For the example under discussion then, a change will occur only, if $CS_1 \vee CS_2 = 1$. In Fig. 11.7a, this signal is used to gate the clock pulse into the clock input of the D flip-flop. Thus, the circuits of Figs. 11.6b and 11.7a accomplish the same function. In this case, the one AND gate and one OR gate saved in the data network of Fig. 11.6b were added back into the clocking network of Fig. 11.7a. A significant savings can be achieved in more complicated systems, particularly when several flip-flops are combined together to form a register. Figure 11.7b is a timing diagram relating the control pulse to the clock and control level. Often the same control signals are used to gate data into all flip-flops of a register. In this case, the clock control logic can be common to all flip-flops of the register, as illustrated in Fig. 11.7c. (The data logic is not shown.) If the approach of Fig. 11.6 were used, this saving could not be achieved.

If the clock is a periodic sequence of narrow pulses, then the signal on the line labeled control pulse in Fig. 11.7b is a pulse synchronized with the clock. This is in contrast to the signals on the control lines entering the OR gate of the same figure, which carry levels that are necessarily stable at the time of a clock pulse. Pulses are allowed as inputs to logic networks, only if the restrictions tabulated in Fig. 11.8 are satisfied.

The case in which a level is ANDed with a pulse is useful, as illustrated in Fig. 11.7. Not allowed, however, is the AND operation in which more than one input

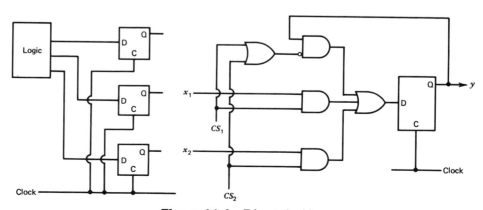

Figure 11.6   Direct clocking.

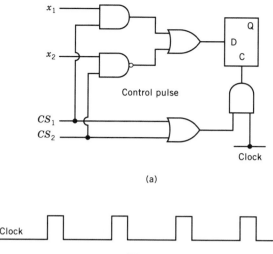

(a)

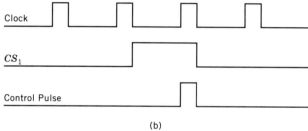

(b)

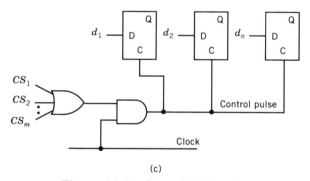

(c)

Figure 11.7 Controlled clocking.

A	B	$A \cdot B$	$A + B$
level	level	level	level
level	pulse	pulse	undefined
pulse	pulse	undefined	pulse

Figure 11.8 Logic on levels and pulses.

line could carry a pulse. Pulses are of short duration relative to the time frame of interest, so two pulses can never be guaranteed to occur simultaneously. It can be seen that the output of an OR gate with one pulse input and one level input might be either a level or a pulse. Every Boolean variable (or network wire) must be classified as either a pulse variable or a level variable. We, therefore, declare the ORing of a level with a pulse to be undefined. The operation of ORing two pulse lines may be useful and is allowed, provided that no more than one pulse input to an OR gate can be active simultaneously.

Using Figs. 11.6 and 11.7, we successfully argued that significant hardware savings could be achieved by the controlled clocking of data registers. In cases where the clock is not a sequence of narrow pulses, controlled clocking carries with it a potential pitfall. This problem must always be considered by the designer of CMOS where the clock is usually a square wave with equal-length high and low half-periods. Assume the memory element shown in Fig. 11.9a is an instance of the CMOS master-slave element given in Fig. 9.27. Figures 11.9b and 11.9c illustrate problems that can develop when the time required for a control signal to stabilize exceeds one half-clock period. In both cases, the control signal **LD** is not stable until the second or positive half of the clock period. In Fig. 11.9b two separate pulses appear at the clock input of the flip-flop. The state of the memory element might change twice in the same period (e.g., a counter), thereby violating the clock-mode assumption. In Fig. 11.9c no pulse should reach the clock input of the memory element, but a narrow one does. Data will be clocked into the memory element when no such activity should take place.

In many systems the logic path with the longest delay, *the critical path*, will always be a data path rather than a control path. It may be possible to establish a clock frequency so that control signals can stabilize in a half-period, whereas the worst-case data path may require nearly a whole period. If such a control/data relationship does not exist, it may be necessary to resort to the less economical structure of Fig. 11.6 for CMOS design.

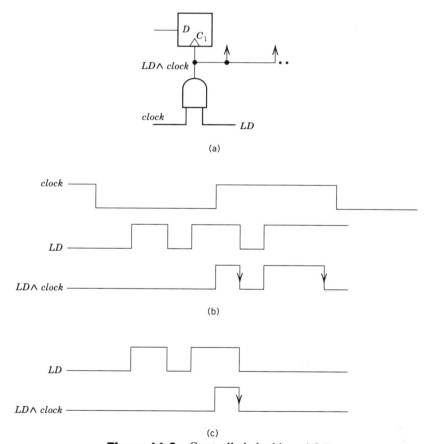

Figure 11.9 Controlled clocking pitfall.

11.4 EXTENDED STATE TABLES

Digital computers and other large digital systems are examples of sequential circuits, but they clearly differ in size and complexity from the circuits discussed up to this point. Such systems typically include many registers of several memory elements each. The number of states in such systems, taken in the formal sense of all possible combinations of values of all flip-flops, can reach into the thousands. Clearly, state diagrams and tables are not practical for the description of such large systems. We must develop more suitable techniques for describing sequential circuits composed primarily of large numbers of registers.

The previous section introduced the concept of controlling the clock signals to memory elements. Let us now explore the distinctions between the control levels and pulses and the logic that develops them on the one hand, and the data networks controlled by these signals on the other. We begin by partitioning the clock-mode model into two separate sequential circuits as shown in Fig. 11.10. The upper portion consists of data registers together with interconnecting combinational logic. The lower sequential circuit provides a sequence of control signals that cause the appropriate register transfers to take place with the arrival of each clock pulse. The numbers of control signal lines and control input lines are normally small in comparison to the numbers of data input and output lines and interconnecting data lines. In Fig. 11.11 is shown a set of three data flip-flops, B_0, B_1, B_2, interconnected by data transfer lines that are controlled by signals C_1, C_2, C_3, C_4, developed by a control sequential circuit. The nature of the control circuit is such that only one of the four control signals can be 1 at any time. A careful examination will reveal that four sets of data transfers can take place in the lower portion of Fig. 11.10. If $C_1 = 1$, the inputs X_0 and X_1 are loaded into flip-flops B_0 and B_1, respectively. The notation we will use for this transfer is given by Equation 11.1. (The contents of B_2 are unchanged.)

$$1. \ B_0 \leftarrow X_0; \qquad B_1 \leftarrow X_1 \tag{11.1}$$

If $C_2 = 1$, the information in the three flip-flops is shifted to the right, with the contents of the rightmost flip-flop shifted around into B_0. For the present, we will denote this transfer similarly as given in

$$2. \ B_1 \leftarrow B_0; \qquad B_2 \leftarrow B_1; \qquad B_0 \leftarrow B_2 \tag{11.2}$$

If $C_3 = 1$, the transfer is

$$3. \ B_1 \leftarrow B_0 \wedge B_1 \tag{11.3}$$

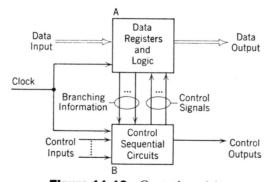

Figure 11.10 Control model.

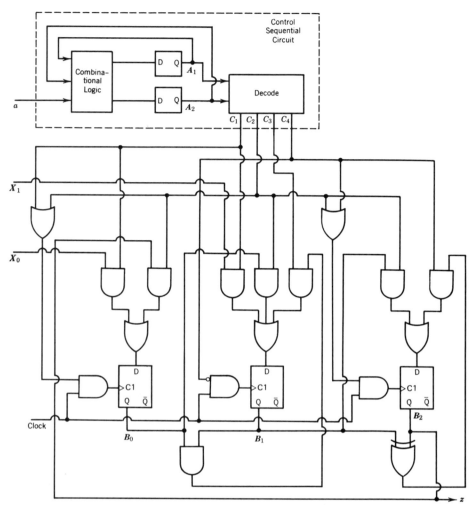

Figure 11.11   Example of control of data transfers.

and if $C_4 = 1$, the transfer is

$$4. \quad B_2 \leftarrow B_1 \oplus B_2 \tag{11.4}$$

Note the use of the symbol $\wedge$ for the AND operation. We will find this symbol convenient as we expand our transfer notation.

The above describes the transfers that can take place, but the sequence of transfers, and therefore the sequence of outputs, will depend on the sequence of control signals. Since there are four possible combinations of control signals, we may characterize the control sequential circuit as a Moore circuit with four states, which we may designate as C_1, C_2, C_3, C_4, each producing a 1 on the corresponding control output line. Let us further assume that the sequencing between these states is controlled by a sequence control signal a, as shown in the state table of Fig. 11.12. This is consistent with Fig. 11.11, where a is shown as an input to this four-state control sequential circuit. To this state table, we add a list of the transfers initiated by the control signals corresponding to each state.

Together with the output equation $z = B_2$, this extended state table provides a complete description of the desired sequential circuit behavior from which the

q^v	q^{v+1}		Transfer corresponding to q^v
	$a = 0$	$a = 1$	
C_1	C_3	C_2	$B_0 \leftarrow X_0; B_1 \leftarrow X_1$
C_2	C_1	C_1	$B_1 \leftarrow B_0; B_2 \leftarrow B_1; B_0 \leftarrow B_2$
C_3	C_4	C_4	$B_1 \leftarrow B_0 \wedge B_1$
C_4	C_2	C_1	$B_2 \leftarrow B_1 \oplus B_2$

Figure 11.12 Extended state table.

complete circuit could be generated. There are three data flip-flops, and at least two control flip-flops will be required. Therefore, a conventional description in the manner of Chapter 10 would require a 32-state table, which would clearly be far less convenient as a design tool than the compact table of Fig. 11.12. We will leave it as a homework problem for the reader to obtain such a table.

11.5 THE EXTENDED ASM CHART AND DESIGN LANGUAGE REPRESENTATION

Even more natural than extending the state table to represent state changes in vectors or registers is to extend the ASM chart for the same purpose. It is only necessary to enter the register transfers in the respective ASM state box as if they were outputs. Effectively, an output from the control sequential circuit causes the register transfers to be executed in the data unit. An ASM chart describing the network of Fig. 11.11 is given in Fig. 11.13. It is important not to be misled by the ordering of transfers and branches in the ASM chart. If conditions are satisfied, outputs in an ASM block are active throughout the clock period in which the state corresponding to that block is active. All transfers in the ASM block and the control state change from the block are triggered by the clock pulse at the end of that period. For emphasis, the ASM block corresponding to control state C_1 is enclosed by dashed lines in Fig. 11.13.

The extended ASM chart representation of a complex sequential circuit suggests that the description might be reduced to program form, just as a flowchart representation of code is reduced to a program in a high-level language. Indeed, for very large digital hardware systems a programlike description is more manageable than an extended ASM chart. A language for expressing a hardware program will be called a *hardware description language*. Many hardware description languages have been defined [5, 6, 8]. Of those still in use, AHPL (a hardware programming language) [3] has the advantage of simplicity, conciseness, and the most direct correspondence to hardware.

The interpretation of timing distinguishes a hardware program or sequential circuit from a software program in a high-level language. In the latter, there is no mechanism and no need for one insisting that sets of statements take place simultaneously. This mechanism is necessary in reducing an extended ASM chart to a program, because each transfer statement in an ASM box is executed at the triggering edge of a single clock pulse while the output statements in a box are active during the clock period preceding that edge. In the next section we will introduce a more compact notation for some simultaneous transfers by collecting flip-flops into registers.

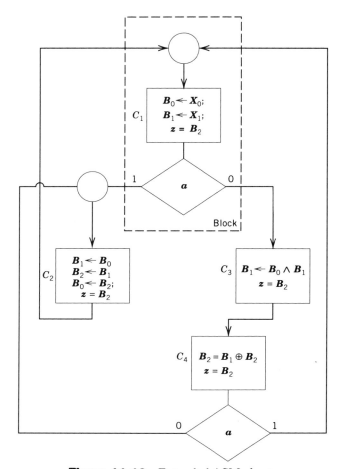

Figure 11.13   Extended ASM chart.

The next control state information represented by decision diamonds in the ASM chart is translated to branch statements in the program representation. Control states C_1 and C_4 include two-way branches, whereas the branches from states C_2 and C_3 are *unconditional*. The unconditional transfer is not a pariah in a hardware description language, as it is in many high-level languages. In AHPL each step or control state consists of a transfer/connection part and a branch part. The former may consist of any number of transfer and/or connection statements. The branch part is a single branch statement, which may be omitted if control always advances to the next step in order.

An unconditional branch has the form

$$\rightarrow (S)$$

where S is the number of the statement to be executed next. The conditional branch has the form

$$\rightarrow (f_1, f_2, ..., f_n)/(S_1, S_2, ..., S_n) \tag{11.5}$$

The functions f_1, f_2, etc., are logic functions of the system variables that take on the values 0 or 1. If a given $f_i = 1$, then S_i, the statement number in the corresponding position to the right of the slash, is the number of the statement to be executed next. Normally, only one of the f_i may be 1 at a time, that is,

$$f_i \wedge f_j = 0 \qquad \text{where } i \neq j \tag{11.6}$$

If none of the $f_i = 1$, the next statement in sequence is executed.

Following the conventions described above, Fig. 11.5 may be translated into the following *control sequence*:

1 $B_0 \leftarrow X_0; \qquad B_1 \leftarrow X_1;$
  $\rightarrow (\bar{a}, a)/(3, 2).$

2 $B_1 \leftarrow B_0; \qquad B_2 \leftarrow B_1; \qquad B_0 \leftarrow B_2;$
  $\rightarrow (1).$

3 $B_1 \leftarrow B_0 \wedge B_1.$

4 $B_2 \leftarrow B_1 \oplus B_2;$
  $\rightarrow (\bar{a}, a)/(2, 1).$

When we have a two-way branch with one destination being the next statement, we can take advantage of the default option (that the next statement is executed if all $f_i = 0$) to simplify the branch statement. For example, statement 1 in the above sequence can be simplified to the form

1 $B_0 \leftarrow X_0; \qquad B_1 \leftarrow X_1;$
  $\rightarrow (\bar{a})/(3).$

This form of branch statement can be interpreted to mean, "IF $\bar{a} = 1$, THEN go to 3, ELSE go to 2." This "IF-THEN-ELSE" form is standard in many programming languages. If there are more than two branch destinations, the default option may not be used.

Throughout the rest of this chapter, we will use the term *control sequence* to identify hardware descriptions of this form. This is a complete description of the circuit of Fig. 11.11 containing all information found in Fig. 11.12. The efficiency of this new method of circuit description is not particularly apparent for the above example. Note that only one branch statement (Step 3) was omitted because the next step (Step 4) is always taken in sequence. This situation will be much more common in larger systems. Similarly, when there are a large number of inputs to the control unit, the f_i at any given step may be a function of only a few of these inputs.

11.6 WRITING THE AHPL DESCRIPTION

A principal advantage of a program description of hardware must necessarily be its use as a synthesis tool. For large systems, it is the only design approach other than trial-and-error manipulation of the logic block diagram. In this chapter, we will show that it can be used effectively over a range of less complex systems that are still too bulky to describe in state table form. We will even find it useful as a means of quickly obtaining unambiguous descriptions of certain smaller sequential circuits for which we obtained state tables in Chapter 10.

Before we move to discard all our state-table-related tools, let us remind ourselves that our new medium is a *language and only a language*. It will have no apparent algebraic structure in terms of which state minimization can be performed. Nor does the language itself provide any constraints that will guide the designer to a minimal circuit. With practice, the designer will readily arrive at an optimal list of program steps for certain types of examples. For other examples, particularly when certain flip-flops can serve a variety of functions, it would be unrealistic to expect to obtain a minimal realization without resorting to state table minimization. Even if the minimal number of flip-flops are used, the fact that "don't care" conditions are not reflected in the language may cause a realization to be more complicated than necessary. The notion of state table minimization should be kept waiting in the wings for occasional use, at least on parts of sequential circuits.

As the first example of translation of an English language description of a problem to AHPL, let us design the three-bit code translator first introduced as Problem 10.22.

■ EXAMPLE 11.2

Obtain the control sequence of a serial translating circuit that performs the translations tabulated in Fig. 11.14. As indicated, there is one data input line x and one output line z in the circuit. The inputs occur in characters of three consecutive bits. One character is followed immediately by another. A reset input, $r = 1$, will cause the translator to begin operation synchronized with the input. A translated output character will be delayed by only two clock periods with respect to an input character.

SOLUTION The output translation cannot be performed until all three input bits have been received, at time t_v. Therefore, as the first two bits arrive, we shift them into a two-bit shift register B consisting of bits B_0 and B_1. At time t_v, x^{v-2} will be in B_1 and x^{v-1} in B_0. When x^v arrives, all three output bits will be computed simultaneously. The immediate output will be z^v, and outputs z^{v+1} and z^{v+2} will be stored in B_1 and B_0, respectively. At time t^{v+1}, z^{v+1} can be read from B_1, z^{v+2} will be shifted from B_0 to B_1, and the first input bit of the next character will be shifted into B_0. At t^{v+2}, the output z^{v+2} will be read from B_1, the previous input bit will be shifted into B_1, and the next input bit will be shifted into B_0, setting the state for the translation of the next character at the following clock time. The operation of this circuit can thus be described in terms of a repetitive three-step sequence, the first two steps being shifts, the third a translation and parallel register load, as shown below. For the first time we have used commas to symbolize *concatenation* of two scalar variables to form a vector.

$$1 \quad B_0, B_1 \leftarrow x, B_0; \qquad z = B_1;$$
$$\rightarrow (r)/(1).$$

$$2 \quad B_0, B_1 \leftarrow x, B_0; \qquad z = B_1;$$
$$\rightarrow (r)/(1).$$

x^v	x^{v-1}	x^{v-2}	z^{v+2}	z^{v+1}	z^v
0	0	0	0	0	0
0	0	1	0	1	1
0	1	0	1	0	0
0	1	1	1	0	1
1	0	0	1	1	0
1	0	1	0	0	1
1	1	0	0	1	0
1	1	1	1	1	1

Figure 11.14 Code translation.

$$3 \quad B_0, B_1 \leftarrow f_2(x, B_0, B_1), f_1(x, B_0, B_1); \qquad z = f_0(x, B_0, B_1);$$
$$\rightarrow (1).$$

The three output functions of Step 3 may be obtained by converting the table of Fig. 11.14 into three K-maps, one for each output, as shown in Fig. 11.15. In correlating these maps to the table, note that $x^{\nu-2}$ is stored in B_1, $x^{\nu-1}$ in B_0. The resulting functions are tabulated in Equations 11.7, 11.8, and 11.9. From Step 3 of the sequence, we note that $z^{\nu+1}$ is initially stored in B_1 and $z^{\nu+2}$ in B_0. A symbol, analogous to the symbol already assigned to AND, will represent OR in AHPL, to avoid the possible use of $+$ for both OR and addition in the same context.

$$f_0 = B_1 \tag{11.7}$$
$$f_1 = (x^{\nu} \wedge \bar{B}_1) \vee (x^{\nu} \wedge B_0) \vee (\bar{x}^{\nu} \wedge \bar{B}_0 \wedge B_1) \tag{11.8}$$
$$f_2 = (\bar{x}^{\nu} \wedge B_0) \vee (B_0 \wedge B_1) \vee (x^{\nu} \wedge \bar{B}_0 \wedge \bar{B}_1) \tag{11.9}$$

The reset mechanism has been incorporated into the three-step sequence. Whenever the input $r = 1$, control is returned to Step 1 to begin the input of a new character. The reader will notice that the single output z is specified separately on each line of the control sequence. This is necessary because the output is not a fixed function of some data register or input line, as was the case for the circuit of Fig. 11.11. Rather, the function is different for different steps in the sequence, that is, it is also a function of the control signals. It is common in large systems for the output to be taken directly from a data register or to be a fixed function of several data registers. In such cases, the outputs need be listed only once in the control sequence. ∎

The above example makes evident a very important difference between the control sequence method of design and the state diagram method presented in Chapter 10. Setting up the state diagram requires only an understanding of the required sequencing of inputs and outputs; no preliminary guesses as to the numbers of state variables or to the internal structure are required. By contrast, the control sequence approach essentially starts with a proposed basic structure. This method thus requires considerable intuitive understanding of possible circuit structures on the part of the designer, and it is very likely that the resulting

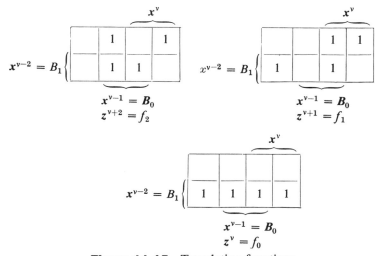

Figure 11.15 Translation functions.

design will not be minimal in the formal sense of Chapter 10. Once a circuit has been obtained by the control sequence method, it can, at least theoretically, be formally minimized, but the number of states usually makes this impractical.

Once a control sequence has been obtained, the creative part of the design process is complete. Translation of an AHPL description to a logic network requires only strict adherence to a few simple rules. In the next chapter we shall see that these rules can be implemented as a hardware synthesis program. At this point in the learning process, we shall find it instructive to follow this translation process in detail. Eventually, we shall consider the design process complete once the AHPL description has been obtained.

We will find it convenient to develop separately the *control* portion and *data storage* and *logic* portions of a circuit realization for the control sequence of Example 11.2. We first construct the data storage and logic portion as shown in Fig. 11.16. Note that Steps 1 and 2 are identical so that one line from the control unit will fuse both these steps to be executed. A second control line will cause the execution of Step 3. These lines are clearly marked as inputs at the left of Fig. 11.16. Note in the control sequence that the contents of both flip-flops are changed at every step. Thus, the clock line may be connected to the $C1$ input of flip-flops B_0 and B_1 without any clock control gating.

The circuit output z will be the output of a multiplexer network gate if it is a function of the control signals. The control step signals and output values corresponding to that step serve as the input to one of the AND gates of the two-level multiplexer network generating z. Only the control signal corresponding to the step being executed is 1, so the output value corresponding to this step is routed through the network to the output z. A similar two-level multiplexer network is placed at the input of each flip-flop so that the new flip-flop values corresponding to the current step are routed to the appropriate D inputs. An AND gate will appear in the input network to a flip-flop for each control step, which actually causes that flip-flop to be updated. In this case, there are only two first-level AND gates, one corresponding to Steps 1 and 2 and the other corresponding to Step 3.

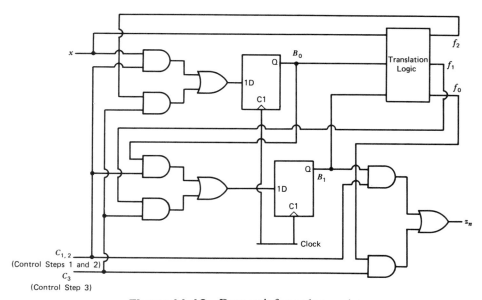

Figure 11.16 Data unit for code translator.

Since these two signals cover all three control steps, we have a check on our earlier observation that clock control was not required. To avoid unnecessarily complicating Fig. 11.16, the Step 3 functions, f_0, f_1, and f_2, are represented by a single block.

Let us now turn our attention to constructing a circuit that will generate the control signals $C_{1,2}$ and C_3. The first step is to obtain a state diagram from the control sequence of Example 11.2. Each control step will correspond to a state, with next-state arrows determined from the branch portion of the step. We begin by representing control Step 1 as state 1 with the appropriate arrows to next states, as shown in Fig. 11.17a. Adding next-state arrows from state 2, as specified by Step 2 of the control sequence, yields Fig. 11.17b. Adding the next-state arrow for state 3 results in the completed state diagram of Fig. 11.17c. The control circuit is a Moore circuit, providing a level output during the duration of each control step, to gate the appropriate signals to the flip-flops at clock time. On the state diagram, the control signals corresponding to each step are shown as outputs for each state, as shown in Fig. 11.17c. Note that only one control signal can be 1 at a time.

Once a state diagram has been obtained, there are two possible ways to obtain the circuit for the corresponding control unit. One is to minimize the corresponding state table and proceed to a circuit design by the formal techniques of Chapter 10.

■ EXAMPLE 11.3

Determine a minimal realization of the control circuit corresponding to Fig. 11.17c.

SOLUTION   The reader can easily verify that Fig. 11.17c, is a minimal state sequential circuit. From this, we can easily obtain the D flip-flop input equations given by Equation 11.10 and the output equations given by Equation 11.11.

$$Y_1^{v+1} = \bar{r} \cdot \bar{y}_2, \qquad Y_2^{v+1} = \bar{r} \cdot \bar{y}_2 \cdot y_1 \tag{11.10}$$

$$C_{1,2} = \bar{y}_2, \qquad C_3 = y_2 \tag{11.11}$$

An implementation of these expressions is given in Fig. 11.18.   ■

The data logic and storage unit of Fig. 11.16, together with the minimal control sequencer of Fig. 11.18, is probably as efficient a realization as can be found for

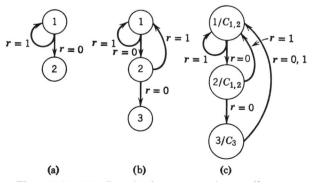

(a)　　　　　(b)　　　　　(c)

Figure 11.17   Developing a control state diagram.

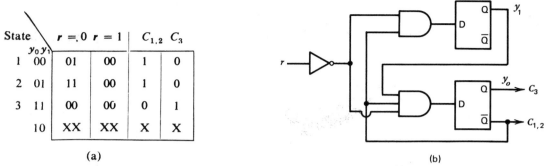

State		$r = 0$	$r = 1$	$C_{1,2}$	C_3
	$y_0 y_1$				
1	00	01	00	1	0
2	01	11	00	1	0
3	11	00	00	0	1
	10	XX	XX	X	X

(a)

(b)

Figure 11.18 Minimal control sequencer for code translator.

the code translator. A minimal state table, determined by the formal techniques of Chapter 10, has 12 states, which agrees with the

$$(3 \text{ control states}) \times (4 \text{ data states})$$

used here. The fact that the number of states agrees does not mean that the complete circuits obtained by the two methods would be identical, but they would be of the same order of complexity.

An alternative state assignment that is sometimes used in the realization of a control unit is the *one-hot code*. That is, the number of memory elements is equal to the number of control states, and only one memory element will assume the value 1 in each state. With this assignment, it is possible to generate the control unit realization directly from the branch statements of the AHPL description, bypassing a control state table. A branching network is used to route control from each step. A branching network corresponding to

$$\rightarrow (r, \bar{r})/(1, 2) \tag{11.12}$$

which represents the branch portion of Step 1 in the control sequence of Example 11.2, is given in Fig. 11.19. Prior to the clock pulse that executes Step 1, the output of the flip-flop is 1. This signal serves as one of the inputs to both AND gates in the branching network. The other input of each is one of the branch functions of Equation 11.12. In general, there will be one gate for each branch function in a branch expression. Since one and only one of the branch functions can be 1, only the corresponding output of the branch network can be 1. In Fig. 11.19a, the output labeled Step 1 will be 1 if $r = 1$ and the flip-flop output $Q = 1$. If $r = 0$ and $Q = 1$, the output labeled Step 2 will be 1. This latter signal will be routed to the D input of the Step 2 flip-flop. If $r = 0$, the clock pulse that executes Step 1 will propagate the signal into the Step 2 flip-flop, thus passing control to Step 2. When the output of more than one branch network can be routed to the input of a control flip-flop, OR gates must be provided at the input of control flip-flops, as illustrated in Fig. 11.19*b*. If any of the illustrated branch signals from Steps 1, 2, or 3 are 1 when a clock pulse arrives, the Step 1 flip-flop is set to 1.

The flip-flops for Steps 1, 2, and 3 are all shown in Fig. 11.19*c*, together with the interconnecting branch networks. Also shown are the output signals C_3 and $C_{1,2}$. The latter signal is formed easily by connecting the outputs of flip-flops 1 and 2 through an OR gate. We will refer to the realization of the control sequential circuit for a control sequence as a *control sequencer* throughout the remainder of the chapter. When the number of states is small and the number of conditional branches few, the one-hot code may result in a savings in combinational logic

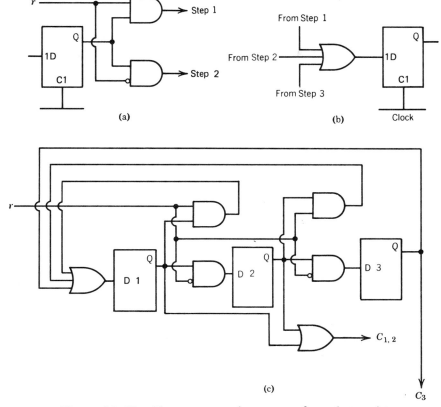

Figure 11.19 Alternate control sequencer for code translator.

and an overall more economical realization or VLSI chip layout using less total area.

11.7 VECTOR OPERATIONS

We have already presented sufficient notation to permit writing a control sequence for any sequential circuit. This notation is convenient when it is necessary to specify the next-state expression for every memory element separately. In larger systems, it is more convenient to treat sets of flip-flops as registers. As we will see, it is usually not necessary to write separate next-state expressions for individual flip-flops within a register. To take advantage of this fact and greatly simplify control sequences for large systems, we will introduce some notation for logical operations on registers and transfers into registers. To remain consistent with *Digital Systems* [3], we will use the vector notation of AHPL.

In the previous section, we denoted scalar variables or flip-flops in two ways: by boldface italic lowercase letters and by boldface italic capital letters with a subscript. We now define a register as a vector variable, that is, a one-dimensional array of flip-flops, to be designated by a boldface italic capital letter (or short string of boldface italic capital letters). Just as is the case in a software program, it is necessary somehow to specify the number of elements or flip-flops in a register. This can be accomplished by a declaration statement such as

MEMORY: $A[3]$, $B[8]$

at the beginning of a control sequence that indicates a sequential circuit consisting of at least 11 flip-flops, 3 in A and 8 in B. In most of the examples in this section, the number of flip-flops in each register will be set forth in the discussion so that the declaration statements are omitted for brevity.

We have used 1's and 0's to represent logical constants throughout the book. Constant vectors whose elements are 1's and 0's will also have a place in the language we are developing. Occasionally, it will be convenient to transfer a constant into a register as illustrated for the four-bit register A in Equation 11.13.

$$A \leftarrow (1, 0, 0, 1) \tag{11.13}$$

Sometimes we want to pick out individual flip-flops within a register and treat them independently. This can be accomplished easily by numbering the flip-flops in the register from *left to right*, starting with 0. We can identify a particular flip-flop in a register by affixing its number as a subscript of the symbol indicating the register. Thus, the leftmost flip-flop would be A_0, the rightmost bit A_3, as shown in Fig. 11.20.

A long segment of a register may be specified by indicating the first and last flip-flops of the segment, separated by a colon. As an example, the string AR_2, AR_3, AR_4, AR_5 could be denoted $AR_{2:5}$.

We may combine vectors into a single vector that may be treated for some operations as a single large vector by writing the individual vector symbols consecutively and separating them by commas. This process will be called *concatenation*. If, for example, A is a four-bit vector and B a three-bit vector, then C as given by Equation 11.14 is an eight-bit vector. A few of the individual element equivalencies are $C_4 = p$, $C_5 = B_0$, and $C_7 = B_2$.

$$C = A, p, B \tag{11.14}$$

As suggested above, it is always possible to form a vector from a subset of the flip-flops of a register by subscripting and catenation. For example, if we wished to transfer the contents of three flip-flops of the six-bit register D into the three-bit register B, we could denote this operation as in Equation 11.15.

$$B \leftarrow D_1, D_4, D_5 \tag{11.15}$$

or as in 11.16

$$B \leftarrow D_1, D_{4:5} \tag{11.16}$$

If the vector notation just introduced is really to be of value in simplifying our program description of hardware, we must be able to express compactly Boolean operations involving registers. One convenient notation is the *reduction operation* over AND and OR, denoted $\wedge / A$ and $\vee / A$, respectively. The operation $\wedge / A$ is defined to mean the ANDing together of all the elements of A. That is,

$$\wedge / A = A_0 \wedge A_1 \wedge A_2 \wedge \cdots \wedge A_{\rho A - 1} \tag{11.17}$$

A_0	A_1	A_2	A_3

Figure 11.20 Numbering of flip-flops in a register.

Similarly,

$$\lor / A = A_0 \lor A_1 \lor A_2 \lor \cdots \lor A_{\rho A-1} \tag{11.18}$$

where ρA is the number of elements in A.

The three logical operators may also be applied to vectors on an element-by-element basis. For example,

$$\bar{A} = \bar{A}_0, \bar{A}_1, ..., \bar{A}_{\rho A} \tag{11.19}$$

and

$$A \land B = A_0 \land B_0, A_1 \land B_1, ..., A_{\rho A-1} \land B_{\rho B-1} \tag{11.20}$$

Clearly, the latter operation is defined only if $\rho A = \rho B$. However, we will allow an $\land$ or $\lor$ operation between a scalar and vector and define it to mean an operation between the scalar and each element of the vector. For example,

$$a \land B = a \land B_0, a \land B_1, ..., a \land B_{\rho B-1}$$

■ **EXAMPLE 11.4**

Compute the vectors to be placed in the register on the left of each of the following transfer statements:

$$A \leftarrow R_{5:6}, R_{0:4}$$
$$B \leftarrow R_{1:6}, 0$$
$$C \leftarrow (\land / R) \lor S$$
$$D \leftarrow R \land S$$

Let $R = (1, 0, 0, 1, 0, 1, 1)$ and $S = (1, 1, 0, 0, 0, 0, 1)$.

SOLUTION

(a) $A = (1, 1, 1, 0, 0, 1, 0)$

(b) $B = (0, 0, 1, 0, 1, 1, 0)$

(c) $(\land/R) \lor S = \land/(1, 0, 0, 1, 0, 1, 1) \lor S$
$$= (1 \land 0 \land 0 \land 1 \land 0 \land 1 \land 1) \lor S$$
$$= 0 \lor S = S$$

(d) $R \land S = \overline{(1, 0, 0, 0, 0, 0, 1)}$
$$= (0, 1, 1, 1, 1, 1, 0)$$

■

The *precedence* rule for operations in AHPL is similar to that for Boolean expressions.

1. Not
2. Selection operators
3. $\land$
4. $\lor$ or $\oplus$
5. Catenate

Parentheses may be used when the precedence rules are insufficient and are sometimes used redundantly for added clarity.

11.8 AHPL MODULES

Clock-level simulation of AHPL descriptions can be accomplished using function-level simulator HPSIM2. This program together with an AHPL network synthesis (see Chapter 12) package may be obtained from the author. These programs require a few keywords including markers of the start and end of a description, together with some declarations in additions to memory elements. Inputs, outputs, and buses must also be declared. A declared bus bit is a wire to which a connection may be established in more than one control state and whose output may fan out to multiple connections. In general, a bus is a vector of bus bits.

An AHPL description may consist of interconnected sequential circuits or AHPL modules. Each module begins with the keyword MODULE followed by the module name and ends with the word END. In between are the declarations followed by the control sequence. The keyword ENDSEQUENCE may follow the control sequence If so, the list of statements between ENDSEQUENCE and END is active every clock period, regardless of the current control state. As is the case with the list of statements in each step, the statements after ENDSEQUENCE are separated by semicolons and terminated by a period. The following might be considered a template for declarations key words, and punctuation syntax in an AHPL module description.

> MODULE: TEMPLATE.
> INPUTS: $XI[8]$; $XJ[4]$.
> MEMORY:
> BUSES:
> OUTPUTS:
> "control sequence"
> ENDSEQUENCE
> "always active".
> END.

■ **EXAMPLE 11.5**

Obtain an ASM chart representation of the system described by the following problem statement:

STATEMENT OF DESIGN PROBLEM

A sequential circuit is to be designed that has a vector of eight input wires that we shall label X, together with two single wire inputs, a and b. The circuit will have a vector of eight output wires labeled Z, and a single wire output labeled out. The circuit may be independently reset to an initial state in which it will wait for the input a to go to 1. When this occurs, a data vector (i.e., an eight-bit unit of information) will be available on lines X. This data vector must be stored by the circuit. Several clock periods after a was 1, line b will go to 1 to indicate that a second data vector is available on lines X. At this time, the circuit must consider the two vectors as binary numbers and compare their magnitudes. It must then make the larger-magnitude vector available on output lines Z for at least four clock periods. During the first four of these clock periods, the output line *out* must be 1. At all other times, *out* = 0. Once *out* goes back to 0, the circuit could return to its initial state. By this time, it may be assumed that both a and b will have returned to 0.

The above problem description is necessarily lengthy, so that the required timing can be specified precisely. The actual function of the system is quite simple. Two successive input vectors on lines X are compared and the largest in magnitude of the two is made available as an output for four successive clock periods. The first of the two input vectors must be stored by the system until the second appears.

SOLUTION   At least one register of eight flip-flops will be required to store the first input vector. Since the eight-bit output vector must be stored for four clock periods, a register is also needed for this purpose. The first input vector will no longer be needed after the output has been determined, so the same register can satisfy both needs. Let us call this register $A[8]$.

Also needed is a counter to count through the four clock periods for which *out* = 1. A two-bit counter that we shall call **COUNT**[2] will serve. It may not be evident to the reader that these are the only two data registers that will be required, but let us proceed as if this is the case. It is not uncommon for a designer to discover, while in the middle of the development of an ASM chart, that not all data storage needs have been anticipated. When this happens, additional registers are declared, and design begins anew. Inputs *a* and *b* are control inputs that will be used in branch statements.

The first block of the ASM chart of Fig. 11.21 represents the initial state in which the system waits for input *a* to go to 1. After *a* goes to 1, the system goes to state Q_2, which causes the values of the input vector X to be transferred into register A. The system remains in state Q_2 for only one clock period, after which it goes to Q_3 to wait for the second control input *b* to go to 1. After *b* goes to 1, the circuit spends a clock period in Q_4 to accomplish two actions. If the second input vector is larger than the one stored in A, this larger vector must be transferred into A. Otherwise, A is left alone. A signal is always generated by state Q_4 that causes the contents of the two counter flip-flops to be reset to 0.

The control circuit will remain in state Q_5 for four clock periods. How this is accomplished should be clear from the ASM chart. When the circuit enters state Q_5, the values stored in **COUNT** will be 00. During each clock period in which the circuit remains in this state, the counter will be incremented. At the same time, the old value of the counter will be tested to determine the next state of the circuit. At the time of the fourth clock period in Q_5, the counter will contain 11. When this is detected, the circuit will return to state Q_1. During the four clock periods in which the circuit remains in Q_5, *out* will be 1 and the contents of A will remain unchanged.   ■

Consider the activity that takes place when the circuit is in control state Q_4 in Example 11.5.

$$\text{If } X > A, \text{ then } A \leftarrow X.$$

This would seem to be a simple transfer statement, except that it is not always executed when the control unit is in state Q_4. That is, the transfer is to happen only if the condition $X > A$ is satisfied. Hence, we call this transfer a *conditional transfer*. If we assume that Step 4 of the description corresponds to control state Q_4 on the ASM chart, we may represent the control and data unit hardware involved in the implementation of this conditional transfer as shown in Fig. 11.22. Of the eight data flip-flops composing the vector A, only one, $A[0]$, is actually shown in the figure. The control pulse line will actually fan out to all eight of the data flip-flops.

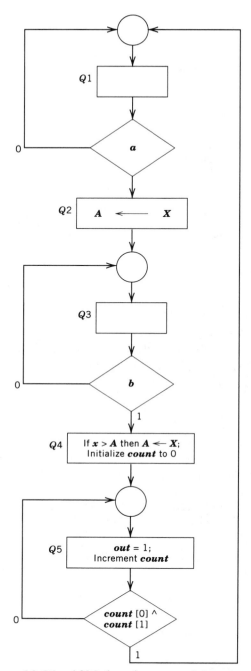

Figure 11.21   ASM chart for sequential comparator.

We now have a hardware implementation of the conditional transfer listed above, but are so far unable to express it in AHPL. To do this, we need, in addition to a notation for conditional transfers, a formal way of expressing $X > A$. A wire labeled $X > A$ will be 1 if the binary number represented by the eight-bit vector X is greater than the eight-bit number stored in A. This line is the output of a combinational logic network with 16 inputs, including all the elements of X and A. A Boolean expression for this network could be written explicitly in the AHPL

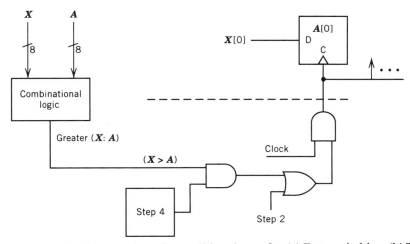

Figure 11.22 Implementation of a conditional transfer. (a) Data switching. (b) Target selection.

module description. The two reasons for not doing so are readability and time and hardware savings, should the same expression be used more than once.

Let us instead use a function-type notation to refer to the output of networks that are known to be strictly combinational logic. For example, $GREATER(X;A)$ will be used to represent a variable that will be 1 whenever $X > A$ and 0 otherwise. The physical realization of this function is symbolized by a box in Fig. 11.22 and the output is connected to the AND gate input, labeled $X > A$. We will not construct the actual combinational logic network represented by $GREATER(X;A)$ at this time, but it can be done easily using repetitive sections much like an adder network. Index operations within AHPL are available to ease the task of writing repetitive combinational logic units. We will defer the introduction of relevant syntax to the final section of this chapter. For now we will simply invoke functions representing logic networks we know to be combinational.

AHPL notation for conditioning a transfer on a signal a (ANDing the clock pulse with a) is

$$R * a \leftarrow X \qquad (11.21)$$

The * when appearing on the left-hand side of a transfer will uniformly mean that the vector at the right of the * will be the target of the transfer (get a clock pulse) only if the scalar variable at the right of the * is 1. We will refer to transfers denoted in this manner as *conditional transfers*. Therefore, the AHPL representation of step Q_4 is

$$A * GREATER(X,A) \leftarrow X$$

■ EXAMPLE 11.6

Translate the extended ASM chart of Fig. 11.21 to an AHPL description of the sequential comparator. Follow the declaration template and punctuate the control sequence as done in Example 11.2.

SOLUTION It has been established that each block of the ASM chart corresponds to a state of the control unit and that each step of an AHPL description corresponds to a control state. Therefore, blocks Q_1, Q_2, Q_3, Q_4, and Q_5 in Fig. 11.21

can correspond to Steps 1, 2, 3, 4, and 5 of the AHPL description. The first three lines of the following AHPL description are declarations of the input and output lines and the memory elements.

Except for the syntax for initializing and incrementing the counter, the remainder of the description was obtained by merely copying the ASM chart into a line-by-line program format. The transfer statements are in the boxes and the branch statements are obtained from the decision diamonds of the same state blocks. In Step 4 we express initialization of the counter as transferring a vector of two 0's into *COUNT*. In Step 5 incrementing the counter is expressed by writing the Boolean expressions for the next (incremented) values of each bit on the right side of a transfer statement. The least significant bit will change values every clock period. The more significant bit, *COUNT*[0], will change values if and only if *COUNT*[1] = 1. GREATER is an instance, as indicated by <:, of the combinational logic unit description GREATERTHAN that would be appended to the module description, prior to processing by a simulator or synthesis program.

> MODULE: SEQUENTIAL COMPARATOR
> INPUTS: X[8]; a; b.
> MEMORY:A[8]; COUNT[2].
> OUTPUTS:Z[8]; *out*
> CLUNITS: GREATER <: GREATERTHAN(X;Y)
> $\quad$ 1 $\rightarrow (\bar{a})/(1)$.
> $\quad$ 2 $A \leftarrow X$.
> $\quad$ 3 $\rightarrow (\bar{b})/(3)$.
> $\quad$ 4 $A * \text{GREATER}(X;A) \leftarrow X$; $COUNT \leftarrow 0,0$.
> $\quad$ 5 $COUNT \leftarrow COUNT[0] \oplus COUNT[1], \overline{COUNT[1]}$;
> $\quad\quad$ *out* = 1;
> $\quad\quad \rightarrow (\wedge/COUNT, \overline{\wedge/COUNT})/(1,5)$.
> ENDSEQUENCE
> CONTROLRESET(1); $Z = A$.
> END. ■

It will be noted that the selection of individual flip-flops or segments of a register was not denoted by subscripts in the above example, as was the case previously. From this point on, we shall regard the meaning of $R_{m:n}$ and $R[m:n]$ as identical. The square brackets must be used in an implementation of the language, whereas subscripts remain convenient for handwritten expressions. Two-dimensional arrays are allowed in AHPL. Rows m through n of array M are denoted $M<m:n>$. Unlike the other statements following the keyword ENDSEQUENCE, which are active every clock period, CONTROLRESET(1) is a special statement intended to assure us that control state 1 is the initial state. It tells the hardware synthesis process that a reset line should be connected to the asynchronous set and reset inputs of the control memory elements, as required to enforce this condition.

The two-bit vector, *COUNT*[0] $\oplus$ *COUNT*[1], $\overline{COUNT[1]}$, in the above example could have been expressed as combinational logic function. The notation

$$COUNT \leftarrow \text{INC}(COUNT)$$

can be used for a counter of any number of bits. INC is the notation for an instance of a combinational logic unit that might be named INCR. INC[8] <: INCR{8} would declare an eight-bit incrementer. The description of the combinational logic unit must be written in terms of a parameter {n} indicating the number of bits. A valid combinational logic unit must result for whatever value of $\{n\}$ is specified in the declaration.

11.9 CONDITIONAL TRANSFERS AND CONNECTIONS

We saw a first example of a conditional transfer in the previous section. In this section we shall generalize this concept to allow multiple targets of a transfer and introduce the notion of a conditional connection. Two distinct control operations are required to specify a register transfer in an AHPL description:

1. A logical selection of the data to be transferred
2. A set of conditions that indicates when and if the data are to be transferred into any of one or more data registers

The controlled selection provides for the routing of the proper data vector to the D inputs of the target memory elements. The second set of conditions causes clock pulses to be routed to the clock inputs of the selected target register at the time of a transfer. Construction of the data busing network as specified by (1) and the clock network as specified by the conditions in (2) can be pursued independently. The general form of a data transfer statement will be as shown in Equation 11.22.

$$target \leftarrow data \qquad (11.22)$$

As a specific example, assume three possible sources of data (*DATA1*, *DATA2*, *DATA3*) that may be connected to an *implied bus* and two possible target registers (*A, B*) to which the data on the implied bus can be transferred. There will be five control variables (p, q, r, f, g) to select specific data sources and targets. The possible actions in this transfer may then be specified as shown in Fig. 11.23.

This figure indicates that the data connected to the implied bus will be the vector *DATA1* if $p = 1$, or will be *DATA2* if $q = 1$, or *DATA3* if $r = 1$. We assume that two or more of p, q, and r cannot be 1 simultaneously. If $p = q = r = 0$, the data vector is a vector of 0's. If $f = 1$, the data are transferred into register *A*. If $g = 1$, the data is transferred into *B*. If $f = g = 0$, no transfer takes place. If $f = g = 1$, the same data are transferred into both *A* and *B*. The implementation of Fig. 11.23 is given in Fig. 11.24 for two-bit data vectors.

The reader should notice the distinction between the interpretation of the unordered IF-THEN clauses defined in AHPL and the IF-THEN-ELSE constructs of most high-level languages. Most of the time, simultaneous 1's on lines controlling separate connections to an implied bus cannot occur. That is, they are "don't care" conditions. If buses were modeled using IF-THEN-ELSE statements, values would be forced on the "don't care" conditions, making optimization of the control logic by a synthesis program impossible.

IF condition, THEN implied bus connection
IF **p**, THEN **DATA1**
IF **q**, THEN **DATA2**
IF **r**, THEN **DATA3**

(a)

IF condition, THEN target register
IF **f**, THEN **A**
IF **g**, THEN **B**

(b)

Figure 11.23 Target and data conditions.

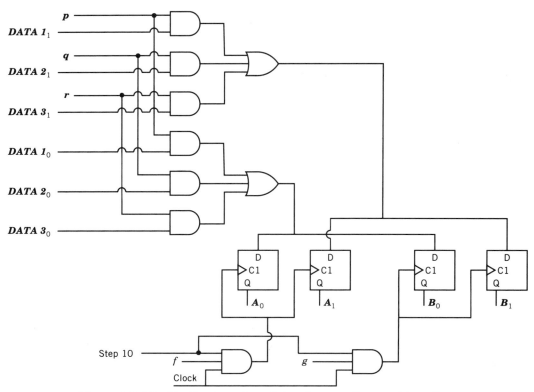

Figure 11.24 Realization of conditional connections and transfers.

The network of Fig. 11.24 can be an implementation of a single step of a control sequence. AHPL provides for a more compact expression of the conditional connection and transfer than that tabulated in Fig. 11.23. An asterisk is used to indicate a condition on either side of the transfer arrow.

$$10 \quad (A!B) * (f, g) \leftarrow (DATA1!DATA2!DATA3) * (p, q, r) \quad\quad (11.23)$$

In the AHPL expression of Equation 11.23, the target registers and data vectors are separated by ! and the conditions that specify the clocking network and implied bus are separated by commas. The target conditions and target registers are specified on the left of the arrow, and the bus network on the right. The transfer actually executed by Equation 11.23 depends on the conditions as given by Fig. 11.23.

We note that in addition to conditions f and g, the transfer specified by Equation 11.24 will take place only when control is at Step 10. This is enforced in Fig. 11.24 by the extra input labeled "Step 10" on the two AND gates that control the clocking of registers A and B.

In certain situations, it is convenient to express the conditional connection of data to a bus independently of a transfer. In that case, the bus must be named in a bus declaration. The bus is now a *declared bus* in contrast to an implied bus. We use the symbol = to denote a connection to a declared bus, a connection that is effective throughout the duration of a clock period. In Equation 11.24, for example, *ABUS* is equal to *DATA1* throughout any clock period for which Step 10 is active and $p = 1$. Similarly, *ABUS* is *DATA2* or *DATA3* if q or r is 1.

$$10 \quad ABUS = (DATA1!DATA2!DATA3) * (p, q, r) \quad\quad (11.24)$$

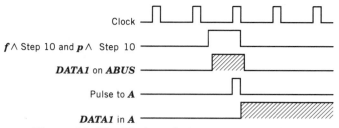

Figure 11.25 Timing of a bused data transfer.

The realization in Fig. 11.24 can also be described by the separate transfer and connection statements given in Equation 11.25.

$$10 \quad ABUS = (DATA1!DATA2!DATA3) * (p, q, r)$$
$$(A!B) * (f, g) \leftarrow ABUS \qquad (11.25)$$

The timing of the execution of Equation 11.25 with $f = p = 1$ is given in Fig. 11.25. The separate declaration of a bus as indicated in Equation 11.25 is helpful in cases where the bus output fans out to a variety of targets that may be involved in separate transfer statements.

A declared bus with fan-out is the focal point of the following example.

■ **EXAMPLE 11.7**
_____

The configuration depicted in Fig. 11.26 includes two sources of data input and two data output channels. Data from either input can be routed to either output. All information passing through this intersection will be in the form of eight-bit characters. A character may be collected serially in register **SERIN** from line **datain**. A character may be shifted out from register **SEROUT** on line z. Four control inputs **a, b, r**, and **s** are provided to allow a system outside the module to route valid characters through the intersection module. Whenever $a = 1$ for one clock period, the module must shift bits in or out for the next eight periods. If $b = 0$, the shift is into **SERIN**. If $b = 1$, the shifting is out from **SEROUT**. If $a, b = 0, 1$, a transfer through **ROUTEBUS** will take place. The source connected to the bus will be

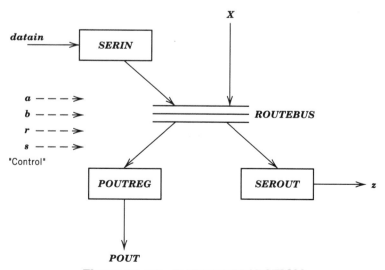

Figure 11.26 DATAINTERSECTION.

specified by *r*, whereas *s* will determine whether the target register is **POUTREG** or **SEROUT**. The contents of **POUTREG** will be observed continuously on lines **POUT**. Write an AHPL description of the intersection module.

SOLUTION   The module description is given below. Note that control leaves step 1 only when *a* goes to 1 to cause a character to be shifted in or out. A three-bit version of the incrementer combinational logic unit is used to count the eight bits of a character when shifting in or out.

MODULE: DATAINTERSECTION.
INPUTS: *X*[8]; *datain*; *a*; *b*; *r*; *s*.
MEMORY: *SERIN*[8]; *SEROUT*[8]; *POUTREG*[8]; *COUNT*[3].
BUSES: *ROUTEBUS*[8].
OUTPUTS: *z*; *POUT*.
CLUNITS: INC[3] <: INCR{3}.

1   *ROUTEBUS* = (*SERIN*!*X*) * (*r*, $\bar{r}$);
     (*SEROUT*!*POUTREG*) * (*s* ∧ $\bar{a}$ ∧ *b*, $\bar{s}$ ∧ $\bar{a}$ ∧ *b*) ← *ROUTEBUS*
     COUNT ← 0, 0, 0;
     → (*a* ∧ $\bar{b}$, *a* ∧ *b*, $\bar{a}$)/(2, 3, 1).

2   *SERIN* ← *datain*, *SERIN*[0:6];
     COUNT ← INC[*COUNT*];
     → (∧/*COUNT*, $\overline{∧/COUNT}$)/(1, 2).

3   *z* = *SEROUT*[7];
     SEROUT ← 0, *SEROUT*[0:6];
     COUNT ← INC (*COUNT*);
     → (∧/*COUNT*, $\overline{∧/COUNT}$)/(1, 3).

ENDSEQUENCE
POUT = *POUTREG*; CONTROLRESET (1).
END.   ■

The final example uses four incrementers, three of which require only a single declaration, because the numbers of outputs are the same.

■ EXAMPLE 11.8

Write an AHPL module description of a stopwatch. The module will be driven by an accurate 1-Mhz clock. The display need consist of one digit for tenths of seconds and two additional digits to display seconds from 0 to 99. There should be four output lines for each digit. Conversion to a seven-segment display will be external to the module and need not be described.

SOLUTION   Two input lines *start* and *stop* are provided. In the following module description, the current count is displayed while control is in Step 1. Control will go to Step 2 for initialization only if the *start* button is pushed while the watch is not currently counting. The 17-bit register, *TCNT*, is included to count from 0 to its maximum value once every 0.1 sec. Because it is driven by a 1-Mhz clock, *TCNT* must count from 0 to 99999 or 11000011010011111 in binary. The line *max*, which is expressed after ENDSEQUENCE, will be 1 when *TCNT* reaches this maximum value. The three remaining registers all count from 0 to 9 to store tenths of seconds and the high and low-order seconds digits. These registers will change value only when **max** = 1 and all lower-order digits are at the maximum values. The notation 17 T 0 is a convenient AHPL representation for a vector of 17 0's. This notation is a mechanism for expressing the binary equivalent of any constant.

MODULE: STOPWATCH.
INPUTS: *start*; *stop*; *go*.
MEMORY: *TCNT*[17]; *TENTHS*[4]; *SECONDSL*[4]; *SECONDSH*[4].
OUTPUTS: *max*; *ZTENTHS*[4]; *ZSECONDSL*[4]; *ZSECONDSH*[4].
CLUNITS: INCCLK[17] <: INCR{17}.
CLUNITS: INC[4] <: INCR{4}.

1   *go* * (*start* ∨ *stop*) ← *start*;
     TCNT * *go* ← (INCCLK(*TCNT*) !17 T 0) * ($\overline{max}$, *max*);
     TENTHS * *go* ∧ *max* ← INC(*TENTHS*) ∧ ($\overline{TENTHS[0] \land TENTHS[3]}$);
     SECONDSL * *go* ∧ *max* ∧ (*TENTHS*[0] ∧ *TENTHS*[3]) ←
                          INC(*SECONDSL*) ∧ ($\overline{SECONDSL[0] \land SECONDSL[3]}$);
     SECONDSH * *go* ∧ *max* ∧ (*SECONDSL*[0] ∧ *SECONDSL*[3]) ∧
                          (*TENTHS*[0] ∧ *TENTHS*[3]) ← INC(*SECONDSH*);
     → ($\overline{start \lor go}$)/ (1).

2   *TCNT* ← T 0;
     TENTHS ← 0, 0, 0, 0;
     SECONDSL ← 0, 0, 0, 0;
     SECONDSH ← 0, 0, 0, 0;
     → (1).

ENDSEQUENCE
max = *TCNT*[0]∧*TCNT*[1]∧*TCNT*[6]∧*TCNT*[7]∧*TCNT*[9]∧(∧/*TCNT*[12:16]);
ZTENTHS = *TENTHS*; *ZSECONDSL* = *SECONDSL*; *ZSECONDSH* = *SECONDSH*.
END.   ■

11.10 MOS REALIZATIONS OF RTL DESCRIPTIONS

Chapter 12 will consider the range of approaches and issues related to computer-aided design, simulation, and realization of clock-mode sequential circuits, with and without the benefit of hardware description languages. Let us complete the background for that chapter by examining the synthesis of MOS VLSI realizations, while still working at the abstract gate and switch levels. Synthesis of a CMOS standard cell or gate array realization from an AHPL description is essentially a reversal of the analysis process considered in Section 11.4. Figure 11.24, for example, is the data unit obtained from synthesis of the single AHPL step given by Equation 11.23. The CMOS standard cell realization is obtained by mapping the individual gates and memory elements (Fig. 9.27*e*) directly to standard cells.

The most efficient NMOS realization can only be obtained by first compiling the AHPL descriptions to switch-level rather than gate-level networks. Our approach will be to present some synthesis examples in the more difficult NMOS environment, leaving the reader to construct the corresponding CMOS networks by inference. The basic building blocks will be the static memory element of Fig. 9.23*b*; the active, passive, and precharged buses of Fig. 9.24; inverters; and logic elements realizable as pull-down networks.

■ EXAMPLE 11.9

Construct a NMOS switch-level diagram of bit 0 of the bus transfer specified in Step 1 of the DATAINTERSECTION module in Example 11.7. Use a precharged bus.

SOLUTION   The resulting realization is given in Fig. 11.27. It is assumed that a 0_2 controlled switch will be in series with each input line to the **ROUTEBUS** network. The AND gates controlling the switches that clock data into the two memory elements will not be duplicated for the other seven elements of the vectors. The control signals will fan out from the points indicated. The ANDing of s with Step 1 and Φ_1 could be accomplished in a single three-input AND gate, if three devices in series in the pull-down network are allowed.   ■

Automatic compilation of an RTL description to a NMOS layout raises several difficult issues. A variety of circuit-level (fan-out, pull-down path length, etc.) decisions were made in generating the switch network diagram of Fig. 11.27. All these degrees of freedom impact the place and route routines necessary in generating a layout. Most of these factors are much more closely controlled in the gate array and standard cell design methodologies. For this reason, gate array and standard cell design are much more likely to be supported by synthesis software.

■ EXAMPLE 11.10

Construct a switch-level diagram of the most significant bit of the network to clock data into the tenths-of-seconds digit, **TENTHS[0]**, of the stopwatch in Example 11.8. Use a passive bus to implement any implied bus in the network.

SOLUTION   The resulting network is shown in Fig. 11.28. An implied bus must be generated by the compiler, because transfers into **TENTHS** are included in both Steps 1 and 2. For convenience in the illustration, part of the logic is represented at the gate level with and the incrementer logic shown as a black box. In this case, **TENTHS[0]** is fed back into the network leading to the D input of that flip-flop. The Φ_2 switch at the memory element output is thus needed to assure that switches controlled by both Φ_1 and Φ_2 are included in series in every feedback loop.

11.10   COMBINATIONAL LOGIC UNIT DESCRIPTIONS*

The separate functionlike representation of combinational logic was included in the original formulation of AHPL in 1972. The concept has proved durable. Whereas other hardware description languages provide for calling processes, the identification of combinational logic units in AHPL allows these units to be singled out for special treatment in syntax for expression, in simulation, and in synthesis. A structured syntax is provided for the description of a combinational logic unit. *What is more structured than combinational logic?* This makes it possible for a synthesis program to process a description step by step and to generate a network wire list for the corresponding combinational logic unit. Each connection statement adds one or more gates to the network. It is the responsibility of the

*This section is not essential for continuity. Refer to the conclusions, Section 11.12, for comment.

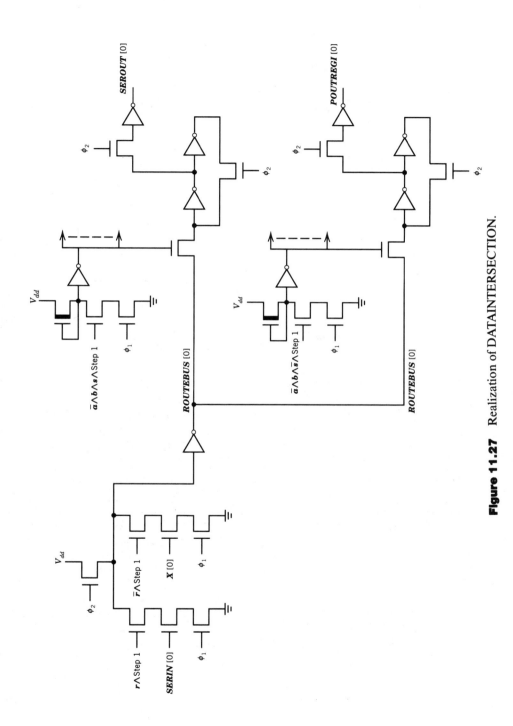

Figure 11.27 Realization of DATAINTERSECTION.

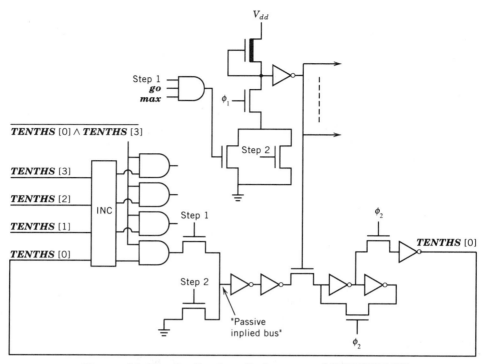

Figure 11.28   Realization of a fragment of STOPWATCH.

designer, the writer of the combinational logic unit description, to assure that the description does indeed represent an implementation of the desired logic function. It is not necessary that connection statements be ordered from input to output.

In combinational logic unit descriptions, only INPUTS, OUTPUTS, nested combinational logic units, and CTERMS are declared. A keyword BODY is included to separate the declarations from the connection statements. A CTERM is a point in the combinational logic network to which the gate-by-gate realization of a single connection statement will be connected. Before presenting the complete syntax for writing and relating connection statements, let us consider a simple example that requires only connection statements and applies a single CTERM.

■ **EXAMPLE 11.11**

Write in AHPL the description of a combinational logic unit that will distinguish the ASCII representation of a decimal digit from other coded characters. The single output will be 1, if the eight input lines represent a digit. The ASCII codes for digits in hexadecimal are 30–39.

SOLUTION   A Boolean expression that will be 1 when the second hexadecimal digit of the coded character is between 0 and 9 is connected to the only CTERM.

CLUNIT: DIGIT(X).
INPUTS: $X[8]$.
OUTPUTS: *yes*.
CTERMS: *lessthan10*.

BODY

> *lessthan10* = $\overline{X[4]} \vee \overline{X[5]} \wedge \overline{X[6]}$;
> *yes* = *lessthan10* $\wedge \overline{X[0]} \wedge \overline{X[1]} \wedge X[2] \wedge X[3]$.

END. ∎

All the variables appearing in the combinational logic unit description of Example 11.11 were physical signals to be connected to the realization of the logic unit. Operations on these signals are limited to the previously defined logic operators. To facilitate the repeated use of connection statements for multiple sets of arguments, it is necessary to introduce the notion of an *index variable* which unlike a signal has no physical significance. As in a high-level language, an index variable is assigned a value to select from a signal vector an element to be used in a connection statement. Index variables are of type Integer whereas signals may assume binary values. Integer expressions and relational expressions are allowed on index variables.

A more detailed syntax of the body portion of a combinational logic unit description is given in Fig. 11.29. The syntax is structured but includes only two of the flow control mechanisms normally found in a high-level language. Because of the natural rigidity of combinational logic, these are all that are necessary. Index variables may appear in any of the expressions in Fig. 11.29. Signals may appear only in connection statements. CTERMS are signals. The evaluation of index expressions is a hardware compile time activity. The purpose of these expressions is to specify the structure of the network, and index variables do not represent points in the final network. The syntax provides for nesting the IF-THEN and FOR-TO constructs by permitting any list of statements within these mechanisms.

Index expressions are integer arithmetic expressions of reasonable complexity. Multiplication, division, and powers of 2, expressed $(2\uparrow k)$, are allowed. Similarly, the relational operators on the integer index variables are restricted to $=, <, \leq, >$, and $\geq$. Index expressions are allowed in row and column selection, that is, within $< >$ and [ ], respectively. A minterm is simply given by the function TERM(i,***BOOLVEC***), where i is an integer index specifying the minterm number, and ***BOOLVEC*** is a vector of the Boolean variables on which the minterm func-

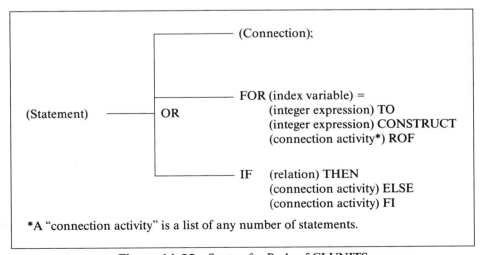

*A "connection activity" is a list of any number of statements.

Figure 11.29 Syntax for Body of CLUNITS

tion is defined. The most significant variable in the binary equivalent of the minterm number is listed first, etc. For example,

$$\text{TERM}(1; x, y, z) = \bar{x} \wedge \bar{y} \wedge z$$

The following is a declaration of a 16-bit adder, with the parameter enclosed in $\{\}$ used to specify the number of bits in each argument.

$$\text{CLUNITS: ADD}[17] <: \text{RIPPLEADDER}\{16\}.$$

Only the number of outputs of a combinational logic network is specified in the declaration. In the case of an adder, this is one more than the value of the parameter enclosed in $\{\}$. The arguments must be specified each time the network is invoked. For the incrementer the parameter enclosed by $\{\}$ is merely the length of the incrementer. The number of inputs and outputs are the same and are specified by this parameter. In general, the relationship between input and output vectors and these numerical parameters is defined in the relevant combinational logic unit description. The designer must be familiar with this description before declaring a combinational logic unit.

The hardware compiler merely executes explicit instructions in generating combinational logic networks. It is worth emphasizing again that the integer index variables will not exist in the resulting hardware network. They serve only as part of a mechanism to provide for compact description. The target networks of the compiler, both combinational and sequential, process only signals of type Boolean. We conclude this section with a description of the incrementer, which illustrates most of the syntax of index expressions.

■ **EXAMPLE 11.12**

Write the combinational logic unit description of an I-bit incrementer.

SOLUTION In the description given below, I is the parameter that specifies the number of bits in the incrementer. The declared inputs, outputs, and CTERMS are all signals. The least significant bit is bit $I - 1$ and bit 0 is most significant. As the index variable J decreases from $I - 1$ to 0, $I - 1$ AND gates and I OR gates are generated by the two connection statements.

```
CLUNIT: INCR(X) {I}
    INPUTS: X[I]
    OUTPUTS: TERMOUT[I]
    CTERMS: TA[I]
BODY
        FOR J=I−1 TO 0 CONSTRUCT
            IF J = I−1 THEN
                    TA[J] = 1
            ELSE
                    TA[J] = X[J+1] ∧ TA[J+1]
            FI
        ROF;
        FOR J = I−1 TO 0 CONSTRUCT
                    TERMOUT[I] = X[J] ⊕ TA[J]
        ROF.
    END.
```

■

The combinational logic unit description of a ripple carry adder is similar to the incrementer, except that there are two rather than one vector(s) of input signals. The first line of a description of a n-bit ($n + 1$ output bits) adder may be given by ADDER(A; B; cin)$\{n\}$, where A and B are both n-bit vectors and cin is one bit. Both the adder and incrementer are examples of iterative networks that will be formally introduced in Chapter 16. Most iterative networks are efficiently represented by combinational logic unit descriptions.

The decoder is one more combinational logic network that is conveniently represented by a combinational logic unit description. If A is an n-bit vector, $DCD(A)$ will have 2^n outputs.

11.12 CONCLUSIONS

Behind this chapter has been the assumption that the ability to design using the notion of an extended state table and a hardware description language is indispensable to any digital hardware designer. This need has been addressed. At the same time, few readers will encounter an AHPL-based design environment in the workplace. Accordingly, the treatment has been limited to presenting a universal design paradigm in terms of a quickly learned language. In deference to the course that will not make use of AHPL-based CAD tools, the definition of this language, particularly combinational logic unit descriptions, was brief.

Software tools are available to readers interested in making further use of AHPL. A function-level simulator, a hardware compiler, and a program capable of translating the output of the hardware compiler to an EDIF netlist are available. The EDIF netlist my be used as input to a commercial CMOS standard cell place and route program or a field programmable gate array development system. The manual [10] available with these programs contains an independent specification of AHPL, including the syntax for combinational logic unit descriptions. Additional information may be obtained from the author at the Department of Electrical and Computer Engineering, University of Arizona. Further definition of AHPL can be found in Appendix B.

PROBLEMS

11.1 Assume that the data unit of Fig. 11.11 is to be redesigned to function at the highest possible clock frequency with a clock with egual 0 and 1 periods. The system clock must, therefore, be connected directly to the clock input of each memory element in the data unit.

    (a) Obtain a logic block diagram of the data unit depicted in Fig. 11.11 with the multiplexer or implied bus networks redesigned to satisfy the constraint just described. Use D flip-flops.

    (b) Repeat the redesign of part **a** using J-K flip-flops in the data unit. Use as few gates in the implied bus network as possible.

11.2 Consider the flip-flop labeled c in Fig. P11.2 with its associated input logic to be a control sequencer, and consider flip-flop a to be the only data flip-flop in the sequential circuit. The control unit is a minimal realization and not a direct translation of the control sequence.

    (a) Determine a state diagram of the control unit.

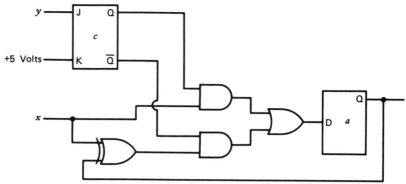

Figure P11.2 Syntax for body of CLUNITS.

(b) Determine a complete control sequence (AHPL description without declarations) for the sequential circuit.

(c) Construct the logic block diagram of a control sequencer that is a one-to-one translation of the above control sequence.

11.3. Familiarize yourself with the function of the above sequential circuit by studying the control sequence determined in Problem 11.2. Now synthesize the complete state diagram of a sequential circuit that will accomplish the same task. Determine a minimal realization of this state diagram using the techniques of Chapter 10, and compare its complexity with that of Fig. P11.2.

11.4 A straightforward realization of a design language step consisting of only a branch is given in Fig. P11.4. This circuit is effectively a two-state sequential circuit. Use the techniques of Chapter 10 to realize this circuit in terms of

(a) A clocked R-S flip-flop

(b) A clocked J-K flip-flop

The inputs are the branch function f and a line from the step $i - 1$ flip-flop. The output leads to step $i + 1$.

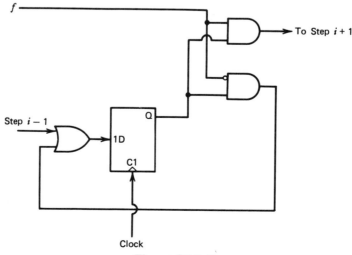

Figure P11.4

11.5 Obtain an alternate realization of the 2-out-of-5 code checker described in Problem 10.7 by first determining a control sequence. Base your approach on a three-bit counting register **BITS** that will count the number of bits of the five-bit code word that have arrived thus far and a two-bit counter **ONES** that will count the number of these bits that have been 1's. Additional memory elements may be added to the data and logic unit, if needed. Consider the problem solved once a complete control sequence has been obtained.

11.6 Construct a control unit realization corresponding to the following AHPL sequence:

1 $Z = X \vee A$;
    $\rightarrow (a, \bar{a} \wedge b, \bar{a} \wedge \bar{b})/(2, 3, 4)$.
2 $A \leftarrow X$;
    $\rightarrow (1)$.
3 $A \leftarrow X_{1:3}, X_0$;
    $\rightarrow (4)$.
4 $A \leftarrow A_{1:3}, A_0$;
    $\rightarrow (1)$.

11.7 Construct a detailed logic block diagram of the hardware realization of both the control and data units specified by the following AHPL description, where a and b are flip-flops, x is an input, and z a single output line:

1 $a \leftarrow x \vee b$.
2 $b \leftarrow x$;
    $\rightarrow (a, \bar{a})/(1, 3)$.
3 $z = 1; b \leftarrow x \oplus b$;
    $\rightarrow (1)$.

11.8 Which of the following are legitimate AHPL action statements or parts of action statements? Register A [2] has two flip-flops, B [3] and C [3] have three, X [2] is a vector of two input lines, a and b are individual inputs, and BUS is a three-bit bus.

(a) $b \leftarrow a$
(b) $B \leftarrow ((A, a) \mathbin{!} (A, b)) * (C_0, \bar{C}_0)$
(c) $(A_0, B_{0:1}) * a \leftarrow C$
(d) $BUS \leftarrow C$
(e) $B \leftarrow BUS + C$
(f) $BUS = 0, 1, 0$;
        $C \leftarrow BUS$
(g) $B \leftarrow (B \oplus (A, a)) \oplus C$
(h) $a * B \leftarrow C$
(i) $B * a \leftarrow (X, b)$
(j) $\wedge/B, A \leftarrow C$

11.9 A digital system has an eight-bit vector X of input lines together with another input line **ready**. The system will have two output lines: z and **ask**. Following each one-period level placed on line **ask**, a new vector will become available on lines X. The availability of such a vector will be signaled by a one-period level on line **ready**. Each time an input vector is exactly the same as either of the two previous input vectors, the output z

should be 1 for four clock periods prior to another signal on line **ask**. The line z should be 0 at all other times. Write an AHPL description for this digital system. Define data registers as required.

11.10 A random-sequence generator is driven by an external source of triggering pulses. This generator provides a level output z that is constant between trigger pulses and may or may not change, on a random basis, when triggered by a clock pulse. A control module is to be designed employing a 1-Mhz clock. This module is to provide a periodic output **trig** to drive the random process at a frequency of 1 Khz. The module must also count the number of level changes in the random process each second. It must also compute and display the average number of level changes per second over the first 2^8 seconds following the depression of its start button. Specify the necessary hardware and write an AHPL description of the module. Accomplish division by shifting.

11.11 The sequential circuit characterized by the following AHPL description will include exactly eight states (control and data units together).

(a) Construct from this description a single Moore model state diagram, representing the control unit and data unit together as a single sequential circuit.

(b) Determine a closed partition of the eight states of this sequential circuit, including four classes of two states each.

INPUTS: x.
MEMORY: a.

1   $a \leftarrow 0; \rightarrow(a)/(1)$.
2   $z = 1. \rightarrow(a, \bar{a})/(1,3)$.
3   $z = 1. \rightarrow(a, \bar{a})/(2,4)$.
4   $a \leftarrow x \vee a; z = 1;$
    $\rightarrow(a, \bar{a} \wedge x, \bar{a} \wedge \bar{x})/(3,4,1)$.

11.12 Refer to a TTL data book for a description of a SN74180 nine-bit odd/even parity checker. Write an AHPL combinational logic unit description of this part.

11.13 AHPL descriptions of a 74194 shift register and a 7485 magnitude comparator follow. A module called COMPARESHIFT is depicted in Fig. P11.13. Write an AHPL description of that module. The line connecting $R[3]$ and input r must be considered part of a constrained version of the shift register module. This module **SHIFTREG** will not be part of the AHPL description of COMPARESHIFT but will be connected to COMPARESHIFT. The IC7485 will be invoked in COMPARESHIFT. The control vector S will be an output of COMPARESHIFT. Assume that a reset line (not shown) is connected so that control flip-flops 1, 2, 3 are initialized to 1, 0, 0 respectively. **COMPARESHIFT** includes only the CLUNIT and the control unit.

CLUNIT: IC7485(A; B; C)
  INPUTS: $A[4]$; $B[4]$; $C[3]$
  OUTPUTS: **CMPR**[3]
BODY
  $D = A \wedge \bar{B}$;
  $E = \bar{A} \wedge B$;

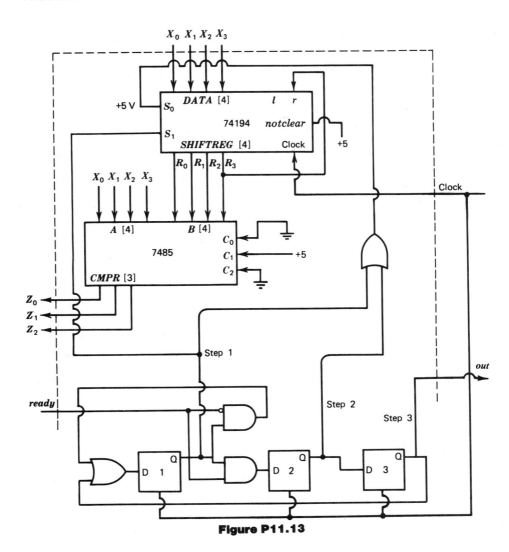

Figure P11.13

$$F = \overline{D \vee E};$$

$$CMPR_0 = D_0 \vee (F_0 \wedge D_1) \vee ((\wedge/F_{0:1}) \wedge D_2) \vee ((\wedge/F_{0:2}) \wedge D_3) \vee$$
$$(\wedge/F) \wedge C_0;$$

$$CMPR_1 = (\wedge/F) \wedge C_1;$$

$$CMPR_2 = E_0 \vee (F_0 \wedge E_1) \vee ((\wedge/F_{0:1}) \wedge E_2) \vee ((\wedge/F_{0:2}) \wedge E_3) \vee$$
$$(\wedge/F) \wedge C_2$$

END

    MODULE: IC74194.
    DATA INPUTS: *l*, *r*, *DATA*[4].
    CONTROL INPUTS: *S*[2], *notclear*.
    OUTPUTS: *SHFTREG*[4].
    MEMORY: *D*[4].
ENDSEQUENCE
    $D*((S_0 \vee S_1) \wedge notclear) \leftarrow ((D_{1:3}, l)! (r, D_{0:2})! DATA)*DCD_{1:3}(S);$
    $D*notclear \leftarrow\circ 0, 0, 0, 0;$ "unclocked"
    $SHFTREG = D.$
END.

11.14 Write a complete AHPL description (including declaration statements) of the sequential circuit described in Problem 10.21.

    (a) Use only a single control step with the two-bit counter and single-bit register included in the data unit.

    (b) Repeat the design with no explicit control steps and all transfers and connections written after ENDSEQUENCE.

    (c) Obtain a D flip-flop realization of the data unit as defined by the AHPL description developed in part **b**. Compare the numbers of components with the J-K flip-flop realization found for Problem 10.21.

11.15 Construct a switch-level logic circuit diagram of a nMOS implementation of the data unit of the following partial AHPL description. A nonoverlapping two-phase clock and a passive bus are to be used. All transistors are to be shown explicitly. No logic gate symbols except simple inverters are allowed.

MODULE: EXAMPLE.
INPUTS: a, c, x, y.
MEMORY: $R[1]$.
OUTPUTS: z.

1   $BUS[0] = (\overline{R[0]} \,!\, x \vee (y \wedge R[0])) * (c, \bar{c});$
   $R[0] * a \leftarrow BUS[0]; z = R[0] \wedge x;$
   $\rightarrow (1).$
END.

11.16 Repeat Problem 11.15 using a precharged bus.

11.17 Repeat Problem 11.15 using an active bus.

11.18 Construct a switch-level (transistor-level) logic circuit diagram of a nMOS implementation of the *data unit* of the following partial AHPL description. A nonoverlapping two-phase clock and *precharged* bus are to be used. All transistors are to be shown explicitly. No logic gate symbols except simple inverters are allowed. Avoid using more devices (transistors) than necessary.

MODULE: ASLIGHTLYDIFFERENTEXAMPLE.
INPUTS: a, c, x.
MEMORY: $R[1]$.
OUTPUTS: z.

1   $BUS[0] = (\overline{R[0]} \wedge x \,!\, x) * (c, \bar{c});$
   $R[0] * a \leftarrow BUS[0]; z = \overline{R[0]} \wedge x;$
   $\rightarrow (1).$
END.

11.19 Construct a logic block diagram of a CMOS standard cell realization of the AHPL description given in Problem 11.15.

11.20 A BCD adder is to be designed as a complex standard cell. It will have nine input bits: eight bits for two BCD digits and one bit for an input carry. It will have five output bits, four bits representing the BCD sum digit and one bit for the output carry. Write a complete AHPL combinational logic unit description of this cell. A four-bit (five output bits) binary adder ADD(A; B; cin) may be invoked in your description.

11.21 An available combinational logic standard cell has four inputs and one output. The output will be 1 if and only if an even number of the inputs are 1 (0 is considered an even number). Write a CLUNIT description of this cell. Invoke this CLUNIT together with exclusive-OR gates in the description of another combinational logic unit that will check for even parity over seven bits (the output should be 1 if the parity is not even).

11.22 Write the combinational logic unit description of a n-bit binary adder.

11.23 An AHPL description is to be written for a module B with registers and communication lines as shown in Fig. P11.23 The function of module B is to facilitate the transfer of information from system A to system C. System A might be a computer, and system C might be some sort of peripheral memory, such as a magnetic tape unit. In that case, system B will perform part of the function of a magnetic tape controller.

The output from system A will be a 32-bit vector, which we will refer to as a word. System C will accept only eight-bit vectors or bytes as input. System B then must sequentially disassemble the 32-bit word into four eight-bit bytes. The principal hardware elements in system B are a 32-bit register *LR* and an eight-bit register *SR*. Also included are an input labeled *ready*, associated with *LR*, and a single flip-flop labeled *full*, associated with *SR*.

We assume that the clock source is provided by system C so that systems B and C are synchronized. During any clock period for which the flip-flop *full* is 1, system C will read a byte from register *SR*. In the clock period immediately after it places a byte in *SR*, system B will set *full* to 1. During the next clock period, system C will utilize the byte in *SR* and simultaneously clear *full* to 0. System B may place another byte in *SR* during the same clock period. Since systems B and C are synchronized, it is possible to read reliably from *SR* and place a new byte in *SR* during the same clock period. The flip-flops in *SR* must be master-slave for this scheme to work. In effect, we have assumed that system C can react in one clock period to simplify the design of system B.

Whenever a word is placed in *LR*, the flip-flop *ready* will be held at 1 by system A. After the fourth byte has been read from *LR*, system B will clear *ready* to 0.

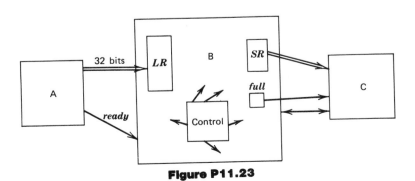

Figure P11.23

BIBLIOGRAPHY

1. Wiatrowski, C. A. and House, C. H. *Logic Circuits and Microcomputer Systems*. McGraw-Hill, New York, 1980.

2. Clare, C. H. *Designing Logic Systems Using State Machines*. McGraw-Hill, New York, 1973.

3. Hill, F. J. and G. R. Peterson. *Digital Systems: Hardware Organization and Design*, 3rd ed., Wiley, New York, 1987.

4. Hines, J. "Where VHDL Fits Within the CAD Environment," in *Proceedings 24th ACM/IEEE Design Automation Conference*. 1987.

5. *IEEE Standard VHDL Language Reference Manual*. IEEE, New York, 1988, Std. 1076.

6. Armstrong, J. R. *Chip-Level Modeling with VHDL*. Prentice-Hall, Englewood Cliffs, N.J., 1988.

7. Barton, D. "Behavioral Descriptions in VHDL," *VLSI Systems Design* (June 1988).

8. Chu, Y. "Why Do We Need Hardware Description Languages?", *Computer*: Vol. 7, No. 12, 18–22 (Dec. 1974).

9. Lipovski, G. J. "Hardware Description Languages," *Computer*: Vol. 10, No. 6, pp. 14–17 (June 1977).

10. Navabi, Z., R. Swanson, and F. J. Hill. *User Manual for AHPL Simulator (HPSIM2) and AHPL Compiler (HPCOM)*. University of Arizona, Tucson, AZ, Jan. 1990.

CHAPTER 12

VLSI REALIZATION OF DIGITAL SYSTEMS

12.1 OVERVIEW

The major premise throughout this book has been to develop the fundamental concepts of logic design needed to support the VLSI realization of digital systems. Having mastered the fundamentals, the reader is no doubt eager to try his or her hand at the layout of an actual chip. A student might be in a position to see the design actually fabricated either by MOSIS [1] or by an ASIC vendor. MOSIS is a National Science Foundation-supported agency that matches chip fabrication resources with the needs of small-volume customers, particularly universities. An ASIC (application-specific integrated circuit) is a chip that will be produced in small numbers, usually to be used in only one particular application. An ASIC contrasts, for example, with an MSI part, a microprocessor, or a support chip that may be designed into numerous applications. Whatever path a student might take to develop a chip design, only a few copies will ever be produced, and the part must, therefore, be classified as an ASIC.

Whatever its classification, the cost of an integrated circuit will include design and test development costs, which are fixed regardless of the number of copies produced, together with incremental production and testing costs. For low-volume parts (ASICS), the minimization of development costs is critical. Vendors have developed several distinct design methodologies, all aimed at reducing "one-time" costs for both the vendor and user of the part. New variations of these methods will no doubt be developed while this edition remains in print.

As production volume increases, so does the importance of incremental costs. In these instances, it becomes economical to allow one-time costs to increase to provide for a design with minimum incremental cost. This will usually mean less design automation and more engineering man-hours focused on design details.

Figure 12.1 shows four possible targets of the two-step digital VLSI design process. The second step consists of translating a structural description or network list (a list of logic elements and interconnections) to a set of masks required for chip fabrication. The leftmost alternative, a "full custom" layout, consists of a complete set of masks, one for each step in the process required by the particular technology used. The data files required to specify full custom mask sets may be created by using one of a large number of available layout editors.

Providing the student with experience with one or more layout editors is a course goal compatible with the use of this textbook. There are enough differences among the features of such editors that it is impractical to attempt to describe an editor within the pages of this book. It is expected that the manual for the editor of the instructor's choice will be available as a supporting document to

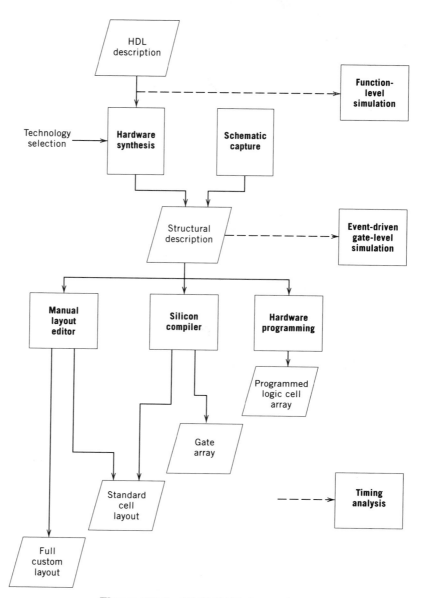

Figure 12.1 VLSI CAD alternatives.

a textbook. Two of the most common descriptive formats for editor outputs are CIF (Caltech intermediate form) and GDSII.

If the designer is willing to sacrifice some flexibility, considerable layout time may be saved by using standard cells. A standard cell may be an already laid out gate, a memory element, or a more complex logic structure. It was pointed out in Chapter 7 that standard cells constitute the most common design approach to CMOS VLSI. A layout editor may be used in placing these cells on the chip and for manually routing connections between the cells. Alternatively, a *place and route* tool, sometimes called a silicon compiler, may be available to translate a structural description to a layout. The silicon compiler will choose an optimal ordering of the standard cells into rows on the chip and then route the connections between the cells. Several iterations of these two activities may be required before the connections fit successfully into the available channel space.

Two other alternative ASIC design targets are given in Fig. 12.1. Unlike standard cells, which require a full mask set for each individual design, gate arrays are partially fabricated independent of the ultimate application. Masks for final metallization layers are generated by a silicon compiler program that routes the interconnections between gates by using a structural description of the user design as input. Sometimes, the wiring process is divided into two steps. One step will interconnect sets of gates to form user-specific cells. The second step of this process resembles the placement and routing in the standard cell method. Placement and routing algorithms have been the subject of intense research activity within the VLSI CALD community. Providing the foundation necessary to engage in this research is beyond the scope of this book. Our point of view is that of the user of CALD tools.

The rightmost ASIC design alternative in Fig. 12.1 uses chips that are completely fabricated, packaged, and tested before the specific application enters the picture. Included in these packages are arrays of logic cells with user controllable interconnections. Also included on these chips are memories in which information specifying these connections is to be written. Embedded in the support software for programmed logic cell arrays are algorithms for mapping the logic network list on the logic cells and best utilizing the transmission lines provided for global interconnections. As with the other ASIC approaches, the worst-case delay and resulting maximum clock rate will depend on a particular layout.

Having discussed the methods for reducing a network list or structural description of a digital system to a realization, let us now consider how the structural description itself might be entered into the database. Two alternatives, hardware synthesis and schematic capture, are represented in Fig. 12.1. The origin of the term schematic capture assumed that a hard-copy description of the desired digital system preexisted the CAD process. With the aid of some variation of a drawing program, this schematic representation is converted to a form that can be displayed on a workstation monitor and then on a network list. Easy to envision but difficult to accomplish would be the on-line design of the system while interacting with the workstation. In this mode, gate-level simulation would alternate with successive refinement of the structural description by schematic capture.

Another design method is to use a hardware description language (HDL), such as AHPL, VHDL, or Verilog, as the medium for the first unambiguous description of a design. In Chapter 11 we observed that a register-transfer-level description accurately represents system behavior and timing, given satisfaction of the clock-mode assumption. It is extremely difficult for the human designer to extract behavioral information from a schematic diagram or network list. Obtaining a network list from an HDL description and agreement on the target technology is straightforward. Bypassing the schematic and translating a higher-level description of a digital system to a structural description is called hardware synthesis.

Figure 12.1 also depicts simulation at each level of description. This will be the topic of Sections 12.6 and 12.7.

12.2 ALTERNATIVE STRUCTURAL DESCRIPTIONS

In the absence of hardware synthesis, the computer-stored representation of a schematic diagram must necessarily be the first structural description of a digital system under design. Additional information, mostly to facilitate the display of the schematic on the computer terminal, is also present in this representation. The particulars of the graphics program and supporting utilities that assist the designer in generating a schematic diagram will be left as a private matter between the reader and the manual for the particular tool. Continuation of the process, once a complete and accurate schematic is stored in a file, does fall within the scope of this book.

Every set of tools will include a utility for extracting the logic element and interconnection information from the schematic file. Almost every tool vendor has a unique format or language for representing this connectivity information. Nearly all such languages may be assigned to one of the three categories shown in Fig. 12.2: *net-list*, *element-list*, and *pin-list*. This figure shows an initial extraction from the schematic file to a net-list. In some CALD systems, a pin-list or an element list might be the target of this extraction process. Similarly, any of the formats may serve as input to a simulation program and a net-list might deliver the network data to a place and route utility. With few exceptions, a description in any of the formats will be sufficiently complete to serve as a source for the silicon

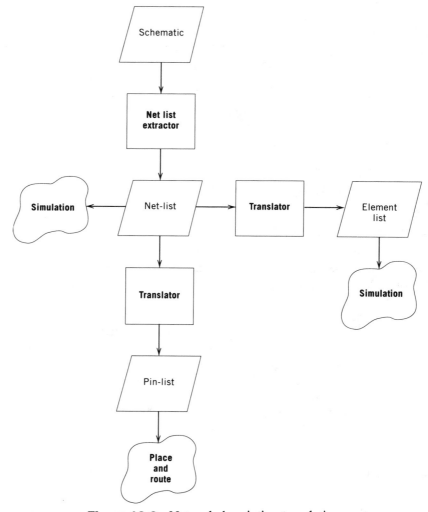

Figure 12.2 Network description translation.

compilation process or to support simulation. Therefore, it should be possible to translate from an example language in one of the three categories to any other example format. An exception might be an element list intended exclusively for simulation, in which information specifying the partition of the list of elements into parts may have been discarded.

Both the schematic diagram and register-transfer-level description are capable of representing the collection of wires, and elements into vectors. Usually, this grouping of objects is lost in the process of translation to a net-list, pin-list, or element-list in which description is on a wire-by-wire, element-by-element, and pin-by-pin basis.

Before we formally define the three language categories, let us clarify terms related to the hardware we wish to describe. We shall use the terms *element*, *pin*, and *part* as generally understood and illustrated in Fig. 12.3*a*. The term part remains important, because it can be used to represent a complex multielement CMOS standard cell, as well as an SSI or MSI part. An element is a logic circuit (combinational or sequential) that can be connected and applied independently of other elements in the part. A pin may be considered an input or output connection to a part or directly to an element within the part. A *net*, as illustrated in Fig. 12.3*b*, is a connection of two or more pins together. Not shown in Fig. 12.3 is a *node*. For any pin a node is the set of all pins that are connected to that pin. Therefore, a node may be defined by one or two or more nets.

To distinguish between the three language categories introduced above, we fall back on the abbreviated Backus–Nauer (*bnf*) representation of each given in Fig. 12.4. Many readers will have had a prior introduction to this notation for defining the syntax of a language. To make the ensuing discussion as self-contained as possible and to adapt the notation to our modest application, we provide the following definitions:

    $< \cdots >$ indicates a nonterminal symbol, one that will not appear in actual text written in the defined language.

    :=   informs us that the more detailed string of terminal and nonterminal symbols at the right of := may replace the nonterminal on the left.

    {**<string>**}   represents any number of copies of <string>.

    (...) and ;   are terminal symbols that may be used as separators in a particular language.

    | indicates OR in a bnf description.

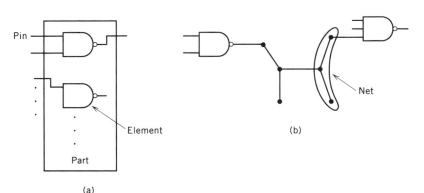

(a)

(b)

Figure 12.3 Definitions.

```
<element_list> := {<element>}
<element>       :=
<output_node_id><element_type><element_attributes>{<input_node>};
<input_node> := <node_id>
```
                                    (a)
```
<pin_list> := {<part>}
<part>       := (<part_id><attributes>;{<pin>})
<pin>        := <pin_number><node_id>;
```
                                    (b)
```
<net_list>          :=  <interface><contents>
<interface>         :=  (interface<port_definition>)
<port_definition>   :=  (define<port_type><port_id>)
<contents>          :=  (contents{<instance>}{<net>})
<instance>          :=  (instance<part_id><element_id>)
<net>               :=  (joined{<pin>})
<pin>               :=  (<element_id><pin_id>) | <IOport_id>
```
                                    (c)

Figure 12.4 Typical Syntax.
(*a*) Element list. (*b*) Pin list. (*c*) Net list.

An identifier, denoted **id**, is a string of symbols that serves as the name of an object in a language. For example, <pin_id> might represent the name or number of a pin. In a particular language, the syntax of an identifier might be controlled so as to include certain semantic information. For example, a pin identifier might always indicate whether a pin is an input or output. This and similar information could alternatively be found in a *type* specifier. An example of <element_type> might be NAND or DFF. Further information that might be of use to a simulator or layout tool could be provided in the form of an *attribute*. Element_attributes of a logic gate might be minimum and maximum delay values for the gate. The interpretation of identifiers, types, and attributes will vary with the syntax and semantics of each individual language. Figure 12.4 includes only enough syntax detail to define the three language categories. Each keyword is printed in boldface to distinguish it from any nonterminal symbol with the same name. More detailed syntax will be illustrated only in examples.

The typical element-list and pin-list syntax in Fig. 12.4 are provided in no more than enough detail to suggest the form of these types of languages. The more detailed syntax of the net-list in Fig. 12.4*c* was deliberately chosen to resemble a simplified form of the industry standard EDIF (electronic data interchange format for net-lists) [20]. The LISP-like language EDIF was developed to provide a mechanism to transport a network description from one design tool environment to another. Many CALD systems have the capability of translating designs expressed in their local format to and from EDIF. Example 12.1 illustrates a translation from the EDIF-like syntax to a pin-list.

■ **EXAMPLE 12.1**

Figure 12.5*a* depicts a simple network, called XORPLUS, generated via schematic capture. The simplified EDIF-like description extracted from the schematic of XORPLUS is given in Fig. 12.5*b*. Translate this net-list description into a pin-list consistent with the syntax of Fig. 12.4*b*.

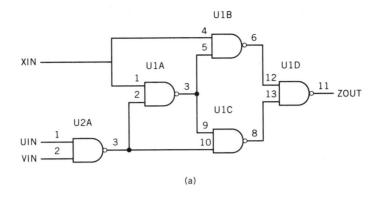

(a)

```
(EDIF a__xorplus__sch
    (interface
        (define input port XIN)
        (define output port ZOUT)
        (define input port UIN)
        (define input port VIN)
        )
    (contents
        (instance   X__74LS00__NET U1A)
        (instance   X__74LS00__NET U1B)
        (instance   X__74LS00__NET U1C)
        (instance   X__74LS00__NET U1D)
        (instance   X__74LS00__NET U2A)
        (joined
            (U1B O) (U1D I0))
        (joined
            XIN (U1B I0) (U1A I0)
            )
        (joined
            (U1A O) (U1B I1) (U1C I0)
            )
        (joined
            (U1D O) ZOUT
            )
        (joined
            (U1A I1) (U2A O) (U1C I1)
            )
        (joined
            UIN (U2A I0)
            )
        (joined
            (U1C O) (U1D I1)
            )
        (joined
            VIN (U2A I1)
            )))
                    (b)
```

Figure 12.5 EDIF description of Xorplus.

SOLUTION   An element or parts library—in this case, the TTL library—was required, as a reference, by the net-list extraction program as it generated the description in Fig 12.5b. Reference to this library is also required by the program that translates from a net-list to a pin-list.

By instance declarations the net-list, like the original schematic, assigns four NAND gates to part U1 and one to part U2. Therefore, the translator lists two 74LS00 parts, one after the other, in the pin-list of Fig. 12.6a. An EDIF like description does not indicate specific pin numbers, so this information must be obtained from the TTL library. That portion of the library information for the 74LS00 is depicted in Fig. 12.6b. The 14 pins of each part may be listed immediately. It remains to determine the identifier of the node connected to each numbered pin, as required by the syntax definition of Fig. 12.4b.

Each pin must be connected to one and only one node, and fortunately the EDIF-like description of Fig. 12.5b includes only one net statement (keyword **joined**) for each node, including input and output ports. The first net statement

```
(I 74LS00.PRT  U1;
    1 XIN
    2 NET00003
    3 NET00002
    4 XIN
    5 NET00002
    6 NET00001
    7 GND
    8 NET00004
    9 NET00002
   10 NET00003
   11 ZOUT
   12 NET00001
   13 NET00004
   14 VCC
)
(I 74LS00.PRT  U2;
    1 UIN
    2 VIN
    3 NET00003
    4 ?
    5 ?
    6 ?
    7 GND
    8 ?
    9 ?
   10 ?
   11 ?
   12 ?
   13 ?
   14 VCC
        (a)
```

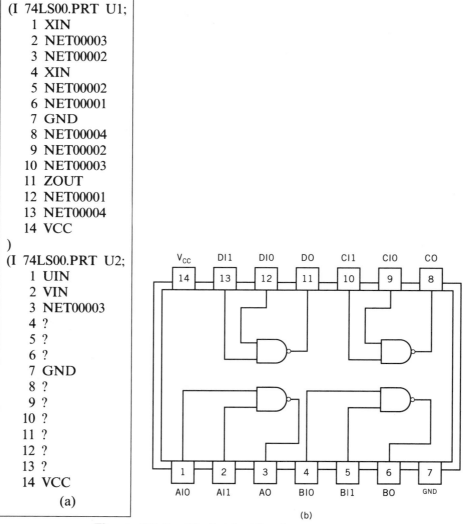

(b)

Figure 12.6   Pin list for Xorplus.

(**joined**) in the description contains no I/O ports, so an internal node__id must be generated by the translator. The node__id, assigned to the node defined by the first net statement, is NET00001. From the library part in Fig. 12.6*b* and the net statement (joined (U1B O) (U1D I0)), the translator is able to determine that NET00001 is connected to pin 6, the output of U1B, and to pin 12, input I0 of U1D. The second net includes the input port XIN, which serves as the node__id. This node is connected to pins 1 and 4, the inputs of U1B and U1A, respectively. There are three more net statements that do not include I/O ports. For these we define internal nodes NET00002, NET00003, and NET00004, respectively. Considering each remaining net statement in order completes the pin-list of Fig. 12.6*a*.

Although not shown in Fig. 12.5, EDIF provides a connection to V_{cc} and ground for each element. The translator realizes that there is just one V_{cc} pin and one ground pin on each part and connects these to the V_{cc} and ground input ports. The symbol ? indicates an unconnected pin on an unused element. ∎

It is unlikely that the reader will ever be expected to write EDIF descriptions manually, but it is much more likely that he or she will have occasion to interpret EDIF descriptions produced by CAD tools. The preceding discussion of a simplified EDIF should provide a helpful introduction to this activity. We shall defer examination of an element list until we open the topic of simulation in Section 12.5.

Every set of schematic capture tools will include utilities for accomplishing a variety of tasks on the stored schematic file once it is generated. Some tools will be used prior to the translation of the schematic to net-list form and others will be used afterward. Utilities might be provided for error checking, cross-referencing, *annotating, back annotating, and flattening*. The schematic capture program will allow the user to annotate manually by entering labels on the screen as the schematic is developed. An *annotation* utility will make the user aware of incomplete labeling and will provide distinct labels for multiple instances of the same part or element. *Back annotation* simply means to change annotation, usually by referring to a text file to update labels and parameters in a net-list. An interesting application of back annotation would be the insertion into a net-list of post-layout capacitance values to facilitate more accurate simulation.

Digital systems are often described in a hierarchical fashion. That is, a schematic description of a design might represent a complex part by only a box with input and output lines. A more detailed description of that part might be developed by the user as a separate schematic or be available in a library. The hierarchy could be extended to more than two levels by invoking a nonprimitive part in the description of a part that is, in turn, invoked in a higher-level description. A description may be extracted to a net list in hierarchical form. *Flattening* is the activity that replaces each instance of a part in a net-list (or pin-list or element-list) with a complete description of that part. The flattening process continues until only *primitive* elements (usually individual logic and memory elements) are invoked in the description. Flattening is often used to prepare a network for simulation.

12.3 LEVELS OF DESCRIPTION

Some of the languages for describing digital systems were introduced in Chapter 11 and the first two sections of this chapter. These descriptive mechanisms are often divided into two categories: structural and behavioral. AHPL is a behavioral language in that the clock-period-by-clock-period behavior of a system is explicit in its AHPL description. An EDIF net-list may be called a structural description, because it explicitly represents only the information needed to specify the physical construction of the circuit. A simulator of an EDIF net-list must refer to a library storing a behavioral description of each circuit element to reproduce the behavior of the circuit over time. The original schematic is also a structural description.

A special feature of AHPL, not found in other behavioral-level languages, is the explicit 1-to-1 hardware correspondence that makes synthesis from AHPL to a structural description a process of direct translation. Representations of digital systems in "C" or Pascal-like languages are behavioral descriptions that do not exhibit a 1-to-1 hardware correspondence. BDsyn, used for the description of combinational logic in the University of California Berkeley OCT tools package [27], is one such behavioral language.

The just discussed sampling of hardware description languages is tabulated in Fig. 12.7. Also included in the table is the IEEE standard language, VHDL (VHSIC hardware description language), whose syntax provides for both behavioral and structural description. Several whole-volume introductions to VHDL [21] are now available. VHDL is less concise than AHPL, and the multitude of VHDL features push that language beyond the scope of this book.

Separate from the structural or behavioral properties of a language is the level of description for which that language is useful. The lowest levels of description most closely reflect the properties of the circuit implementing that description, and low-level languages are consequently least efficient for describing large complex systems or algorithms. The highest-level or system behavioral-level languages do not reflect circuit properties and are the most efficient descriptive mechanisms. Figure 12.8 depicts five additional levels of description between the highest and lowest. After the system behavioral level comes the instruction set level, intended for the simulation of computer instruction sets, and the register transfer level. ISP [28] is the only well-known language at the instruction set level. DDL [26] is a *register-transfer-level* (RTL) language that appeared just prior to AHPL.

Whenever an implementation of a digital system includes a custom LSI or VLSI part, the system must "work" when it is first connected. It is not feasible to make a series of design modifications on the circuit while a prototype is under test on the bench. Each such design change would require the replacement of at least one costly diffusion or metallization mask. In addition, a time delay of days or weeks would be associated with the fabrication of each new part. Consequently, a

Behavioral	Structural
AHPL	EDIF net list
"C" or Pascal-like languages	pin list
	schematic
VHDL	VHDL

Figure 12.7   Dichotomy of hardware description.

Level of Description	Language			Simulation Precision	Vectors
System behavioral	↑			Data types	Yes
Instruction set			ISP	2-value	Yes
Register transfer			AHPL, DDL	3-value	Yes
Gate behavioral	VHDL		Verilog	3-value	Yes
Gate structural	↓		↑	3-value	No
Switch			EDIF	3-value	No
				(5 or more signal strengths)	
Circuit		↓		Continuous	No

Figure 12.8   Hardware description language levels.

trial-and-error design on the bench could take the designer weeks and bankrupt the company in the process. Since design errors are inevitable, some means must be provided for checking out a system before it is actually fabricated. The answer, of course, is *simulation*.

Simulation precision is the feature that distinguishes descriptions at the gate, switch, and circuit levels. There are many approaches to the simulation of sequential circuits. The approach chosen will be a function of the level of detail of circuit performance that must be monitored. As a general rule, the greater the level of detail provided or the greater the precision of the simulation, the greater the computer time consumed. This computer time can be significant when one considers the simulation of a circuit containing 10,000 or even 100,000 gates. Some of the alternative approaches are tabulated in Fig. 12.9, which includes a hypothetical plot of the useful information that might be obtained from each type of simulation versus the amount of computation time consumed. The plot is only an intuitive estimate, so no scale is provided.

The highest degree of precision of interest in digital design may be obtained by applying an analog simulator, such as SPICE, to circuit-level structural descriptions (point *A* of Fig. 12.9). By accurately simulating the internal function of individual circuit elements, the precision of the output of all gates as functions of time is sufficient. Unfortunately, the cost of running such programs on more than one or two critical paths or for longer than single clock periods is prohibitive for large logic networks.

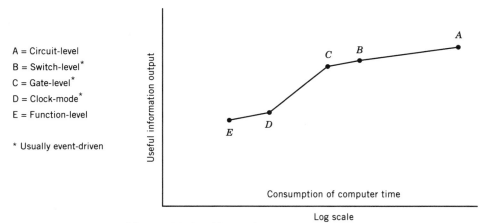

A = Circuit-level
B = Switch-level[*]
C = Gate-level[*]
D = Clock-mode[*]
E = Function-level

[*] Usually event-driven

Figure 12.9   Alternative simulation models.

At the opposite end (points *D* and *E*) of the curve in Fig. 12.9 is the clock-mode simulation. This type of simulation is the least costly and should be used whenever it will provide the insight required. Point *D* represents a clock-mode simulation based on a structural description. The *function-level* simulation, point *E*, is based on the clocked execution of statements in a hardware description language representation of a system. The function-level simulator for **AHPL**, discussed in References [15, 19], is available from the authors. By definition, synthesis must guarantee that a function-level simulation of the HDL representation will produce the same output as a clock-mode network-level simulation of the synthesized structural description. Any discrepancy between these two simulations would indicate failure of the synthesis process. Still the network-level simulation will almost always consume more computer time.

A graphic comparison of the computer time required by clock-mode and gate-level network simulations (represented by point *C* in Fig. 12.9) is given in Fig. 12.10. By definition, a clock-mode simulation implies that one iteration through the network will be accomplished each clock period. That is, output and next-state values are computed from the combinational logic networks and new values stored in the memory elements once each clock period. Many more iterations are required in a clock period if a clocked system is simulated in terms of delays in individual gates. Several iterations must take place between each input change to allow signals to propagate through multilevel gating. The clock itself must be regarded as a circuit input; hence, several iterations must be accomplished during the duration of a clock pulse. More network elements must be considered in a gate-level simulation. A flip-flop that can be treated as a unit in a clock-mode simulation must be treated as eight or ten separate gates in a gate-delay simulation. Thus, each iteration might be more time-consuming. Overall, in the absence of some yet to be discussed time-saving features, one should expect at least a 10-to-1 (conceivably 100-to-1) increase in computer time when resorting to a gate-delay as opposed to a clock-mode simulation.

In Chapters 7 and 11 we observed that networks of pass transistors in **NMOS** and, less commonly, transmission gates in CMOS could be used to realize certain logic functions using less chip area than would be used by logic gate realizations. These networks can readily be modeled at the circuit level but often not so easily at the gate level. Point *B* in Fig. 12.9 represents switch-level modeling and simulation. We shall see in Section 12.8 that this technique will provide some of the data to be obtained from a circuit-level simulation at a significant savings in computer time.

Gate- and switch-level simulations are very often *event-driven*. (Indeed, the clock-mode simulation of structural descriptions could be event-driven as well.) That is, a circuit element will be simulated during a particular iteration, if and only if an *event* has occurred at one of its inputs during a previous iteration. An event is simply a change in value propagated from the driving circuit element. Because large segments of any system are inactive at any particular point in time, the event-driven approach will result in a savings in computation time in comparison to processing every network element once each iteration. The algorithm for event-driven simulation will be the subject of Section 12.7.

Computations in gate-level simulators are usually accomplished using integer arithmetic. Time and delay values and capacitance values must be scaled to make this possible. With the exception of emitter-coupled logic and gallium arsenide logic, gate delays are usually an integral number of nanoseconds. The iteration interval (as depicted in Fig. 12.10) and the unit delay are often set at 1 nsec. The unit delay is established as a model and is defined for a particular

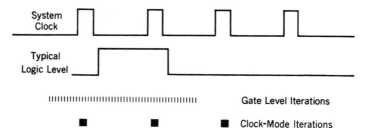

Figure 12.10 Timing in clock-mode and gate-level models.

logic element in a particular technology and cannot be changed later by the tool user.

■ EXAMPLE 12.2

Figure 12.11a depicts a segment of a timing trace of four lines, as they might appear on the output display of a gate-level simulator. Similar information must be provided by the user as input to the simulator. Devise a compact notation for representing values on the three input lines *clock*, *x*, and *D*. Although the marked intervals in Fig. 12.11a are 5 nsec, use 1 nsec, as the unit interval in the input notation.

SOLUTION Specification of the values at each nanosecond interval for the first 30 nsec is given in Fig. 12.11b. Alternately, the values could be specified only for 1-nsec intervals in which, at least, one of the lines changes value. In this case, it is necessary to explicitly provide the time at which the new set of values is applied. Figure 12.11c is a translation of the entire trace interval of Fig. 12.11a into this format.

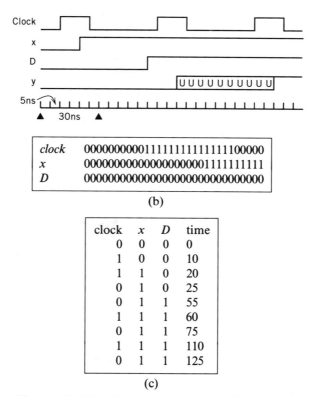

clock	0000000000111111111111111100000
x	00000000000000000000001111111111
D	00000000000000000000000000000000

(b)

clock	x	D	time
0	0	0	0
1	0	0	10
1	1	0	20
0	1	0	25
0	1	1	55
1	1	1	60
0	1	1	75
1	1	1	110
0	1	1	125

(c)

Figure 12.11 Alternative timing specifications.

12.4   UNSYNCHRONIZED INPUTS, SETUP TIME, AND LOGIC VALUE U

Design in the clock mode (or controlled clock mode) is by far the easiest and most convenient approach to the design of sequential circuits. The large majority of existing digital systems have been designed this way. The purpose of this section is to begin to identify the role of CAD tools in verifying that a digital system actually operates in the clock mode as designed. The clock-mode assumption may be restated from Chapter 10 as follows.

In a circuit operating in the clock mode, all input transitions and state-variable transitions must be synchronized by a clock pulse, and all the variables must be stable before the next clock pulse. Usually, these transitions are in synchronism with either a rising or falling edge of the clock.

The designer is able to ensure that transitions on state variables are clock-synchronized. There is often no such control of inputs from other systems. Suppose that system D to be designed is to receive inputs from system S. There are many reasons why an input from system S might not be synchronized with the internal clock of system D. (1) System S might be driven by an internal clock not synchronized with the clock of system D. (2) System S might not be a clock-mode system. (3) There might be a significant transmission line delay between systems S and D, so that the inputs would be delayed and, therefore, unsynchronized even if it were possible to drive the two systems with the same clock.

Even with the above reasons in mind, the designer of system D should probably not give up without a fight. It may be possible to adjust specifications or physical location so that the two systems can be driven by the same clock.

Try as one might, the system designer may finally have no choice but to deal with some unsynchronized primary inputs. These inputs might be connected anywhere within the combinational logic, but let us assume that the designer tried to eliminate the problem by isolating each unsynchronized input using what he or she hoped was a synchronizing flip-flop. If the system were described in AHPL and x were the only unsynchronized input, a synchronizing flip-flop r might be added to the description as shown below.

$$\text{ENDSEQUENCE}$$
$$. \ . \ . \ .$$
$$r <- x; \ . \ . \ . \ .$$
$$\text{END.}$$

The output of r would now replace x throughout the description of the system. That this measure will solve the synchronization problem might be argued as follows.

Argument for a Synchronizing Flip-Flop   If a transition of x from 0 to 1 occurs very near a clocking transition, this transition may or may not set r to 1. No matter! If r is set by these almost simultaneous transitions, the impact of the change on x will be felt during the next clock period. If not, r will be set one period later, and the impact of the input transition will be felt one period later. In many applications, where x is a control signal, the difference will be immaterial.

The reader may be tempted to breathe more easily at this point, but this would be unwise! In some technologies, the flip-flop r may not behave precisely, as assumed in the above argument. A number of researchers have verified empirically and argued theoretically that simultaneous transitions on the clock and data inputs of a flip-flop can leave the output gates of the device in the linear or *metastable* region. See, for example, Reference [24]. The result is that the output may oscillate for longer than a clock period without settling to either the old or a new logical value.

For an illustration of this behavior, consider Fig. 12.12 that shows the pair of cross-coupled gates that connect to the flip-flop output. Every type of bipolar flip-flop will have this pair of gates at the output. The rest of the flip-flop that will vary with the type is represented by the box. Suppose the flip-flop output is initially 0. Point *a* in the circuit will be normally 1, but must be driven to 0 for a period of time to cause the output *z* to go to 1. The third waveform of Fig. 12.12 shows without verification what can happen when the *D* input goes to 1 too close to the triggering clock transition. This tiny zero-going pulse may not have sufficient energy to drive *z* to +5 V or even past the threshold of about 2.5 V for logical 1. In this case, the output of both gates in the cross-coupled pair may be between 1 and 2.5 V when the pulse on *a* has disappeared. The result is the oscillation of *z* shown in the last waveform, which can persist until the next triggering clock edge.

The master-slave CMOS memory element of Fig. 9.27 with a transmission gate in the feedback loop of the slave enjoys an advantage over its bipolar counterpart. The transmission gate will interrupt oscillations in the slave (repeated in Fig. 12.13) when the memory element is clocked again. If the unsynchronized input *x* always remains stable for multiple clock periods, the CMOS circuit could function as a synchronizer as argued above. A technology-independent system-level approach to synchronization is discussed in Section 9.5 of Reference [15].

Long delays in combinational logic, unconnected to a primary input, can cause the *D* input of a memory element to change value too close to the triggering clock transition, resulting in oscillation as discussed. The minimum time interval, prior to the clocking transition during which changes on *D* must not occur, was defined in Chapter 9 as the *setup time*. The vendor's data sheet,

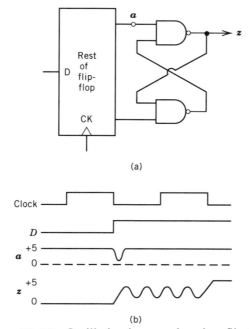

Figure 12.12   Oscillation in a synchronizer flip-flop.

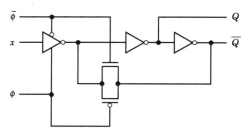

Figure 12.13   Output stage of a CMOS memory element.

whether for SSI or VLSI, will prominently display the setup time in a form similar to Fig. 12.14.

Although it is the job of the designer to avoid violating the setup or hold time of any memory elements in the system under design, the role of a gate-level simulator is to uncover design errors. A simulator may react to a setup or hold time violation by delivering an error message, by setting the memory element output to a U value, or both. *Unknown* or U values may be introduced into a simulation in other ways as well. By initializing all memory elements to U, for example, the simulation may reveal whether the system will be driven to a known state following power-up.

AND gates and OR gates with one or more U values as inputs may be simulated by consistently treating U as meaning unknown. Truth-table representations of AND and OR gates whose inputs may take on values of 0, 1, or U are given in Fig. 12.15. In general, there may be more than one source of U values. Therefore, when two U values arrive at both inputs to an AND or OR gate, these values are independent. It is tempting to designate $\bar{U}$ as the output of a NOT gate with input U. To do so would be misleading, because the actual value represented by that $\bar{U}$ would not necessarily be the complement of values represented by U

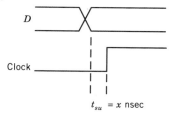

$$t_{su} = x \text{ nsec}$$

Figure 12.14   Setup time t_{su}.

x	y	$x \wedge y$	$x \vee y$		x	$\bar{x}$
0	0	0	0		0	1
0	U	0	U		U	U
0	1	0	1		1	0
U	0	0	U			
U	U	U	U			
U	1	U	1			
1	0	0	1			
1	U	U	1			
1	1	1	1			

Figure 12.15   Three-value logic.

elsewhere in the circuit. Instead, we say that the output of the NOT gate with unknown input is unknown. That is, $\overline{U} = NOT(U) = U$.

■ EXAMPLE 12.3

Determine the output of an exclusive-OR network for input combinations 0U, 1U, and UU. Use this information to determine the outputs of a three-bit adder with inputs $A = U1U$ and $B = 110$. The sum bits of the full adders may be expressed in terms of exclusive-OR gates.

SOLUTION   We simply substitute values in the Boolean definition of exclusive-OR.

$$A \oplus B = A\overline{B} \vee \overline{A}B$$
$$0 \oplus U = 0\overline{U} \vee \overline{0}U = 0 \vee 1U = U$$
$$1 \oplus U = 1\overline{U} \vee \overline{1}U = 1U \vee 0U = U$$
$$U \oplus U = U\overline{U} \vee \overline{U}U = U \vee U = U$$

For the least significant bit of the three-bit adder, we have

$$S[2] = A[2] + B[2] = U + 0 = U$$
$$C[2] = A[2] \wedge B[2] = U \wedge 0 = 0$$

For the next bit,

$$S[1] = (A[1] + B[1]) + C[2]) = (1 + 1) + 0 = 0$$
$$C[1] = A[1] \wedge B[1] \vee A[1] \wedge C[2] \vee B[1] \wedge C[2]$$
$$= 1 \wedge 1 \vee 1 \wedge 0 \vee 1 \wedge 0 = 1$$

and, finally,

$$S[0] = (A[0] + B[0]) + C[1] = (U + 1) + 1 = U$$
$$C[0] = A[0] \wedge B[0] \vee A[0] \wedge C[1] \vee B[0] \wedge C[1]$$
$$= U \wedge 1 \vee U \wedge 1 \vee 1 \wedge 1 = 1$$

Thus, the four-bit output is 1 U 0 U.   ■

The oscillations depicted in Fig. 12.12 can be modeled only in a circuit-level simulator. The following example assigns U values to signals that would be driven into the metastable region so that simulation can continue. Sometimes, the U values will disappear before reaching a primary output.

■ EXAMPLE 12.4

Let the timing trace of Fig. 12.11a represent the inputs to a leading-edge-triggered D flip-flop. Assume that the setup time is 6 nsec and the clock-to-output delay time is 10 nsec. Construct a timing trace of the flip-flop output y generated by a gate-level simulator that will assign a U value whenever a possibility exists that a line might be in the linear region.

SOLUTION   The change on D occurs 5 nsec prior to the clock transition. This is within the 6-nsec setup time interval, causing the simulator to report an unknown output. The timing trace of y was also included in Fig. 12.11a. Note that y is unknown for only one clock period, because D remains 1, allowing the next clock pulse to drive y to a stable value. ∎

12.5 STANDARD CELL CMOS LAYOUT AND DELAY MODEL

As background for our discussion of gate-level simulation in the next section, let us examine briefly the sources of logic propagation delay on a VLSI chip. We will restrict our discussion to standard cell CMOS, in many ways the easiest technology to model and the likeliest medium for the beginning designer. With some modification, the concepts discussed here are applicable to other technologies as well.

Figure 12.16 shows connections between two cells on a tiny chip standard cell layout. For emphasis, the switch-level diagrams of the inverter and NAND gates are shown in place of the actual layouts of the two cells. The location of these two cells, their relation to other cells not shown, and the positioning of the interconnecting lines were determined by an automatic place and route program. As suggested earlier in Fig. 12.2, it is the place and route program that translates the net-list or pin-list to a final chip layout. Because most readers will be users rather than developers of these programs, we choose to regard the underlying place and route algorithms as beyond the scope of this book. Consistent with most CMOS processes, two metal layers are included to facilitate minimal capacitance row-to-row connections. One metal layer (M1) including V_{dd} and ground are shown as solid lines. The second metal (M2) layer is represented by bold dashed lines. The two layers of metal are separated by a deposit of insulating material to make possible the crossover of interconnecting metal lines. Vias or connections between layers are represented by solid squares. It is common practice to use polysilicon to route both inputs an outputs to cell boundaries. The narrow dashed lines representing polysilicon are consistent with Fig. 8.1.

A gate-level simulation must model the time delay between the time a signal value change reaches the input of cell 1 until the change is observable at the input of cell 2 or 3. The resistance and capacitance of the interconnecting metal lines, the gate capacitance of cells 2 and 3, and the circuit within cell 1 will all impact this delay. A circuit-level simulation might use the RC network of Fig. 12.17b to approximate the transmission line. With the goal of determining a single number to assign as the delay of cell 1, we resort immediately to the less exact approximation of Fig. 12.17b, which lumps the total line capacitance and input gate capacitance as a single circuit element.

The slightly more complicated situation of Fig. 12.16 in which cell 1 drives two separate cells may be modeled as shown in Fig. 12.18. The dashed line in Fig. 12.17 denotes a separation of the polysilicon within the cell from the cell-to-cell metal lines. If the resistances R_2 and R_3 were unequal but significant, then the cell-1-to-cell-2 delay might differ significantly from the cell-1-to-cell-3 delay. The goal is to derive a single simple expression for delay at the output of cell 1. We take advantage of the fact that most CMOS technologies accomplish interconnection using low-resistance metal lines. Table 12.1 lists the resistivity of the various layers of a typical CMOS process. The Ω/square values will vary with layer thickness, but are close to values reported for 2μ, 3μ, and 5μ processes.

Because the resistance of a section of metal interconnect will be small with respect to the combined impact of the channel of the turned-on transistor and R_{poly}, we replace the metal part of the resistive network by single resistance R_{SM}. To avoid a separate approximation of R_{SM} for each gate prior to simulation, a

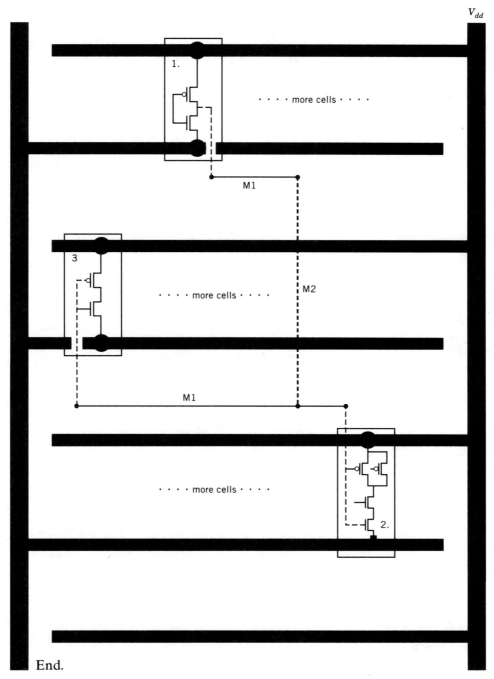

Figure 12.16 Layout fragment.

single average or "almost minimal" value is often used for all cells on a chip. A 200-Ω nominal R_{SM} would allow for a 2 λ wide interconnect 13,333 λ long. This value will always be less than the resistance of the polysilicon line within the cell.

Ignoring the distributed resistance makes it possible to calculate a single total load capacitance C_L for each gate in a simulation. In Fig. 12.18 this would be $C_1 + C_2 + C_3$. In general, C_L is given by Equation 12.1.

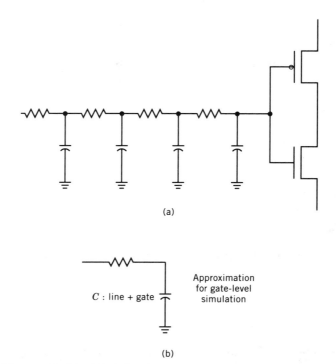

(a)

(b)

Figure 12.17   Modeling interconnection transmission lines.

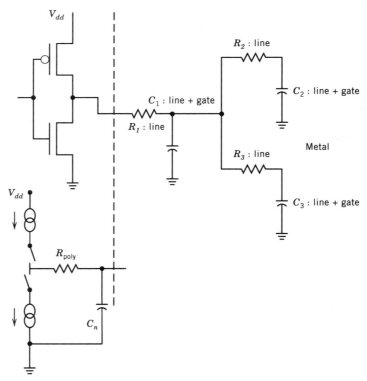

Figure 12.18   Circuit-level model of Fig. 12.15.

Table 12.1 Resistivity Values

Layer	Resistivity in Ω/square
Metal	0.03
Diffusion	10–50
Polysilicon	20–100
Transistor channel	10,000[a]

[a]Representative value: See I_{ds} expressions in Chapter 4.

$$C_L = \sum_{\substack{\text{connected}}} C_{\text{line}} + \sum_{\substack{\text{connected}}} C_{\text{gate}} \qquad (12.1)$$

In the cell output model shown in Fig. 12.19, only C_L varies with each instance of a particular cell. R_{poly} and C_n are functions of the cell type and R_{SM} is taken as constant. Expressions for delay as a function of C_L may be obtained by repeated detailed circuit-level simulation of the cell output model using Equations 4.1 and 4.2 as expressions for the current sources. This was the approach used to compile delay parameters for the National Security Agency 3-μ CMOS standard cell library described in Reference [11].

Simulations were carried out for C_L = 1.5 and 4.0 pF. Equation 12.2 is the resulting linear approximation of the expression for a (0$\rightarrow$1) propagation delay for a two-input NAND gate at a worst-case operating temperature of 125° C. Delays are less at lower temperatures. Equation 12.3 is a similar expression for the (1$\rightarrow$0) propagation delay. Capacitance values are in picofarads. Both equations are based on an input rise time of 10 nsec and R_{SM} = 200 Ω.

$$T_D(0\rightarrow1) = 3.11\,C_L + 3.03 \text{ nsec} \qquad (12.2)$$

$$T_D(1\rightarrow0) = 3.01\,C_L + 3.61 \text{ nsec} \qquad (12.3)$$

It is unnecessary to tabulate delay expressions for any standard cell library here or in the appendix. As we shall see in the next section, the appropriate expressions will be found built into the simulator support library for the reader's chosen logic family.

The simulation for the two-input NAND gate was repeated by using a series R_{SM} = 10K Ω to determine the impact of very long interconnecting lines on delay for that gate. The result was the term given in Equation 12.4 that may be added to

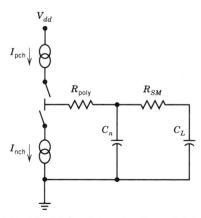

Figure 12.19 Model for determination of delay expressions.

the delay computed by either Equation 12.2 or 12.3, if the series resistance exceeds 200 Ω. In Equation 12.4 R_{SM} is in K ohms and C_L in picofarads.

$$T_{D\text{corr}} = 0.62 \, (R_{SM} - 0.2) \, C_L \text{ nsec} \tag{12.4}$$

Table 12.2 is provided to permit computation of actual capacitances of interconnect lines.

■ EXAMPLE 12.5

The output of a two-input CMOS (2-μ) NAND gate is connected by two unusually long interconnect lines to two cell inputs. The dimensions and gate capacitances are as depicted in Fig. 12.20. Use Equations 12.2 and 12.4 to estimate the gate A delay (0→1) observed at point B and the gate A (0→1) delay observed at point C. The 3-μ capacitance values in Table 12.2 may be used.

SOLUTION To begin with, we compute the resistances of the three separate segments. In each case, we multiply the resistivity (0.03 for metal) in Ω/square by the length divided by the width of the line, cancelling the λ's in the numerator and denominator.

$$R_1 = 0.03 \, (15{,}000/2) = 225$$

Similarly,

$$R_2 = 300 \quad \text{and} \quad R_3 = 150$$

The significance of each resistor must yet be decided, but the entire capacitance of all line segments and both gates will be required. We, therefore, compute a single C_L. A 2μ technology tells us that $\lambda = 1\mu$. We were not told how much of each line was metal 1 and how much was metal 2. We accept a common approximation

Table 12.2 Capacitance in Picofarads per Square Micron

Area Type	5μ CMOS	3μ CMOS
Gate to channel	4×10^{-4}	5.7×10^{-4}
Polysilicon	0.4×10^{-4}	0.65×10^{-4}
Metal 1 to substrate	0.3×10^{-4}	0.52×10^{-4}
Metal 2 to substrate	0.2×10^{-4}	0.3×10^{-4}
Metal 2 over metal 1	0.4×10^{-4}	0.5×10^{-4}
Metal 2 over poly	0.3×10^{-4}	0.4×10^{-4}

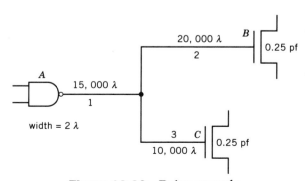

Figure 12.20 Delay example.

of equal lengths of metals 1 and metal 2 and, therefore, simply average the values of capacitance per unit area, using 0.4×10^{-4} picofarads/μ^2. Now

$$C_L = (2\mu)(15 + 20 + 10) \cdot 10^3\mu \cdot (0.4 \times 10^{-4}) + 0.25 + 0.25$$
$$= 9\,(0.4) + 0.5 = 4.1 \text{ pF}$$

From Equation 12.2, we compute the nominal delay for the NAND gate.

$$T_D(0 \to 1) = 3.11\,(4.1) + 3.03 = 15.78 \text{ nsec}$$

Now approximation must begin in earnest. We assume that the resistance applicable to path AB is $R_{AB} = R_1 + R_2 = 525\,\Omega$ and that $R_{AC} = R_1 + R_3 = 375\,\Omega$. Next we insert these values (converted to KΩ) into Equation 12.4 and add the appropriate correction factors to 15.78 nsec.

$$T_{AB} = 15.78 + 0.62\,(R_{AB} - 0.2)\,(4.1)$$
$$= 15.78 + 0.62(0.525 - 0.2)\,(4.1)$$
$$= 15.78 + 2.54\,(0.305) = 16.55 \text{ nsec}$$
$$T_{AC} = 15.78 + 2.54\,(0.175) = 16.22 \text{ nsec}$$

∎

Three important observations may be made based on the calculations in Example 12.5:

1. Even for the very long interconnects (long enough to connect diagonal corners of a reasonable-size chip), the value of R_{SM} will be such that the correction value calculated from Equation 12.4 is less than 10% of the gate delay. Neglecting R_{SM} will allow the delay used by a simulation to be associated with each gate rather than each gate-to-gate path.
2. The capacitances of the interconnecting lines themselves are not negligible and the accuracy of a simulation will suffer, if this capacitance is ignored.
3. Load capacitance will accumulate with fan-out from a cell as well as with line length. A fan-out limit must be imposed at some point in the design or synthesis process.

A simulation of gate delay may use values computed from Equations 12.2 and 12.3. The use of Equation 12.1 that includes line capacitance in the computation of C_L must be based on an already existing layout. The process of determining the total capacitance driven by each cell output and entering this information in the element list used by a gate-level simulator is a form of *back annotation*. A capacitance file for this purpose may be generated by the place and route program and then entered in the element list by a net list translator.

12.6 TIMING ANALYSIS AND SIMULATION

After a potential VLSI design has been functionally tested by simulation at the clock level and a chip layout developed by whatever means, it is necessary to verify that the chip design will actually meet all specifications. If the original design was 100% clock mode and no layout design rules were violated, the problem reduces to verifying that the chip will work at the expected clock frequency. If circuits such as synchronizers or nonclock-mode arbitrators are present, then more elaborate timing relationships must be verified. In any case, this verification must be accomplished by analysis and simulation before a chip is actually fabricated.

Once the layout is complete, there must exist some path in the combinational logic, the delay through which will determine the maximum clock frequency. Figure 12.21a depicts such a *worst-case path*, and the timing diagram of Fig. 12.21b illustrates the delay in that path. The clock is single-phase, with all memory elements triggered on the trailing edge of that clock. (Adjusting the timing diagram to account for complementary clocking or a two-phase clock will be left as a problem for the reader.) Control pulse CP_a will cause a change in the value of Y, the origin of the worst-case path. Control pulse CP_b will trigger the value f at the termination of the path into a memory element one clock period later. To maximize the clock frequency, it must be set so that the output of this worst-case path will change value just barely more than the setup time interval ahead of the triggering edge of CP_b.

Definition 12.1 A *worst-case delay path* in the combinational logic originating at memory element output Y and terminating at point f must satisfy the following two conditions:

1. The sum of the applicable delay values (alternating $0\rightarrow1$ and $1\rightarrow0$) of all gates in the path must not be less than that sum for any other path through the combinational logic.
2. There must exist a vector $W = W_1, W_2, ..., W_{n-1}$ of memory element and primary input values such that (A) for every point h on the path, the logical value of the output f depends on the logical value at point h, and (B) a clock pulse CP_b to the target memory element will be enabled.

A path that satisfies condition 1 but not condition 2 is called a *false path*. Condition 2 is included, because the time required for a value to change is irrelevant if that final value itself does not matter. The function $f(Y, W_1, W_2, ..., W_{n-1})$ will depend on the value at a point h if and only if Equation 12.5 is satisfied. The functions $f|_{h=0}$ and $f|_{h=1}$ represent the function f with h forced to 0 and 1, respectively.

$$f|_{h=0} \oplus f|_{h=1} = 1 \tag{12.5}$$

The left side of Equation 12.5 is a function of $W_1, W_2, ..., W_{n-1}$.

If the design and layout are for some reason not subject to change, then the maximum clock frequency is established as less by a suitable safety margin, than the reciprocal of the sum of the delays in a worst-case path. If change is allowed and the clock determined in this way is too slow, then the circuit must be somehow modified to eliminate the worst-case delay path. Of course, once a worst-case delay path is eliminated, another will appear to take its place. The design must be repetitively modified and the layout must be reevaluated, eliminating worst-case paths individually or in groups, until the calculated clock frequency meets specifications.

Occasionally, redundant elements are added to a design for the express purpose of reducing delay. The original maximum delay path may then become a false path. In the carry skip adder, introduced in Section 8.4, special logic paths are introduced to permit carries to bypass propagate stages, that is, stages for which one input bit is 1 and the other is 0.

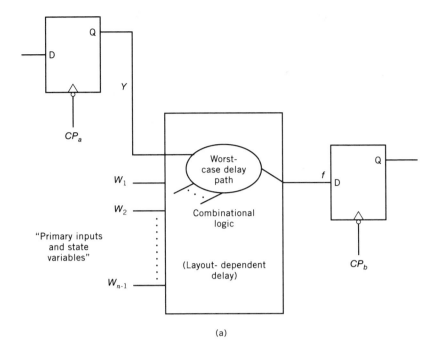

(a)

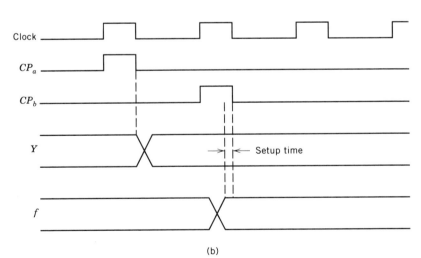

(b)

Figure 12.21   Worst-case delay path.

■ **EXAMPLE 12.6**

The carry skip around two full adder stages with all gates shown explicitly is depicted in Fig. 12.22. The input carry C_2 is routed to the output of the multiplexer, if both stages are propagate stages. Otherwise, the normal carry output of the second stage is connected through the multiplexer. We use the AHPL convention of labeling the most significant bit as bit 0. To simplify the notation in the following expressions, we identify the false path in the final (most significant) two stages of the adder.

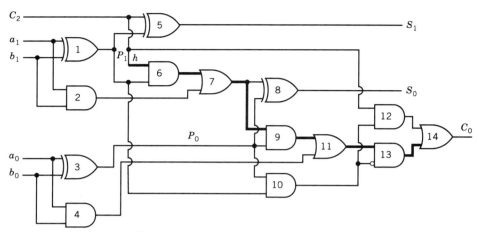

Figure 12.22 Two-bit carry skip adder.

SOLUTION Although the carry skip path has been added, the original worst-case path through gates 6, 7, 9, 11, 13, and 14 still exists and is marked by a heavy line. For convenience, P_1 and P_0, whose values indicate whether stages 1 and 0 are propagate stages, are marked in Fig. 12.22. The expression for the final carryout C_0, taken directly from Fig. 12.22, is given in Equation 12.6.

$$C_0 = ((hP_1 + a_1b_1)P_0 + a_0b_0)\overline{P_0P_1} + C_2P_0P_1 \tag{12.6}$$

Manipulating the exclusive-OR expressions, arising from Equation 12.5, can be tedious, so we first simplify Equation 12.6.

$$C_0 = hP_1P_0\overline{P_1P_0} + a_1b_1P_0\overline{P_0P_1} + a_0b_0\overline{P_0P_1} + C_2P_0P_1 \tag{12.7}$$
$$= a_1b_1P_0\overline{P_0P_1} + a_0b_0\overline{P_0P_1} + C_2P_0P_1$$

We see that C_0 is not a function of point h on the marked path. Therefore,

$$C_0|_{h=0} = C_0|_{h=1} \quad \text{and}$$
$$C_0|_{h=0} \oplus C_0|_{h=1} = 0 \tag{12.8}$$

Because there is no set of input values for which the expression on the left of Equation 12.8 is 1, we conclude that the path through 6, 7, 9, 11, 13, and 14 is a false path. ∎

The false path in Example 12.6 could have been eliminated by connecting the output of gate 2 rather than the output of gate 7 as an input to gate 9. Doing so would have made no difference in the total number of gates. One could argue that a false path is always nonessential and should have been eliminated by the designer. An alternative to asking the designer to guarantee the existence of no false paths is to include a mechanism for false path checking in a CAD tool for timing analysis.

We offer no algorithm for efficiently identifying worst-case paths or making corresponding design modifications, but the resources available are suggested by the design flow for CMOS standard cells depicted in Fig. 12.23. This diagram should prove a useful overview, whatever the set of design tools available to the reader. Synthesis from a RTL description is the design methodology of Fig. 12.23. Adapting the diagram for the case where design input is done through schematic capture will be left as a problem for the reader.

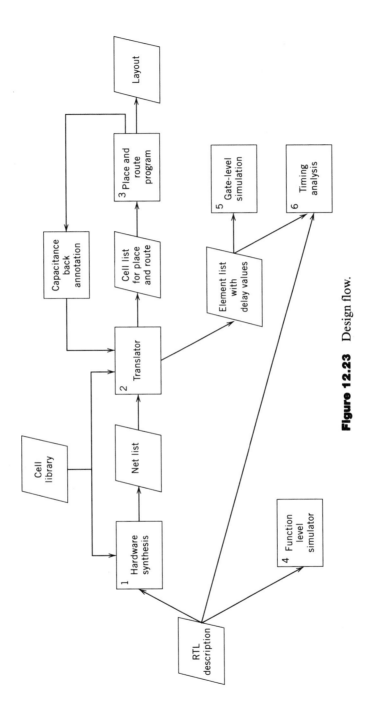

Figure 12.23 Design flow.

The information necessary to compute path delay is included in the "element-list with delay values" generated by the translator in block 2 of Fig. 12.23. The first translation of the net-list generated by hardware synthesis will use only the gate values to compute C_L from Equation 12.1. Once the place and route process is complete, the line capacitance will be provided to the translator by back annotation. The revised version of the element-list will reflect accurate delay values consistent with the actual circuit layout. The revised element-list may now be used to support accurate gate simulation. Any further increase in precision would probably require simulation at the circuit level.

In many design environments, the gate-level simulator is the only tool available for use in identifying and determining the delay of the worst-case delay path. What are the sequences of primary inputs to be simulated? For some smaller designs, it may be possible to determine an input sequence that constitutes a complete functional test of the design. Somewhere in the process of simulating that sequence at the specified clock frequency, the worst-case path will be excited. If the output is as anticipated and U values do not appear, no design modification or reduction in clock frequency is required.

Simulating a complete functional test at the gate level might be prohibitive for a complex circuit. As indicated in Section 12.3, a function-level simulator is usually more efficient than a gate-level simulator. It may be possible to use a function-level simulator together with engineering judgment based on layout to identify sequences that will excite a worst-case path when applied to the gate-level simulation.

One analytical approach to identification of a worst-case delay path consists of tracing all paths in the combinational logic from output to input using the element list. The resulting candidate for worst-case path might be a false path. The timing analysis block 6 in Fig. 12.23 represents a more comprehensive approach. Once a candidate worst-case path (a path satisfying condition 2 of Definition 12.1) is identified, a search over time on the RTL description is carried out to verify that the path can indeed be excited (condition 1 of Definition 12.1). The search routine of the timing analysis block would share many of the characteristics of the function-level simulator.

12.7 EVENT-DRIVEN GATE-LEVEL SIMULATION

The timing analysis algorithms referred to in the previous section are beyond the scope of this book. Although not necessarily easy to implement, the basic concept of function-level simulation is easy to understand. The steps of the hardware description are simply executed in sequence as in a program. Event-driven simulation may seem less transparent. Because so much of the gate-level simulation on which design evaluations are based is event-driven, this subject merits a closer look.

Most of the time most memory elements of any digital system are at rest in stable states. When memory elements do change, the impact of each change flows through a narrow path in the combinational logic impacting one gate at a time, not all at once. Why evaluate every element in the network each time step? An event-driven simulation evaluates only those elements, one of whose inputs experiences a value change or an event.

A simplified flow chart characterizing an event-driven simulation is shown in Fig. 12.24. The principal data structure is an *event list* that includes all future

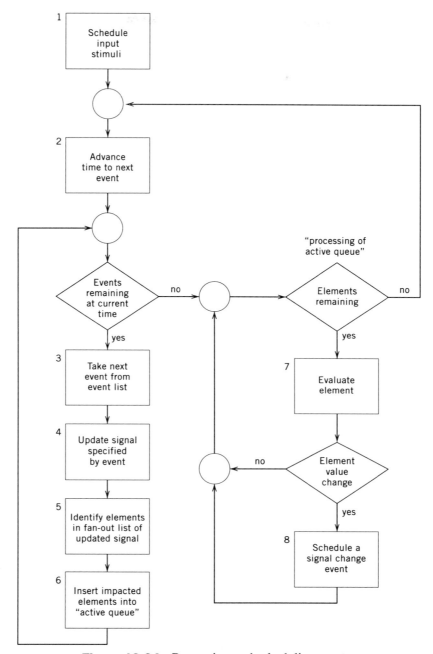

Figure 12.24 Processing and scheduling events.

scheduled output changes. As provided by block 2, the simulation advances past time steps in which no events are scheduled. All events scheduled in a given time step are processed by blocks 3, 4, 5, and 6. The processing of an event will ordinarily cause the output signal of some element to change value. Block 5 determines from the fan-out list of that element the list of all elements with an input connected to the changing signal. These elements with new input values are inserted into the *active queue*.

Once processing of all events scheduled for the current time step is complete, activity passes to block 6 on the left in Fig. 12.24. Now logic and memory elements in the active queue are evaluated. If an input change will affect the output of an element, an event is scheduled into the event list by block 8. The scheduled event is an output change that will occur at some future time as dictated by the 0→1 or 1→0 delay of the element. The event is entered in the list at the time step in which the event is to occur, not necessarily at the end of the list. Delay values are computed, as discussed in Section 12.5. Once all elements in the "active queue" have been evaluated, time is advanced to the time step of the next scheduled event; and the whole process is repeated.

The reader may recognize that structuring the event list as just discussed implies that delay will be treated as transport delay in contrast to inertial delay. The transport delay model of standard cell layout is least accurate in the presence of narrow pulses. In the physical system, the onset of a pulse will begin the charge of a capacitor at the input of an element. At the end of a sufficiently narrow pulse, the capacitor will begin to discharge before any impact appears at the element output. An inertial delay model will swallow the narrow pulse exactly as occurs in the physical system. Suppression of pulses narrower than some minimum value can be accomplished in the event-driven simulation at the price of the significant complication of Fig. 12.24 and resultant slowing of the simulation.

■ **EXAMPLE 12.7**

The initial values within a simple sequential circuit are depicted in Fig. 12.25a. Also shown are the delay parameters needed by an event-driven simulation of the circuit. Suppose that B goes to 1 after 6 nsec and that A goes to 1 eight nsec after

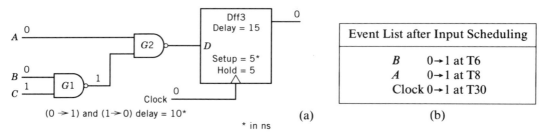

Event List after Input Scheduling
B   0→1 at T6
A   0→1 at T8
Clock 0→1 at T30

(a)   (b)

Time	Event List Activity	Active Queue Activity (Scheduling of Events)
6	B 0→1; pass input change to gate 1	schedule G1 1→0 at T16
8	A 0→1; pass input change to gate 2	schedule G2 1→0 at T18
16	G1 1→0; pass input change to gate 2	schedule G2 0→1 at T26
18	G2 1→0; pass input change to DFF3	
26	G2 0→1; pass input change to DFF3	
30	Clock 0→1; pass input change to DFF3	schedule DFF3 0→U at T45

(c)

Figure 12.25 Example of event-driven simulation.

the beginning of the simulation. We assume the memory element to be leading-edge-triggered and that the clock will go to 1 30 nsec into the simulation. Determine the results of an event-driven simulation of one clock period of circuit activity.

SOLUTION The first step is to schedule the input transitions into the event queue, as shown in Fig. 12.25b. This initial status of the event queue is the only instantaneous snapshot of the queue shown. The left column of Fig. 12.25c reflects the history of the processing of events in the queue. In this simple example, there is never more than one event at a time step. Shown at the right is the subsequent processing of the active list each time step. Notice that consecutive changes on both inputs to G2 cause two events affecting the output of this element to simultaneously reside in the event queue. This will cause an 8-nsec pulse to appear at the output to this gate. Because this pulse width is 80% of the gate delay, the pulse would probably not be suppressed by a simulation with that feature.

We note that the change on the D input to DFF3 at T26, followed by a clock transition at T30, is a setup time violation. Therefore, the output of the memory element is scheduled to go to the unknown value U at T45. ∎

Example 12.7 focuses on a small circuit fragment in which activity exists throughout the period of interest. It is easy to imagine the existence of additional gates that would not change the contents of the event queue over this period. We are able to observe that the processing of the queue is required at only six time steps over a 30-step period.

12.8 SWITCH LEVEL SIMULATION

The reason for the existence of switch-level simulators is to more closely model arbitrary networks of NMOS and PMOS transistors while consuming less computation time than circuit-level simulators. As a digital designer, the reader will likely use one or more of the existing switch-level simulators [2, 6, 10] or perhaps one yet to be developed. For CMOS standard cell technology, it is likely that the simulation level of choice for complete systems will be the gate level. Switch-level simulation may be helpful in the design of complex cells. In the case of NMOS, switch-level simulation may be used for complete system simulation. The purpose of this section is to look briefly at the theoretical foundation on which most switch-level simulation algorithms are based. Unlike what can be done mathematically for circuit-level simulation algorithms, this analysis will not establish quantitative bounds on the level of accuracy of a switch-level simulation.

Any n-channel or p-channel transistor will be considered a switch. A switch may be conducting or nonconducting, depending on the value of its gate terminal. The connecting point between two or more switches will be considered a *node*. The response of a network of switches to a change on one of its primary inputs will depend on charges previously stored at nodes in the network and the logical values of all the input gates. The set of possible input and output and node values will be $\{0, 1, U\}$, as discussed in Section 12.5. If a sequential network has been properly designed, node values will eventually stop changing in response to a single input change. This final set of values is called the *network steady state*. Determination that a network has failed to reach a steady state after some fixed large number of iterations, or determination of that network steady state will be called a *switch network evaluation*.

Elapsed time in response to a stimulus is of interest at the switch level, just as it is at the gate level or circuit level. Quantitative timing information is not obtained as part of a switch network evaluation. Some "so-called" switch-level simulators superimpose gate-level time delay information on the network evaluation. Others simply accomplish successive network evaluations in response to successive primary input changes. In this way, accuracy is reduced to a "yes or no" question. The worth of a particular evaluation algorithm becomes the size of the class of circuits for which that accuracy question is answered yes. The following discussion will focus on a single evaluation.

Throughout each evaluation primary inputs, most of which will drive gate terminals in the network, will be fixed. We may apply the notion of "divide and conquer" to an evaluation by allowing all internal gate values to remain fixed throughout what we shall call a component iteration.

Definition 12.2   A switching network is partitioned into *switch-level components* by cutting the lines leading to all NMOS or PMOS transistor gate terminals.

When we use Definition 12.2, switch-level operations may be confined to individual components.

■ **EXAMPLE 12.8**

Realize in a NMOS network that portion of the data unit of Fig. 11.11 leading to flip-flop output B_2. Use a precharged implied bus and pass network realization of the exclusive-OR function. Partition the network according to Definition 12.2.

SOLUTION   As shown in Fig. 12.26, Definition 12.2 partitions the resulting circuit into seven components A through G. The NAND-NAND realization of $\Phi_1 \wedge (C_2 \vee C_4)$ is not shown.   ■

The results of each component iteration are passed to other components as gate inputs. If the inputs to some component j change following a component k iteration, the component j is entered in the component event queue. Initially, the components connected to the changed primary input are in the component event queue. When the component event queue is empty, the evaluation is complete. Figure 12.27 summarizes the switch network evaluation algorithm, neglecting the possibility that gate inputs may have the value U.

In the absence of time delays, the lists in the algorithm are simpler than the event lists of the previous section. The former are merely first-in first-out lists. The function **firstin** merely picks one of the entries that have been waiting longest in the queue. A component iteration consists of redetermining the value of each node in the component until node values are stable, that is, until the cevent queue is empty. When a node changes value, it is entered in the cevent queue as the source node of each switch path connected to that node. The node at the other end of the switch is termed the target.

The capacitance to ground of each node makes it possible to store a value at that node. The potential of this value to be passed to or to affect the value of nearby nodes is directly related to the magnitude of capacitance of that node. When turned on, a switch acts as an attenuating resistor. Both effects are modeled directly in a circuit-level simulation. In most switch-level modeling schemes, capacitance and resistance effects are jointly modeled in terms of a single set of

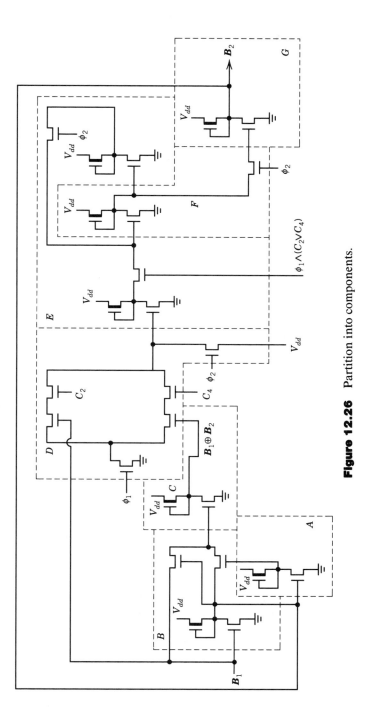

Figure 12.26 Partition into components.

```
place components with changed primary input in network event list;
while   (network event list not empty) {
        current component = firstin (components in
                                     network event list);
        place all nodes of current component in cevent queue;
        while (cevent queue not empty) {
              source node = firstin (nodes in cevent queue);
              for (each path from source node) {
                  resolve target node;
                  if (value of target node changes)
                        place or update target node in cevent queue;
              }
              pass current component outputs to inputs of other components;
                    for (each component whose inputs change)
                          place or update component in network event queue;
        }
```

Figure 12.27   Switch network evaluation algorithm.

strength values. That will be our position here. Increasing the number of strength values will enhance the ability of the simulation to identify increasingly subtle design errors. We will restrict our attention to five strength values and leave it to the reader to propose design errors that will not be detected. (**Hint: Consider charge-sharing in a pull-down path of several switches in series.**)

The highest strength value 5 will be associated with a terminal connected directly to ground or V_{dd}. The strength value 1 will be assigned at the beginning of an evaluation to nodes with the least capacitance. The value 2 will be assigned to nodes with greater capacitance, such as bus wires and gate inputs. The value 4 will be used to express the conductance of n-channel devices in pull-down paths or p-channel pull-up devices. The effects of the attenuation through all other devices including depletion-mode loads and pass transistors will be expressed by strength value 3. A strength value 0 will be used only for unconnected nodes and to represent the impact of an open switch. A node value will be represented by the pair $<s,v>$, where s is the strength and v the logical value (0, 1, U) of the node.

The innermost activity of the algorithm in Fig. 12.27, *resolve target node*, will consist of two steps. The first will use the source node and state of the connecting switch to compute a *propagated target pair*, which will be used together with the old value of the target node to determine the new pair for that node. The rule for determination of the propagated target pair is given in Fig. 12.28.

Following application of the rule in Fig. 12.28, there exist two competing target node pairs, neither of which is necessarily the correct final pair. The last step is to use these two values to determine the best possible approximation of that final pair. For this purpose, we introduce in Equation 12.9 the compact and mathematically satisfying # operator, first defined by John Hayes [3].

$$\text{new target pair} = (\text{propagated target pair}) \, \# \, (\text{old target pair}) \qquad (12.9)$$

The lattice of Fig. 12.29 defines the # operator for the special case of six strength values. The # of two pairs is merely the least upper bound of the two corresponding points in the lattice. We note that the largest strength value will always prevail. If two logical argument values with the same strength conflict, the resulting logical value will be U. These statements are consistent with the propagation of values in the actual circuit. Whether a given simulation using the # operator, as

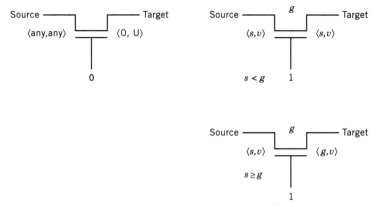

Figure 12.28 Computation of propagated target value.

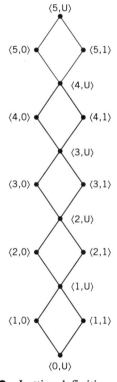

Figure 12.29 Lattice definition of # operator.

defined here, will converge to the proper set of steady-state pairs will depend on component complexity, the number of strength values used, and the assignment of strength values to nodes and switches.

■ **EXAMPLE 12.9**

The user is usually asked by a simulator for only the initial values of primary inputs and memory elements. The switch-level simulator will then assume U for the remaining logical values and execute until a steady state is reached. Assume that component A of Fig. 12.26 was evaluated first during initialization so that the

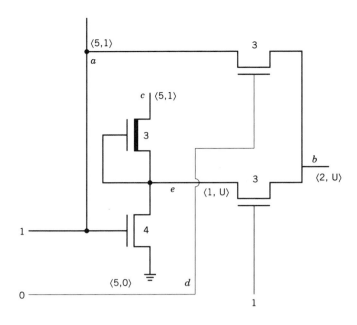

Source Node	Target Node	Former Target Pair	New Pair
a	b	<2,U>	<2,U>
b	e	<1,U>	<2,U>
c	e	<2,U>	<3,1>
d	e	<3,1>	<4,0>
e	b	<2,U>	<3,0>
b	e	<4,0>	<4,0>

(b)

Figure 12.30 Initialization example.

output of the inverter in that component is now 1. Execute the first initialization evaluation of component *B* using six strength values and the # operator defined in Fig. 12.29.

SOLUTION The initial node values and switch strength values for component *B* are given in Fig. 12.30*a*. The computation of new node values in the order that source-target pairs emerge from the cevent queue are given in Fig. 12.30*b*. For each line the propagated target value is first determined using the rules in Fig. 12.28 and then the new target pair is determined using the # operator. Note that the fixed nodes are sources but never targets. ∎

■ **EXAMPLE 12.10**

Because Fig. 11.11 is a clock-mode circuit, the input nodes to component *E* of Fig. 12.26 do not change as $\Phi_2 \Phi_1$ pass within one sequence of values 00, then 10, then 00, then 01. Execute three successive evaluations of component *E* for these sets of values, taking advantage of this observation.

SOLUTION During initialization both clock phases will be 0, and both internal nodes of the component will be isolated from external values. Therefore, the U values will not disappear during initialization. The node pairs and switch strengths following initialization are shown in Fig. 12.31a, and the results of the three successive evaluations are given in Fig. 12.31b. Notice that the strength values of nodes b and c are reduced to node strength values as the evaluation for $\Phi_2\Phi_1 = 00$ begins. ∎

Thus far we have neglected occurrences of U values on gate inputs. Unknown gate inputs must be faced in the design of a simulator to allow for lack of control of the order of component evaluation and for the possibility that initial values of memory elements might be assigned as U. The two switches in Fig 12.32 characterize the propagation step for switches with unknown gate values. In each case, there are two possible propagated target pairs, but only one new node pair can actually be assigned in the simulation. One approach [4] that has been implemented with reasonable efficiency calls for two evaluations for each component.

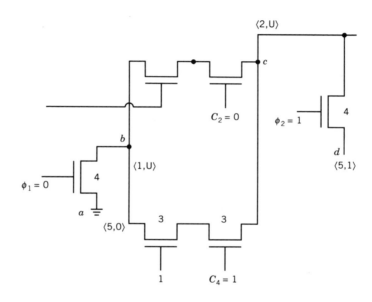

Source Node	Target Node	Old Target Pair	New Pair
	$\Phi_2 = 1, \Phi_1 = 0,$		
d	c	$<2,U>$	$<4,1>$
c	b	$<1,U>$	$<3,1>$
b	c	$<4,1>$	$<4,1>$
	$\Phi_2 = 0, \Phi_1 = 0$		
c	b	$<1,1>$	$<2,1>$
b	c	$<2,1>$	$<2,1>$
	$\Phi_2 = 0, \Phi_1 = 1$		
a	b	$<2,1>$	$<4,0>$
b	c	$<2,1>$	$<3,0>$
c	b	$<4,0>$	$<4,0>$

(b)

Figure 12.31 Three evaluations of Component E.

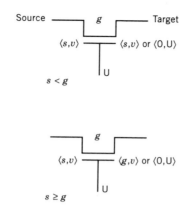

Figure 12.32   Propagated pair alternatives for gate = U.

Both cases use the lattice of Fig. 12.29 to determine two alternative new target node pairs. One simulation will choose logic pairs according to the priority of logic values—first 1, then U. The other simulations choose pairs with a logic value priority of first 0, then U. Once both simulations have stabilized, nodes with logic value 1 for both cases will be assigned 1, and nodes with value 0 for both cases will be assigned 0. All other nodes are assigned U values.

12.9 PROGRAMMABLE LOGIC DEVICES AND PROGRAMMABLE GATE ARRAYS

Sometimes design cost and development time are more critical than the number of transistors per package and even than the clock frequency at which that package can operate. This issue was first raised in the discussion of Fig. 4.18. In that figure every alternative, to full custom VLSI or realization with MSI parts at the other extreme, were lumped under the label *semi-custom VLSI*. There are, at least, five distinct approaches to semi-custom VLSI. Five are listed in Table 12.3 between the full custom and MSI part extremes.

Standard cell and gate array realization have already been considered in Chapter 9 and earlier in this chapter. The three remaining alternatives, field programmable gate arrays (FPGAs), logic cell arrays (LCAs), and programmable logic devices (PLDs), share one common feature. They are all in some sense user programmable. The first user programmable logic element to become available was the PAL, which we introduced in Chapter 6. When memory elements are included within a PAL, these parts can be used to realize sequential circuits and, therefore in theory, any digital system. In practice PALs and similar PLDs are usually quite limited with respect to the complexity of a system that can be realized in a single package. This notion is represented in Table 12.3 by assignment of a relatively low density of usable logic elements per package.

CALD tools for PLD design are provided by vendors and are independently available. Economics in conjunction with the limited-device density in a PLD force the tailoring of designs to efficiently utilize available parts. As suggested by the leftmost design flow in Fig. 12.33, CALD tools must reflect the features of the PLD to the user. This may be done in a specialized schematic capture program or by asking for input in the form of Boolean equations. Logic expressions are optimized by the CALD tool as it generates a program to be passed to the PLD programmer.

Table 12.3

	Pre-manufacture	Field Programming Mechanism	Usable Element Density	Performance	CALD Tool Sophistication
Full custom	NO	—	▮▮▮▮▮▮▮▮	▮▮▮▮▮▮▮	▮▮▮▮▮▮
Standard cells	NO	—	▮▮▮▮▮▮	▮▮▮▮▮▮▮	▮▮▮▮▮▮▮
Gate arrays	Partial	—	▮▮▮▮▮	▮▮▮▮▮▮	▮▮▮▮▮▮▮
FPGAs[a]	Yes	Fuses; antifuses	▮▮▮	▮▮▮▮▮	▮▮▮▮▮▮▮
Logic[a] Cell arrays	Yes	On-chip memory elements	▮▮	▮▮▮▮▮	▮▮▮▮▮▮▮
PLDs[a]	Yes	Fuses	▮	▮▮▮▮▮	▮▮▮▮
MSI	Yes	—	▮	▮▮▮▮▮	▮▮▮▮▮

[a] See Figure 12.33

An FPGA is just that, a gate array in which internal cell connections and connections between cells can be established by programming. Although the final steps of gate array manufacture are accomplished after a design is specified, an FPGA is completely manufactured and packaged with no hint of what its ultimate application might be. At least one vendor, Actel, establishes connections during programming by using an antifuse technique rather than destroying all unwanted connections by blowing fuses. This implies that a surplus of horizontal and vertical interconnecting metal lines must be provided during manufacture, with some lines left unconnected after programming.

Another programming part similar to an FPGA will be called an LCA. Xilinx is one manufacturer of LCAs. In the LCA technology programming is not accomplished by antifuses but by loading a set of special on chip memory elements that control connections within the user design. Thus a large number of "overhead" gates, tri-state drivers, and memory elements must be located on the chip in addition to those available to the user. The CALD tools will initially write the LCA program in a separate RAM (or a PROM may be programmed with the same information). The program will be down loaded from the RAM or ROM to the LCA as part of each "power on" process. One advantage of the LCA, like the EPROM, is that the LCA can be reprogrammed. Redesign after prototype LCAs have been programmed and tested is less costly than similar activity in the other technologies of Table 12.3.

The largest FPGA and LCA chips marketed at the time of this writing are competitive in terms of elements available to the designer. Table 12.3 indicates a lower usable element density for the LCA, because significant chip space is occupied by the overhead elements. Typical chips include 8000 user available gates and 1000 memory elements. In a given application the routing limitations will restrict utilization to 80% to 90% of the available elements.

The "front end" CALD tools may be the same for the LCA and FPGA. A technology-independent netlist may be generated by schematic capture or syn-

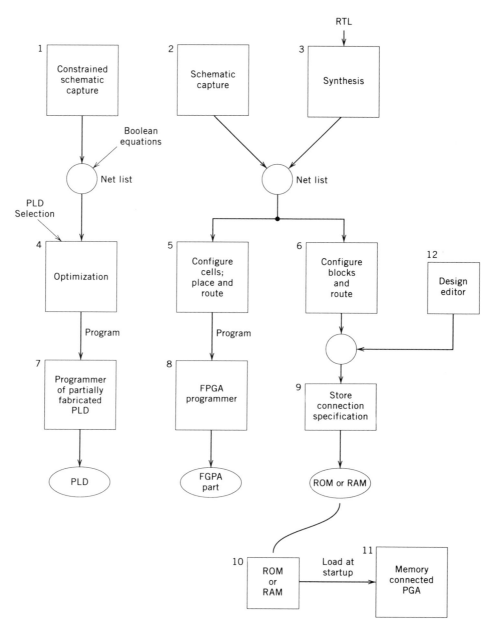

Figure 12.33 PLD and programmable gate array development.

thesis. The technology sensitive processes of partitioning the design among modules defining the cells and accomplishing place and route may be hidden from the designer. The LCA approach as defined in Fig. 12.33 may allow the designer to edit (block 12), the final layout, in an attempt to increase gate and memory element utilization. This requires more specialized knowledge on the part of the designer and is likely to lengthen the design time.

The performance column in Table 12.3 reflects the clock frequency at which final parts may operate. The programmable devices are typically at a performance disadvantage with respect not only to full custom but to standard cells and gate arrays as well. The notation in the table was devised to permit relative comparison. Linear scales are not implied.

12.10 SUMMARY

The goal of this section has been to identify the primary issues of interest to the user of any set of computer-aided logic design tools. For the potential tool developer, the chapter should provide an introduction and incentive for further study. For this reason the reference list for Chapter 12 is more extensive than that for the other chapters. The selection of topics was not biased toward any set of tools, to a hardware description language or even to synthesis. The reader who now sees synthesis from a hardware description language to the preferred alternative may refer back to the conclusion of the previous chapter for information on the available set of AHPL-based synthesis tools.

PROBLEMS

12.1 Write a simplified EDIF-like net list description of the sequential circuit given in Fig. 10.52. Regard the outputs of the two memory elements as primary outputs. Use instances of the 74LS136 (quad EXOR), 74LS74 (dual DFF), and 74LS04 (hex inverter). Unconnected pin lists for these parts are given in Fig. P12.1.

12.2 Write a simplified EDIF-like net list description of the full adder as given in Fig. 8.12. Replace the AND-OR network with three NAND gates. The 74LS136 (quad EXOR) is characterized in Fig. P12.1. Refer to Fig. 12.6 to deduce the pin list for the 74LS00.

12.3 An EDIF-like description of a simple combinational logic network is given below. Translate this description to a pin list of the form given in Fig. 12.6. On the 74LS02 VCC is pin 14 and ground is pin 7.

```
(EDIF a__prob3__sch
    (interface
        (define input port A)
        (define input port B)
        (define output port Z)
        (define input port C)
        (define input port D)
    )
```

(74LS74.prt U?)	(74LS136.prt U?)	(74LS04.prt U?)
1 U?A clock	1 U?A I0	1 U?A I
2 U?A D	2 U?A I1	2 U?B O
3 U?A $\overline{\text{clear}}$	3 U?A O	3 U?B I
4 VCC	4 U?B I0	4 VCC
5 U?B $\overline{\text{clear}}$	5 U?B I1	5 U?C I
6 U?B D	6 U?B O	6 U?C O
7 U?B clock	7 gnd	7 U?D I
8 U?B $\overline{\text{preset}}$	8 U?C O	8 U?D O
9 U?B Q	9 U?C I1	9 U?E I
10 U?B $\overline{\text{Q}}$	10 U?C I0	10 U?E O
11 gnd	11 U?D O	11 gnd
12 U?A $\overline{\text{Q}}$	12 U?D I0	12 U?F O
13 U?A Q	13 U?D I1	13 U?F I
14 U?A $\overline{\text{preset}}$	14 VCC	14 U?A O

Figure P12.1

```
(contents
    (instance X__74LS02__NET U1A)
    (instance X__74LS00__NET U2A)
    (instance X__74LS00__NET U2B)
    (joined
      A (U1A I0)
    )
    (joined
      (U2A I0) (U1A O)
    )
    (joined
      B (U1A I1)
    )
    (joined
      (U2A O) (U2B I0)
    )
    (joined
      Z (U2B O)
    )
    (joined
      (U2B I1) D
    )
    (joined
      C (U2A I1)
)))
```

12.4 An element list (suitable for the clock-mode simulation) of a sequential circuit is given below. This structural-level description could be the result of the application of the AHPL synthesis program to an AHPL description of a three-bit linear feedback shift register.

Element Output	Type	I0 (or D)	I1 (or clock)	I2 ...
1	DFF	4	clock	
2	DFF	1	clock	
3	DFF	2	clock	
4	EXOR	1	3	

(a) Assume the translator is told that the 74LS74 and 74LS136 will be used in the realization of this element-list. What additional information is necessary to allow the translation of this description to a pin list of the form given in Fig. 12.6? Which of this missing information is supplied arbitrarily by the translation program? Where must the translator look for the remaining data?

(b) Construct a pin list of a realization of the above element list.

(c) Translate the element list given above to a simplified EDIF-like description.

12.5 Consider the following AHPL fragment in which x is a primary input and Y a two flip-flop register. The setup time of the *falling-edge-triggered* memory element is 6 nsec, the clock-to-output delay 10 nsec, and the unit time 1 nsec.

ENDSEQUENCE
$$Y \leftarrow x \vee Y[1], x;$$

The initial value of Y is 1, 0 prior to application of the following input sequence. Construct a timing trace of $Y[0]$ and $Y[1]$ from 0–150 nsec.

clock	0	1	1	0	0	1	0	0	1	0
x	0	0	1	1	0	0	0	1	1	1
time	0	10	20	25	40	60	75	90	110	125

12.6 Prove that the three-value logic system $\{0, 1, U\}$, as defined in Section 12.4, is not a Boolean algebra.

12.7 Manually execute six periods of a clock-mode simulation of the partial AHPL description of the code translator developed in Example 11.2. Assume that the initial control state is Step 1 and $B[0:1]$ is initialized to U,U. Use the input sequence 101011. Assume $r = 0$ throughout.

12.8 Determine the outputs $f_\lambda, f_\beta,$ and f_γ of the combinational logic network given in Fig. 6.30, if the inputs are $ABCD = 110U$.

12.9 Determine the outputs $S_i, S_{i+1},$ and C_{i+2} of the carry skip adder network of Fig. 8.14
   (a) For inputs $C_i = U, X_i Y_i = 01,$ and $X_{i+1} Y_{i+1} = 10.$
   (b) For inputs $C_i = U, X_i Y_i = 11,$ and $X_{i+1} Y_{i+1} = 10.$
   (c) For inputs $C_i = U, X_i Y_i = 01,$ and $X_{i+1} Y_{i+1} = 11.$

12.10 Manually execute a three-value (0, 1, U) function-level simulation of the sequential comparator module described in Example 11.4. That is, execute the language description for 10 clock periods, with the machine initially in control state 1 and the values of all data memory elements initially U. The 10 consecutive sets of input values are listed below. List the clock-period-by-clock-period values of *COUNT*; *A*; *Z*; and *out* in the same format.

a	b	$X[0:7]$
0	0	00000000
0	0	00000001
1	0	00000010
1	1	00000100
1	1	00001000
1	1	00010000
1	1	00100000
1	1	01000000
1	0	10000000
1	0	00000011

12.11 Suppose the CONTROLRESET statement were removed from the sequential comparator description of Example 11.4. This will dictate that the simulation begin with U assigned initially to all control memory elements. Would it be possible to execute a function-level simulation under these conditions?

12.12 A simple inverter with input a and output z is simulated by a simulator that models both minimum and maximum delay values. In response to an input change, the output is U in the interval between minimum and maximum delays. For the inverter, the minimum delay is 5 nsec and the maximum 10 nsec. Construct the timing trace of z resulting from the application of the following input specification. The initial value of z is 1.

a	0	1	0
time	0	20	40

12.13   What, if any, modifications of the timing diagram of Fig. 12.21b are required to accommodate master-slave memory elements of the form given in Fig. 9.27b. Add a timing trace at a point in the flip-flop representing the output of the master element (Fig. 9.27c). Is the entire clock period less a fixed setup time (independent of clock frequency) available for signal propagation through the combinational logic?

12.14   Suppose the control unit of the DATAINTERSECTION module of Example 11.5 is implemented with one flip-flop per control state and that the entire module is to be realized on a 2-μ CMOS standard cell chip. Delay equations 12.2 and 12.3 apply. Assume that the gate input capacitance is 0.25 pF for all standard cells that will appear in the realization. Assume that the width of all metal lines is 2 λ and the average length of all leads fanning out from control flip-flop 1 (and from the gate realizing Step 1 and r) is 400 λ. Use a capacitance value of $0.3 \cdot 10^{-4}$ pF/μ^2.

(a)   Determine the 0→1 delay at the output of control flip-flop 1.

(b)   Determine the 0→1 delay at the output of the AND gate realizing *step1* $\wedge$ *r*.

(c)   On the basis of only the calculation in parts a and b, state an upper bound on the frequency of the clock of the DATAINTERSECTION module.

12.15   Use the data set out in Problem 12.14, including average lead lengths.

(a)   Determine the 1→0 delay at the output of an OR gate implementing one of the bits of **ROUTEBUS**.

(b)   By simply adding delay values, determine the delay from input to output of the combinational logic unit INC, invoked in Steps 2 and 3 of the DATAINTERSECTION module. Make use of the combinational logic unit description developed in Example 12.10.

12.16   Repeat Example 12.5, assuming that nothing is changed except that the dimensions of the metal lines conform to 1 μ technology.

12.17   Adapt the flow diagram of Fig. 12.22 to reflect schematic capture as the design input mechanism. Is clock-mode simulation still possible? Will it be less time-consuming than gate-level simulation? Would it be possible to make the clock-mode simulation event-driven? Explain. Accommodate a clock-mode simulation block in your diagram. Discuss any disadvantages of your just developed schematic-capture-based design methodology as contrasted to Fig. 12.22. Discuss any advantages.

12.18   Sometimes, false paths exist because the circuit will never reach a control state that will excite the path. Without recourse to a net list, describe a false worst-case delay path that exists in the AHPL description given below.

MODULE: FALSEPATH.
INPUTS: *C*[4].
BUSES: **OBUS**.

MEMORY: $A[4]$; $B[4]$.
OUTPUTS: $Z[4]$.

1   **OBUS** = ADD$[1:4](A;B)$;
    $A \leftarrow$ **OBUS**; $Z = B$.

2   **OBUS** = A;
    $B \leftarrow$ ADD$[1:4]($**OBUS**$;C)$;
    $\rightarrow (1)$.
END.

12.19 In the carry skip adder of Fig. 12.22, suppose that the output of gate 6 is chosen as point h on the false path. Does there exist a set of input values for which Equation 12.5 is satisfied? If so, determine that set of values.

12.20 Repeat Problem 12.19 with point h chosen at each of the outputs of gates 7, 9, 11, and 13.

12.21 Assume that the turn-on and turn-off delays of all gates in the sequential circuit of Fig. 10.1 are 10 nsec, the delays at the output of the memory elements are 15 nsec, and the setup and hold times are 5 nsec. Let the initial values of both memory elements be U and x be initially 0. Repeat the process of Example 12.7, given that the event queue after input scheduling is the following. Assume rising-edge-triggered memory elements.

clock	$0 \rightarrow 1$	T20
clock	$1 \rightarrow 0$	T40
clock	$0 \rightarrow 1$	T60
x	$0 \rightarrow 1$	T70
clock	$1 \rightarrow 0$	T80
clock	$0 \rightarrow 1$	T100
clock	$1 \rightarrow 0$	T120
x	$1 \rightarrow 0$	T125
clock	$0 \rightarrow 1$	T140
clock	$1 \rightarrow 0$	T160

12.22 Repeat problem 12.21 with one change in the input schedule. Let x: $1 \rightarrow 0$ occur at T118 rather than T125.

12.23 Replace the inverter symbols by their switch-level equivalents and partition the network of Fig. 11.27 as specified by Definition 12.2. Consider the component that includes the inverter whose output is **ROUTEBUS**$[0]$. Assign strength values to nodes and devices consistent with the six-value (0–5) system of Fig. 12.29. Suppose that initially $\overline{ROUTEBUS[0]} = step1 = b = 1$ and $a = s = \Phi_1 = \Phi_2 = 0$. Execute the switch network evaluation algorithm on just this component of the network until all nodes are initialized. Assuming that $\overline{ROUTEBUS[0]}$, $step1$, and s do not change, execute the algorithm on this component for $\Phi_1 \Phi_2 = 0\ 1$, then for $\Phi_1 \Phi_2 = 0\ 0$, and then for $\Phi_1 \Phi_2 = 1\ 0$.

12.24 Examples 12.9 and 12.10 only partially execute the switch network evaluation algorithm for two components of the network of Fig. 12.26. The following tasks call for executing the entire algorithm on all components of the network until all node values stabilize. Assign strength values to nodes and devices consistent with the six-value (0–5) system of Fig. 12.29.

(a) Initially $B_1 = B_2 = C_4 = 1$ and $\Phi_1 = \Phi_2 = 0$. Execute the algorithm until initial values have stabilized at all network nodes.

(b) Following the initialization of part (a) and while other input values remain unchanged, execute the algorithm for $\Phi_1 \Phi_2 = 0\ 1$, then for $\Phi_1 \Phi_2 = 0\ 0$, and then for $\Phi_1 \Phi_2 = 1\ 0$.

BIBLIOGRAPHY

1. Bouldin, D. *Mosis User Group Newsletter*. University of Tennessee, Knoxville, Tenn., July 1988, Issue 3.
2. Bryant, R. "A Switch Level Model and Simulator for MOS Digital Systems," *IEEE Trans. Computers*, Vol. C-33, 160–177 (Feb. 1984).
3. Hayes, J. "Pseudo-Boolean Logic Circuits," *IEEE Trans. Computers*, Vol. C-35, 602–612 (July 1986).
4. Bryant, R. "A Survey of Switch Level Algorithms," *IEEE Design & Test*, Vol. 4, 26–40 (August 1987).
5. Sunblad, R. and C. Svensson. "Fully Dynamic Switch Level Simulation of CMOS Circuits," *IEEE Trans. CAD*, Vol. CAD-6, pp. 282–289 (March 1987).
6. Chen, C. F., C.-Y. Lo, H. N. Nham, and P. Subramaniam. "The Second Generation MOTIS Mixed Mode Simulator," in *Proceedings of the 21st Design Automation Conference*. Albuquerque, N.M., June 1984, pp. 10–17.
7. Tsao, D. and C.-F. Chen. "A 'Fast Timing', Simulation for Digital MOS Circuits," in *Proceedings of IEEE International Conference on Computer-Aided Design*. Nov. 1985, pp. 185–187.
8. Szygenda, S. A. and E. W. Thompson. "Digital Logic Simulation in a Time-Based, Table-Driven Environment, Part I, Design Verification," *Computer*, **8:** 24–36 (March 1975).
9. Hwang, H., Y. H. Kim, and A. R. Newton. "An Accurate Delay Modeling Technique for Switch-Level Timing Verification," in *Proceedings of the 23rd Design Automation Conference*. IEEE, Las Vegas, Nev., June 1986, pp. 227–233.
10. Adler, D. "A Dynamically Directed Switch Model for MOS Logic Simulation," in *Proceedings 1988 Design Automation Conference*. IEEE, Anaheim, Calif., pp. 506–511.
11. Heinbuch, D. V. *The CMOS3 Cell Library*. Addison-Wesley, Reading, Mass., 1988.
12. *OrCAD/SDT III User Manual*. OrCAD Systems Corp., Hilsboro, Oregon, 1987.
13. *L-Edit User Manual*. Tanner Research Inc., Pasadena, Calif., 1989.
14. *Gatesim User Manual*. Tanner Research Inc., Pasadena, Calif., 1989.
15. Hill, F. J. and G. R. Peterson. *Digital Systems: Hardware Organization and Design*, 3rd ed. Wiley, New York, 1987.
16. *ACT 1 Family Gate Arrays Product Guide*. Actel Corp., Sunnyvale, Calif., 1989.
17. *The Programmable Gate Array Design Handbook*. Xilinx Corp., San Jose, Calif., 1988.
18. Lipsett, R., C. Schaefer, and C. Ussery. *VHDL: Hardware Description and Design*. Kluwer, Boston, Mass., 1989.

19. Navabi, Z., R. Swanson, and F. J. Hill. *HPSIM/HPCOM User Manual*. Dept. ECE, University of Arizona, Tucson, Az., revised 1990.

20. *Electronic Data Interchange Format Version 2.00*. ANSI/EIA-548-1988, Electronics Industries Assoc., Washington, D.C., 1988.

21. *IEEE Standard VHDL Language Reference Manual*. IEEE, New York, 1988, Std. 1076.

22. Chaney, T. and C. E. Molnar. "Anomalous Behavior of Synchronizer and Arbiter Circuits," *IEEE Trans. Computers*, **C-22**: 421, 422 (April 1973).

23. Couranz, G. R. and D. F. Wann. "Theoretical and Experimental Behavior of Synchronizers Operating in the Metastable Region," *IEEE Trans. Computers*, **C-24** (June 1975), pp. 604–616.

24. Liu, B. and N. C. Gallagher. "On the 'Metastable Region' of Flip-Flop Circuits," *Proc. IEEE*: 581–583 (April 1977).

25. Marino, L. R. "General Theory of Metastable Operation," *IEEE Trans. Computers*, **C-30**: 2, 107–115 (Feb. 1981).

26. Duley, J. R. and D. L. Dietmeyer. "A Digital System Design Language (DDL)," *IEEE Trans. Computers*, Vol. C-17, 850–861 (Sept. 1968).

27. Segal, R. B. *BDSYN User's Manual*. University of California, Berkeley, March 1990.

28. Siewiorek, D. "Introducing ISP," *Computer*, Vol. 7; No. 12, 39–44 (Dec. 1974).

29. Armstrong, J. R. *Chip-Level Modeling with VHDL*. Prentice-Hall, Englewood Cliffs, N.J., 1988.

30. Hines, J. "Where VHDL Fits Within the CAD Environment," in *24th ACM/IEEE Design Automation Conference Proceedings*. 1987.

31. Barton, D. "Behavioral Descriptions in VHDL," *VLSI Systems Design* (June 1988).

CHAPTER 13

INCOMPLETELY SPECIFIED SEQUENTIAL CIRCUITS

13.1 INTRODUCTION

Throughout Chapter 10, we assumed that the output and next state were specified for every possible combination of input and present state. Thus, no don't-care conditions arose due to the circuit specifications, except somewhat indirectly when the number of internal states required was not a power of 2. Although such don't-cares are very useful in simplifying excitation functions, they are more a result of the form of realization chosen than of the actual circuit specifications. Because of this special character, such don't-cares, due to extra internal states, are sometimes known as *incidental* don't-cares [3].

Also important are "don't cares" that result because certain combinations of inputs or sequences of inputs can never occur. Consider, for example, a clock-mode circuit with three input lines A, B and C, such that no more than one of the inputs may be 1 during a given clock period. It becomes even more interesting if this restriction holds and only certain sequences are possible. It might be stated that subsequent to each reset of the circuit each line will go to 1 for only one clock period. This would imply only the sequences of inputs ABC, ACB, BAC, BCA, CAB, and CBA.

Just as don't-cares offered the possibility of simple realizations for combinational circuits, it is reasonable to expect that don't-cares in state tables may make possible further simplification than might otherwise be the case. For combinational circuits, the designer in effect specifies the don't-cares as 1's or 0's in whatever manner provides the simplest realization, and the appropriate choice is generally quite obvious. Sequential circuits being generally more complex, however, the problems of assigning don't-cares are correspondingly much more difficult.

First, there are several types of don't-cares in sequential circuits. The most common is that described above, resulting from input sequences that cannot occur, in which case both the next state and output are unspecified. The justification is the same as for combinational don't-cares; if the event will never occur, then we don't care what the circuit would do in response to it. Second, we may have circuits that receive a continuous series of inputs, to which they respond with appropriate sequences of states, but the outputs are sampled only at specified times, not continuously. When the outputs are not being sampled, they may be unspecified even though the next states are specified. Third, we have situations in which a given input is the *last* event of a sequence in the sense that it will always be followed by a general reset signal, which can reset the circuit from any state. For a Mealy circuit, the output will then be specified, but the next state will not.

In the case of combinational don't-cares, there are only two ways to assign the don't cares (0 or 1), but for don't-cares in a state table, there are generally many choices. When two rows of a state table are identical in every column when *both* are specified, the obvious step is to assign the don't-cares in such a manner as to make the two rows identical, in which case they are equivalent by inspection. This is what was done in Example 11.1 to combine states q_3 and q_4 (see Fig. 11.12). Note that when a given column is unspecified in *both* rows, it may be left unspecified, resulting in don't-cares in the resultant transition table.

In Example 11.1, only two rows could be combined in this manner, and the resultant four-state circuit was obviously minimal. In most cases, however, there are a number of ways that the don't-cares might be assigned, and it may not be obvious which choice will lead to the minimal state table. If the number of choices is not too large, we might assign all don't-cares in all possible ways and minimize the resulting completely specified state tables (by the methods of Chapter 10) to see which is minimal.

■ **EXAMPLE 13.1**

A clock-mode sequential circuit has an input x and output z. Starting in a reset state, it stays in reset as long as $x = 0$ and the output is unspecified. An input of $x = 1$ will move the circuit from reset state R. This will be followed by a string of 0 inputs of arbitrary length, producing $z = 1$ outputs at the time of the second, fourth, sixth, etc., 0 inputs. When the circuit is not in state R, a 1 input may occur at any time and will return the circuit to state R with an output of $z = 0$. Develop a minimal state table.

SOLUTION   A state diagram corresponding to the above operation and the corresponding state table are shown in Fig. 13.1. There is only one unspecified entry, the output for $x = 0$ and state R.

At first glance, it might seem that the safe way to proceed would be to form two state tables by first specifying the single don't care as a 0 and then as 1. These tables could then be minimized, finally choosing the one with the fewest states. This results in the two completely specified state tables of Fig. 13.2. For Fig. 13.2*a*, we see that only states R and 1 have identical outputs, but their equivalence would imply the equivalence of R and 2. Thus, the table of Fig. 13.2*a* is minimal, and by similar argument, the table of Fig. 13.2*b* is also minimal.

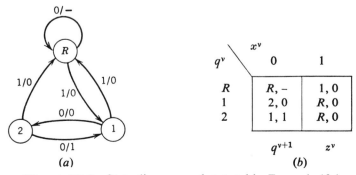

Figure 13.1   State diagram and state table, Example 13.1.

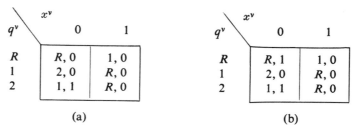

Figure 13.2   Equivalent completely specified state tables, Example 13.1.

It appears that Example 13.1 cannot be realized with fewer than three states. But now consider the two-state circuit specified by the state diagram and table of Fig. 13.3, in which states A and B are equivalent to states 1 and 2 of Fig. 13.1, and the function of state R is "shared" between them. First, consider the circuits in either states 1 or 2, or states A or B, with a string of 0 inputs applied. Both circuits will generate as outputs an alternating string of 0's and 1's. For a string of 1's applied, both circuits will produce a string of 0's. Next, consider the circuit in state 2 or state B, with a 1 applied, followed by a string of 0's. The two-state circuit will produce a 0 followed by an alternating string of 0's and 1's. This behavior satisfies the specifications for the three-state circuit since response to a string of 0's beginning in state R is a string of don't care outputs. A similar analysis applies to a 1 followed by a string of 0's from states 1 or A. Thus, the circuits described by these two state diagrams are equivalent in the sense of generating the same required response to any sequence of inputs.   ■

The above example illustrates that the arbitrary specification of don't-care entries in rows of a state table that are not identical may prevent minimization of the state table. As in Example 11.1, such simple and obvious specification may result in a table that is obviously minimal or so close to minimal that more complex and time-consuming efforts at minimization may not be justified. However, if absolute minimization is deemed important, more powerful techniques must be developed. For this purpose, we will need some new definitions and theorems.

13.2   COMPATIBILITY

We have seen that in some cases incompletely specified states can be combined to reduce the total number of states in a circuit. However, we cannot say that such states are equivalent because the formal definition of equivalence requires that both outputs and next states be defined for all inputs. Instead, we will say that incompletely specified states that can be combined are *compatible* with one

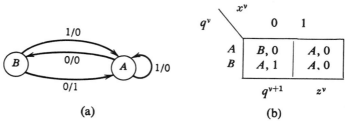

Figure 13.3   Two-state equivalent, Example 13.1.

another. When two states are identical wherever both are specified, as in the partial state table of Fig. 13.4, they can be made equivalent by assigning the unspecified entries the same values in both states.

States that are identical whenever both are specified are thus clearly compatible, but in the last section we saw a case in which states that were not identical could be combined. In such cases, we cannot simply specify the states so as to make them equivalent. To define compatibility, we must go back to the basic concept of states having the same function in a circuit if they produce the same outputs in response to the same inputs. Since circuit behavior is not specified for all possible input sequences, we cannot require identical outputs for all possible sequences, as was the case for equivalent states.

Definition 13.1 Let a sequence of inputs be applied to an incompletely specified circuit $S*$ in initial state q. The sequence is said to be *applicable* to q if no unspecified next-state entries are encountered, except possibly as a result of the last input of the sequence.

Definition 13.2 Two states p and q of a circuit S are said to be *compatible* if and only if

$$\lambda_S[\delta(p, X_1 X_2 \cdots X_K), X_{K+1}] = \lambda_S[\delta(q, X_1 X_2 \cdots X_K), X_{K+1}] \qquad (13.1)$$

whenever both of these outputs are specified, for every input sequence $X_1 X_2 \cdots X_{K+1}$ applicable to both states. The expression $\delta(q, X_1 X_2 \cdots X_K)$ denotes the state of the circuit after the sequence of inputs $X_1 X_2 \cdots X_K$ is applied to the circuit, originally in state q.

Suppose the output sequences resulting from some applicable input sequence are computed for a circuit initially in compatible states p and q, leaving blank those outputs that are not specified. Each of these output sequences can be computed only as far as the next states are specified. Definition 13.2 says that at no point will specified outputs in these two sequences differ.

Theorem 10.1 was used as the basis for eliminating pairs of states that cannot be placed in the same equivalence classes. Theorem 13.1 will be used similarly to eliminate pairs of states that are not compatible. The proofs of most theorems in this chapter will be omitted. They are similar to but in general more tedious than the proofs presented in Chapter 10.

THEOREM 13.1 If two states q_i and q_j of circuit S are compatible, then the following conditions must be satisfied for every input X:

q^v	A	B	C	K
3	2, –	3, –	4, 0	1, –
4	2, 1	–, –	–, 0	1, 0

$$q^{v+1}, z^v$$

Figure 13.4

*Strictly speaking, once constructed, any deterministic (nonrandom) circuit is completely specified. In Definition 13.1, we refer to the analytic determination of outputs and next states from the state table.

1. $\lambda(q_i, X) = \lambda(q_j, X)$ whenever both are specified.

2. $\delta(q_i, X)$ and $\delta(q_j, X)$ are compatible whenever both are specified.

The processing of the implication table proceeds in exactly the same fashion as for the completely specified case. We first cross out any squares corresponding to pairs of states for which the *specified* outputs differ, that is, for which condition 1 of Theorem 13.1 is not satisfied. In the remaining squares, we enter the numbers of pairs of states whose compatibility is implied. For example, if $\delta(q_i, X) = q_2$ and $\delta(q_j, X) = q_3$ in a certain state table, we will enter 23 in square i, j of the implication table. We enter checkmarks in squares corresponding to pairs of states that imply only their own compatibility. Once the initial implication table has been obtained, the elimination of further incompatible pairs of states by successive "passes" through the table proceeds in exactly the same manner as the elimination of nonequivalent states in completely specified circuits.

■ **EXAMPLE 13.2**

A clock-mode sequence checker has four input lines A, B, C, and K, and two output lines Z_1 and Z_2. No two of the inputs will be 1 during the same clock period. At any time, the circuit may be in the *check* or *noncheck* mode. In the noncheck mode, these 1's on these input lines may arrive in any order, but only $K = 1$ will move the circuit into the check mode. Following this K input, the 1 inputs on lines A, B, C, will occur exactly once in any order followed by another K input. At the time $K = 1$ following the sequence on lines A, B, C, an output will appear for one period on line Z_2, if the inputs arrived in the order A, B, C. The output Z_1 will be 1 for one clock period, if the order was C, B, A. If the inputs on A, B, C, arrive in any other order, there will be no output. The same $K = 1$ period that can generate an output will return the circuit to the noncheck mode, requiring another K to get it back to the check mode. The circuit outputs will be observed only at the time of a K input. Therefore, the output must be 00 when a K input drives the circuit to the check mode. The circuit should not change state unless one of the four input lines is 1.

SOLUTION With four separate input lines, we would normally expect 16 arrows leaving each state in the state diagram and 16 next-state columns in the state table. The restriction that the circuit will not change state unless some input is 1 and that no two inputs will be 1 simultaneously reduces the number of arrows in the state diagram to 1 for each input line. A state diagram and state table are given in Fig. 13.5. Because there is not a column in the state table for each input combination, care must be taken in translating the final state table to Karnaugh maps obtain realization.

State q_1 constitutes the noncheck mode, from which a K input will take the circuit to q_3, the initial state of the check mode. Since only two of the A, B, or C inputs are needed to identify a sequence, the second of these inputs leads to one of three possible states q_7, q_8, q_9, corresponding to the three output possibilities. The output-producing K input sends the circuit back to q_1. It is seen that there are many don't-cares, because of the restriction to exactly one A input, one B input, and one C input per sequence and the fact that the output is specified only for the input K.

None of the missing columns in the state table of Fig. 13.5*b* will contribute to the incompatibility of states. Because the next states would be the same as the present states in a column corresponding to all 0 inputs, next states will always be

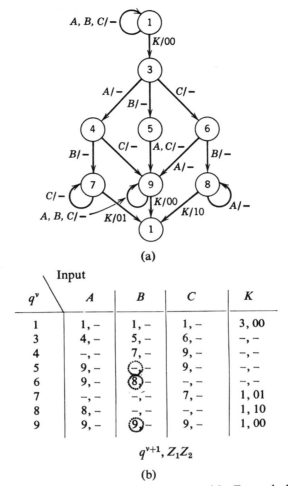

(a)

Input				
q^v	A	B	C	K
1	1, –	1, –	1, –	3, 00
3	4, –	5, –	6, –	–, –
4	–, –	7, –	9, –	–, –
5	9, –	⊙–	9, –	–, –
6	9, –	⑧–	–, –	–, –
7	–, –	–, –	7, –	1, 01
8	8, –	–, –	–, –	1, 10
9	9, –	⑨–	9, –	1, 00

$$q^{v+1}, Z_1 Z_2$$

(b)

Figure 13.5 State diagram and state table, Example 13.2.

compatible in that column, if present states are compatible. In all the other omitted columns, all next states and outputs are "don't cares."

The initial and final implication tables are shown in Fig. 13.6. The squares that contain checkmarks correspond to pairs whose compatibility is not dependent on the compatibility of any other states. The remaining squares, not crossed out in the final implication table, correspond to sets that are compatible because the implied states satisfy condition 2 of Theorem 13.1. ∎

The reader will recall that, in Chapter 10, equivalent pairs determined by the implication tables were combined to form equivalence classes. A similar process can be applied to compatible pairs. Although not formally proved, it should be clear that *pairs of states corresponding to squares not crossed out in the final implication table are compatible pairs.*

Definition 13.3 A set of states S_i of a circuit S will be said to form a *compatibility class* if every pair of states in S_i is compatible.

Definition 13.4 A *maximal compatible* is a compatibility class, which will not remain a compatibility class if any state not already in the class is added to the class. A single state that is compatible with no other state is a maximal compatible.

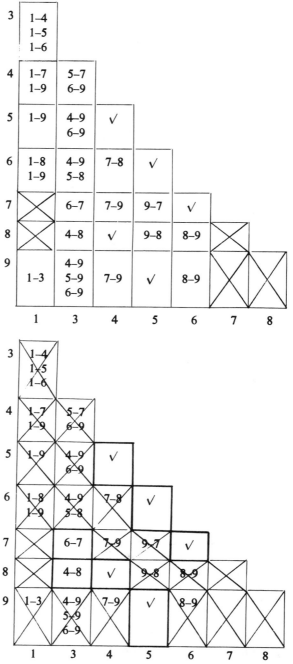

Figure 13.6 Initial and final implication tables, Example 13.2.

■ EXAMPLE 13.2 (continued)

Determine the set of all maximal compatibles for the sequence checker.

SOLUTION   The process is exactly that utilized to determine equivalence classes, as shown in Fig. 13.7. There are no compatibles indicated in columns 7 and 8. In column 6, we find that q_6 and q_7 form a compatible pair. Next we find q_5 compatible with q_6 and q_9, but here we note a situation we have not seen before. For com-

```
8  –
7  –
6  (6, 7)
5  (6, 7) (5, 6) (5, 9)
4  (6, 7) (5, 6) (5, 9) (4, 5) (4, 8)
3  (6, 7) (5, 6) (5, 9) (4, 5) (4, 8) (3, 7) (3, 8)
1  (6, 7) (5, 6) (5, 9) (4, 5) (4, 8) (3, 7) (3, 8) (1)
```

Figure 13.7 Determination of maximal compatibles, Example 13.2.

pletely specified circuits, if a state is equivalent to two other distinct states, then these states are equivalent to each other. But here we find that q_6 and q_9 are not compatible, so instead of having a class (5, 6, 9), we have two compatible pairs, (5, 6) and (5, 9). The completion of the process follows in the same manner. ∎

In this case, the maximal compatibles all happen to be pairs, but that is a peculiarity of this problem and not the general case. What is significant is the fact that several states appear in more than one maximal compatible, as was the case with (5, 6) and (5, 9) as discussed above. It is easy to see why these pairs are compatible while states 6 and 9 are not from an inspection of the circled next-state entries in column B of Fig. 13.5b. Note that the next state for state 5 in this column is unspecified so that (at least with respect to column B) (5, 6) and (5, 9) are compatible regardless of the next states specified for 6 and 9. However, the next states for 6 and 9 are 8 and 9, respectively, and 8 and 9 are, in turn, incompatible since their outputs differ in column K. Therefore, 6 and 9 are not compatible. Similarly, state 3 is compatible with states 7 and 8, while these two states are not compatible with each other.

The above are two examples of a more formal observation that compatibility is *not* a transitive relation. Thus, compatible states cannot be grouped neatly into disjoint classes, and the formal basis of the procedures of Chapter 10 slips away.

This fact accounts for such puzzling situations as we saw in Example 13.1 and also makes the Huffman partition method inapplicable to incompletely specified circuits.

Nevertheless, our basic simplification problem is still the same: to find a circuit with fewer states that will produce the same output sequence. Since the simplified circuit need have the same output as the original circuit only for those outputs specified, slightly different criteria than specified by Definition 10.1 must be satisfied by the states of the two circuits.

Definition 13.5 A state p_j of a state table T is said to *cover* a state q_i of a state table S, written $p_j \geq q_i$, if for any input sequence applicable to q_i and applied to both T and S initially in states p_j and q_i, respectively, the output sequences are identical whenever the output of S is *specified*.

The distinction between covering and compatibility is illustrated in Fig. 13.8. Note that, at least for the single input sequence 0112031, q_1 and q_2 are compatible. In addition, the state p covers *both* q_1 and q_2 for the input sequence shown, and (q_1, p) and (q_2, p) are compatible pairs. Neither q_1 nor q_2, however, covers p. Thus, we observe that requiring a state r to cover a state s is more restrictive than requiring the two states to be compatible.

We further note that q_2 and q_3 are compatible, while q_1 and q_3 are not. Thus, as we have observed from Example 11.1, compatibility is not a transitive property.

Input Sequence	0	1	1	2	0	3	1
Initial States	Output Sequences						
q_1	X	1	X	X	0	1	1
q_2	0	1	1	X	0	X	X
q_3	X	1	X	1	0	0	0
p	0	1	1	0	0	1	1

Figure 13.8

Definition 13.6   A state table T is said to cover a state table S if for every state q_i in S there exists some state p_j in T that covers q_i.

A circuit T that covers S necessarily performs the function of S. Given S, then, our goal is to find such a circuit T having a minimum number of states. Toward this goal, we present the following theorems.

THEOREM 13.2   If an internal state p of T covers both internal states q_j and q_i of S, then states q_i and q_j must be compatible.

PROOF   If p covers q_i and q_j, then for any applicable input sequence $X_1 X_2 \cdots X_{K+1}$ for which

$$\lambda_S[\delta(q_j, X_1 \cdots X_K), X_{K+1}] \qquad \text{and} \qquad \lambda_S[\delta(q_i, X_1 \cdots X_2), X_{K+1}]$$

are both specified, we must have

$$\lambda_T[\delta(p, X_1 \cdots X_K), X_{K+1}] = \lambda_S[\delta(q_j, X_1 \cdots X_K), X_{K+1}] \qquad (13.2)$$

and

$$\lambda_T[\delta(p, X_1 \cdots X_K), X_{K+1}] = \lambda_S[\delta(q_i, X_1 \cdots X_K), X_{K+1}] \qquad (13.3)$$

Therefore,

$$\lambda_S[\delta(q_j, X_1 \cdots X_K), X_{K+1}] = \lambda_S[\delta(q_i, X_1 \cdots X_K), X_{K+1}] \qquad (13.4)$$

and q_i and q_j are compatible.

The following is an immediate corollary of Theorem 13.2 and the definition of a compatibility class, Definition 13.3.

COROLLARY 13.3   If a state p_i of T covers a set of states S_i in S, these states must form a compatibility class.

For completely specified machines, it was only necessary to include each state of S in one equivalence class. A machine T with a state corresponding to each such equivalence class would then be equivalent to S. In the incompletely specified case, however, it may be necessary to include a state in more than one compatibility class.

Definition 13.7   A collection of compatibility classes is said to be *closed* if, for any class $\{q_1, q_2, ..., q_m\}$ in the collection and every input X_i, all specified next states, $\delta(q_1, X_i)$, $\delta(q_2, X_i)$, ..., $\delta(q_m, X_i)$, fall into a single class in the collection.

We are now in a position to state the critical theorem of this section.

THEOREM 13.4   Let the n states of an incompletely specified sequential circuit S form a collection of m compatibility classes such that each of the n states is a member of at least one of the m compatibility classes. Then the circuit S may be covered by a circuit T, having exactly m states, $p_1, p_2, ..., p_m$, such that each compatibility class of S is covered by one of the states of T, if and only if the collection of m compatibility classes of S is closed.

From this theorem, it can be seen that our basic problem is to find a minimal closed collection of compatibility classes that includes all the states of the original circuit. It can be shown that the set of all maximal compatibles is a closed collection. Therefore, this set of maximal compatibles could be used to form the basis for an equivalent circuit. However, the resulting circuit will seldom be minimal, since there will usually be considerable overlap. Indeed, the number of maximal compatibles will often be greater than the number of states in the original circuit.

It is obvious that we will wish to delete redundant classes whenever possible. However, we must be careful not to violate the closure conditions by removing a class that is necessary to satisfy the implications of a class that is retained. Since any subset of a compatibility class is also a compatibility class, we may delete sets from classes in order to reduce closure requirements.

To facilitate discussion, we will refer to a closed collection that contains each state of a machine S in at least one class as a *cover collection*. Once we find a cover collection, the construction of a machine covering S is straightforward.

■ **EXAMPLE 13.2 (continued)**

All the compatibility classes determined in Fig. 13.7 are listed in Fig. 13.9 together with any implied classes. We first note that state q_9 appears in only one compatibility class, so this class $(5, 9)$, along with the state that did not combine at all, q_1, will be included in the cover collection. The selection of these classes leaves states q_3, q_4, q_6, q_7, and q_8 to be covered. Four of these states can be covered by including classes $(4, 8)$ and $(6, 7)$ in the cover collection, which can be done without implying the inclusion of any other classes. It remains to cover state 3. This can be accomplished by including any of the three classes $(3, 7)$, $(3, 8)$, and (3) in the cover collection. We note that $(3, 7)$ implies $(6, 7)$ and $(3, 8)$ implies $(4, 8)$. Both of these implied classes have already been included, so any of the following can be used to form the final cover collection:

$$
\begin{array}{lllll}
(1) & (5, 9) & (4, 8) & (6, 7) & (3, 7) \\
(1) & (5, 9) & (4, 8) & (6, 7) & (3, 8) \\
(1) & (5, 9) & (4, 8) & (6, 7) & (3) \qquad (13.5)
\end{array}
$$

The third of these closed collections is probably preferable, since it will result in more don't-cares in the final state table.

Once a closed collection of compatibility classes has been obtained, the output entries in the table are specified as required to satisfy Definition 13.5. That is,

Comp. Classes	(6, 7)	(5, 6)	(5, 9)	(4, 5)	(4, 8)	(3, 7)	(3, 8)	(1)
Impl. Classes	√	√	√	√	√	(6, 7)	(4, 8)	√

Figure 13.9 Compatibility classes and implied classes, Example 13.2.

for each state p_i of T and each input, the output must agree with any outputs that are specified for the states covered by p_i. Theorem 13.4 assures us that there will be no conflicts.

Similarly, each next-state entry for present state p_i must be the state covering the class that includes all the next-state entries for the states covered by p_i. A minimal state table covering our machine is given in Fig. 13.10. As shown, states a, b, c, d, and e are assigned to cover classes (1), (3), (4, 8), (5, 9), and (6, 7), respectively. The outputs and next-state entries are obtained in much the same way as for completely specified circuits. Consider the circled next-state entry for present state d and input B. In Fig. 13.5, $\delta(5, B)$ is unspecified while $\delta(9, B) = 9$. Since state 9 is covered by class (5, 9), we enter the corresponding state d as $\delta(d, B)$. The remaining entries in Fig. 13.10 may be obtained in a similar manner. ∎

Comp. Class	q^v	Inputs			
		A	B	C	K
(1)	a	$a, -$	$a, -$	$a, -$	$b, 00$
(3)	b	$c, -$	$d, -$	$e, -$	$-, -$
(4, 8)	c	$c, -$	$e, -$	$d, -$	$a, 10$
(5, 9)	d	$d, -$	$ⓓ, -$	$d, -$	$a, 00$
(6, 7)	e	$d, -$	$c, -$	$e, -$	$a, 01$

Figure 13.10 Final minimal state table, Example 13.2.

Let us next apply our modified minimization techniques to the three-state table of Example 13.1 to see that they lead to the two-state cover circuit that we determined by trial-and-error methods.

■ **EXAMPLE 13.1 (continued)**

Minimize the three-state table of Fig. 13.1b (repeated in Fig. 13.11a) by obtaining a closed collection of compatibility classes.

SOLUTION The simple implication table of Fig. 13.11b leads to the list of maximal compatibles and implied classes of Fig. 13.11c.

As might be expected since they form a collection of maximal compatibles, the two classes provide a cover collection. The two classes imply each other so that both must be used to satisfy closure. Thus, it is the closure requirement that prevents us from arbitrarily combining q_R with one or the other of q_1 or q_2, as was originally suggested. It may seem strange compatible classes that overlap can be used at the same time, but a comparison of the final state table of Fig. 13.11d with the original (Fig. 13.1b) shows that q_A satisfies all specifications of q_1 and q_R, and q_B does the same for q_2 and q_R. ∎

In both the above examples, the closure problems were trivial, and the choice of closed covers was fairly obvious. In some cases where there are a large number of interacting closure requirements, it may be desirable to reduce the size of some of the compatibility classes. It is apparent that any subset of a maximal compatible is also a compatibility class, and deleting one or more states from a maximal

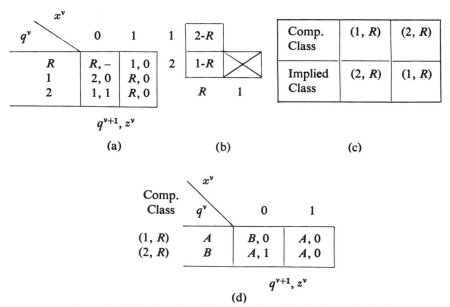

Figure 13.11 Minimization of state table of Example 13.1.

compatible may reduce the closure requirements. One such situation is illustrated by the following example.

■ **EXAMPLE 13.3**

Find a minimal state table equivalent to the state table of Fig. 13.12a.

SOLUTION The final version of the implication table is shown in Fig. 13.12b and the determination of maximal compatibles in Fig. 13.12c.

The maximal compatibles are listed in Fig. 13.12d, with the implied pairs. The reader may easily verify that the smallest closed cover obtainable from this listing is the set

$$[(abc), (ae), (bd), (de)] \tag{13.6}$$

If we delete state c from the class (abc), we remove the implication of class (ae), thus obtaining the three-state closed cover

$$[(ab), (bc), (de)] \tag{13.7}$$

Similarly, deleting state b from (abc) leads to the three-state cover

$$[(ac), (ae), (bd)] \tag{13.8}$$

The minimal state table corresponding to the cover of Equation 13.7 is shown in Fig. 13.12e. In setting up this table, we find that the overlapping classes create a situation we have not seen before. Consider the next-state entries for state q_1, representing the class (ab). In the first three columns of the original state table, the next-state entries are (bc), (de), and (e), corresponding to states q_2, q_3, and q_3, respectively. In the fourth column, the next-state entry is (b), which appears in both q_1 and q_2. Thus, the next-state entry can be either q_1 or q_2, as indicated by the entry 1/2. Such entries may be considered partial don't-cares and usually provide extra flexibility in design. The other optional entries in the table arise in the same manner. ■

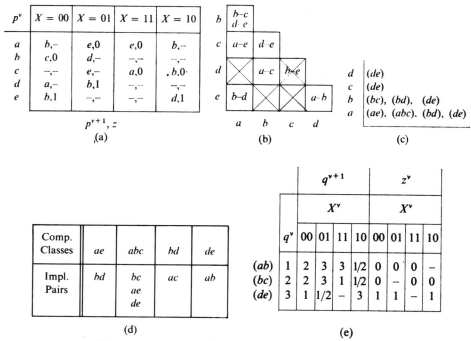

Figure 13.12 Minimizing state table, Example 13.3.

13.3 REALIZATION OF SEQUENTIAL CIRCUITS WITH MANY "DON'T CARES"

Three state variables will be required to accommodate the five states in the final state table of Fig. 13.10. There are four inputs. Thus, the expressions for the outputs and next values of state variables will be Boolean expressions in seven variables. Thus far we have not considered a seven-variable Karnaugh map and have generally avoided six-variable maps. Indeed, we have made a case that logic optimization programs should be applied when the number of variables exceeds six. To use such a program, one must first compile truth tables for the functions to be realized or alternatively (for Espresso) "not necessarily minimal" Boolean expressions for the functions, together with expressions for the "don't care" conditions.

If minimal expressions are to be obtained by whatever means, "don't care" conditions must be interpreted properly. To do so requires state assignment and consideration of the relationship between column-compressed state tables like Fig. 13.10 and the transition tables needed to obtain a realization. As we do this for Example 13.2, we shall see that Boolean expressions can be obtained for this "don't-care" rich example by using a slight extension of the Karnaugh map approach. The effort expended is probably less than would be required to prepare the input for a logic optimization program.

■ EXAMPLE 13.4

Obtain a minimal-cost D flip-flop realization of the state table of Fig. 13.10.

SOLUTION Without actually identifying partitions, our experience with partition-based state assignment in Chapter 10 suggests that using one state variable to distinguish the noncheck mode (state a) from the check mode (all other states) may lead to a near-optimal realization. We make that assignment, resulting in the transition table of Fig. 13.13. "Don't care" conditions in explicit columns of this table are left blank. The column corresponding to all zero inputs, which was omitted from the state table, because it did not impact on compatibility, has reappeared. It must be included in the transition table to assure that no state change will take place in the absence of a 1 input. The all "don't care" columns are omitted, because they may or may not be included in selected cubes at will.

Two rules may be stated for interpreting K-maps applicable to the very special case in which no two inputs can be 1 simultaneously.

(1) In the formation of cubes, each column (except the *KCBA* = 0000 column) may be lumped with all "don't care" columns to cover half the map. If the all inputs = 0 column of the map is 0 in the rows to be included in the cube, cubes are formed from individual columns. (2) The all inputs = 0 column should be included in all cubes formed by partially covering rows for which this column is 1. The column coverage of these cubes is limited by ANDing the complements of all variables corresponding to columns with 0's in these rows.

The values for the function y_1^{v+1} were extracted from the transition table to form the K-map in Fig. 13.14. Consider first the group partially covering four rows of the map. It appears to be an incomplete cube covering only three columns. Actually, it covers the all "don't care" column $KCBA = 0101$ as well. According to rule 2, we AND the complements of the variables in the uncovered columns, leading to the product $\bar{K}\bar{B}y_1$. Including the remaining four cubes leads to Equation 13.9. The last two products are formed from individual columns.

$$y_1^{v+1} = \bar{K}\bar{B}y_1 + \bar{K}y_1y_0 + By_2\bar{y}_1 + Cy_2\bar{y}_1 \tag{13.9}$$

The expressions for the next values of the remaining two state variables are similarly determined to form Equations 13.10 and 13.11.

$$y_0^{v+1} = \bar{K}\bar{B}y_0 + Ay_1 + By_1 + Ay_2\bar{y}_0 + By_2\bar{y}_0 \tag{13.10}$$

$$y_2^{v+1} = K\bar{y}_2 + \bar{K}y_2 \tag{13.11}$$

Translating the output values from the state table yields the two-function Karnaugh map of Fig. 13.15. Because all but one column contains all "don't care" values, the output expressions are easily found to be

$$Z_2 = \bar{y}_1y_0 \quad \text{and} \quad Z_1 = y_1\bar{y}_0$$

KCBA $y_1y_2y_0$		0000	0001	0010	0100	1000
a	000	000	000	000	000	100
	001					
	011					
	010					
b	100	100	101	111	110	
c	101	101	101	110	111	000
d	111	111	111	111	111	000
e	110	110	111	101	110	000

$$y_2^{v+1}y_1^{v+1}y_0^{v+1}$$

Figure 13.13 Abbreviated transition table for the sequence checker.

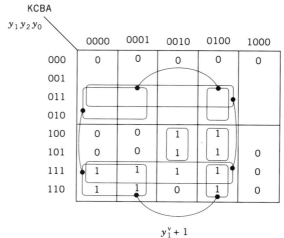

Figure 13.14   Karnaugh map for y_1^{v+1}.

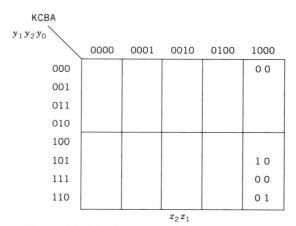

Figure 13.15   Output map for Example 13.4.

13.4 CONCLUSIONS

In the predecessor of this book, *Introduction to Switching Theory and Logical Design*, pulse-mode sequential circuits were introduced in advance of this chapter on incompletely specified sequential circuits. Inputs and outputs were divided into two categories: level and pulse. In a sense, a clock-mode sequential circuit is a special case of a pulse-mode circuit with the clock considered a pulse input. A pulse-mode circuit may have more than one pulse input line. Multiple-pulse lines are difficult to manage when the realization is to be a VLSI chip. For this reason, we have deleted the discussion of pulse mode from the present volume. Pulse-mode sequential circuits are almost always incompletely specified, as are level-mode circuits to be introduced in the next chapter.

PROBLEMS

13.1   A single input line *X*, on which no more than two consecutive 1 or 0 bits may occur, is to serve as the input to two distinct sequential circuits. Both circuits are synchronized by the same clock.

(a) Determine the incompletely specified state table of a circuit with a single output Z_1, such that $Z_1^{v+2} = X^v X^{v+1} + X^{v+2}$ (logical-OR).

(b) Determine the incompletely specified state table of a circuit with a single output Z_2, such that $Z_2^{v+2} = (X^v + X^{v+1}) \cdot X^{v+2}$. Do not use overlapping notation for the states of (a) and (b).

(c) Using the implication table, show that three-state covers can be obtained for each of the circuits of parts (a) and (b).

(d) Obtain the state table of a three-state circuit with four outputs, 00, 01, 10, and 11, that will simultaneously accomplish the function of both circuits.

13.2 (a) Determine the set of all compatible pairs of states of the incompletely specified sequential circuit given in Fig. P13.2.

(b) Determine the set of all maximal compatibles.

(c) Verify that compatibility of states is not an equivalance relation for this circuit.

13.3 A *strongly connected* sequential circuit S is a circuit such that, for every pair of states p and q of S, there exists a sequence of inputs $X_1 X_2 \cdots X_K$ such that $\delta(q, X_1 X_2 \cdots X_K) = p$.

Although the result may not serve a useful function, form a circuit simply by writing the two state tables of Problem 13.1 (a) and (b) as a single table. The outputs and initial states of each row of the new table remain the same as in the original tables.

(a) Show that the circuit formed as above is not strongly connected.

(b) Using the Paull–Unger implication table, obtain a five-state equivalent of the above circuit, which is strongly connected.

13.4 Determine a minimal state equivalent of the sequential circuit described by the flow table in Fig. P13.4.

	q^{v+1}, Z		
q^v	$X = 0$	$X = 1$	$X = 2$
q_1	$q_3, 0$	$q_5, 1$	
q_2	$q_3, 0$	$q_5,$	
q_3	$q_2, 0$	$q_3, 0$	$q_1,$
q_4	$q_2,$	$q_3,$	$q_5,$
q_5		$q_5, 0$	$q_1,$

Figure P13.2

	q^{v+1}, Z			
q^v	$X = 0$	$X = 1$	$X = 2$	$X = 3$
A	$B,$	$C, 0$		$D,$
B		$E, 0$		
C	$D,$	$F,$	$C,$	
D	$E,$	$C, 0$		$A,$
E		$F, 0$		
F		$, 1$	$D,$	$B,$

Figure P13.4

13.5   Determine a minimal state table for a circuit with a single input and a single output that is only sampled once every five clock times. When sampled, the output must indicate whether three or more of the last five input bits have been 1 (including the current bit). Also provided are a clock and synchronizing input that resets the circuit to an initial state following each sampling instant. The latter need not be shown in the state table.

13.6   Determine the state table of a circuit that will accomplish the same function as the circuit of Problem 10.7. The only change is that input X_2 will be 1 only for the one-clock pulse immediately preceding the input of each coded character. The state table is now incompletely specified. Show that 14 states are still required in the minimal circuit.

13.7   One information input line X_1, a clock, and the outputs of a modulo-4 counter serve as the inputs to a sequential circuit. Binary-coded decimal (BCD) numbers appear sequentially in four-bit characters (lowest-order bit first) on line X_1, synchronized with the output of the counter. That is, $c_2 c_1 = 00$ when the lowest-order bit arrives, and $c_2 c_1 = 01, 10$, and 11, respectively, as the next three bits arrive. Derive the incompletely specified state table for a circuit that will translate the BCD code to excess-3 code with a delay of three clock pulses.

13.8   Obtain a minimal state table of a synchronous sequential circuit that will perform the translation shown in Fig. P13.8 with only a two-period delay. The process may be assumed to be synchronized by a reset signal not described in the state table. As shown, only the first two bits of the output character are sampled, so the third is redundant. Only the four valid characters of Fig. P13.8 will occur as inputs.

13.9   Determine the maximal compatibles of the circuit described by the state table in Fig. P13.9. Determine a four-state cover of this circuit.

13.10   Determine a minimal covering collection for the sequential circuit partially described by the implication table of Fig. P13.10.

13.11   Repeat 13.10 for the implication table of Fig. P13.11.

13.12   A clocked detector circuit is connected to a controller, as shown in Fig. P13.12. Only when $x_1 = 1$ are the outputs z_1, z_2 gated to the controller. x_1 and x_2 are level signals changing only between clock pulses. When x_1 goes to 1 for two successive clock periods, then x_2 for the next four clock periods will represent a decimal digit in the Gray code of Fig. P13.12b, bit b_0 first. At the time of the fourth bit $(b_3), z_1 = 1$ if the digit received is even, and $z_2 = 1$ if it was odd. At all other times that $x_1 = 1$, $z_1 z_2 = 00$. If x_1 remains 1 for the two clock periods following receipt of a digit, another digit will follow, and the process will repeat until x_1 goes to 0, which may

X^{v+2}	X^{v+1}	X^v	Z^{v+4}	Z^{v+3}	Z^{v+2}
0	0	1	–	1	0
0	1	0	–	1	1
1	0	0	–	0	1
1	1	1	–	0	0

Figure P13.8

q^v	q^{v+1}, Z						
	$X = 1$	$X = 2$	$X = 3$	$X = 4$	$X = 5$	$X = 6$	$X = 7$
q_1	$q_1,0$	$q_3,1$	$q_5,$		$q_5,$	$q_5,1$	$,$
q_2	$q_1,0$	$q_3,1$	$q_5,1$				$q_8,0$
q_3		$q_4,$		$q_1,$	$q_1,0$		$q_5,$
q_4	$q_1,$	$q_4,$	$q_5,$		$q_1,$	$q_4,$	$q_5,1$
q_5	$q_5,0$		$q_3,0$	$q_4,1$	$q_1,0$	$q_5,$	
q_6	$q_5,1$	$q_4,0$	$q_3,1$	$q_1,0$	$q_1,$	$q_4,$	$q_5,1$
q_7	$q_1,0$	$q_2,$	$,1$	$q_6,1$	$q_7,1$	$q_8,0$	
q_8		$q_2,1$		$q_4,1$		$q_8,0$	$q_7,0$

Figure P13.9

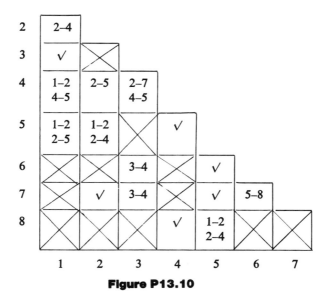

Figure P13.10

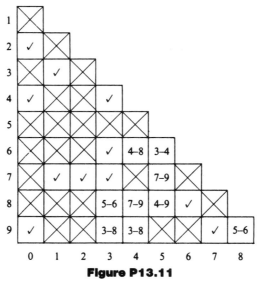

Figure P13.11

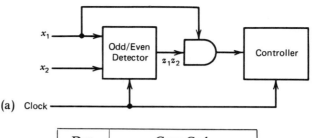

Dec. Digit	Gray Code			
	b_0	b_1	b_2	b_3
0	0	0	1	0
1	0	1	1	0
2	0	1	1	1
3	0	1	0	1
4	0	1	0	0
5	1	1	0	0
6	1	1	0	1
7	1	1	1	1
8	1	1	1	0
9	1	0	1	0

(b)

Figure P13.12

occur only between digits. Design the odd/even detector using *J-K* flip-flops.

13.13 In the sequence detector of Example 13.4, the next state on the final input of the third sequence could be a don't-care since the state of the detector once the car has started is unimportant. Determine whether this additional don't-care would provide any further simplification of the design.

13.14 Redesign the sequence detector of Example 13.4, removing the restriction that each button can be pushed only once.

13.15 Write a complete AHPL module description of the sequence checker in Examples 13.2 and 13.4.

(a) Write the description based on the final state table in Fig. 13.10. Consider the state table to be a description of the control unit. Let each state correspond to a step in the AHPL sequence.

(b) Write the entire description after ENDSEQUENCE, assuming only one control state. Base the description on the transition table of Fig. 13.13 and the resulting realization.

BIBLIOGRAPHY

1. Grasselli, A. and F. Luccio. "A Method of Minimizing the Number of Internal States in Incompletely Specified Sequential Networks," *IEEE Trans. Electronic Computers*, **EC-14**: 3, 330–359 (June 1965).

2. Paull, M. C. and S. H. Unger. "Minimizing the Number of States in Incompletely Specified Sequential Switching Functions," *IRE Trans. Electronic Computers*, **EC-8**: 3, 356–357 (Sept. 1959).

3. Hartmanis, J. and R. E. Stearns. *Algebraic Structure Theory of Sequential Machines*. Prentice-Hall, Englewood Cliffs, N.J., 1966.

4. Harrison, M. A. *Introduction to Switching and Automata Theory*. McGraw-Hill, New York, 1955.
5. Miller, R. E. *Switching Theory, Sequential Machines*, Vol. 2. Wiley, New York, 1965.
6. Moore, E. F. *Sequential Machines, Selected Papers*. Addison-Wesley, Reading, Mass., 1964.
7. Ginsburg, S. *Introduction to Mathematical Machine Theory*. Addison-Wesley, Reading, Mass., 1962.
8. Givone, D. D. *Introduction to Switching Circuit Theory*. McGraw-Hill, New York, 1970.
9. Kohavi, Z. *Switching and Finite Automata Theory*. 2nd ed. McGraw-Hill, New York, 1978.
10. Sheng, C. L. *Introduction to Switching Logic*. Intext, Scranton, Pa., 1972.
11. Hill, F. J. and G. R. Peterson. *Introduction to Switching Theory and Logical Design*. 3rd ed. Wiley, New York, 1981.

CHAPTER 14

LEVEL-MODE SEQUENTIAL CIRCUITS

14.1 INTRODUCTION

In this chapter we shall continue the discussion of sequential circuits, but eliminate the requirement that all state changes be triggered by clock pulses. In the process, we shall eliminate the clock as a special and necessary input to every sequential circuit and not insist on identifiable memory elements. What will remain is combinational logic with feedback. Is sequential activity still possible under these conditions?* In the following paragraphs, we shall argue that it is. This more general class of sequential circuits will be called *level-mode sequential circuits*. In many quarters, these circuits are casually called *asynchronous*, but this is a term assigned many other interpretations. We will avoid its use in this chapter.

The timing chart of Fig. 14.2 is intended to show that the simple configuration in Fig. 14.1 is sequential. As a starting condition, we assume both inputs and the output are at 0. Because of the feedback, the output is a function of itself and these

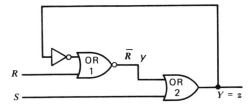

Figure 14.1 Combinational circuit with feedback.

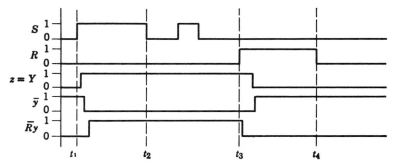

Figure 14.2 Timing chart for circuit of Fig. 14.1.

*Not all combinational circuits with feedback are sequential. See p. 141 of Reference 1.

assumptions are not necessarily consistent. In this case, if $Y = 0$, the output of NOR gate 1 is $\bar{R}y = 0$, so OR gate 2 implements $Y = S + \bar{R}y = 0 + 0 = 0$, which checks.

At time t_1, S goes to 1. This causes the output Y of OR gate 2 to go to 1 after a short propagation delay. In turn, $\bar{y}$ goes to 0 and $\bar{R}y$ to 1, again after short propagation delays. OR gate 2 is now implementing $1 + 1 = 1$, so when S returns to 0 at time t_2, there is no further change. Indeed, the $\bar{R}y = 1$ term has "locked in" the output at 1, so that S has no further effect.

Next, with $S = 0$, R goes to 1 at t_3. This drives $\bar{R}y$ to 0 and Y to 0 in turn. Then $\bar{y}$ goes to 1, locking in $\bar{R}y$ at 0. Thus, when R returns to 0 at t_4, there is no further change and we are back at the starting condition.

Consider the condition of the circuit immediately before t_1 and immediately after t_2. At both times, the inputs are the same, $S = R = 0$, but the output is different. Thus, the output is not dependent solely on the inputs, and the circuit is sequential. Thus, we do not have to use flip-flops or other memory devices to produce sequential circuits. Indeed, the reader who compares Fig. 14.1 with Fig. 9.3 will recognize that this circuit is the basic S-R flip-flop, redrawn to emphasize the form of the basic model.

The general model for the level-mode circuit is repeated in Fig. 14.3. Recall that the delay elements are not as a rule specific elements inserted in the feedback path for that purpose, but rather represent a "lumping" of the distributed delays in the combinational logic elements into single delay elements, one for each feedback variable. These delay elements may be considered as providing short-term memory. When an input changes, the delay enables the circuit to "remember" the present values of the variables, $y_1, y_2, ..., y_r$, long enough to develop new values of $Y_1, Y_2, ..., Y_r$, which in turn, after the delay, become the next-state values of the y's. Note the distinction between the y's and Y's. In the steady-state condition, they are the same, but during transition they are not. The y's will be referred to as the *secondaries* and the Y's as the *excitations*. Both correspond to the state variables in the clocked circuits, and either may be referred to by that name when the distinction between them is not important.

In the timing chart of Fig. 14.2, we note that changes in the secondaries and excitations, that is, changes in state, may occur in response to any change in the value of any input. Furthermore, in that analysis, we allowed only one variable to change at a time and no further changes to occur until all variables had stabilized. When such constraints on the circuit inputs can be enforced, we will say that the

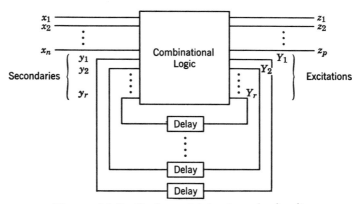

Figure 14.3 Basic model, level-mode circuit.

circuit is operating in the *fundamental mode*. This assumption about input changes considerably simplifies our design problems. Although the fundamental mode is a method of operation, not a special type of circuit, we will refer to circuits designed with this assumption as fundamental-mode circuits. In the next sections, we will consider only fundamental-mode circuits. In Section 14.9, we will consider the special problems introduced by the elimination of fundamental-mode restrictions.

14.2 ANALYSIS OF A FUNDAMENTAL-MODE CIRCUIT

We will now see that a fundamental-mode circuit can be described by transition and flow tables in essentially the same manner as clocked and pulse circuits. The transition table for a fundamental-mode circuit is a tabulation of the excitations and outputs. For the circuit of Fig. 14.1, the equation for the excitation is

$$Y = S + \bar{R}y \tag{14.1}$$

Since the excitation is the same as the output in this case, this is also the equation for z. The map corresponding to this equation is shown in Fig. 14.4. Just as with the clocked and pulse circuits, we list the inputs across the top and the secondaries (state variables) down the side, one row for each of the 2^r combinations of secondaries. Since the excitations are the next values of the state variables in fundamental-mode circuits, no distinction will be made between excitation maps and transition tables. The excitation maps will differ from the transition tables only in that they will be maps of single-state variables, whereas the transition tables will show all state variables. For only one state variable, they are identical.

Since we must assume that both inputs will not change simultaneously from their unexcited or 0 states, don't-care conditions are in the $SR = 11$ column. This is reflected in the specification that both inputs of an S-R flip-flop must not be pulsed simultaneously. The circled entries represent *stable* states, that is, states for which $y = Y$. For example, for entry $SRy = 000$, $Y = 0$. Since Y represents the next value of y, this is a stable condition. This corresponds to the starting condition in the timing table of Fig. 14.2. Similarly, $SRy = 010$ is a stable entry since a 0 output will be unaffected if R goes to 1. The square $SRy = 100$ corresponds to the case in which S goes to 1 while the flip-flop output is 0. As specified by Equation 14.1, the resulting value of Y is 1. Since this is not equal to the present value of y, it represents an unstable value and is not circled. After a delay, y changes to 1 and the circuit moves to the circled stable entry at $SRy = 101$. As S returns to 0, the circuit assumes the stable state indicated in square $SRy = 001$ without a further change in output. Finally, R going to 1 moves the circuit through unstable entry $SRy = 011$ to stable entry $SRy = 010$ and then to $SRy = 000$ when R returns to 0.

y $\diagdown$ SR	00	01	11	10
0	⓪	⓪	–	1
1	①	0	–	①

Y and *z*

Figure 14.4 Transition table for circuit of Fig. 14.1.

We have seen that more than one stable state may be found on the same row of a fundamental-mode transition table. It is convenient, in fact, to let each stable combination of inputs and secondary states be numbered as a separate state. Doing this to the transition table of Fig. 14.4 results in the *flow* (state) table of Fig. 14.5. Note the distinction between this flow table and those of clock-mode circuits, for which the states are only a function of the state variables. Also note that the circled entries indicate the final destination when the state variables must change, that is, the stable state into which the circuit will settle when all state variables have stabilized.

14.3 DEVELOPING A FLOW TABLE

We have just seen how an existing circuit can be analyzed in terms of transition and flow tables. In the design process, we must first go from a verbal statement of the desired circuit performance to a flow table and then to a transition table and actual circuit. The likelihood of error in the process of translating the problem statements into a flow table is minimized by initially going to a primitive flow table.

Definition 14.1 A *primitive flow table* is a flow table in which only one stable state appears in each row.

The primitive flow table corresponding to the flow table of Fig. 14.5 is shown in Fig. 14.6. This table describes the same response as the reduced flow table of Fig. 14.6. Starting in stable state ① with $SR = 00$, S going to 1 takes us through 2 to ②. When S returns to 0, the transition is through 3 to ③. Again, changes in S will move the circuit back and forth between ② and ③. When R goes to 1, the transition is through 4 to ④, and R going to 0 takes the circuit through 1 to ①. The outputs corresponding to each stable state are the same as in Fig 4.5.

SR	00	01	11	10	z
	①	④	–	2	0
	③	4	–	②	1

Figure 14.5 Flow table for circuit of Fig. 14.1.

SR	00	01	11	10	z
	①	4	–	2	0
	3	–	–	②	1
	③	4	–	2	1
	1	④	–	–	0

Figure 14.6 Primitive flow table for circuit of Fig. 14.1.

It is often convenient in fundamental-mode problems first to represent the problem as a timing chart rather than a state diagram. This procedure is followed in the following example.

■ EXAMPLE 14.1

Design a fundamental-mode circuit with two inputs x_1 and x_2 and a single output z. The circuit output is to be 0 whenever $x_1 = 0$. The first change in the input x_2, occurring while $x_1 = 1$, will cause the circuit output to go to 1. The output is then to remain 1 until x_1 returns to 0. Changes in inputs will always be separated sufficiently in time to permit the circuit to stabilize before a second input transition takes place. Some typical input output sequences are illustrated in Fig. 14.7.

SOLUTION   The first step of the design process is to determine a primitive flow table, for which we must assume some convenient starting stable state. In this case, it is most convenient to choose a state for which $x_2 x_1 = 00$. For this state, the output will be 0, regardless of which input was the last to change. This initial state is indicated as ①in Figs. 14.7 and 14.8. We also indicate the output corresponding to the stable state. For now, we indicate the outputs only for stable states on the flow table. The outputs for unstable states will be considered later.

Since both inputs are not allowed to change simultaneously, we enter a dash in the $x_2 x_1 = 11$ column of the first row of the flow table. This will eventually result in a don't-care condition in the circuit realization. Dash marks will similarly be entered in each row of the primitive-flow table in that column, differing in both inputs from the inputs for the stable state of that row.

If x_2 goes to 1 while x_1 is still 0, there will still be no circuit output, so we let the circuit go to a stable state ②with $z = 0$. If x_1 goes to 1 while $x_2 = 0$, the circuit goes to stable state ③also with $z = 0$. From state ②x_2 returning to 0 will take the circuit back to ①as will x_1 going to 0 from state ③From state ②x_1 going to 1 will take the circuit to a new state ④also with $z = 0$. We have so far determined the

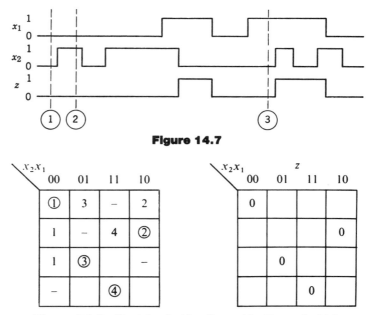

Figure 14.7

Figure 14.8   Partial primitive flow table, Example 14.1.

partial flow table of Fig. 14.8. The uncircled entries, or unstable states, merely represent the circuit destination following an input transition.

Both states ③ and ④ correspond to the last change having been from $x_1 = 0$ to $x_1 = 1$, so that a change in x_2 should cause the output to go to 1. Thus, from state ③, x_2 going to 1 takes us to ⑤, with $z = 1$. Similarly, from state ④ x_2 going to 0 takes us to ⑥, with $z = 1$. From state ④ x_1 going to 0 will return the circuit to the previous state, ②. These additions to the primitive flow table are shown in Fig. 14.9.

From either state ⑤ or ⑥, the output should remain 1 as long as x_1 remains 1, regardless of further changes in x_2. So we provide states ⑦ and ⑧, both with $z = 1$, such that the circuit will cycle between ⑤ and ⑦ or ⑥ and ⑧ for consecutive changes in x_2. A change of x_1 to 0 will take ⑥ or ⑦ to ① and ⑤ or ⑧ to ②. The completed primitive flow table is shown in Fig. 14.10. ∎

14.4 MINIMIZATION

It would, of course, be possible to implement Fig. 14.10 as is, utilizing eight internal states corresponding to the eight possible combinations of three secondaries.

x_2x_1	00	01	11	10
①	3	–	2	
1	–	4	②	
1	③	5	–	
–	6	④	2	
		⑤		
	⑥			

x_2x_1	00	01	11	10	z
0				
			0	
	0			
		0		
		1		
	1			

Figure 14.9

x_2x_1	00	01	11	10
①	3	–	2	
1	–	4	②	
1	③	5	–	
–	6	④	2	
–	7	⑤	2	
1	⑥	8	–	
1	⑦	5	–	
–	6	⑧	2	

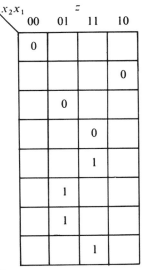

x_2x_1	00	01	11	10	z
0				
			0	
	0			
		0		
		1		
	1			
	1			
		1		

Figure 14.10

If the primitive flow table were identical to the state tables of Chapter 10 or 11, we might consider application of the various techniques developed in those chapters. However, if we compare the primitive flow table of Fig. 14.10 with a typical state table, we see a significant difference—namely, the absence of a present-state column in the flow table.

This difference arises from the fact that a clocked circuit is stable only between clock pulses, and the state of the circuit at these times is independent of the present inputs. Therefore, all the stable states may be grouped in a single column, corresponding to no clock input. A fundamental-mode circuit will be stable any time all the inputs are stable, and the state will depend on the values of the inputs. Thus, a present-state entry may appear in any column of the flow table.

The distinctions may be further clarified by considering a partial-state table and a partial primitive-flow table, as shown in Figs. 14.11a and 14.11b. First, consider that the circuit of Fig. 14.11a is in state q_1, and the inputs are 10 when a clock pulse arrives. The transition is from q_1 to unstable 3 during the clock interval and then to stable q_3 when the clock interval is complete, as shown by the dotted arrow in Fig. 14.11a. Next assume that the circuit of Fig. 14.11b is in stable ① with inputs 00. If the inputs go to 10, the circuit moves across to unstable 3 and then down to stable ③, as shown.

Since there is only one stable state in each row of the primitive flow table, we can regard these as the present states for the individual rows and group them in a present state column, as shown in Fig. 14.12a. Now the transition from ① to ③ may be regarded as a transition from a present state to a next state. Thus, we see that there is little real distinction between the state table for a clocked circuit and the primitive-flow table for a fundamental-mode circuit. The only distinction is that for *every row of the flow table there must be one input combination for which the next state is the same as the present state*. This condition could be imposed on clocked circuits if the designer so desired. This would only limit flexibility, having no effect on the determination of a covering circuit.

We also see that states may be combined in the same manner on both state and flow tables. In Fig. 14.11a, states q_1 and q_3 are obviously compatible and could be

q^v	q^{v+1}				z				States				Output			
	00	01	11	10	00	01	11	10	00	01	11	10	00	01	11	10
1	1	2	=	3	0	–	–	–	①	2	=	3	0	–	–	–
2	5	2	4	–	–	0	–	–	5	②	4	↓	–	0	–	–
3	1	–	8	3	–	–	–	0	1	–	8	③	–	–	–	0

(a) State Table	(b) Primitive Flow Table

Figure 14.11   Partial-state and primitive-flow tables.

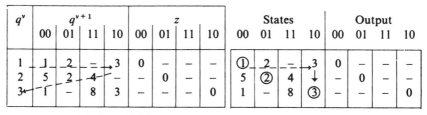

Present State	Next State				Output				Present State	Next State				Output			
	00	01	11	10	00	01	11	10		00	01	11	10	00	01	11	10
①	①	2	=	3	0	–	–	–	⑨	⑨	2	8	⑨	0	–	–	0
②	5	②	4	–	–	0	–	–	②	5	②	4	–	–	0	–	–
③	1	–	8	③	–	–	–	0									

(a)	(b)

Figure 14.12   Primitive flow table with present-state column added.

combined. Similarly, states ① and ③ in Fig. 14.12a are compatible and could be combined. Should they be combined, they would be replaced by a state, say, ⑨, which would be stable in both columns 00 and 10, as indicated in Fig. 14.12b. That is, the next state would be the same as the present state in both columns.

We have established a precise analogy between the minimization problems for clocked circuits and fundamental-mode circuits. We may proceed then to utilize all the techniques of Chapters 10 and 11 on Example 14.1.

■ EXAMPLE 14.1 (Continued)

We now proceed to reduce the number of rows in the primitive-flow table of the earlier section of Example 14.1. We repeat Fig. 14.10 as Fig. 14.13 in the present-state, next-state format justified by the above discussion.

The implication table corresponding to this flow table is shown in Fig. 14.14a. It is obtained in exactly the same way as for the clocked circuits. We enter an X for any pair with conflicting outputs, a ✔ for any pair compatible without implication, and the implied pairs in the remaining squares. We then eliminate pairs that imply incompatible pairs to produce the final implication table of Fig. 14.14b.

The maximal compatibles are easily determined to be

		x_2x_1				x_2x_1		
q^v	00	01	11	10	00	01	11	10
①	①	3	–	2	0			
②	1	–	4	②				0
③	1	③	5	–	0			
④	–	6	④	2			0	
⑤	–	7	⑤	2			1	
⑥	1	⑥	8	–		1		
⑦	1	⑦	5	–		1		
⑧	–	6	⑧	2			1	
		q^{v+1}				z		

Figure 14.13

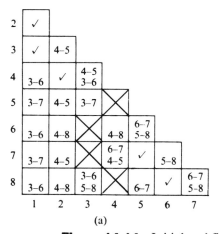

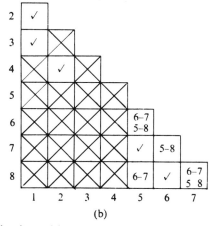

Figure 14.14 Initial and final implication tables, Example 14.1.

$$(12), (13), (24), (5678) \tag{14.2}$$

The smallest closed collection may be found by inspection to be

$$(13), (24), (5678) \tag{14.3}$$

The three-state table corresponding to this closed collection is shown in Fig. 14.15. Note that the combining of states may result in more than one stable state in a single row. In the primitive-flow table, we see that, for present state ① and input $x_2 x_1 = 00$, the next state is also ①. Similarly, in state ③, for input $x_2 x_1 = 01$, the next state is also ③. Thus, when states ① and ③ are combined into state @, the next state must be the same as the present state, for both $x_2 x_1 = 00$ and $x_2 x_1 = 01$. The reasoning is similar in arriving at the pairs of stable states in the other rows of the minimal table.

In row ① of the primitive table, the entry for $x_2 x_1 = 11$ is a don't-care, as this would correspond to a forbidden double transition from the stable ① entry. In row ③, however, the entry in this column is an unstable 5, corresponding to the transition from ③ to ⑤. When states 1 and 3 combine into state @, the entry in this column must be an unstable c, corresponding to a transition to the class containing state ⑤. This does not mean that a double transition, from $x_2 x_1 = 00$ to $x_2 x_1 = 11$, is any more "legal" than it was before. If the circuit is operated properly in the fundamental mode, the only transition through this unstable c will be from the stable @ corresponding to $x_2 x_1 = 01$. The other cases in which the circuit is to go to a new stable state are handled in a similar manner, as shown in Fig. 14.15.

Now let us reconsider our earlier assumption that the outputs are of no concern during the transitions. If an output variable is to change value as a result of a state change, then this variable is a don't-care during the transition. Assume an output variable is to change from 0 to 1 in a certain transition. If a 0 is entered as the value during the transition, then the change of variable will not take place until the end of the transition. If a 1 is entered, the change will take place at the start of the transition. Since it makes no difference exactly when in the transition the output change occurs, the entry is a don't-care. Referring to the final flow table of Fig. 14.15, we see that all transitions to or from state ⓒ involve a change in output. The outputs during these transitions thus remain unspecified.

If an output variable is *not* supposed to change as the result of a transition, the situation becomes a bit more complicated. It may be that the outputs are sampled by whatever other circuits may be involved only between transitions, in which case the transition outputs are don't-cares. However, if the outputs during tran-

q^ν	$x_2 x_1$ 00	01	11	10	$x_2 x_1$ 00	01	11	10
(13)	ⓐ	ⓐ	ⓐ	c	b	0	0	--	0
(24)	ⓑ	a	c	ⓑ	ⓑ	0	–	0	0
(5678)	ⓒ	a	ⓒ	ⓒ	b	–	1	1	–
	$q^{\nu+1}$				z			

Figure 14.15   Minimal flow table, Example 14.1.

sitions may affect other circuits, we usually must specify them. Refer again to Fig. 14.15 and consider the transition from ⓐ to b to ⓑ, as the input goes from 00 to 10. If the output were a 1 for unstable b, as might occur if we leave it unspecified, a 1 pulse might appear on the output line during the transition. If the circuit driven by z were, for example, a pulse circuit, this pulse might cause some undesired transition. Thus, the output corresponding to unstable b must be specified as 0. Similarly, the output for unstable a in row ⓑ is specified as 0.

Several more problems must be considered before an adequate circuit can be constructed for Example 14.1. Before considering these difficulties, let us develop another fundamental-mode circuit of some practical interest. ∎

∎ EXAMPLE 14.2

Over the years the control units [11] of certain computers have made liberal use of a fundamental-mode sequential circuit called a control delay. This circuit has a control pulse input x, a clock input C, and a control pulse output z, as illustrated in Fig. 14.16a. Pulses on line x will always be separated by several clock periods. Whenever a pulse occurs on line x, it will overlap a clock pulse and be of approximately the same width as a clock pulse. That is, line x will only go to 1 after the clock has gone to 1 and will return to 0 only after the clock has returned to 0. For each input pulse, there is to be an output pulse on line z coinciding with the next clock pulse following the x pulse. Thus, each x pulse results in a z pulse delayed by approximately one clock period, as shown in Fig. 14.16b.

The input line x originates at the output of a similar control delay with perhaps an additional gate and line delay in between. The control delay output will be a very slightly delayed version of the clock due to internal gate delay. Thus, we assume that no two inputs will change simultaneously so that the *fundamental-mode* constraint is satisfied. Develop a minimal state table for this circuit.

SOLUTION State ① representing a condition between clock pulses and at least two clock periods after the last pulse on line x, is a natural starting state. From here, the next input change can only be from 0 to 1 on line C. We let this change take the circuit to state ② as shown in Fig. 14.17a. From here, the next change may be the leading edge of a pulse on line x, in which case the circuit goes to state ③ If no x pulse occurs, the circuit returns to state ① when C goes to 0. We have assumed that the clock will always return to 0 ahead of a coincident x pulse. Thus, the circuit must go from state ③ to state ④, as shown in Fig. 14.17. Next, x will return to 0, taking the circuit to state ⑤ We have assumed that another control pulse will not occur until at least one clock period after an output has been generated. Therefore, the circuit will go from state ⑤ to state ⑥ with the arrival of

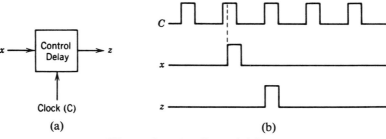

Figure 14.16 Control delay.

the next clock pulse and remain in that state for the duration of that clock pulse. The output $z = 1$ while the circuit is in state ⑥. A complete tabulation of outputs for all stable states is given in Fig. 14.17c. Following the output pulse, the circuit returns to state ①.

When we use the implication table of Fig. 14.18a, the primitive flow table is readily minimized to form the flow table of Fig. 14.18b. The outputs correspond-

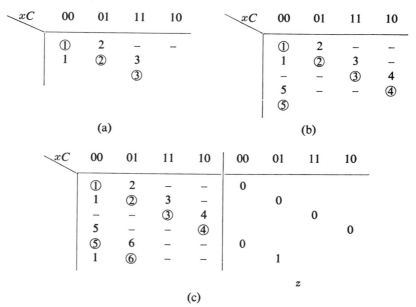

xC	00	01	11	10
	①	2	–	–
	1	②	3	
			③	

(a)

xC	00	01	11	10
	①	2	–	–
	1	②	3	–
	–	–	③	4
	5	–	–	④
	⑤			

(b)

xC	00	01	11	10	00	01	11	10
	①	2	–	–	0			
	1	②	3	–		0		
	–	–	③	4			0	
	5	–	④				0	
	⑤	6	–	–	0			
	1	⑥	–	–		1		

z

(c)

Figure 14.17 Developing flow table for control delay.

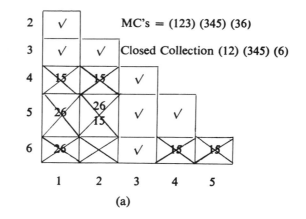

MC's = (123) (345) (36)

Closed Collection (12) (345) (6)

(a)

xC

	00	01	11	10	00	01	11	10
(12)	ⓐ	ⓐ	b	–	0	0	0	–
(345)	ⓑ	c	ⓑ	ⓑ	0	–	0	0
(6)	a	ⓒ	–	–	–	1	–	–

(b) z

Figure 14.18 Minimal flow table for control delay.

ing to unstable states are specified in this table to prevent unwanted output transitions. ∎

14.5 TRANSITION TABLES, EXCITATION MAPS, AND OUTPUT MAPS

The next step in the design is to assign specific combinations of the secondaries to the rows of the reduced flow table. Here we have all the state assignment problems of the clock- and pulse-mode circuits, plus some special problems peculiar to the fundamental mode. Let us proceed for now as though there were no new problems. As an example, let us construct a transition table corresponding to the flow table just developed for the control delay. We first assign the combinations of secondary values $y_2 y_1 = 00$ to ⓐ, $y_2 y_1 = 01$ to ⓑ, and $y_2 y_1 = 11$ to ⓒ Substituting these values for the stable states in Fig. 14.18b results in the partial transition table of Fig. 14.19a. This figure is a tabulation of the excitations, $Y_2 Y_1$, as a function of the inputs and present values of the secondaries, $y_2 y_1$. The excitations for unstable states must be the same as those for the corresponding stable states. For example, the inputs 11 cause a transition from state ⓐ to state ⓑ through unstable state b. To cause this change to take place in the physical circuit, we enter $Y_2 Y_1 = 01$ in place of unstable state b. Thus, when xC takes the value 11, the excitations are logically specified as 01. After a circuit delay, the secondaries take on the values $y_2 y_1 = 01$. Now the secondaries and excitations are the same (01), and the circuit is stable.

Excitation values corresponding to all unstable states are entered in the completed transition table of Fig. 14.19b. In succeeding problems, we will obtain the transition table in one step by making no distinction between stable and unstable

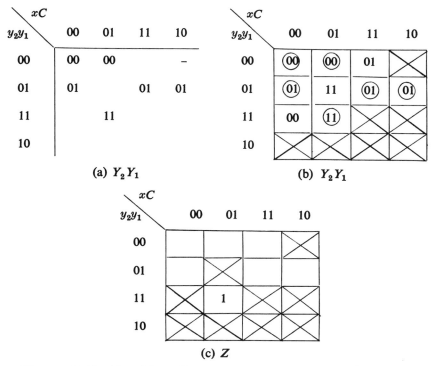

(a) $Y_2 Y_1$

(b) $Y_2 Y_1$

(c) Z

Figure 14.19 Transition table and output map for control delay.

states in the assignment of excitation values. The output table of Fig. 14.19c is a direct translation of the minimal state table.

For convenience in reading off the excitation equations, we may separate the transition table into separate *excitation maps* for the excitation variable, as shown in Fig. 14.20a. From these maps and the output map, we read the equations

$$Y_2 = \bar{x}Cy_1 \tag{14.4}$$

$$Y_1 = x + \bar{y}_2 y_1 + Cy_1 \tag{14.5}$$

$$z = y_2 \tag{14.6}$$

The corresponding circuit is shown in Fig. 14.20b.

If this circuit were clock or pulse mode, we would also check the other two possible assignments to see which is minimal. However, as mentioned above, fundamental-mode circuits have some special problems, so let us look at the functioning of this circuit in some detail before worrying about an absolute optimum.

Let us assume that the circuit is in the stable state $xCy_2 y_1 = 0111$, corresponding to the stable ©state in the second column, third row of the state table. The cor-

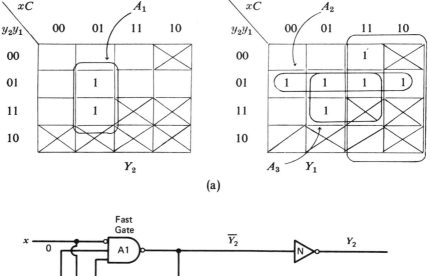

(a)

(b)

Figure 14.20 Excitation maps and circuit for control delay.

responding values at various points in the circuit are shown in Fig. 14.20b. Note that the 1 values for Y_2 and Y_1 are produced by gates $A1$ and $A3$, respectively. Now assume that C goes from 1 to 0, as shown. This should take the circuit through the unstable a entry in the third row to the stable ⓐ in the first column, first row, with both state variables changing to 0.

Recalling that a NAND gate will have a 1 out if any input is 0, we see that the C = 0 signal will drive the outputs of $A1$ and $A3$ to 1. Providing $A2$ does not change too fast, the 1 at the output of $A3$ will then drive Y_1 to 0 at the output of $A4$. This value of Y_1, fed back to the input of $A2$, will lock its output at 1, in turn stabilizing Y_1 at 0, as required.

However, now assume that the gates have unequal delays, as shown in Fig. 14.20b. Then the change in C will first drive the output of $A1$ to 1, which will in turn drive $A2$ to 0 *before* the output of $A3$ goes to 1. As a result, Y_1 will not change and the circuit will make an erroneous transition to $Y_2 Y_1 = 01$, the stable ⓑ state in the first column.

The difficulty just described occurred because both secondaries were required to change simultaneously following a single input change, but did not change at the same time due to unequal circuit delays. Such a situation is termed a *race* because the nature of the transition may depend on which variable changes fastest. Sometimes, races may be "fixed" by introducing extra delays, but this approach is expensive and doesn't always work. In the next section, we will investigate the nature of these races more carefully and show that they can be eliminated by a better state assignment.

14.6 CYCLES AND RACES

It is possible for a circuit to assume more than one unstable state prior to reaching a new stable state. If for a given initial state and input transition, such a sequence of unstable states is unique, then it is termed a *cycle*. For example, in the table and corresponding Y map of Fig. 14.21, the circuit will cycle through three unstable states during a transition from ① to ④. Note in the Y map that the next state for $y_2 y_1 x_2 x_1 = 0001$ is not $y_2 y_1 = 10$. Instead, the machine proceeds from 00 to 01 to 11 to 10. Each of the transitions involves a change in only one secondary. Thus, the circuit proceeds reliably through each of the unstable states to the final stable state. The cycle may be entered from stable states ② and ③ as well.

When a change of more than one secondary is specified by the Y map, the resulting situation is termed a *race*. The race illustrated in Fig. 14.22 is a *noncritical*

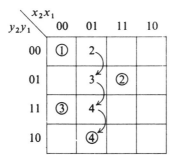

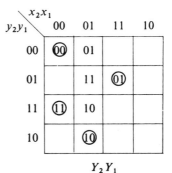

Figure 14.21

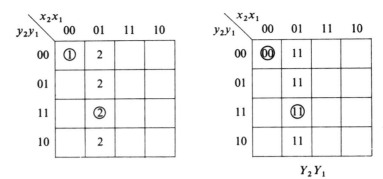

Figure 14.22

race. Suppose the inputs change to 01 while the machine is in state $y_2 y_1 = 00$. If y_2 changes first, the machine goes to unstable state 10. If y_1 changes first, the machine goes to 01. For either of these unstable states, the Y map shows the next state to be stable ②. Thus, no matter what the outcome of the race, the stable state ②, as designated by the Y map, is reached.

The situation encountered in the control delay, as discussed in the last section, is an example of a *critical race*, a situation in which there are two or more stable states in a given column, and the final state reached by the circuit depends on the order in which the variables change. In other words, the desired result might or might not occur, depending on the actual circuit delays.

Another critical race is illustrated in Fig. 14.23. There the circuit is supposed to move from stable ②($y_2 y_1 = 01$) in the $x = 1$ column to stable ③($y_2 y_1 = 10$) in the $x = 0$ column. If Y_1 changes first, the excitations will be $Y_2 Y_1 = 00$, leading to an erroneous transition to stable ①. If Y_2 changes first, the excitations will be $Y_2 Y_1 = 11$, leading to unstable 1 in the third row and then to stable ①. Thus, no matter what the delays, the wrong transition will occur.

It might seem that the first type of critical race would be less troublesome since it can, in theory, be eliminated by the proper adjustment of delays. In practice, however, constraining delays is difficult and expensive so that both types of critical races are to be avoided. In some problems, this may be accomplished by a judicious assignment of state variables. In others, it becomes necessary to use more than a minimal number of states.

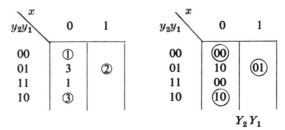

Figure 14.23

■ **EXAMPLE 14.2 (continued)**

We are now in a position to construct a more satisfactory version of the control delay. It is only necessary to eliminate the critical race in column 00 of Fig. 14.19*b* and replace it with a cycle through the spare state. A new transition table is given in Fig. 14.24. The only changes are in the lower two squares of column $xC = 00$. Thus, from stable state ⑪, the circuit will cycle as indicated by the arrows when C goes to 0. In terms of the excitation maps, the only change is an additional 1 in the third row. The new equation for Y_2 is

$$Y_2 = y_2 y_1 + \bar{x} C y_1 \tag{14.7}$$

There is no change in the equation for Y_1. To avoid a possible output pulse during the cycle, an additional 1 in the output map must be specified, as shown in Fig. 14.24*b*, but this does not change the equation for z. The revised circuit for the control delay shown in Fig. 14.25 will function properly as a control delay, provided the input pulses satisfy the fundamental mode constraints as previously assumed.

∎

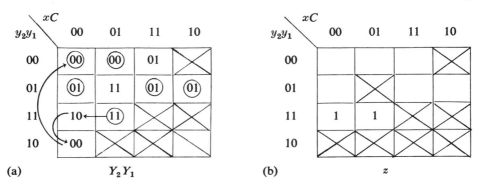

(a) $Y_2 Y_1$ (b) z

Figure 14.24 Revised transition table for the control delay.

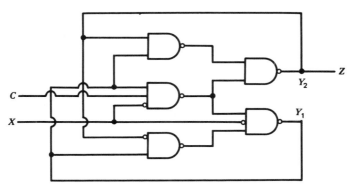

Figure 14.25 Control delay.

14.7 RACE-FREE ASSIGNMENTS

To eliminate critical races, we must assign the state variables in such a manner that transitions between states require the change of only one variable at a time. To this end, it is desirable that stable states between which transitions occur be given adjacent assignments, that is, assignments differing in only one variable. To find such assignments, it is convenient to make a *transition diagram* on the Boolean hypercube. Consider the state table shown in Fig. 14.26*a* and the transition diagram in Fig. 14.26*b*. To set up the transition diagram, we start by assigning ⓐ to 00. Noting that ⓐ has transitions to ⓒ, we give ⓒ an adjacent assignment at 01 and indicate the transition by an arrow from ⓐ to ⓒ. Next, the transition from

ⓒback to ⓐis indicated by an arrow from ⓒto ⓐ. Since there is a transition from ⓒto ⓑ, ⓑis given an adjacent assignment at 11, leaving 10 for ⓓ. The diagram is then completed by filling in arrows for all the other transitions. In this case, all transitions are between adjacent states, so this assignment is free of critical races.

A slightly different situation is illustrated in Fig. 14.27. Now ⓓhas a transition to ⓒ, resulting in a diagonal transition, that is, a change of two variables. Note that this diagonal transition cannot be eliminated since ⓓ has transitions to three other states and cannot be adjacent to all of these with only two state variables. However, this diagonal transition is seen to represent a noncritical race and will cause no problems. Whenever there is only one stable state in a column, transitions into that column need not be considered in choosing a state assignment since critical races cannot occur unless there are at least two stable entries in a column. Thus, the same assignment will be valid for this state table.

In Fig. 14.28, we see a more difficult situation. Here we have several critical races, as indicated by the diagonal transitions, and it is clear that no permutation of the assignment can eliminate all diagonal transitions. For example, the first row requires both Ⓑ and Ⓓ to be adjacent to Ⓐ, whereas the second row requires Ⓑ and Ⓓ to be adjacent. Clearly, not all these requirements can be satisfied.

In situations such as the above, the critical races can be eliminated only by the use of spare states. If the number of states in the original state table is not a power of 2, spare states will automatically be available. This was the case in the control delay, in which there were only three states and the spare state was used for a cycle

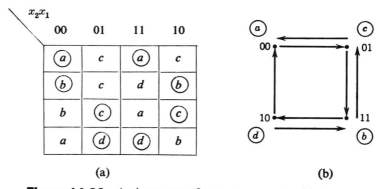

(a)   (b)

Figure 14.26   Assignments of states to prevent critical races.

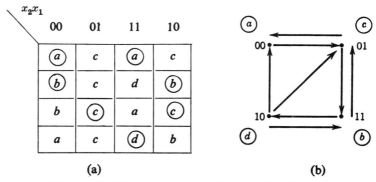

(a)   (b)

Figure 14.27   Assignment with noncritical race.

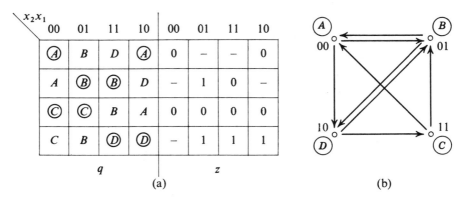

Figure 14.28 State assignment with critical races.

to avoid the critical race. If the original number of states is a power of 2 or a satisfactory assignment is not found using available spare states, spare states must be created by adding extra state variables.

There are two basic techniques for using the spare states: *shared-row* assignments and *multiple-row* assignment. In multiple-row assignment, as the name implies, each state is assigned two or more combinations of state variables, that is, two or more rows in the transition table. Figure 14.29 shows a universal multiple-row assignment for four-state circuits. The extra rows are created by using three state variables. The four states are arbitrarily numbered 1 to 4, and their assignments are indicated by their location on the K-map of the state variables. There are two state-variable combinations assigned to each state, with the two assignments for each state being logical complements.

The structure of this assignment is shown more clearly in Fig. 14.30. The two combinations assigned to each state are on diagonally opposite corners of the cube. Both assigned combinations for a state are adjacent to one of the combinations for each of the other states. This is illustrated by the arrows in Fig. 14.30, indicating adjacent transitions from either ① state to any of the other states.

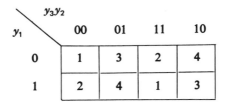

Figure 14.29 Universal assignment for four-state tables.

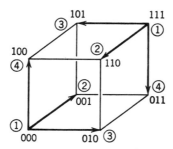

Figure 14.30 Universal four-state assignment on hypercube.

The development of the transition table for the universal assignment for the state table of Fig. 14.28 is shown in Fig. 14.31. We start by setting up an augmented state table (Fig. 14.31a) with the eight-state variable combinations listed and the corresponding state assignment. Since state numbers are arbitrary, we let $1 \rightarrow A, 2 \rightarrow B$, etc. Although there is no logical difference, we denote the two representations of each state as A_1, A_2, B_1, B_2, etc., for ease of reference. For each row assigned to a given state, the stable entries are the same as in the original state table. For example, state $\textcircled{A}$ is stable in the first and fourth columns, so rows $\textcircled{A_1}$ and $\textcircled{A_2}$ are stable in the same columns. Similarly the outputs are the same for each row assigned a given state. For state $\textcircled{A}$, the output row is $0, -, -, 0$, so the output rows for $\textcircled{A_1}$ and $\textcircled{A_2}$ are both $0, -, -, 0$.

The transitional entries are filled in to provide adjacent transitions. State $\textcircled{A}$ makes transitions to $\textcircled{B}$ and $\textcircled{D}$. $\textcircled{A_1}$ is adjacent to $\textcircled{B_1}$ and $\textcircled{D_2}$, and $\textcircled{A_2}$ is adjacent to $\textcircled{B_2}$ and $\textcircled{D_1}$, so the transitional entries are filled in accordingly. We can now see why the external behavior of the circuit is not changed by the use of multiple states. Consider the circuit to be in state $\textcircled{A}$, with a stable output of 0 and either $x_2 x_1 = 00$ or $x_2 x_1 = 10$. It makes no difference whether the circuit is actually in $\textcircled{A_1}$ or $\textcircled{A_2}$. From state $\textcircled{A}$, it should go to $\textcircled{B}$ or $\textcircled{D}$. From $\textcircled{A_1}$, it will go to $\textcircled{B_1}$ or $\textcircled{D_2}$; from $\textcircled{A_2}$, it will go to $\textcircled{B_2}$ or $\textcircled{D_1}$. Either way, the new stable state will correspond to $\textcircled{B}$ or $\textcircled{D}$ of the original state table. The same argument holds for any other state transitions. The final transition table, Fig. 14.31b, is obtained by replacing the state entries with the assigned combinations of state variables.

The multiple-row assignment has the advantage of minimal delay. There are no cycles. Each transition is direct from one stable state to another. The disadvantage is that the number of rows is at least twice the number of states. At least one extra state variable will always be required. When minimal delay is not the principal design criterion and unused rows are available (the number of states is not a power of 2), the shared-row approach will result in a more economical realization.

When unused rows are insufficient, it is usually possible to find a shared-row assignment by adding a single state variable. For example, Fig. 14.32 depicts an assignment that will always work for eight or fewer stable states. It will be left to the reader to verify that nonconflicting cycles may be found between all pairs of states assigned to squares marked X. The reader is referred to [12] for similar universal shared-row assignments for larger numbers of variables.

Although universal assignments require an extra variable to allow for all possible cases, shared-row assignments that do not require extra variables can often be found for specific tables, particularly if there are several spare states available without adding extra variables. A useful concept in finding such assignments is the *destination set*. In any column of a state table, a destination set consists of any stable state and any rows that make transitions into that state. For example, in the state table of Fig. 14.33, in the first column, the stable states are $\textcircled{1}, \textcircled{3}, \textcircled{5}$, and row 4 leads to $\textcircled{1}$, row 2 to $\textcircled{5}$. The destination sets for this column are thus $(1, 4)(2, 5)(3)$. Similarly, the destination states for the second, third, and fourth columns are $(1, 3)(2, 4)(5)$, $(1, 4)(2, 3)(5)$, and $(1, 3)(2)(4, 5)$, respectively. The destination sets are important because, to avoid critical races, the members of each set must either be adjacent or so located in relation to the spare states that cyclic transitions for all sets in a given column may be made without interference.

$y_3y_2y_1$ \ x_2x_1		00	01	11	10	00	01	11	10
A_1	000	A_1	B_1	D_2	A_1	0	–	–	0
B_1	001	A_1	B_1	B_1	D_1	–	1	0	–
D_1	011	C_1	B_1	D_1	D_1	–	1	1	1
C_1	010	C_1	C_1	B_2	A_1	0	0	0	0
D_2	100	C_2	B_2	D_2	D_2	–	1	1	1
C_2	101	C_2	C_2	B_1	A_2	0	0	0	0
A_2	111	A_2	B_2	D_1	A_2	0	–	–	0
B_2	110	A_2	B_2	B_2	D_2	–	1	0	–

States z

(a)

$y_3y_2y_1$ \ x_2x_1	00	01	11	10
000	000	001	100	000
001	000	001	001	011
011	010	001	011	011
010	010	010	110	000
100	101	110	100	100
101	101	101	001	111
111	111	110	011	111
110	111	110	110	100

$Y_3Y_2Y_1$

(b)

Figure 14.31   Development of transition table for state table of Fig. 14.28.

A standard assignment for five-state tables is shown in Fig. 14.34a. We term it *standard* rather than *universal* because it will work for almost every table but will fail for a few rare cases, as we will see. For this case, we arbitrarily assign the five states as shown in Fig. 14.34b on the K map, and Fig. 14.34c on the Boolean hypercube. In the first column of the state table, the destination sets are (1, 4)(2, 5)(3), requiring transitions from 4 to 1 and from 2 to 5. We see in Fig. 14.34c that this is impossible with this assignment as the two cycles must use the same spare rows

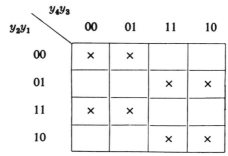

Up to 8 states

Figure 14.32 Universal shared-row assignment.

	x_2x_1 00	01	11	10	00	01	11	10
1	①	3	4	①	0	–	0	0
2	5	4	②	②	–	1	1	0
3	③	③	2	1	1	1	1	–
4	1	④	④	④	–	1	0	1
5	⑤	⑤	⑤	4	0	1	1	1

States

Figure 14.33 Example five-state table.

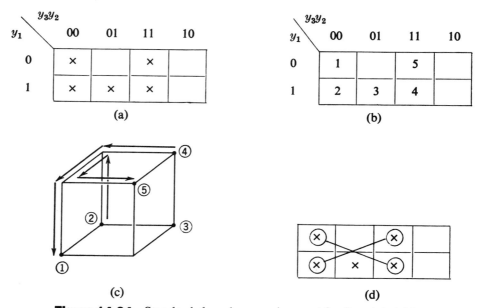

(a)

(b)

(c)

(d)

Figure 14.34 Standard shared-row assignment for five-state tables.

enroute to different destinations. Rows may be *shared* only in the sense that they may be used for different cycles in different columns. In any single column, a transitional row may be part of a cycle to only one destination.

This problem occurs only in columns having (2, 2, 1) destination sets, that is, columns having 2, 2, and 1 members in three destination sets. It can be eliminated by permuting the assignment so that the two two-member sets do not occupy diagonally opposite vertices of the hypercube. In terms of the assignment map, the two-member sets must not occupy the connected rows as shown in Fig. 14.34d. There are 15 possible (2, 2, 1) column configurations and, as long as all 15 do not occur in the same table, it is possible to avoid assigning any (2, 2, 1) sets in the "forbidden" pattern. This assignment will thus fail only in the unlikely event of a state table with 15 or more columns, including all 15 possible (2, 2, 1) columns [12].

The easiest way to use this assignment is to construct any arbitrary (2, 2, 1) group of destination sets that do *not* occur in the destination sets of the state table and assign them to the "forbidden" pattern. This guarantees that none of the (2, 2, 1) columns which are in the state table can be assigned to that pattern.

For the table of Fig. 14.33, we see that the (2, 2, 1) pattern (1, 2)(3, 5)(4), for example, does not appear among the destination sets, so we assign (1, 2) and (3, 5) to the diagonal locations, as shown in Fig. 14.35a. For ease of reference, the spare rows are labeled α, β, γ. To set up the augmented state table (Fig. 14.33b), we start by filling in the stable states and the corresponding outputs, just as for a multiple-row assignment.

The transitional entries are filled in with reference to the map of the assignment (Fig. 14.35a) to determine if the transition is direct or whether a cycle through spare states must be provided. State ① makes transitions to ③ in column 01 and to 4 in column 11. The transition to ③ is direct, but the transition to ④ uses the cycle ① → α → 4 → ④, as seen in the third column of Fig. 14.35b. Transitions ② → ④ and ② → ⑤ are both direct. Transition ③ → ① is direct, but ③ → γ → 2 → ② cycles through γ. The other entries are filled in similarly. We note here an advantage of the shared-row method. It is rare that all the spare row entries are needed, resulting in many don't-cares. By contrast, the multiple-row method does not provide any don't-cares other than those in the original state table.

Up to now, very little has been said with regard to the assignment of outputs corresponding to unstable states. Some rules must be followed, and we will introduce them now. The output entries for the transitional states must be chosen to eliminate any spurious pulses. If the output is to be the same before and after the transition, it must stay constant throughout the cycle. If the output is to change, it must change only once during a cycle. Therefore, it can be a don't-care for only one transitional state during the cycle. Consider the transition from ④ in the second column to ① in the first column, in Fig. 14.35b, a transition that cycles through x. If we let both transitional outputs be don't-cares, as shown in Fig. 14.36a, the result, depending on groupings of don't-cares, might be a momentary 1 pulse, as shown in Fig. 14.36b. To prevent this, either the first transitional state must be specified as 1, or the second as 0, as shown in Figs. 14.36c and 14.36d, respectively. Then the choice between the latter two should be made on the basis of which seems likely to result in the simplest output equations.

The transition table follows directly from the augmented state table by replacing each state with the corresponding state-variable combination, as shown in Fig. 14.37c.

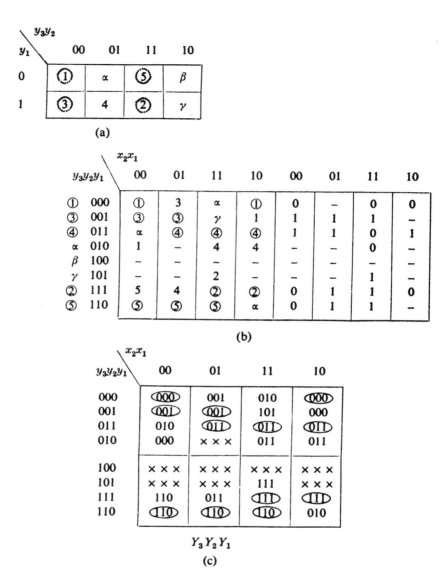

Figure 14.35   Valid five-state assignment for Fig. 14.33.

There are no generally applicable three-variable assignments for 6 and 7 state tables, but suitable assignments can often be found through careful analysis of the destination sets. This may be illustrated by the following example.

Figure 14.36   Transitional output assignment in cyclic transition.

■ **EXAMPLE 14.3**

Obtain a critical-race-free state assignment for the table of Fig. 14.37.

SOLUTION The destination sets are $(1, 3, 4)(2, 5)(6), (1, 6)(2, 3)(4)(5), (1, 2)(3)(4)(5, 6)$, and $(1, 4)(2, 6)(3, 5)$. First, we note a destination set with more than two members, $(1, 3, 4)$, in the first row. In such cases, it is often possible to cycle within the destination set, for example, ①→ 4→ ③ or ④→ 1→ ③. In view of the flexibility offered by such cycles, it is usually best to consider the two-member destination sets first. In this case, we note that state 2 is a member of four such sets $(2, 5), (2, 3)$, $(1, 2)$, and $(2, 6)$. In a three-variable assignment, a single state can be adjacent to only three other states, so these four destination states require that ② be adjacent

	$x_1 x_2$			
	00	01	11	10
	3	6	①	①
	②	3	1	②
	③	③	③	5
	3	④	④	1
	2	⑤	6	5
	⑥	⑥	⑥	2

Figure 14.37 Example six-state table.

to one of the spare states, which we shall designate α. Of the four states paired with ② in destination sets, ⑤ and ⑥ are not part of the $(1, 3, 4)$ set, so we place them adjacent to the $(2, α)$ pair, giving the initial assignment shown in Fig. 14.38a. The remaining state paired with ②, ①, and ③ must be adjacent to ② or α, so we locate them as shown in Fig. 14.38b. Finally, noting that state ④ can cycle to ③ through state ①, we place 1 and the remaining spare state β as shown in Fig. 14.38c. We then check the destination sets for all columns to see that all transitions can be made without critical races with the assignment of Fig. 14.38c. The development of the augmented state table and transition table is shown in Fig 14.39. ■

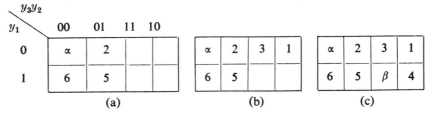

Figure 14.38 Development of race-free assignment for Fig. 14.37.

14.8 HAZARDS IN SEQUENTIAL CIRCUITS

The difficulty encountered in dealing with critical races in the last two sections has probably cooled the enthusiasm of the reader for designing in the level mode. Races are not a problem in clock-mode circuits, because the values of all state variables are assumed to have stabilized before a subsequent clocking transition appears. If not sooner, the reader should have been convinced by Section 12.6 that these state variables will indeed stabilize, given a sufficiently slow clock. In spite of the relative simplicity of clock-mode design, there are a few applications in which level-mode design cannot be avoided.

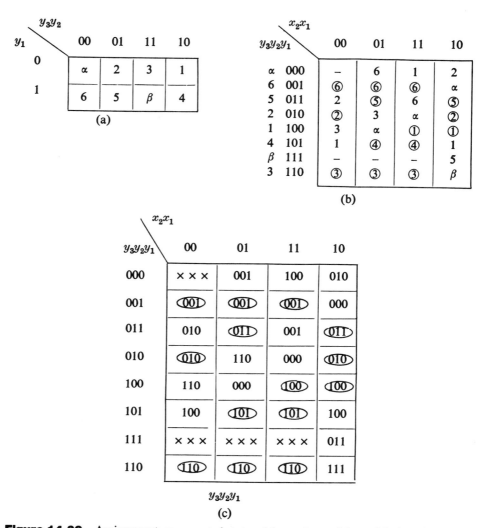

Figure 14.39   Assignment, augmented state table, and transition table for Fig. 14.37.

In Section 5.9 we observed that logic hazards can be a problem at the output of clock-mode circuits. As with critical races and for the same reason, hazards are not a problem internally within a clock-mode circuit. A hazard within a level-mode circuit, however, can cause a malfunction, that is, an actual next state different than predicted by the flow table. In Fig. 14.40, we show the flow table and transition table for a sequential circuit. The circuit is free from critical races and so, presumably, will function properly. The minimal equations for the state variables are

$$Y_2 = y_2 x_2 + y_2 \bar{y}_1 x_1 + \bar{y}_1 x_2 \bar{x}_1 + y_2 y_1 x_1 \tag{14.8}$$

and

$$Y_1 = \bar{y}_2 \bar{x}_2 y_1 + y_2 y_1 x_1 + y_2 y_1 x_1 + y_1 x_2 \bar{x}_1 \tag{14.9}$$

The logic circuit for Y_1 is shown in Fig. 14.41. The logic for Y_2 is only indicated symbolically since it is not pertinent to the present discussion.

Let us assume that the circuit is in the stable state Ⓒ at $y_2 y_1 x_2 x_1 = 1111$. Under this condition, both gates A2 and A3 will develop 1 at their outputs, producing Y_1

	x_2x_1			
y_2y_1	00	01	11	10
00	ⓐ	ⓐ	ⓐ	d
01	ⓑ	ⓑ	a	ⓑ
11	d	b	ⓒ	ⓒ
10	a	ⓓ	c	ⓓ

	x_2x_1			
y_2y_1	00	01	11	10
00	⓪⓪	⓪⓪	⓪⓪	10
01	⓪①	⓪①	00	⓪①
11	10	01	⑪	⑪
10	00	⑩	11	⑩

$$Y_2Y_1$$

Figure 14.40 Flow table and transition table for hazard example.

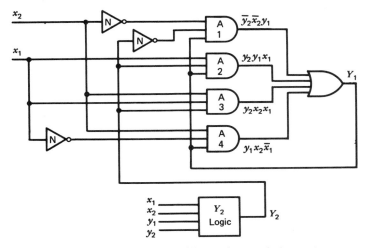

Figure 14.41 Circuit illustrating static hazard.

= 1 at the output of the OR gate. Now let x_1 change to 0. We see that the circuit should move to the stable state ⓒ at $y_2y_1x_2x_1 = 1110$, that is, neither Y_1 nor Y_2 should change. But let us consider the circuit action in detail. When x_1 goes to 0, the outputs of gates A2 and A3 will go to 0, and the output of gate A4 will go to 1 due to $\bar{x}_1$ going to 1. However, because of the delay in the inverter, $\bar{x}_1$ will not go to 1 as fast as x_1 goes to 0, so that there will probably be a short interval when the outputs of all AND gates will be 0. The output of the OR gate will thus momentarily go to 0. Since y_2 does not change, the circuit will momentarily see the state-variable input $y_2y_1 = 10$ and may make an erroneous transition to stable state ⓓ at $y_2y_1x_2x_1 = 1010$.

Next, let us assume the same starting state, $y_2y_1x_2x_1 = 1111$, and now let x_2 go to 0. Now the circuit should go through unstable b to stable ⓑ at $y_2y_1x_2x_1 = 0101$. Again, Y_1 should not change. As before, gates A2 and A3 are 1 initially. When x_2 goes to 0, gate A3 goes to 0, and at the same time, the transition in Y_2 is started. After a delay through the Y_2 logic, y_2 goes to 0, taking A2 to 0, and A1 goes to 1 due to $\bar{y}_2$ going to 1. However, as before, the delay through the inverter will cause all gates to be 0 momentarily, in turn causing Y_1 to go to 0 momentarily. The circuit may thus make an erroneous transition to stable a at $y_2y_1x_2x_1 = 0001$.

The above are examples of malfunctions caused by *static hazards*.

Definition 14.2 Let there be a state transition in a fundamental-mode circuit such that one or more of the secondary variables is required to remain constant.

If it is possible, due to unequal delays in the circuit, for one or more of the expected constant state variables to change momentarily, then a *static hazard* is said to exist.

A static hazard is a function of the combinational logic in the realization and may be removed using the method as given in Section 5.10.

■ EXAMPLE 14.4

Remove the static hazard from state variable Y_1 in the circuit of Fig. 14.41.

SOLUTION   Using the technique of Section 5.10, we obtain the cover given in the Karnaugh map of Fig. 14.42. Note that the minterms $y_2 y_1 x_2 x_1 = 0110$ and 1110 are not covered by a 1 cube. This extra cube is unnecessary, because both minterms correspond to stable states in Fig. 14.40. No transition will occur between stable states.   ■

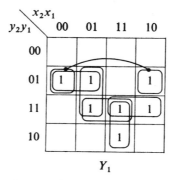

Figure 14.42   Removal of static hazards.

The situation illustrated above, in which the output is supposed to remain 1, is known as a *static hazard in the 1's*. It should be obvious that the same thing could happen when the output is supposed to remain 0, in which case we have a *static hazard in the 0's*. It is also possible to have a hazard when the output is supposed to change. When certain patterns of delay occur, the circuit may go momentarily to the new value and then back to the initial value before making the permanent transition, that is, $0 \to 1 \to 0 \to 1$ instead of $0 \to 1$. This situation is known as a *dynamic hazard*. Like the static hazard, the dynamic hazard is seldom of real concern in combinational circuits but may cause malfunctions in fundamental-mode sequential circuits.

The detection of all possible types of hazards in a given circuit is very complicated and has been the subject of extensive investigation [3]. However, as designers, all we need is some rule that is sufficient to ensure freedom from hazards. The following theorem (stated without proof) provides this rule.

THEOREM 14.1   A second-order sum-of-products circuit that is free of all static hazards in the 1's will be free of all static and dynamic hazards.

Like races, the static and dynamic hazards are due to unequal delays in various signal paths. Still another type of hazard arises when delays result in one excitation changing before the circuitry generating another excitation is even aware of the input change. In Fig. 14.43, we show the transition table, excitation maps, and output map for a sequential circuit designed to count modulo-4, the number of changes in input x.

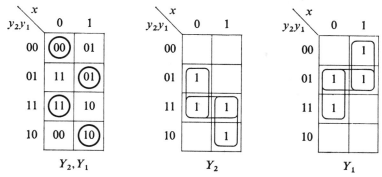

Figure 14.43 Transition table and design maps for counter.

The excitation equations with static hazards eliminated are given by Equations 14.10 and 14.11.

$$Y_2 = y_1\bar{x} + y_2 x + y_1 y_2 \tag{14.10}$$

$$Y_1 = y_1\bar{x} + \bar{y}_2 x + \bar{y}_2 y_1 \tag{14.11}$$

Since we have carefully avoided all critical races and all static and dynamic hazards, the resulting circuit should work as intended. Or will it?

Consider the realization of these equations as shown in Fig. 14.44. Assume the circuit is in the stable state (11), for which the signals on the various lines are as shown. Now let x go to 1, which should change Y_1 and take the circuit to stable (10). The input $x = 1$ ($\bar{x} = 0$) drives AND 3 to 0 and thus Y_1 to 0. The change in Y_1, in turn, takes the output of AND 1 to 0. If the circuit is functioning properly, AND 2 will change to 1 at the same time AND 3 changes to 0, so that the subsequent change of AND 1 will not affect Y_2. But suppose that the delay in AND 2 is very long so that its output is still 0 when the output of AND 1 goes to 0. Then OR 1 will see all 0 inputs and Y_2 will go to 0. Instead of seeing itself at the stable combination, $xy_2y_1 = 110$, the circuit now sees itself at the unstable condition, $xy_2y_1 = 100$, and thus makes an erroneous transition to stable (01).

At first, this might look very much like the static hazard discussed in connection with Fig. 14.40, but in that case, there was only one excitation involved, so

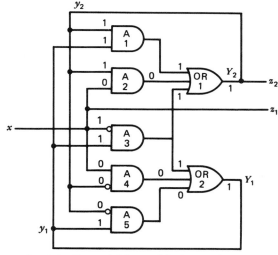

Figure 14.44 Essential hazard in input-change counter.

that the problem could be corrected by adding gates in the circuitry generating that excitation variable. Here the problem involves an interaction between two excitation variables. The variable Y_1 has changed and propagated back through the circuitry for Y_2 before that circuitry has completed responding to the initial input change. This type of hazard, which is peculiar to sequential circuits, is known as an *essential hazard*. As the name suggests, an essential hazard is created by the basic logical structure, or specification, of the circuit and is not affected in any way by the particular form of realization used. The hazard can be detected on the original state table.

Definition 14.3 An *essential hazard* exists in a state table if there is a stable state from which three consecutive changes in a single input variable will take the circuit to a different state than the first change alone [4].

This behavior will occur if any two adjacent columns of the state table exhibit either of the two patterns shown in Fig. 14.45a and 14.45b. In both cases, if the circuit starts in ① a single change of variable takes the circuit to ② and two more changes take it to ④ The reader should note that the numbering of the states and ordering of the rows in these two tables are quite arbitrary; the patterns may appear quite different in a larger flow table. For example, the transition indicated in the table of Fig. 14.45c, ③→ ⑤→ ⑥→ ②, corresponds to the pattern of Fig. 14.45b.

The reader will note that for the hazard described in connection with Fig. 14.44 to occur, the delay through AND 2 would have to be longer than the cumulative delay along the path AND 3, OR 2, AND 1. The reader may feel that such a combination of delays would be very unlikely, but it is not. In typical IC gates, the standard production tolerances on delay specification allow a range of three-or-four-to-one between minimum and maximum delay times.

Essential hazards can be found in many circuits of considerable practical importance, including almost all counters. Since the effect of these essential hazards cannot be eliminated by modifying logic, they can be handled only by carefully controlling the number of gates in each feedback loop. Very often, realizations can be worked out with pairs of cross-coupled gates making up loops that must respond very rapidly. Gates on a single chip will usually have less variance in delay times than typical gates on different chips. Thus, essential hazards are less likely to cause trouble in a level-mode circuit implemented in a single IC package.

Occasionally, it will be necessary to insert a pair of inverters in a feedback loop so that the secondaries do not change until the input change has propagated to all parts of the circuit. The deliberate insertion of delay is not at all desirable, since speed is generally of the essence, but there is often no other remedy for the essential hazard.

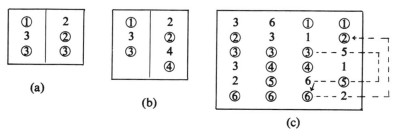

Figure 14.45 Flow tables indicating essential hazards.

As with races and logic hazards, essential hazards are not a problem in clock-mode circuits. Changes in state-variable values will not take place until a triggering edge of the clock appears. By then the impact of all input changes will have propagated through the combinational logic and stabilized.

14.9 ELIMINATE THE FUNDAMENTAL-MODE RESTRICTION AT YOUR OWN RISK

Over the years many hard lessons have been learned in the process of attempting level-mode design with more than one input allowed to change simultaneously. As we shall see, final verification that such circuits will work is *outside the scope of logic design*. Careful design with flow tables and attention to hazard-free and race-free realizations is not enough. Designers whose work resulted in expensive product revisions were not necessarily aware that they were designing in the level mode. They may have assumed that they were designing in the clock mode, but gave insufficient consideration to some input that could occasionally change very close to the time of the triggering edge of the clock.

The problems discussed thus far—races, hazards, and essential hazards—all exist in the more general level-mode case, just as they do for the fundamental mode. There are at least two more potential problems—one manageable within the scope of flow table design, the other not. To illustrate the first issue, consider the flow table of Fig. 14.46. Assume that it is possible for both inputs a and b to change from 0 to 1 at or approximately at the same time. Because of unpredictable delays within the ultimate realization, it cannot be guaranteed that the circuit will react differently, when the two inputs change at once, than when b goes to 1 just slightly before a goes to 1. For some reason, the flow table is asking for distinct behavior for these two cases. This distinct behavior is unrealizable. The ultimate realization cannot be expected to reliably distinguish between the case in which a and b change at the same time and the case in which b changes slightly sooner.

What about the class of level-mode circuits in which simultaneous input changes are possible, but where the flow table behavior for any dual input change would be the same as if either of the two inputs changed slightly before the other? If general level-mode design offered no new problems in addition to the one discussed in the above paragraph, we could still use flow tables to address this class of circuits. Research on synchronizing circuits [14, 15] uncovered the second pitfall of general level-mode design.

A simple attempt at designing a synchronizing circuit using an edge-triggered D flip-flop was discussed in Section 12.4. It was argued that a change on the unsynchronized D input to a memory element too close to the triggering edge of the clock could drive the circuit to a metastable state. The resulting oscillation

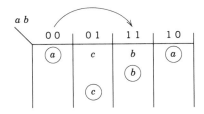

Figure 14.46 Impossible-to-guarantee behavior.

within the linear regions of the two cross-coupled gates in Fig. 12.12 could last for several clock periods. Circuit parameters that sustain the oscillations are not necessarily those that determine the setup, hold, and delay times of the memory element. The simulation of this phenomenon requires a circuit-level rather than gate-level simulator.

It was not established that synchronization is impossible. We must simply conclude that phenomenon that cannot be modeled and simulated at the gate level cannot be treated by level-mode design methods. Circuit design and system design techniques can be used to reduce the probability of synchronization failure. Reference [11] considers a method of trading time for synchronization. It may be conjectured that a possibility for oscillations will exist within any level-mode circuit in which two inputs can change simultaneously. The following example will find the same problem in a slightly more complex synchronizer.

■ **EXAMPLE 14.5**

Because of gate delays, line delays, and communication between units with separate clocks, a control pulse to a section of a computer system may arrive out of synchronism with the clock for that section. In order to avoid unpredictable operation in the clocked circuits of the computer, such a control pulse must be routed through some type of pulse synchronizer. This pulse synchronizer will have two inputs: the computer clock line C and the control pulse line P. Pulses on line P may arrive at random times relative to clock pulses.

Corresponding to every pulse on line P, the synchronizer is to emit a pulse on output line Z, synchronized with a clock pulse. Pulses on line P are of approximately the same duration as clock pulses but will occur less frequently, that is, always separated by several clock periods. This is an example of a situation in which some, but not all, multiple-input changes may occur.

SOLUTION As with fundamental-mode design, it is possible to start with a primitive-flow table, but in this case, we find it convenient to proceed directly to a table with more than one stable state per row. Let state ①be a quiescent stable state following an output pulse, that is, a situation in which no further output should occur until another input appears on P. As illustrated in Fig. 14.47a, the circuit merely moves back and forth between the stable states labeled ①as the clock goes off and on in the absence of an input control pulse. When a pulse appears on line P, the circuit goes to stable state ②, as indicated in Fig. 14.47b. Whether the ultimate circuit senses single or multiple changes on row ①, it will eventually stabilize in row ②when a clock pulse arrives.

Note that the output, although not shown, must remain 0 for all stable states in the first two rows of the final flow table in Fig. 14.48c. Continuing as shown in the figure assures that only one possible change of state is specified for any input change while in the first two rows of the flow table. Figure 14.48c specifies that the output will be 1 for the duration of the first clock pulse after an interval in which $PC = 00$ following a pulse on P. This table assures that there will be no output pulse of shorter duration than a clock pulse.

Figure 14.48 depicts the only possible multiple-input change ($PC = 10$ to 01, as shown by the arrow in Fig. 14.48c) where behavior might depend on circuit delay. Accepting the flow table tells us that an output pulse of clock-pulse width will occur immediately, if the circuit reacts as if P changed first. Otherwise, the output pulse will coincide with the next clock pulse. Either behavior satisfies the problem specification. ∎

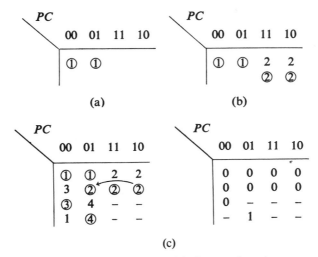

Figure 14.47 Flow table for synchronizer.

We now turn our attention to obtaining a circuit realization. States are assigned without difficulty, yielding the transition table of Fig. 14.49a. Individual Karnaugh maps are included to illustrate the minimal multiple-output realization for Y_2 and Y_1 as given by Equations 14.12 and 14.13. Note the inclusion of the term $y_2 y_1$ to avoid a static hazard in Y_2. Equation 14.14 for output z is taken directly from Fig. 14.48.

$$Y_2 = \bar{P}\bar{C}y_1 + y_1 y_2 + y_2 C \tag{14.12}$$

$$Y_1 = \bar{P}\bar{C}y_1 + \bar{y}_2 y_1 + P \tag{14.13}$$

$$z = y_2 \bar{y}_1 \tag{14.14}$$

We have already asserted that a multiple-input change can lead to instability and oscillation, not predicted by the flow table. Consider the input transition PC = 10 to 01 marked by the arrow in Fig. 14.47c. The critical term is $\bar{P}\bar{C}y_1$ in Equation 14.12, the expression for Y_2. If delays are such that the circuit reacts as if $\bar{C}$ goes to 1 before $\bar{P}$ goes to 0, Y_2 will be logically 1 for a short interval, possibly driving the gate into the linear region. Once $\bar{C} = 0$, Equation 14.12 reduces to $Y_2 = y_2$. *With Y_2 already in the metastable or linear region, oscillation is possible.*

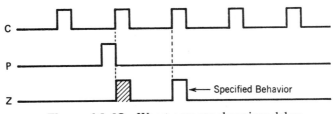

Figure 14.48 Worst-case synchronizer delay.

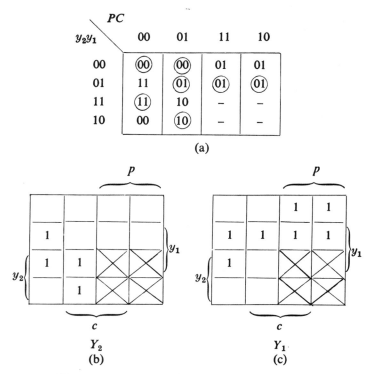

Figure 14.49 Transition table for synchronizer.

PROBLEMS

14.1 Determine the primitive flow table of a fundamental-mode sequential circuit with two inputs x_2 and x_1. The single output z is to be 1 only when $x_2x_1 = 10$, provided that this is the fourth of a sequence of input combinations 00 01 11 10. Otherwise, the output is to be 0. Include a tabulation of outputs assuring that no spurious 1 outputs will occur during transitions between two states with 0 outputs. Both inputs will not change simultaneously.

14.2 A controller is to be designed for an automatic parking lot gate of the type in which the driver inserts a coin in a slot and the gate then opens long enough to let the car enter. When the driver inserts the coin, a coin detector will emit a pulse P, which will cause the gate control signal z to go to 1 to open the gate. As the car passes through the gate, it will pass by a photocell that will cause a signal x to go to 1. When the car is past the gate, x will go back to 0, at which time z should go to 0, closing the gate. To prevent persons from defeating the system by blocking off the photocell to keep the gate open, there will be a treadle switch in the pavement immediately in front of the gate that will emit a pulse T whenever car wheels pass over it. The car passing through will trip this switch twice. If it is triggered a third time before x has gone back to 0, z should go to 0 to close the gate. Assume that the switch is located relative to the photocell so that x will go to 1 before the front wheels hit the switch, and the rear wheels will hit the switch before x goes back to 0. Determine a primitive-flow table of a circuit to develop the control signal z.

14.3 A circuit has a single-input line on which pulses of various widths will occur at random. Construct the primitive-flow table of this circuit with a single output z, on which a pulse will occur coinciding with every fourth input pulse. Because this circuit is part of a random-process generation scheme, the power on state is of no interest.

14.4 A sequential circuit has two level inputs and two level outputs. The outputs are coded to count in the straight binary code, modulo-4, as shown in Fig. P14.4. The count is to increase by one for each 0-to-1 transition of either input occurring when the other input is at the 0 level. The two inputs will not change at the same time. Determine a primitive-flow table of a circuit to meet this specification.

z_2z_1	Count	Next Count
00	0	1
01	1	2
10	2	3
11	3	0

Figure P14.4

14.5 Derive the primitive flow table of a fundamental-mode circuit with inputs a and b and single output z. The output z will be 1 if and only if both inputs a and b are currently 1 and there has been a time interval in which both inputs have been 0 since the last interval in which $z = 1$. The output z must also be 1 following power up (when the circuit first turns on).

14.6 Determine a minimal state equivalent of the primitive-flow table given in Fig. P14.6. Suppose that the don't care outputs were all specified such that unstable states always have the same output as similarly lettered stable states. Thus, the state table becomes completely specified. Repeat the minimization after this modification and compare the two solutions. Would this approach in general prevent the determination of a minimal table? Why not?

	x_2x_1				z x_2x_1			
	00	01	11	10	00	01	11	10
ⓐ	f	i	k		0	–	0	–
ⓑ	f	g	j		0	–	0	–
ⓒ	d	h	j		0	–	0	–
c	ⓓ	i	k		–	1	–	1
a	ⓔ	i	l		–	1	–	–
a	ⓕ	h	j		–	1	–	1
a	e	ⓖ	k		0	–	0	–
c	f	ⓗ	k		0	–	0	–
a	d	ⓘ	j		0	–	0	–
c	f	i	ⓙ		–	1	–	1
a	d	h	ⓚ		–	1	–	1
b	d	h	ⓛ		0	–	0	0

Figure P14.6

14.7 Determine a minimal row equivalent of the flow table of Fig. P14.7.

14.8 Determine minimal state equivalents of the primitive-flow tables determined in:

(a) Problem 14.1.

(b) Problem 14.2.

14.9 Assign secondaries to the minimal flow table of Fig. P14.9. Construct a transition table that avoids critical races. Obtain Boolean expressions for each excitation variable.

14.10 Assign the eight combinations of values of three secondaries to the six rows of the flow table in Fig. P14.10. Determine a transition table that avoids critical races. Do not minimize the flow table.

14.11 Obtain a minimal flow table covering the table of Fig. P14.10. Assign secondary states to rows of the minimal state table using only two secondaries. If possible, obtain a transition table for this assignment that avoids critical races.

	q				z			
	x_2x_1				x_2x_1			
00	01	11	10	00	01	11	10
①	2	–	5	0	0	–	0
–	②	3	–	–	0	0	–
–	4	③	9	–	–	0	0
1	④	–	–	..	1	–	–
–	–	6	⑤	–	–	0	0
–	7	⑥	8	–	0	0	0
1	⑦	–	–	0	0	–	–
1	–	–	⑧	0	–	–	0
1	–	–	⑨	0	–	–	0

Figure P14.7

	q				z_2z_1			
	x_2x_1				x_2x_1			
00	01	11	10	00	01	11	10
①	2	①	3	01	—	01	—
1	②	②	②	—	00	00	00
1	③	2	③	—	10	—	10

Figure P14.9

	q				z			
	x_2x_1				x_2x_1			
00	01	11	10	00	01	11	10
①	2	–	5	0			
②	②	4	6	1	1		
–	2	④	6			1	
1	2	–	⑤			0	
1	⑥	8	⑥	0		1	
–	6	⑧	5		0	0	

Figure P14.10

14.12 Make a secondary assignment for the flow table of Fig. P14.12. Construct a transition table without critical races.

14.13 For the circuit and secondary assignment of Fig. P14.13, obtain expressions for Y_2, Y_1, and z that are free from all static and dynamic hazards.

14.14 For the flow table shown in Fig. P14.14, determine a state assignment free of critical races, using three state variables.

	q $x_2 x_1$				z $x_2 x_1$			
	00	01	11	10	00	01	11	10
	ⓐ	ⓐ	d	b	0	1		
	a	ⓑ	c	ⓑ		0		0
	ⓒ	a	ⓒ	d	1		1	
	c	b	ⓓ	ⓓ			1	1

Figure P14.12

$y_2 y_1$	q $x_2 x_1$				z $x_2 x_1$			
	00	01	11	10	00	01	11	10
00	ⓐ	ⓐ	ⓐ	b	1	1	0	0
01	a	ⓑ	c	ⓑ	1	1	1	0
11	ⓒ	ⓒ	ⓒ	d	0	0	1	1
10	c	–	a	ⓓ	0	–	0	1

Figure P14.13

	$x_1 x_2$			
0 0	0 1	1 1	1 0
Ⓐ, 0	Ⓐ, 0	E, 0	D, 0
D, 0	E, 0	Ⓑ, 0	Ⓑ, 0
A, 0	Ⓒ, 0	F, 0	Ⓒ, 0
Ⓓ, 0	G, 0	Ⓓ, 0	Ⓓ, 0
D, 0	Ⓔ, 0	Ⓔ, 0	F, 0
Ⓕ, 0	A, 0	Ⓕ, 0	Ⓕ, 0
Ⓖ, 1	Ⓖ, 0	B, 0	C, 0

q, z

Figure P14.14

14.15 For the flow tables shown in Fig. P14.15, determine assignments free of critical races. Assume the tables are minimal.

14.16 For the flow tables of Fig. P14.16, determine assignments free of critical races and excitation equations free of combinational hazards.

14.17 Write hazard-free excitation and output expressions for the following flow tables. Make certain that there are no transient output pulses.

(a) Figure P14.10.

(b) Figure P14.12.

x

0	1
(a)	f
h	(b)
(c)	f
(d)	b
d	(e)
g	(f)
(g)	e
(h)	i
c	(i)

q

(a)

x_1x_2

00	01	11	10
(a)	c	–	b
j	(b)	–	(b)
h	(c)	–	(c)
e	(d)	–	(d)
(e)	f	–	c
a	(f)	–	(f)
(g)	d	–	(g)
(h)	i	–	g
h	(i)	–	c
(j)	(j)	–	c

q

(b)

Figure P14.15

x_1x_2

00	01	11	10
(a), 0	(a), 0	d, 0	e, 0
(b), 0	(b), 0	d, 0	(b), 0
b, 0	(c), 0	(c), 0	e, 0
e, –	(d), 0	e, 0	b, 0
(e), 0	(e), 0	c, 0	e, 0

q, z

(a)

x_1x_2

00	01	11	10
(1), 0	2, –	(1), 1	5, –
1, –	(2), 0	(2), 0	4, –
(3), 0	5, –	4, –	(3), 1
3, –	5, –	(4), 0	(4), 0
(5), 0	(5), 0	1, –	(5), 1

q, z

(b)

Figure P14.16

14.18 Inspect the flow tables of Fig. P14.18 for essential hazards. If any exist, indicate them by specifying the single and triple transitions fulfilling the condition of Definition 14.3. For example, for the flow table of Fig. 14.45b, the essential hazard would be indicated as

$$① \rightarrow 2 \rightarrow ② \rightarrow 3 \rightarrow ③ \rightarrow 4 \rightarrow ④$$

x_2x_1

00	01	11	10
(a)	b	–	(a)
(b)	(b)	c	a
–	–	(c)	d
b	–	–	(d)

(a)

x_2x_1

00	01	11	10
(a)	–	b	(a)
–	c	(b)	d
d	(c)	b	–
(d)	–	f	(d)
–	(e)	f	(e)
–	c	(f)	e

(b)

Figure P14.18

14.19 Obtain an alternate flow table for the rising-edge-triggered D flip-flop in which the circuit changes state on any rising edge of the clock. For that clock transition, it must always go to the same next stable state. Depending on D, the circuit may go to two different states at the time of the falling edge of the clock. Obtain the most economical possible realization of this flow table and compare the result to Fig. 9.19.

14.20 A fundamental-mode circuit has two inputs and two outputs. The two outputs should indicate which of the inputs has changed most recently and whether the number of times it has changed is odd or even.

(a) How many secondaries will be required?

(b) Obtain a minimal flow table of such a circuit.

(c) Obtain a race-free transition table.

(d) Design a circuit to realize the transition table of (c) that is free of static and dynamic hazards.

(e) If any essential hazards exist, indicate them on the flow table.

14.21 A fundamental-mode circuit is to be designed that will identify a particular sequence of inputs and provide an output to trigger a combination lock. The inputs to the circuit are three switches labeled X_3, X_2, and X_1. The output of the circuit is to be 0 unless the input switches are in the position $X_3 X_2 X_1 = 010$, where this position occurs at the conclusion of a sequence of input positions $X_3 X_2 X_1 = 101\ 111\ 011\ 010$. Note that two switches are not required to be switched simultaneously.

Using only NAND gates, design a fundamental-mode circuit that will have an output of 1 only at the conclusion of the above-described sequence of switch positions. A new correct sequence may begin every time the switches are set to $X_3 X_2 X_1 = 101$. Be sure that the circuit is free from hazards and critical races.

14.22 Complete the design of the circuit of Example 14.1, starting with the final flow table of Fig. 14.15. Make an assignment free of critical races, determine hazard-free equations, and draw the final circuit.

14.23 Complete the design of the controller for the parking lot gate, previously considered in Problems 14.2 and 14.8(b).

14.24 Let us now consider a more elaborate version of the parking lot gate of Problem 14.2. Assume that there are two lanes entering a parking garage and that two gates, as described in Problem 14.2, are arranged side by side. After these gates, the two lanes merge into a single-lane ramp to the parking floors. The operation of the gates will be just as described in Problem 14.2, except that we wish to control the spacing of cars as they go up the ramp. For this purpose, we locate another treadle switch a short distance up the ramp. After a car passes through either of the gates, its front wheels should pass over the ramp treadle before either gate can open again. Let the output pulse of the ramp treadle be R. If there are two cars at the gates at the time of a R pulse, they should be released in alternate order. If two cars arrive at the gates at the same instant, they may be released in either order, but take care in your design to ensure that both cannot be released at the same time. Design a circuit to meet these requirements.

The following three problems consider systems that are not fundamental mode. Assume that low-frequency operation and careful circuit design will eliminate oscillation in the linear region as a problem. Avoid instances of inconsistent behavior between multiple-input changes and consecutive single transitions in the same inputs.

14.25 A detector circuit is to be designed for a reaction tester. At random intervals, a pulse X_1 will be emitted by a pulse generator. This X_1 pulse will turn on a red light for the duration of the pulse. When the person being tested sees the light come on, he or she is to push a button that emits a pulse X_2, which may be assumed to be shorter in duration than the X_1 pulse. The output of the detector Z is to be 0 unless a complete X_2 pulse occurs during the X_1 pulse, in which case the output is to go to 1 and remain there until a pulse R appears on a reset line. Design a detector to meet these requirements. Include the reset line in your design.

14.26 On a television quiz show, three contestants are asked a question, all at the same time. A contestant who thinks he or she knows the answer pushes a button. Whoever pushes a button first gets to answer the question. Design a detector circuit to turn on a light in front of the contestant who pushed a button first. Let the signals from the three push buttons be X_1, X_2, X_3 and the outputs controlling the corresponding lights be Z_1, Z_2, Z_3. Include a provision for a reset line controlled by the host. Since the contestants are racing each other, two or more signals will occasionally coincide. The circuit may resolve these "ties" in a random fashion, but the circuit must be designed so that no two lights can ever come on at the same time.

14.27 Repeat Problem 14.26, except now assume that we wish an indication of the order in which the contestants pushed their buttons. There will be three lights in front of each contestant, numbered 1, 2, 3. The appropriate light should come on in front of each contestant, indicating whether the person was first, second, or third to push a button.

BIBLIOGRAPHY

1. Miller, R.E. *Switching Theory*, Vol. 1. Wiley, New York, 1965.
2. McCluskey, E.J. "Fundamental and Pulse Mode Sequential Circuits," in *Proceedings of IFIP Congress, 1962*. North Holland Publ. Co., Amsterdam, 1963.
3. McCluskey, E.J. *Introduction to the Theory of Switching Circuits*. McGraw-Hill, New York, 1965.
4. Unger, S.H. "Hazards and Delays in Asynchronous Sequential Switching Circuits," *IRE Trans. Circuit Theory*, **CT-6:** 12 (1959).
5. Huffman, D.A. "The Design and Use of Hazard-Free Switching Networks," *J. ACM*, **4:** 47 (1957).
6. Huffman, D.A. "The Synthesis of Sequential Switching Circuits," *J. Franklin Inst.*, **257:** 161–190, 275–303 (March–April 1954).
7. Huffman, D.A. "A Study of the Memory Requirements of Sequential Switching Circuits." *Tech. Rept. 293*, Res. Lab. of Electronics, M.I.T., Cambridge, Mass., April 1955.
8. Caldwell, S.H. *Switching Circuits and Logical Design*. Wiley, New York, 1958.

9. Eichelberger, E.B. "Hazard Detection in Combinational and Sequential Switching Circuits," *IBM J. Res. and Dev.*, **9:** 2 (1965).

10. Maley, G.A. and J. Earle. *The Logic Design of Transistor Digital Computers.* Prentice-Hall, Englewood Cliffs, N.J., 1963.

11. Hill, F.J. and G.R. Peterson. *Digital Systems: Hardware Organization and Design*, 3rd ed. Wiley, New York, 1987.

12. Saucier, G.A. "Encoding of Asynchronous Sequential Networks," *IEEE Trans. Computers*, **EC 16:** 3 (1967).

13. Unger, S.H. *Asynchronous Sequential Switching Circuits.* Wiley, New York, 1969.

14. Chaney, T.J. and C.E. Molnar. "Anamalous Behavior of Synchronizer and Arbiter Circuits," *IEEE Trans. Computers*, **EC 22:** 4, 421–422 (April 1973).

15. Couranz, G.R. and D.F. Wann. "Theoretical and Experimental Behavior of Synchronizers Operating in the Metastable Region," *IEEE Trans. Computers*, **C-24:** 604–616 (June 1975).

CHAPTER 15

TEST GENERATION FOR VLSI

15.1 FAULT DETECTION AND DIAGNOSIS

Testing is an integral part of any manufacturing process. A high probability that some failure in that process will cause a manufactured item to be defective will dictate comprehensive testing of each individual item. Usually, the probability (expressed as percent) that items **will not** be defective is called the *yield* for a particular process. For VLSI chips, yields range from 10% to 90%, depending on the process, circuit complexity, and lambda (λ). Yields in this range dictate the comprehensive testing of each item. If a yield were 99.99% or if the cost of repair in the field were less than the testing cost, items might be shipped without testing. VLSI chips are not repaired and the replacement of defective chips on printed circuit boards is unattractive. Therefore, every chip must be thoroughly tested.

Allowing for modest literary license, Fig. 15.1 depicts the essential features of the testing process. A sequence of input combinations or test vectors must be applied to input pins, and these must cause observable outputs for a defective chip to in some way differ from the sequence of outputs expected from a good chip. A defective chip *detected* by the application of a sequence of test vectors is then discarded. Chip output pins are observable outputs. The set of inputs and observable outputs may be enlarged by pressing tiny probes against metal pads on the chip itself. In contrast to the "wished for" arrangement suggested by Fig. 15.1, *probing can only be incorporated in the test process, if testing is done before chips are packaged.* Testing at the chip stage in the fabrication process is more difficult to automate and, therefore, significantly more costly than package testing. Tests, which are only required to *detect* defective parts, usually involve only pins on the package.

Somewhat less important in integrated circuit development is *diagnostic testing*, a term borrowed from medicine where discarding a patient after the detection of a problem is not the normal procedure. The diagnostic test must identify a defect by stimulating output behavior in the defective item that differs not only from the expected behavior of the good item but also from the behavior caused by any other defect. Diagnostic testing is important at the system level, where a failure may occur during operation. At the chip level, diagnostic testing might be used to isolate a frequently occurring defect, so that its likelihood of occurrence may be reduced through design or the fabrication process modification. Probes are more likely to be available for the diagnostic testing of VLSI chips. The more difficult subject of diagnostic testing will be considered only briefly in this chapter.

Before fault detection tests can be applied, these tests must be devised or *generated*. Testing is part of manufacturing. *Test generation* is properly part of the

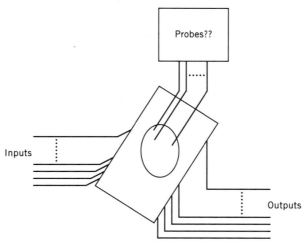

Figure 15.1 Failure detection testing.

design process and, therefore, a suitable subject for a book with the word "design" in the title.

The part shown in Fig. 15.1 may fall into one of three categories: (1) combinational logic only, (2) a sequential circuit that does not execute programs, or (3) a programmable processor with or without on-chip memory. Typically (but not necessarily), parts in the third category are by far the most complicated. Test generation for a processor package would be extremely difficult, if we were unable to take advantage of its programmability to enlist the assistance of the part in testing itself. The definitions in the next section will apply to all three categories. Following that section we will consider the categories of parts in order. Test generation for combinational logic must be understood before sequential circuits can be considered. Similarly, sequential circuit test generation is a prerequisite to the brief treatment of category 3 found at the end of the chapter.

15.2 STUCK-AT FAULT MODEL

One approach to testing a part is to apply a sequence of test vectors that will "prove it works." For a combinational logic circuit, this may be accomplished by applying all possible input combinations and determining whether the outputs expected from the good network are observed for each input vector. In this case, 2^n input vectors must be applied to a circuit with n inputs. We shall argue shortly that fewer test vectors will adequately test a typical combinational logic circuit with large n. The behavior of sequential circuits is characterized by response to input sequences rather than to independent input vectors. For some special cases of sequential circuits, upper bounds have been established [1] on the length of input sequence necessary to verify the proper function of the circuit. These upper bounds are typically the product of the number of input combinations times some exponential order function of the number of states. Finding a test sequence by using the algorithms on which these upper bounds are based is impractical for all but the simplest sequential circuits.

An alternative to the "prove it works" strategy is to prove the "absence of failure." This strategy requires a list L_f of all possible failures that might occur in the parts to be tested. A test sequence that will determine whether any of the failures in L_f actually exists may be used to identify good parts. For each member of L_f, the "absence of failure" test must cause output behavior in some way different from the behavior it would stimulate in the good part.

How is the list of failures L_f compiled? Before trying to answer, let us define the most often considered failures in digital integrated circuits.

Definition 15.1   *A single stuck-at-0 (stuck-at-1) fault* is a failure in which **one** input or output line of one gate (or memory element) in the network is permanently connected to logical 0 (1) with the network otherwise unchanged. A stuck gate input is assumed to be disconnected from its driving source.

Definition 15.2   A stuck-at fault is said to be *detected* by a vector of input values, if and only if a circuit output value for that single-fault network differs from the output value for the good network. A fault is *excited* at any point in the network, if the good and single-fault network values differ at that point.

Clearly, single stuck-at-0 (SA0) and stuck-at-1 (SA1) faults are not the only failures that can occur within a real integrated circuit. Studies [2] have been made of real failures that may occur within VLSI parts in various technologies, their causes, and the probabilities of occurrence. Included are combinations of single stuck faults, short circuits between adjacent points in the layout, shorts between layers, stuck open conditions, parameter variations, and combinations of occurrences of any or all of these conditions.

In short, compiling the list L_f is impractical for a part of any complexity. This would be a major concern in a world governed by the following rule, deeply ingrained by the experience with the diagnostic testing of almost anything:

Murphy's Law of Testing   If a hard-to-detect failure can occur, it will occur, and it will occur independently.

If such a law were really applicable to failure detection testing in VLSI, there would have been little exploitation of these technologies. Although not always admitted, the *silver spoon law of VLSI* usually applies instead.

Silver Spoon Law of VLSI   If a failure occurs on a VLSI chip, the probability is high that it will occur in conjunction with a fault, detectable by a test sequence generated to detect all single SA0 and SA1 faults.

We offer the silver spoon law without proof. We merely conjecture that this has been the "real world" experience in any VLSI technology appearing to date. Closest to an exception is the impact of stuck-open faults in CMOS, where two consecutive input vectors must be applied to excite a fault. This class of failures will be given special attention in Section 15.7.

Acceptance of the silver spoon law allows us to ignore the very-difficult-to-compile L_f and instead find a sequence of test vectors, which will detect all distinct single stuck-at faults for the network to be tested. This task can be approached by using gate-level search and simulation techniques. A starting point for a list of all single stuck-at faults is all gate (or memory element) inputs or outputs in the network. Clearly, the number of single faults is directly proportional to the overall complexity of the network. It is sometimes possible to model members of $[L_f]$ by using multiple faults, combinations of more than one single fault. If it were necessary to consider multiple faults, the number of multiple faults of interest would surely grow with network complexity much faster than the number of single faults. If n is the number of single faults, there are a total of 2^n multiple-fault combinations.

The number of single faults for which tests must be found will always be less than the obvious upper bound of twice the number of gate inputs and outputs. Figure 15.2 depicts two observations applicable to all combinational logic networks. In Fig. 15.2a we notice that all stuck-at-0 faults in an AND gate are really the same fault. All are *excited* only by driving all gate inputs to 1. We represent the excited fault by 1/0 at the gate output, indicating that the good network is 1 and the single-fault A SA0 network is 0. For an OR gate, all SA1 faults are the same. In Fig. 15.2b we see that the SA1 fault at the output of an AND gate need not be included in the list of distinct faults. It will always be tested by any test found for a SA1 input fault. Likewise, the SA0 output fault of an OR gate may be ignored. These observations are summarized as Theorem 15.1.

THEOREM 15.1 When n is the number of gate inputs, $n + 1$ single faults must be listed for an AND, OR, NAND, or NOR gate. For each AND (OR) gate, there are n SA1 (SA0) input faults and one SA0 (SA1) input fault.

Further reduction in the list of stuck-at faults depends on the properties of particular logic networks.

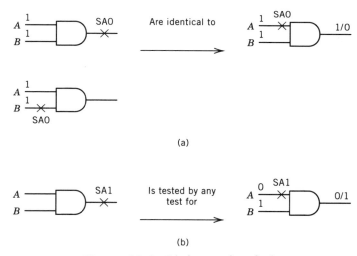

Figure 15.2 depicts diagrams.

(a)

(b)

Figure 15.2 Listing stuck-at faults.

■ EXAMPLE 15.1

Verify that the input vector $A = B = 1$ is a test for input B SA0 in the exclusive-OR implementation in Fig. 15.3a. Proceed as if the exclusive-OR gate were separately packaged, with A and B as package inputs and Z as the package output.

SOLUTION The values of the good and faulty networks are shown at each gate output. The generation of these values is accomplished by simultaneous simulation of the good and single-fault networks. The package output for the single-fault network differs from that of the good network, so the fault is considered detected.

Notice in Fig. 15.3b that the SA0 fault at the input to the OR gate is not detected by $A = B = 1$. We conclude that gate input and gate output faults connected at a fan-out point must be considered separate faults. ■

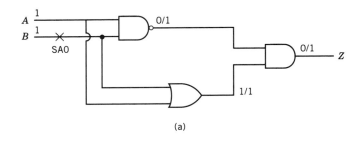

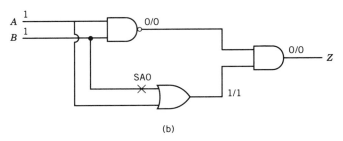

Figure 15.3   Faults connected to a fan-out point.

■ **EXAMPLE 15.2**

Develop a set of tests for all single faults in a separately packaged exclusive-OR gate. Assume that failures cannot occur at the input connections to the package.

SOLUTION   The assumption in the problem statement allows us to ignore the input points prior to fan-out and to regard the AND and OR gate inputs as primary inputs, as given in Fig. 15.4. Each gate input and output is assigned a distinct label in what is now a *fan-out-free* network.

Consider a SA0 at point $5'$. This will only be excited if the good network value at point 5 is 1. This, of course, is the good network value that will excite the SA0 fault at point 5. Therefore, a test for point 5 SA0 must be a test for point $5'$ SA0. The same argument is applicable to point $5'$ SA1 and to single faults at points 6 and $6'$. Theorem 15.1 tells us that faults at point 7 are detected by tests for faults at points $5'$ and $6'$, and that faults at points 5 and 6 are detected by tests for faults at points 1–4. Therefore, the list of distinct faults for the network will consist of the six faults at the inputs to the NAND and OR gates.

Figure 15.4*a* depicts a simulation of the good network and the point 1 SA1 network for inputs $A = 0, B = 1$. We see that the fault is detected. In this case, all four input combinations are required to detect the six distinct faults. Verification of this conjecture can be accomplished by simulation of each of the distinct faults for all four input combinations. We leave this computation, the results of which are summarized in Fig. 15.4*b*, to the reader. Notice that each fault is detected by only one test vector.

Although fan-out at the package inputs was ignored in this example, we may still conclude that modeling any exclusive-OR gate with the six faults identified above will force any test generation process to find a complete set of tests for that gate. That is, tests will be found involving all four combinations of input values. This will be true wherever the gate might be located in the network.   ■

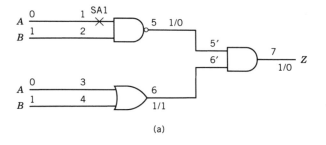

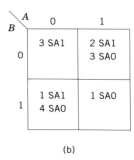

(b)

Figure 15.4 Test generation for an exclusive-OR gate.

The argument used in establishing the fault list in Example 15.2 may be generalized to form a proof of the following theorem.

THEOREM 15.2 A set of tests for all primary input faults will detect all single stuck-at faults in a fan-out-free combinational logic network.

In a combinational network with fan-out, it is only necessary to find tests for all primary input faults together with tests for all gate inputs connected to fan-out points.

15.3 TEST GENERATION STRATEGY

A test vector to detect a fault in a particular logic element must excite that fault at the output of the element. The test vector must also cause the effect of that fault to propagate forward to a network output. Figure 15.5 depicts the conditions that cause a fault, excited in a previous gate, to propagate through AND, OR, and exclusive-OR gates. We note that an excited fault will propagate through an exclusive-OR gate, regardless of the value on the other input. A zero at an AND gate input or a 1 at an OR input will block the path of a fault.

Once a set of tests has been found for detecting all faults in a combinational logic network, the Quine–McCluskey algorithm could, in theory, be used to

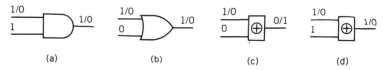

Figure 15.5 Fault propagation.

determine a minimal subset of these tests that will still detect all faults. The faults would take the place of minterms to be covered, and the tests would be entered in place of prime implicants. Unfortunately, this approach is computationally impractical for all but the simplest networks. Usually, one must settle for an approximation of a minimal test set. Occasionally for a special case, a unique argument that a particular test set is minimal can be found.

■ EXAMPLE 15.3

Determine a minimal set of tests for all stuck-at faults in the odd-parity network shown in Fig. 15.6a.

SOLUTION   We cannot argue that the circuit is fan-out-free, because there is fan-out within the exclusive-OR elements. We learned in Example 15.2 that all four combinations of input values are necessary to excite the internal faults of an exclusive-OR gate at the output of that element. The test set must include at least four vectors. Figure 15.5c and d tells us that once a fault is excited at the output of either G1 or G2, it will be propagated to the output of G3. This is illustrated in Fig. 15.6a for a fault excited at the output of G1.

The four test vectors given in Fig. 15.6b excite all internal faults at the outputs of G1 and G2 and will, therefore, cause them all to be detected. It must be emphasized that only single-fault networks are under consideration. No more than one fault will appear when the test vector is actually applied. Also tabulated in Fig. 15.6b for each of these test vectors are the values at the outputs of G1 and G2, when no internal faults are present. We see that the combinations of values of G1 and G2 constitute all possible input combinations to G3. These four test vectors will also cause the internal faults of G3 to be excited at the output of G3 and thus be detected.   ■

We can, at least, generalize from the above example that a minimal set of test vectors for a combinational logic network can be significantly less than the total number of input combinations required by the "prove it works" approach. We cannot expect to find a simple analytical argument leading to that minimal test

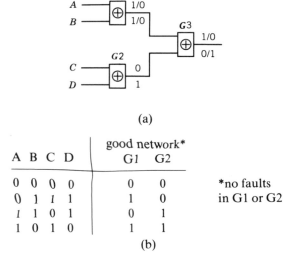

(a)

A	B	C	D	good network* G1	G2	
0	0	0	0	0	0	*no faults
0	1	1	1	1	0	in G1 or G2
1	1	0	1	0	1	
1	0	1	0	1	1	

(b)

Figure 15.6   Minimal set of test vectors for four-input odd-parity network.

set. The usual practice is to construct an approximately minimal test set by using some variation of the procedure depicted in Fig. 15.7. For combinational logic networks, a test vector is found for a single fault. Then the set of all single stuck-at faults detected by this same test vector is collected. One simple but time consuming approach to this step is simulation of all networks corresponding to undetected single faults. All faults newly detected by that vector are then deleted from the list of undetected faults. The process continues until all faults are detected.

With the following qualifications, the same approach can be used for sequential circuits. (1) Input sequence segments rather than individual test vectors are submitted for simulation. (2) The simulator must store the memory element values for all faulty networks following the simulation of a sequence segment. (3) The process is often terminated before all faults are detected, perhaps with 90% or 95% fault coverage. (4) Trial input sequences are not always determined by searching for a test for a single fault. Sequence segments that are known to exercise the circuit or even random sequences are sometimes used.

15.4 BOOLEAN DIFFERENCES

This section will present the first of two approaches to finding tests for a single fault. *Boolean differences* will find an expression for all possible tests for a fault. This Boolean algebraic approach is analytically useful, but will not lead to a computationally efficient implementation. Before defining a Boolean difference, let us obtain one preliminary result.

Shannon's expansion theorem from Section 5.11 may be used in the proof of a similar theorem for exclusive-OR. As defined in that section, $f_{x_i}(1)$ refers to the function resulting when 1 is substituted for x_i in f. Similarly for $f_{x_i}(0)$.

THEOREM 15.3

$$f(x_1, x_2, ..., x_n) = x_i f_{x_i}(1) \oplus \bar{x}_i f_{x_i}(0)$$

PROOF $x_i f_x(1) \oplus \bar{x}_i f_x(0)$

$$= [x_i f_{x_i}(1) + \bar{x}_i f_{x_i}(0)] \overline{[x_i f_{x_i}(1) \bar{x}_i f_{x_i}(0)]}$$
$$= [x_i f_{x_i}(1) + \bar{x}_i f_{x_i}(0)] (\bar{0})$$
$$= [x_i f_{x_i}(1) + \bar{x}_i f_{x_i}(0)]$$

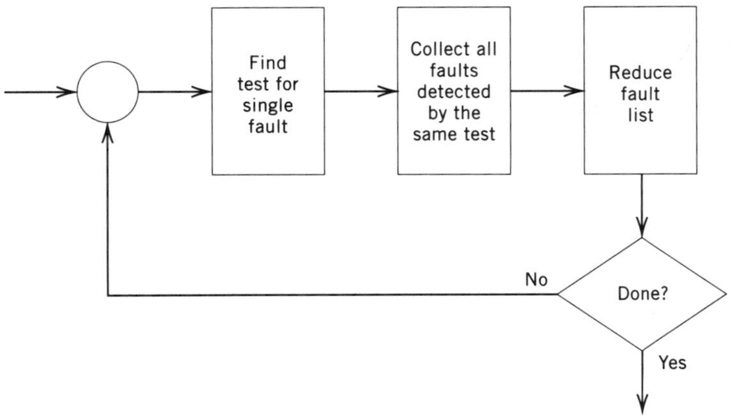

Figure 15.7 Generation of an approximately minimal test set.

and by Theorem 5.4 (Shannon's expansion theorem)

$$= f(x_1, x_2, ..., x_n)$$

If a fault at some point y in a combinational logic network is to be detectable at a particular output, the value of the output must be made to depend on the value at point y. That is, the value of the output when $y = 0$ must differ from the output value when $y = 1$. As shown in Fig. 15.8, point y may be a node within the network, or it may be desired that the value of f depend on a primary input x_i. We define the function

$$df/dx_i = f_{x_i}(1) \oplus f_{x_i}(0) \tag{15.1}$$

as the Boolean difference of f with respect to x_i. This Boolean difference will be a function of all the inputs except x_i and will be 1 if and only if f depends on x_i. The solutions to Equation 15.2 will cause f to depend on internal node y. df/dy will be a function of all variables.

$$df/dy = 1 \tag{15.2}$$

A SA0 fault at point y will be detected if Equation 15.2 is satisfied and $y = 1$. A SA1 fault at point y will be detected for input vectors that satisfy Equation 15.3.

$$(df/dy)\bar{y} = 1 \tag{15.3}$$

Chapter 8 of Reference [1] lists several Boolean difference identities. Of these we shall have occasion to use only Equation 15.4 that follows directly from $\bar{a} \oplus \bar{b} = a \oplus b$ (see Problem 15.3).

$$d\bar{f}/dx = df/dx \tag{15.4}$$

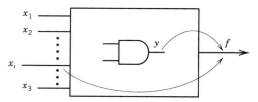

Figure 15.8 Defining Boolean differences.

■ **EXAMPLE 15.4**

Use Boolean differences to determine a test for input C SA1 in the combinational logic network of Fig. 15.9.

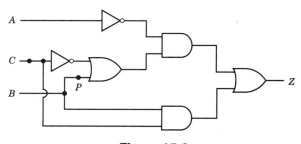

Figure 15.9

SOLUTION   The Boolean expression realized by the network is given in Equation 15.5.

$$Z = \bar{A}(B + \bar{C}) + BC \qquad (15.5)$$

First, we compute the Boolean difference dZ/dC.

$$
\begin{aligned}
dZ/dC &= f_C(0) \oplus f_C(1) \\
&= [\bar{A}(B + 1) + B0] \oplus [\bar{A}(B + 0) + B1] \\
&= \bar{A} + (\bar{A}B + B) = \bar{A} + B \qquad (15.6)
\end{aligned}
$$

The tests for C SA1 are given by

$$(dZ/dC)\,\bar{C} = (\bar{A} + B)\,\bar{C}$$

The three possible test vectors are $A, B, C = (0\ 1\ 0)$, $(1\ 1\ 0)$, and $(0\ 0\ 0)$. ∎

■ EXAMPLE 15.5

Find the set of all tests for point P SA0 in the network of Fig. 15.9.

SOLUTION   Although P is connected directly to the primary input B, the fan-out from that input makes it necessary to treat P as an internal network node. Accordingly, we insert P into the expression for output Z.

$$
\begin{aligned}
Z &= \bar{A}(P + \bar{C}) + BC \\
dZ/dP &= (\bar{A}\bar{C} + BC) \oplus (\bar{A} + BC) \\
&= (\bar{A} + BC)(\bar{C} + BC) \oplus (\bar{A} + BC) \\
&= (\overline{\bar{C} + BC})(\bar{A} + BC) \text{ (see Problem 15.3)} \\
&= C\,(\overline{BC})(\bar{A} + BC) = \bar{A}C(\bar{B} + \bar{C}) = \bar{A}\bar{B}C
\end{aligned}
$$

To excite the SA0 fault at point P, we must have $P = B = 1$. Therefore, the tests for P SA0 are given by

$$(dZ/dP)\,B = (\bar{A}\bar{B}C)\,B = 0$$

The existence of no tests for P SA0 tells us that this fault is undetectable. The reason is that this particular realization of Z is redundant. That is,

$$Z = \bar{A}(B + \bar{C}) + BC = \bar{A}\bar{C} + BC \qquad (15.7)$$

∎

The redundant product in Equation 15.7 eliminates a static hazard. Redundancies or undetectable faults are a problem in any implementation of a search to find tests for single faults. The search must continue until all possibilities are tried, before concluding that a fault is undetectable. An exhaustive search is time-consuming for combinational logic and cannot necessarily be accomplished for sequential circuits. This observation reinforces the importance of logic minimization, even though the payoff in reduced chip area may be unimportant. It is also another argument for clock-mode design, in which hazards may usually be ignored.

A chain rule applies in Boolean differences, extending the analogy with calculus to more than notation. The chain rule can be used in some instances to shorten algebraic manipulation.

THEOREM 15.4 Let $f(y,...)$ be a Boolean function of inputs to a combinational logic unit and some node y (perhaps an input). Let $G = g(f,h)$ be a composite function, where h is not a function of y. Then

$$dG/dy = (df/dy) \cdot (dg/df) \tag{15.8}$$

PROOF Using Theorem 15.3, we express G as

$$G = \bar{f}g_f(0) \oplus fg_f(1)$$

Now, because g_f is not a function of y

$$dG/dy = [\bar{f}_y(0)g_f(0) \oplus f_y(0)g_f(1)] \oplus [\bar{f}_y(1)g_f(0) \oplus f_y(1)g_f(1)]$$

Using the associative and distributive laws for exclusive-OR (Problems 4.16 and 4.17), we obtain

$$
\begin{aligned}
dG/dy &= [\bar{f}_y(0) \oplus \bar{f}_y(1)]\, g_f(0) \oplus [f_y(0) \oplus f_y(1)]\, g_f(1)] \\
&= d\bar{f}/dy\, g_f(0) \oplus df/dy\, g_f(1) \\
&= df/dy\, [g_f(0) \oplus g_f(1)] \qquad \text{(Equation 15.4)} \\
&= (df/dy)\,(dg/df)
\end{aligned}
$$

■ EXAMPLE 15.6

The four-input odd-parity network of Example 15.3, including the explicit realization of one of the exclusive-OR gates, is repeated in Fig. 15.10. Use Boolean differences to find expressions for all tests for input A SA0 and point P SA0.

SOLUTION In preparation for application of the chain rule, we compute dZ/dQ after expressing $Z = Q + (C + D)$.

$$
\begin{aligned}
dZ/dQ &= [1 \oplus (C \oplus D)] \oplus [0 \oplus (C \oplus D)] \\
&= (1 \oplus 0) \oplus [(C \oplus D) \oplus (C \oplus D)] \\
&= 1 + 0 = 1
\end{aligned}
$$

This reinforces our previous observation that any value on the other input will propagate a fault through an exclusive-OR gate. Now with $Q = (\overline{A B})(A + B)$

$$dZ/dA = dZ/dQ\, dQ/dA = dQ/dA = B \oplus \bar{B} = 1$$

Therefore, A is the expression for all tests for A SA0. Similarly, with $Q = (\overline{A B})(A + P)$

$$
\begin{aligned}
dZ/dP &= dZ/dQ\, dQ/dP = dQ/dP = (\overline{A B})\, A \oplus (\overline{A B}) \\
&= (\overline{A B})\, (A \oplus 1) = (\overline{A B})\, \bar{A} = \bar{A}
\end{aligned}
$$

The tests for P SA0 are given by $\bar{A} B$. ■

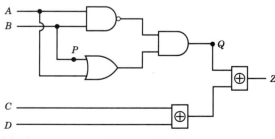

Figure 15.10

15.5 TEST GENERATION BY EVALUATION AND SEARCH

Algebraic methods like Boolean differences are often not well suited to implementation. Usually, one test vector for each fault is sufficient. Boolean differences yield an expression for all tests. In this section we will develop a test generation based on the determination of gate input values that will excite a fault and propagate that excited fault to the output. It will consist, insofar as possible, of explicit evaluation followed by a branch-and-bound search as required.

The most obvious network-based method for test generation, path sensitization, was demonstrated not to work [9] for certain circuits with reconvergent fanout. After this discovery, path sensitization was replaced by the D-algorithm [8], which for many years was the most commonly implemented test generation method. More recently, forward search methods, such as PODEM [6, 7] and FAN [10], have been found to be faster than the D-algorithm for many networks. The algorithm to be described below is new [4]. It is less complex than FAN and the D-algorithm and is easily shown to be faster than PODEM.

A three-value logical calculus was introduced in Section 12.3 to allow for unknown values in gate-level simulation. Here we shall use this same three-value logic, as summarized in Fig. 15.11, to provide for *unspecified* or not yet specified values at points within the network. We now have third symbols other than 0 or 1 to represent "don't care," unknown, and unspecified values. Regardless of the overlapping symbology, these three concepts are distinct, usually *but not always** appearing in different contexts.

Figure 15.5 suggests propagating a fault by assigning single 0 or 1 values to gate inputs on which the fault is not excited. This approach is adequate for simulation, but destroys important information in a test generation context. The gates in Fig. 15.5 actually display values for the good and single-fault networks. It is not necessary to assign the same values to a point in both networks. This notion is illustrated in Fig. 15.12a with a SA0 fault injected at the output of the OR gate. A 1 at AND gate input q is necessary to propagate the 1 value through that gate in the good network. However, the 0 on line r of the faulty network will propagate through the AND gate, regardless of the value of q. We indicate that q in the faulty network need not yet be specified by assigning value X. There are now nine possible combinations of good and faulty network values that may appear at any point in the network. The use of a nine-value calculus in this context first appeared in [5]. The development to be presented in the rest of this section is a simplification of an approach originated by Wang [4].

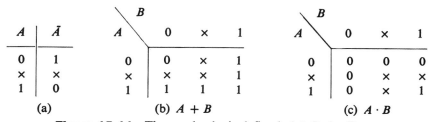

Figure 15.11 Three-value logic defined. $A + B. A \cdot B.$

*An interesting example of interaction between these concepts occurs in the simulation of hardware descriptions written in a high-level language. When combinations of values that are application "don't cares" appear as inputs, the simulation will compute and assign unknown values to points within the network. This situation will often occur in fault simulations that employ both high- and low-level hardware descriptions.

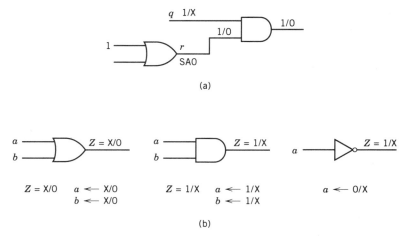

Figure 15.12   Forward and reverse propagation.

On occasion in the test generation process to be described, the output value of a gate will be determined before the inputs are specified. In instances where the output value forces values on all inputs, that value is propagated back through the gate. One such case is shown for each gate type in Fig. 15.12*b*.

The Wang algorithm for single-fault test generation for combinational logic is summarized in Fig. 15.13. Block 2 determines gate input values, leaving unspecified values whenever possible, to propagate an excited fault to an output. One path is selected, and the fault is propagated along this path only. Thus far, no gate input values have been specified other than those on the propagation path. Next gate input values are propagated backward, only insofar as values of all inputs to a gate are forced by the output value.

When good (faulty) network primary inputs have been specified by the backward propagation process of block 3, the faulty (good) ones are specified to agree. Next, the impact of these now specified primary input values is propagated forward. If a conflict arises, the test path is rejected. Once the forward propagation of established primary inputs is completed, block 5 accomplishes a branch-and-bound tree search over the remaining primary inputs. As each input value is specified, a node is created in the search tree and values are propagated forward through the network. If a conflict with already established values arises, the node is rejected; and network values, unique to that node, are erased. If not already considered, the other possible value of the same input is tried. When both nodes for the most recently specified input are rejected, the search backtracks to the predecessor node. The second value of the corresponding input is tried, if not already considered, and the search continues. Otherwise, the search backtracks again. When backtrack is impossible, the test path is rejected. The search will be illustrated in Example 15.7.

A node corresponding to a completely specified vector of input values, which does not conflict with values on the propagation path, is a test for the selected fault. Except for the criteria for rejecting a node, the forward search just described is similar to PODEM [6, 7]. The prior establishment of values on a path significantly shortens the time for evaluation of each node. In Reference [4], Wang extends the backward propagation step, thereby enlarging the footprint of the test path and further reducing evaluation time for each node. He also introduces a mechanism for estimating the best propagation path prior to application of the algorithm in Fig. 15.13. The five-value D-algorithm [8] was the first test genera-

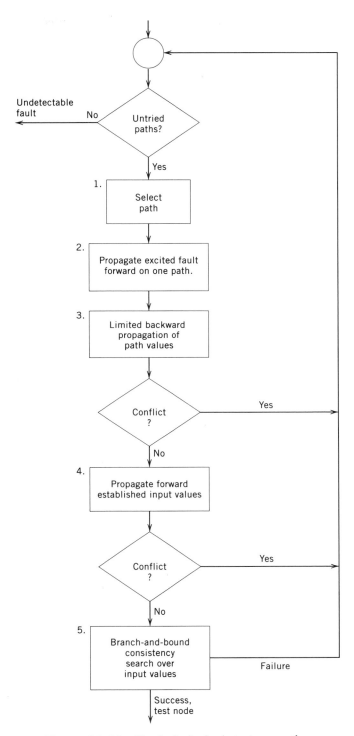

Figure 15.13 Single-fault single test generation.

tion procedure to eliminate problems with single-path propagation in circuits with reconvergent fan-out. Wang's method avoids these problems through separate specification of good and single-fault network values at each point. Solutions requiring fault propagation on multiple paths will appear during the search procedure.

■ EXAMPLE 15.7

Use the Wang algorithm to generate a test for point *P* SA0 in the combinational logic network of Fig. 15.14.

SOLUTION To excite the fault, one of two inputs to gate 5 must be 1. The lower of the two paths from gate 5 is selected. The excited fault is propagated along this path to the output. The values enclosed in rectangles in Fig. 15.14*a* are the result of the propagation step, together with the backward propagation of 1/X through gate 4 and X/0 through gate 11. Forward propagation of the established primary input values (*A* and *D*) determines the output of gate 7 to be consistent with the input to gate 11.

The starting node of the branch-and-bound search given in Fig. 15.14*b* is *ABCDEF* = 1XX1XX. Arbitrarily, it is decided to assign values to variables in the

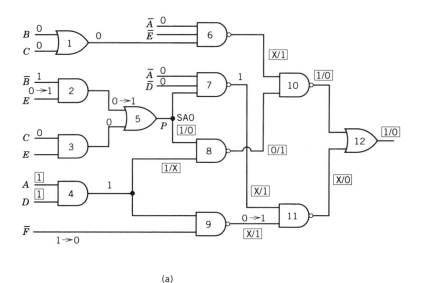

(a)

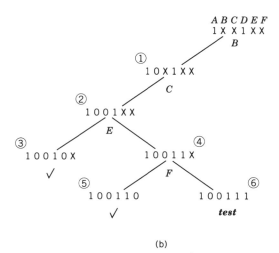

(b)

Figure 15.14 Test generation search.

order $B\,C\,E\,F$ and to try the value 0 first in each case. Other choices could be made on the basis of the heuristic evaluation of the network. Letting $B = 0$ to form search node (1) does not impact the network. The search continues forward, letting $C = 0$ to form node (2). Now the outputs of gates 1 and 3 evaluate to 0. This causes no conflict, so the search continues forward. Next E is set to 0 to form node (3). For this node, the output of gate 2 evaluates to 0. Both inputs to gate 5 are now 0, conflicting with the 1 output of that gate needed to excite the fault. Thus, node (3) is a terminal node (marked with a ✔) and the search must backtrack to node (2). Letting $E = 1$ (denoted $0 \rightarrow 1$) form node (4) resolves this conflict. The search continues forward, letting $F = 0$ generate node (5). Now the output of gate 9 evaluates to 0, conflicting with the input to gate 11 established by backward propagation. Node (5) is terminal. The search backtracks to node (4) to let $F = 1$. Now the values of all inputs are specified without conflict. The resulting node 100111 is a test vector.

There may be other tests for point P SA0; the process is required to find only one. If node (6) had been terminal, the search would have backtracked to node (1) to let $C = 1$. If the search had backtracked to the starting node with B already 1, the search fails, and the alternative propagation path is tried. ∎

Example 15.8 is the classic example used to show the inadequacy of five-value single-path propagation (see Problem 15.8b). The Wang algorithm is successful in this example, even though the fault is initially propagated forward along a single path.

■ EXAMPLE 15.8

Use the Wang algorithm to determine a test for point h SA0 in the combinational logic network of Fig. 15.15.

SOLUTION Again, the values established by propagating the excited fault through gate 9 to the output are enclosed by rectangles. Critical is the fact that three inputs to gate 12 are assigned 0/X rather than 0/0. Primary inputs x_2 and x_3 must be 0 to excite the fault. By slightly extending the capability of block 4 of Fig. 15.13, it is possible to avoid a search. After propagating $x_3 = 0$ to the input of gate 11, it is possible to uniquely determine that the other input to that gate must be 1/X. We assume that block 4 is extended to allow the backward propagation of an output value on a single-gate input once the other inputs are appropriately

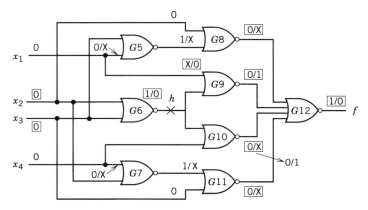

Figure 15.15 Multiple-path fault propagation.

specified. This backward propagation continues through gate 7, establishing that x_4 must be 0. Similarly, x_1 is fixed at 0 by backward propagation through gate 8 and then gate 5. Propagation of these newly established inputs forward is accomplished without conflict. In particular, 1/0 is propagated through gate 10. The output of gate 10 becomes 0/1, consistent with the required 0/X. ∎

We come away from Example 15.8 with the key observation that inputs 0, 0, 0, 0 propagate the excited fault forward on two paths that reconverge at gate 12. There is no other test for the fault (see Problem 15.10).

In implementing the simplified Wang algorithm, the same capability for the backward propagation of values discussed above must also be included in the branch-and-bound search. This will have the effect of reducing the time required by subsequent node evaluations.

15.6 MODELING CMOS STUCK-OPEN FAULTS

CMOS presents some special testability problems not found in other logic families. In MOS circuits, faults typically arise from stuck-open or stuck-short FET devices. In NMOS gates, these conditions may be interpreted directly as traditional stuck-at faults on inputs and outputs. In CMOS circuits, stuck-short faults result in intermediate output voltage levels that may or may not be distinguished from the corresponding voltages in the good network. Baschiera [11] and others have argued that tests for stuck-short faults can be accomplished only by monitoring power-supply current during the testing process or with the addition of overhead FET devices.

Test vectors that excite stuck-open device faults cause the output of the CMOS gate to retain its previous value. The test process must, therefore, cause the output of the gate in the fault-free network to assume opposite values over two consecutive clock periods. A test for a stuck-open fault in a combinational logic network will consist of two successive test patterns.

Complex networks of devices in the source with dual networks in the drain are often used to realize AND-OR-INVERT logic functions in CMOS standard cells. It has been observed that some tests for stuck-open faults in these networks are dependent on the order of input changes. Moritz [12] also suggests that charge-sharing between network nodes in a marginal CMOS AND-OR-INVERT network can invalidate a static test. The fault would be detectable only if increasing the delay associated with the gate would cause a malfunction. In the discussion that follows we assume that circuits with more than two devices in series are buffered by an output inverter so that charge-sharing will not mask a stuck-open device. Test pairs for stuck-open faults in general CMOS circuits can be found by modeling these circuits with gate-equivalent networks.

Before considering the more complex CMOS AND-OR-INVERT networks, let us focus on exciting the set of all stuck-open faults in the simple NAND structure of Fig. 15.16a. Clearly, $N1$ stuck open and $N2$ stuck open are indistinguishable. The test $(T1, T2)$ of Table 15.1 is a test for this fault. That is, $T1$ initializes the output of the gate to 1 and $T2$ causes the output of the good network to go to 0, while leaving the faulty network at the original value, 1. Similarly, the vector pair $(T3, T4)$ excites the fault $P1$ stuck open, and the pair $(T5, T6)$ excites $P2$ stuck open. If the good network output was 0 before application of $T2$, this value would be stored at the output node when this node was disconnected by the stuck-open

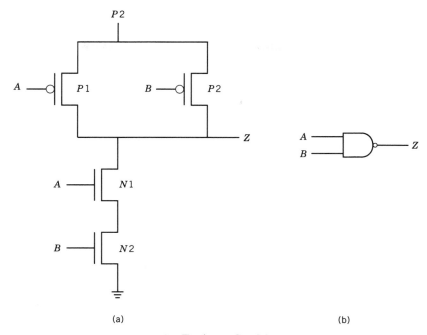

Figure 15.16 Testing a CMOS NAND gate.

Table 15.1 Fault Analogies

	$T1$	$T2$	$T3$	$T4$	$T5$	$T6$
A	0	1	1	0	1	1
B	X	1	1	1	1	0
Stuck-open fault		$N1$		$P1$		$P2$
Stuck-at fault	Z	SA1	A	SA1	B	SA1

fault. Thus, the faulty network and good network would have the same value after the application of $T2$. Without initializing the output to 1, the fault $N2$ stuck open would not be detected by $T2$.

Now consider the stuck-at faults of the generic NAND gate of Fig. 15.16b. Note that $T2$ is the test for the output of this gate stuck-at-1 (SA1). $T4$ is the test for input A SA1, and $T6$ the test for input B SA1. Both $T4$ and $T6$ are tests for Z SA0, so this fault may be disregarded. The argument is immediately extendible to NAND gates with any number of inputs, with the result expressible as Theorem 15.5.

THEOREM 15.5 There exists a 1-1 mapping of stuck-open faults in a CMOS NAND gate onto the set of all stuck-at faults in an arbitrary n-input NAND gate. Each test for a stuck-open fault in the n-input CMOS NAND gate constitutes a test for the corresponding stuck-at fault in any n-input NAND gate. Every test vector for a stuck-at fault in an arbitrary n-input NAND gate can be extended to form a test for the corresponding stuck-open fault in a CMOS NAND gate by prefacing that test vector with a vector that will initialize the output of the fault-free gate to the opposite value.

Theorem 15.5 is easily extended to other gate configurations and can be applied in a variety of ways in adapting tools for test generation for stuck-at faults to handle stuck-open faults. It eliminates the need for reference to a transistor-level diagram for enumerating the stuck-open faults or referring to a particular fault. Here, we shall use the term ST1 in place of SA1 to indicate that a fault listed at a gate input or output represents an internal stuck-open. This notation will emphasize that in addition to the SA1 test vector, an initializing vector is necessary to excite the fault.

■ **EXAMPLE 15.9**

In the more complex network of Fig. 15.17a, the vectors $T7 = (1, 1, 0, 0)$ followed by $T8 = (0, 1, 1, 0)$ would be expected to be a test for device PD stuck open. Show that there is an order of input changes in the transition from $T7$ to $T8$ that will invalidate the test.

SOLUTION $T7$ will drive the circuit output Z to 1, after which $T8$ will cause the output of the good network to go to 0. For the faulty network with PD stuck open, the output is expected to retain its previous value. However, if A goes to 0 before C goes to 1, a momentary path through PC and PA is created, charging the capacitance at the input to the inverter and invalidating the test. Regardless of the timing of input changes, the test could be invalidated, if sufficient charge could be stored at point Q while $T7$ is active. A substantial charge at point Q could flow to the input of the inverter once $T8$ is applied, again driving the out-

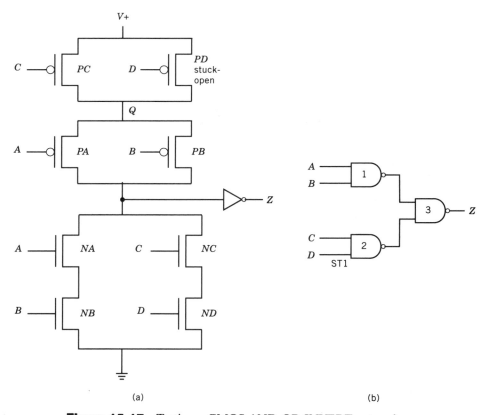

(a)   (b)

Figure 15.17   Testing a CMOS AND-OR-INVERT network.

put to 0 irrespective of the fault. This is an example of charge-sharing, as introduced in Chapter 7. ■

The all-NAND network of Fig. 15.17*b* is the "gate equivalent" (realizes the same function) of the AND-OR-INVERT network of Fig. 15.17*a*. We shall verify that, with the addition of an initializing vector, every test for the ST1 fault on input *D* of gate 2 is a test for *PD* stuck open in Fig. 15.17*a*. However, the vectors (*T*7, *T*8) that were the problem test for *PD* stuck open would not be identified as a test for the analogous fault in Fig. 15.17*b*. The first vector *T*7 will drive the output *Z* to the correct initial value, but will not initialize the output of gate 2 to 0 as required. The vector *T*7′ = (X, X, 1, 1) will properly initialize gate 2 to 0. Therefore, (*T*7′, *T*8) is a test for *D* ST1. This approach may be used to find tests for all stuck-open faults in an AND-OR-INVERT. The second vector of each test must test for the analogous stuck-at fault in the gate-equivalent network, and the *initializing vector must initialize the opposite value at the output of the individual gate in that gate-equivalent network*.

15.7 FAULT-SIMULATION IN SEQUENTIAL SYSTEMS

We have completed our discussion of test generation in combinational logic and are now ready to widen our focus to include sequential circuits. Algorithms for automatic test generation in sequential circuits are either suitable for only unrealistically small circuits or are nondeterministic and insatiable in their consumption of computer time. Very often, explicit effort to search for test sequences over sequential models is simply omitted from the design process. Frequently, an input sequence, found during the design process to *exercise* a system, is used to drive a simulation of all faulty networks. This approach is often called *fault-grading*. In other cases, tests are automatically generated for complex combinational logic portions of sequential systems, with these tests applied through *scan paths*, a technique discussed in the next section.

Even if we accept the silver spoon law articulated in Section 15.2 and consider only stuck-at faults, the time required for fault simulation is often enormous and increases with the cube of the number of gates. Simulation time for the good network typically increases as the square of the size of the design. In fault simulation, every single-fault network must be simulated separately and the number of faults is proportional to the number of gates.

Several approaches have been proposed to reduce the time consumed by fault simulation. The first to appear, parallel fault simulation [19], used each individual bit of a data word to represent a separate single-fault network. This method was eventually found wanting [16], because the size of the database could not be reduced as the simulation continued after most faults were detected. Subsequent methods have included deductive simulation, concurrent simulation, reverse fault collection in clock-mode simulation [15], and hardware acceleration (special-purpose parallel architecture). In the end, most readers will care only that all single-fault networks are individually simulated. Few will resent that the details of the simulator are hidden from view.

The deductive approach [18] is the most convenient platform for discussing issues common to all approaches to clock-mode fault simulation. Simulating only in the clock mode is certainly a first step in minimizing the time required for fault simulation. In deductive simulation, the good network is simulated for a particular gate and then faults local to that gate are examined to identify faults for which the gate output will differ from the good network value. These will be included in a fault list assigned to a gate output. Fault lists will also appear at inputs to the gate. Depending on good network values, these lists may or may not propagate through the gate, to be included in the output fault list. The simulation of a clock period is concluded by storing the fault list at each D input in the respective memory element.

Figure 15.18 depicts the separate deductive simulation of two gates. The reader should be able to infer the handling of other gate types and other input combinations from the following discussion of these two cases. During the clock period depicted in Fig. 15.18a, the AND gate inputs are 0, 1 and the good network output is, therefore, 0. The fault list adjacent to input $I1$ includes all single-fault networks for which that input value differs from the good network input (is 1). The fault list adjacent to $I2$ includes all networks for which $I2 = 0$. The gate output will be 0 regardless of the value of $I2$; hence, none of the faults in this list will propagate through the gate or be included in the output fault list. With one exception, the fault list for input $I1$ will propagate through the gate. The exception is the network, gate15I1SA0 because gate 15 is the gate being simulated. The fault must have been stored in a memory element after a previous clock period and fed back to the originating gate. For this network, the gate output will be 0, the same value as the good network, so that the presence of gate15I1SA0 in the incoming list has no effect. This phenomenon is called *reconvergent fan-out over time*. Particular care must be taken to assure correct simulation when this type of reconvergence occurs. Fault networks faultxx3 and faultxx4 are joined in the output fault list by local fault gate15I1SA1, which will cause the gate output to be 1.

As we asserted in Section 15.3, faults propagate through an exclusive-OR gate, regardless of good network values. The exception as noted in Fig. 15.18b are faults that simultaneously appear on both inputs, as is true for fault network faultxx3 in Fig. 15.18b. When both inputs are changed, the outputs remain the same, so faultxx3 is not included in the output fault list. Again, reconvergent fan-out has occurred. If one of the fault lists had propagated from the output of a memory element, it would be *reconvergent fan-out over time*.

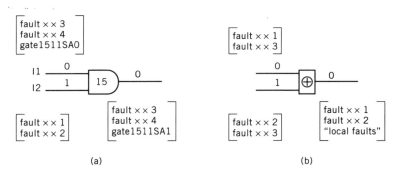

Figure 15.18 Deductive fault simulation prototypes.

■ EXAMPLE 15.10

Execute a three-clock-period clock-mode deductive fault simulation for the sequential circuit of Fig. 15.19. Assume that the initial values of $Q1$ and $Q2$ are 0 and 1, respectively. The values of b for three consecutive clock periods are 1, 1, 0.

SOLUTION   Theorem 15.2 allows us to ignore the internal faults of AND gate $A2$, expecting them to be detected if faults at the inputs to NAND1 and OR4 are detected. The internal faults of the inverter are also ignored, because they are equivalent to faults at the input of gate $A3$. Internal faults within the memory elements may invalidate the clock-mode simulation [19], but clock-mode simulation is specified in the problem statement. By modeling memory element faults as stuck-at outputs only, we are relying on a slight extension of the "silver spoon law," that all faults will be detected by tests for clock-mode stuck-at faults. Assigning letters to gate inputs to shorten the fault notation yields the following list of all distinct faults: pSA1, rSA1, pSA0, sSA0, uSa0, sSA1, vSA1, wSA1, vSA0, $Q1$SA1, $Q1$SA0, $Q2$SA1, $Q2$SA0.

The three clock periods of simulation are shown as the three rows of Table 15.2. Values for each gate output in the good network are shown above the fault list deduced for the same gate. At the start of the simulation, no faults have been driven to the D inputs and stored in the memory elements. The stored faults represent fault networks corresponding to memory element outputs stuck opposite the stored values for the good network. The short fault list from $Q2$ is fed back to the logic network to begin the simulation. The fault list from $Q1$ propagates to the input of flip-flop $Q2$ into which it is clocked at the end of the first period. The fault list from gate $A2$ is simultaneously clocked into $Q1$. Only one fault is detected prior to the third clock period, because the fault list on line v is blocked by the 0 on line w. Notice the faults Q1SA1 and Q2SA0 appeared to have propagated back into memory elements $Q1$ and $Q2$, respectively. These faults are not actually stored because the values of the good and faulty networks are the same. The two self-masked faults are lined out in Table 15.2. This is another example of reconvergent fan-out over time.   ∎

It is less than obvious how the initial state of 0, 1 might actually be established in the sequential circuit of Fig. 15.19. In the absence of preset and clear lines not shown in the figure, it is impossible to establish any known initial state. Consider a three-value simulation of the good network with $Q1$ and $Q2$ initialized to the unknown value U. The logic network will effectively exclusive-OR the U in $Q2$ with a known input, to generate another U to be clocked into $Q1$. The U in $Q1$ is passed to $Q2$; the network remains in an unknown state indefinitely. We say that the circuit of Fig. 15.19 lacks a *synchronizing sequence*. **A synchronizing sequence is a sequence of inputs that will drive a given sequential circuit from any initial state to a**

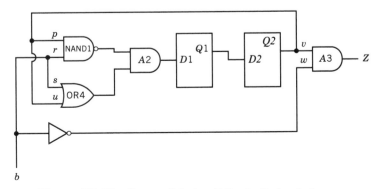

Figure 15.19   Sequential circuit for fault simulation.

Table 15.2 Deductive Fault Simulation

	$Q1$	$Q2$	b	NAND1	OR4	$A2$	$A3$ (detected)
Good	0	1	1	0	1	0	0
Fault	$Q1$SA1	$Q2$SA0		$Q2$SA0	none	$Q2$SA0	wSA1
list				pSA0		pSA0	
Good	0	0	1	1	1	1	0
Fault	$Q2$SA0	$Q2$SA1		$Q2$SA1	sSA0	sSA0	none
list	pSA0	$Q1$SA1		$Q1$SA1		$Q2$SA1	
	$Q1$SA1			pSA1		$Q1$SA1	
						pSA1	
Good	1	0	0	1	0	0	0
Fault	sSA0	~~$Q2$SA0~~		none	sSA1	sSA1	vSA1
list	$Q2$SA1	$Q1$SA1			$Q1$SA1	$Q1$SA1	$Q1$SA1
	~~$Q1$SA1~~	pSA0			pSA0	pSA0	pSA0
	pSA1	$Q2$SA1			$Q2$SA1	$Q2$SA1	$Q2$SA1
	$Q1$SA0						

single known state. Problem 15.16 introduces a simple modification of Fig. 15.19 to provide for a synchronizing sequence.

15.8 OBSERVABILITY, CONTROLLABILITY, AND SCAN PATHS

A test for a single stuck-at fault in a sequential circuit can be considered the concatenation of two sequences. The first or *state justification* sequence will take the circuit to a state that will allow the single-fault version to malfunction. The result can be interpreted as *the single-fault network reaching the wrong state* or equally well the *storage of the excited fault in a memory element*. A measure of the average length of a state justification sequence may be defined as *controllability*. In a circuit with low controllability, state justification sequences are long and difficult to find. High controllability implies the existence of a short state justification sequence.

Once a circuit has been driven to a wrong state or the fault has been stored, a *fault propagation* sequence, which will cause the stored fault or state transition malfunction to impact on the output, must be found. If a typical propagation sequence is short, the observability in the circuit under test is high. An easily testable circuit would have both high controllability and high observability.

Consider an eight memory element digital system, defined primarily by the following AHPL statements. Included are two combinational logic units, the details of which we will leave for conjecture.

$$R \leftarrow \text{NSLOGIC}(X, R); \tag{15.9}$$

$$z = \text{OUTLOGIC}(R).$$

Controllability and observability will depend on the complexity of the combinational logic units. Observability is limited by the existence of only one output line (z is a scalar.). Indeed, a fault stored in a single memory element of R might not impact the output. In that case, it would be necessary to find a propagation sequence to drive the good and single-fault networks to states for which the outputs will differ.

If a sequential system is considered too difficult to test, controllability and observability may be increased by a simple redesign to add a scan path. Statements 15.9 with a scan path added are given as 15.10.

$$R \leftarrow \text{NSLOGIC}(X, R) \,!\, (td, R[0:6]) * (\overline{tc}, tc); \tag{15.10}$$
$$z = \text{OUTLOGIC}(R); \; ztest = R[7].$$

Two additional input lines, *td* and *tc*, and a single test output line, *ztest*, have been added. If *tc* is 1, the circuit is in the test mode. If *tc* = 0, the circuit will function normally. When operating in the test mode, values may be shifted to the right through register *R*, which is now a scan path. In the process, the original contents of *R* are shifted serially out on line *ztest*. Now the two internal combinational logic units may be treated as if they were independent combinational logic networks. Test vectors may be shifted into *R* along the scan path and the values of the eight output lines of NSlogic may be shifted serially to *ztest*. Tests found for the separate combinational logic units will thus test the sequential circuit. A realization of statements 15.10 is depicted in Fig. 15.20. The simple busing network at the input to each memory element and the three additional input/output wires are hardware overhead, the price of enhancing testability.

The addition of scan paths at the RTL level allows consideration of the relative testability of a design for different overhead configurations. Evaluation can be accomplished by fault simulation with a minimum of design time committed to a particular approach. In some design environments, scan paths remain transparent to the user until the design has reached the net list or even the layout level. This implies forwarding a design to a separate organization for test generation, a deceptively inefficient approach. *The designer should be responsible for test.*

■ EXAMPLE 15.11

The many important VLSI chips that include counters are among the least controllable and most difficult to test configurations. Evaluate the testability of the stopwatch introduced in Section 11.5 and refine the design with the addition of scan paths to achieve testability with minimal hardware overhead.

SOLUTION   Controllability is minimal in the stopwatch, as given in Example 11.6. Each individual counter must increment to its maximum value before triggering the count for the next digit. The *TENTHS*, *SECONDSL*, and *SECONDSH* are observable, because they are connected directly to chip outputs. We take the position that it is only necessary to cause these registers to increment each clock period independent of the value in *TCNT*. In order to test the logic associated with *TCNT*, a scan path is introduced in that register. The AHPL description of the stopwatch, as modified for testability, is given below. Again, we include the test lines *tc*, *td*, and *ztest*. The signal *go* is replaced by *tgo* = *tc* ∨ *go* in the clock control

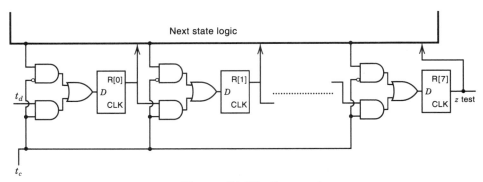

Figure 15.20   Scan path.

expression of each transfer. Only the expression transferred into *TCNT* is further modified to form the scan path activated by *tc*.

MODULE: TESTABLESTOPWATCH.
INPUTS: *start*; *stop*; *tc*; *td*.
MEMORY: *TCNT*[17]; *TENTHS*[4]; *SECONDSL*[4]; *SECONDSH*[4].
OUTPUTS: *tgo*; *ztest*; *max*; *ZTENTHS*[4]; *ZSECONDSL*[4]; *ZSECONDSH*[4].
CLUNITS: INCCLK[17] <: INCR{17}.
CLUNITS: INC[4] <: INCR{4}.

1   *go* * (*start* ∨ *stop*) ← *start*;
    TCNT * *tgo* ← (INCCLK(*TCNT*) ! 17 T 0 ! *TCNT*[1:16],*td*)
                    * ($\overline{tc}$ ∧ $\overline{max}$, $\overline{tc}$ ∧ *max*, *tc*);
    TENTHS * *tgo* ∧ *max* ← INC(*TENTHS*) ∧ ($\overline{TENTHS[0] \land TENTHS[3]}$);
    SECONDSL * *tgo* ∧ *max* ∧ (*TENTHS*[0] ∧ *TENTHS*[3]) ←
                    INC(*SECONDSL*) ∧ ($\overline{SECONDSL[0] \land SECONDSL[3]}$);
    SECONDSH * *tgo* ∧ *max* ∧ (*SECONDSL*[0] ∧ *SECONDSL*[3]) ∧
                    (*TENTHS*[0] ∧ *TENTHS*[3]) ← INC(*SECONDSH*);
    → ($\overline{start}$ ∨ *tgo*)/ (1).

2   *TCNT* ← 17 $ 0;
    TENTHS ← 0, 0, 0, 0;
    SECONDSL ← 0, 0, 0, 0;
    SECONDSH ← 0, 0, 0, 0;
    → (1).

ENDSEQUENCE
tgo = *go* ∨ *tc*;
ztest = *TCNT[0]*;
max = *TCNT*[0]∧*TCNT*[1]∧*TCNT*[6]∧*TCNT*[7]∧*TCNT*[9]∧(∧/*TCNT*[12:16]);
ZTENTHS = *TENTHS*; *ZSECONDSL* = *SECONDSL*; *ZSECONDSH* = *SECONDSH*.
END.   ■

The hardware overhead introduced in the above example is less than would be necessary to provide an end-to-end scan path and to allow reliance on test generation for combinational logic. We offer no argument that TESTABLE-STOPWATCH satisfies any measurable criteria for controllability or observability. A design will usually be approved, if fault simulation verifies the existence of an input sequence that will test 90% to 95% of the stuck-at faults. The sequence might be devised analytically or be generated by single-fault searches. The authors' paper [21] describes an effective method for single-fault search over time. The first to appear and most interesting approach to sequential circuit fault search reduces the sequential circuit to combinational logic via an iterative network analogy. This method is not computationally efficient, but will follow almost immediately once iterative networks have been introduced in Section 16.7. We, therefore, defer the topic to Section 16.8.

15.9   BOUNDARY SCAN

The testing of VLSI packages to be placed on a printed circuit board does not eliminate the need to test the finished board. Board testing might be confined to testing failures, introduced in the process of bonding VLSI packages to the board. Alternatively, this testing might extend to less than exhaustive testing for failures internal to the chip. If the board manufacturing process or subsequent *burn-in* causes a chip to fail, this failure will usually be easy to detect. An array of probes (*bed of nails*) to make contact with strategic interconnections on a board has traditionally been a critical feature of the board-testing process. With the continued increase in package density and the acceptance of new package configurations, the bed-of-nails approach has become more difficult and more expensive.

The scan register technique discussed in the previous section has been accepted with enthusiasm. With the ever-increasing number of gates realizable on a chip, the resistance to allocating significant portions of the chip area to facilitate testing has diminished to the vanishing point. How would it be possible to use an analogous concept to ease the problem of board-testing? The answer has been the allocation of still more area on individual VLSI chips to facilitate board-testing. The mechanism, called *boundary scan*, is illustrated in Fig. 15.21.

Notice that two packages in Fig. 15.21 are mounted on a board in very close proximity, making it difficult to probe the connection lines between the packages. The internal logic of the rightmost chip is partially exposed. The boundary scan registers shown at the left and right on the chip are provided to aid the board-level test. If the core logic of the chip were originally testable by application of test vectors by way of the normal input/output pins, it may now be tested by scanning these vectors in and out on only two pins, which may be probed at less cost. In the extreme, the boundary-scan registers of several chips might be connected in series and routed to the board edge connector, completely eliminating probing.

Boundary scan can only be made to work if everyone, including chip suppliers and systems manufacturers, "play the same game." Consequently, a boundary-scan standard usually referred to as JTAG/P1149.1 [20, 24] has been established. The JTAG protocol is more complex than suggested above and incorporates dedicated on-chip test control. The precise JTAG protocol is properly classified

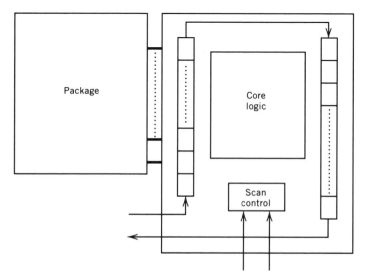

Figure 15.21 Boundary scan.

as "technology" and cannot be included here. The most readily available documentation is a primer [24] made available by Texas Instruments Corporation.

15.10 BUILT-IN SELF-TEST

The notion that a VLSI chip might test itself after connection and power-up must be considered attractive. If the chip to be tested were a processor, it could be asked to execute a program that has been determined in advance to test a satisfactorily large percentage of chip faults. The test results might be collected on chip, reducing the function of the tester to making a yes or no decision.

Storing long sequences of predetermined test vectors on chip is not an attractive alternative for testing chips that are not processors. If one is willing to buy the self-test at the price of longer test time, testing can be accomplished by generating on chip all possible input vectors to isolated blocks of combinational logic. Suppose a particular block of combinational logic (perhaps a data unit) had 16 inputs. The exhaustive set of test vectors could be generated by a 16-bit counter, but the maximal-length linear feedback shift register (LFSR) depicted in Fig. 15.22 will generate all 16-bit vectors (except 16 T 0), while requiring substantially less chip area. The reader with experience in error-correcting codes will realize that not every network of 16 memory elements with exclusive-OR feedback will generate a sequence of $2^{16} - 1$ distinct vectors. Available space will not accommodate the luxury of a proof, but the sequential circuit in Fig. 15.22 is indeed the maximal-length LFSR.

It is likewise undesirable to store on chip the expected output vectors necessary to evaluate each vector as it appears at the output of a combinational logic unit under test. A *signature register* may be used to compress this data into a single vector. With modest modification, a maximal-length LFSR may be configured as a signature register. The AHPL description of one such multipurpose register is given below.

$$R[1:15] \leftarrow (\mathbf{a} \wedge X[1:15]) \oplus (\mathbf{b} \wedge R[0:14]);$$

$$R[0] \leftarrow (\mathbf{a} \wedge X[0]) \oplus \mathbf{b} \wedge (R[3] \oplus R[12] \oplus R[14] \oplus R[15]);$$

For $\mathbf{a},\mathbf{b} = 0, 0$, the register is reset. For $\mathbf{a}, \mathbf{b} = 0, 1$, the register is a maximal-length vector generator. For $\mathbf{a},\mathbf{b} = 1, 0$, it is loaded with X. For $\mathbf{a},\mathbf{b} = 1, 1$, the register is a signature register. The signature register exclusive-ORs the vector X with the vector that would normally be shifted along the LFSR. At the start of the test, the signature register is initialized to 16 T 0. Each clock period, the vector X is the output from a combinational logic unit under test. A particular signature should be found in the signature register at the end of a $2^{16} - 1$ vector test process. If any other vector is found in R following the test, the logic unit includes a failure. Unlikely combinations of faults can cause the test procedure to fail. If in any period, the error pattern on lines X is the same as the stored cumulative

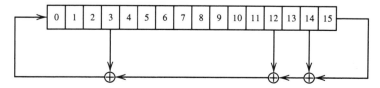

Figure 15.22 Maximal-length FLSR.

error pattern, the patterns will cancel and the register will assume an error-free pattern. This cannot happen for a single fault that is always excited on the same output line.

Figure 15.23 depicts the use of a signature register in two self-test applications. In Fig. 15.23b, signature registers collect values from both the address and data buses as the microprocessor executes a program. A somewhat more detailed analysis of these applications may be found in Miczo [6].

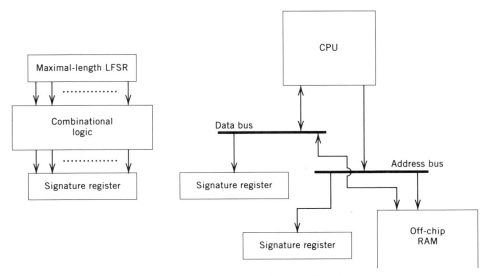

Figure 15.23 Signature register applications.

PROBLEMS

15.1 The set of tests $S_a = \{000\,(A,B,C = 0, 0, 0), 111\}$ is a test for all SA0 and SA1 faults on the three primary input lines in the two EXOR gate circuits shown in Fig. P15.1.

(a) If S_a will not test a SA0 fault at point p, form S_b by adding a test (for which $A = 1$) for this fault to S_a.

(b) Form S_c by adding to S_b any tests needed to test for p SA1, z SA0, or z SA1. Add as few tests as possible and indicate the purpose of each test.

(c) Let g be the Boolean expression for the good network, f' be the expression with multiple fault (A SA0, C SA0) imposed, and f'' the expression with (A SA0, B SA0). Choose f' or f'', as appropriate, and show by algebraic manipulation that S_b will **not** test the respective multiple fault.

(d) Do the above results conflict with Theorem 15.2? Why or why not?

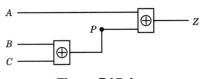

Figure P15.1

15.2 For some reason, a logic function has only six possible disjoint fault conditions: f_a, f_b, f_c, f_d, f_e, and f_i. There are six tests that test for various combinations of these six faults as tabulated below. A 1 at ij indicates that the test in row i tests for the fault in column j. Use the generalized Quine–McCluskey algorithm to determine a minimal set of fault detection tests for the faults. At each step, indicate why a test is found to be essential or eliminated from the list. Consider set B to be the initial set of tests and that including a test for a particular fault is analogous to covering a minterm.

	f_a	f_b	f_c	f_d	f_e	f_i
$T1$	1		1			
$T2$		1				1
$T3$				1	1	
$T4$		1			1	
$T5$		1	1			
$T6$			1	1		

15.3 Prove each of the following minor results involving exclusive-OR.
(a) $ab \oplus b = \bar{a}b$
(b) $(a + b) \oplus b = \bar{b}a$
(c) $(a + b) \oplus (c + d) = (c \oplus a)\bar{b}$
(d) $\bar{a} \oplus \bar{b} = a \oplus b$

15.4 Use Boolean differences to find a Boolean expression representing all possible tests for point x **stuck-at-1** in the logic circuit Fig. P15.4.

15.5 Construct the logic block diagram of a combinational logic network in which an undetectable stuck-at-1 fault could occur. Indicate the line on which that undetectable fault would appear.

15.6 Consider a two-level sum-of-products network realizing the Boolean expression

$$f(A,B,C,D) = AB + ACD + BD$$

(a) Use Boolean differences to find the set of all tests for input c SA0.
(b) Use Boolean differences together with the procedure described in Fig. 15.7 to find the set of tests for all stuck-at faults in the network. Use Theorem 15.2 to reduce the initial fault list.

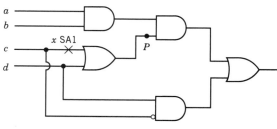

Figure P15.4

15.7 Use Boolean differences to prove that the fault point p SA1 in Fig. P15.7 is undetectable. Make use of the theorem

$$X \oplus (X \oplus Z) = \bar{Z}X$$

15.8 Use Boolean differences to verify that $0, 0, 0, 0$ is the only test for h SA0 in Fig. 15.15a.

(b) It would seem that $x_1 = 0$ and inputs to $G12$ of $0, 0/1, 0, 0$ would constitute a test for h SA0. Prove that no set of input values is consistent with these values. This is the counter example to simple *path sensitization* as a means of test generation.

15.9 Use Boolean differences to find a set of tests for input c SA0 in a direct implementation of the Boolean expression

$$g = \overline{[(a\,b)(\overline{b\,h})(\overline{e\,h})]}, \qquad \text{where } h = cd$$

15.10 One interpretation of the Wang algorithm calls for assigning values to input lines in alphabetical order and always trying the value 0 before 1. In the logic circuit shown in Fig. P15.10, use the Wang algorithm to find a test for point P stuck-at-0. Include a block diagram showing network values immediately following forward propagation of the fault and presearch backward propagation of path values. Show the search tree in complete detail. Indicate only nodes that would actually be generated by this version of the algorithm. Show terminal nodes with a checkmark,

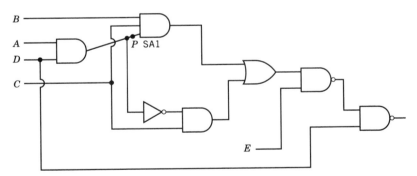

Figure P15.7

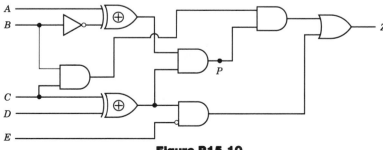

Figure P15.10

and in each such case cite the reason the node is terminal. Indicate the success node on the tree.

15.11 Repeat Problem 15.9 for point P SA1 in Fig. P15.7. What observation will the algorithm pass to the user to indicate that the fault is undetectable?

15.12 Repeat Problem 15.10 for point P SA1 in the network of Fig. P15.12.

15.13 Find the two consecutive sets of values for A and B that will constitute a test for device D stuck open in the CMOS EXOR gate given in Fig. P15.13.

15.14 Suppose the AND gates in Fig. P15.4 are realized in CMOS. Use whatever methods are appropriate to find a pair of value vectors (a,b,c,d) that will constitute a test for point P SA0.

15.15 Use the Wang algorithm to find tests for the following two additional points in the network of Fig. 15.14a. Show the values of all points in the network determined by the forward fault propagation and initial backward propagation. Show the consistency search tree.

(a) gate 4, output SA0

(b) gate 10, input 2 SA0

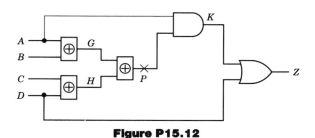

Figure P15.12

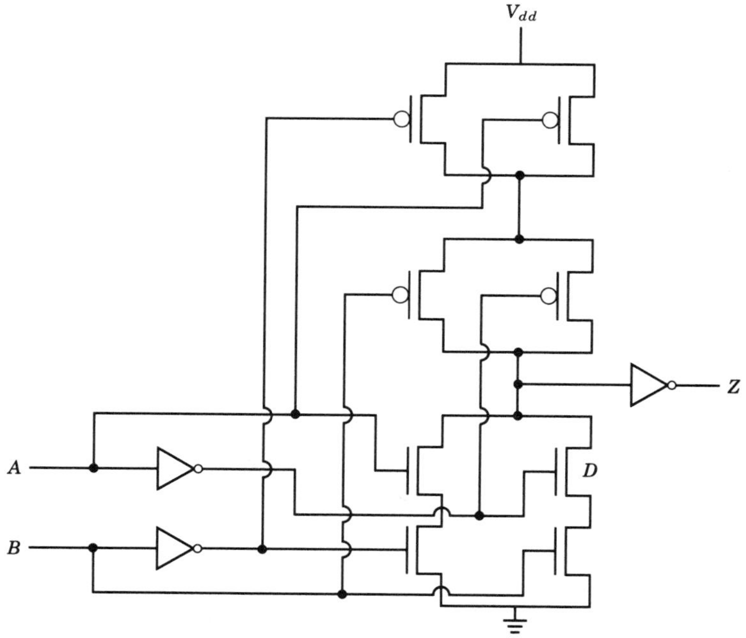

Figure P15.13

15.16 A complete reduced fault list for the sequential network given in Fig. P15.16 is

a-SA1, b-SA1, b-SA0, c-SA1, c-SA0, d-SA1, e-SA1, e-SA0, f-SA0, g-SA1, g-SA0, h-SA0, j-SA1, j-SA0, k-SA0, k-SA1, p-SA1

The initial state of the circuit is $Y0 = 0$, $Y1 = 1$. Perform manually three clock periods of deductive simulation, beginning with this initial state and assuming that no faults are initially stored in memory elements. For the first period of the simulation, $X0 = 0$, $X1 = 1$, and for the second and third periods, $X0 = X1 = 0$. Use the format of Table 15.2 with the stored faults and detected faults for each clock period.

15.17 A complete list of faults of interest for the sequential network given in Fig. P15.17 is

a-SA1, b-SA1, b-SA0, c-SA1, c-SA0, d-SA1, e-SA1, e-SA0, f-SA1, f-SA0, g-SA0, h-SA1, h-SA0, p-SA1, q-SA1, q-SA0

The initial state of the circuit is $Y0 = 0$, $Y1 = 1$. Repeat Problem 15.16 for this network. For the first period of the simulation, $x = 0$, and for the second and third periods, $x = 1$.

15.18 Modify the AHPL description given below to add a single scan path from input to output to support the testing process. A single control line *test* is to be added. When *test* = 1, inputs on line *a* are to be shifted through all data unit memory elements to output z. When *test* = 0, the circuit is to function as described below.

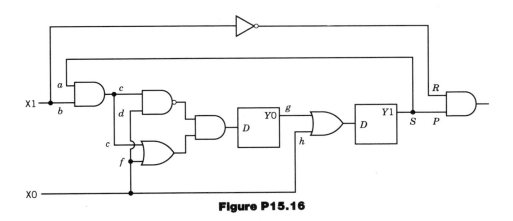

Figure P15.16

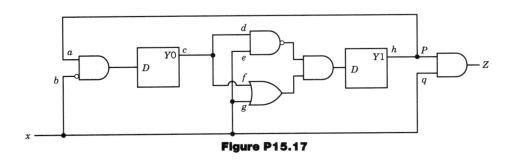

Figure P15.17

MODULE: HARDTOTEST
INPUTS: $X[4]$; a.
MEMORY: $CNT[4]$; $R[4]$.
OUTPUTS: z.

1   $CNT \leftarrow 4T0$;
2   $CNT \leftarrow \text{INC}(CNT)$.
3   $R * \wedge/CNT \leftarrow (R \oplus X!4T0)*(a, \bar{a})$.
4   $z = R[0] \wedge R[1] \wedge R[2] \wedge R[3]$;
    $\rightarrow (a, \bar{a})/(1, 2)$.

ENDSEQUENCE
CONTROLRESET(1).
END.

15.19   Add test control lines as needed to modify the design of the DATAIN-TERSECTION module of Section 11.9 so that all memory elements are connected into a single scan path for testing purposes. Add no more input/output lines than necessary. Write an AHPL description of the revised module that must function the same as the original, when the test control line is 0.

BIBLIOGRAPHY

1. Kohavi, Z. *Switching and Finite Automata Theory*, 2nd ed. McGraw-Hill, New York, 1978.
2. Shen, J. P., W. Maley, and F. J. Ferguson. "Inductive Fault Analysis in MOS Integrated Circuits," *IEEE Design and Test*: Vol. 2, 14–26 (Dec. 1985).
3. Breuer, M. and A. D. Fredman. *Diagnosis and Reliable Design of Digital Systems*. Computer Science Press, Woodland Hills, Calif., 1976.
4. Wang, X. L., M. Karunaratne, and F. J. Hill. "Test Generation Based on Dynamic Search Space Reductions," in *Proceedings 9th Phoenix Conference on Computers and Communications*. IEEE Computer Society Press, Los Alimitos, Calif., March 1990, pp. 622–629.
5. Muth, P. "A Nine Value Circuit Model for Test Generation," *IEEE Trans. Computers*: **C-25**: 630–636 (June 1976).
6. Miczo, A. *Digital Logic Testing and Simulation*. Harper and Row, New York, 1986.
7. Goel, P. "An Implicit Enumeration Algorithm to Generate Tests for Combinational Logic Circuits," *IEEE Trans. Computers*, **C-30**: 215–222, 3 (March 1981).
8. Roth, J. P. "Diagnosis of Automata Failures: A Calculus and a Method," *IBM J. Research and Development*, **10**: 4, 278–291 (July 1966).
9. Chang, H. Y., E. Manning, and G. Metze. *Fault Diagnosis of Digital Systems*. Wiley, New York, 1970.
10. Fujiwara, H. *Logic Testing and Design for Testability*. MIT Press, Cambridge, Mass., 1985.
11. Baschiera, D. and B. Courtois. "Testing CMOS: A Challenge," *VSLI Design*: 58–62 (Oct. 1984).

12. Moritz, P. S. and L. M. Thorsen. "CMOS Circuit Testability," *IEEE J. Solid-State Circuits*, **SC-21:** 2, 306–309 (April 1986).

13. Hong, S. J. "Fault Simulation Strategy for Combinational Logic Networks," in *Proceedings of 8th Annual International Conference on Fault Tolerant Computing*, 1978, pp. 96–99.

14. Ozguner, F., et al. "On Fault Simulation Techniques," *J. Design Automation and Fault Tolerance Computing*, Vol. 3, No. 2, pp. 83–92.

15. Nishida, T., et al. "RFSIM: Reduced Fault Simulation," *IEEE Trans. Computer Aided Design*, Vol. CAD-3, No. 3, pp. 392–403 (May 1987).

16. Chang, H. Y., et al. "Comparison of Parallel and Deductive Fault Simulation Methods," *IEEE Trans. Computers*, **23:** 11, 1132–1138 (Nov. 1974).

17. Ulrich, E. "High Speed Concurrent Fault Simulation with Vectors and Scalars," in *Proceedings of 17th Design Automation Conference*, 1980, IEEE, New York, pp. 374–380.

18. Armstrong, D. B. "A Deductive Method for Simulation Faults in Logic Circuits," *IEEE Trans. Computers*, **C-21:** 464–471 (May 1972).

19. Seshu, S. "On an Improved Diagnosis Program," *IEEE Trans. Electronic Computers*, **EC-14:** 76–81 (Feb. 1965).

20. *IEEE Standard Test Access Port and Boundary Scan Architecture*. New York, May 1989, Std. 1149.1-1989/D4.

21. Hill, F. J. and B. Huey. "SCIRTSS: A Search System for Sequential Circuit Test Sequences," *IEEE Trans. Computers*, **C-26:** 490–502 (May 1977).

22. Leet, D., P. Shearon, and R. France. "A CMOS LSSD Test Generation System, *IBM J. Research and Development*, **28:** 5, 625–635 (Sept. 1984).

23. Komonytsky, D. "LSI Self-Test Using Level Sensitive SCAN Design and Signature Analysis," in *Proceedings of 1982 IEEE Test Conference*. IEEE, New York, pp. 414–424.

24. *Testability, Test, and Emulation Primer*. SSYA002, Texas Instruments Corp., Richardson, Tex., 1989.

CHAPTER 16

COMBINATIONAL LOGIC FUNCTIONS WITH SPECIAL PROPERTIES

16.1 INTRODUCTION

In Chapters 5 to 8, we dealt with general techniques applicable to the design of combinational circuits of all types, with emphasis on two-level, or second-order, design. In this chapter, we return to combinational design to consider some special classes of logic functions that do not lend themselves readily to second-order design and to explore some special techniques for dealing with such functions.

One particular type of function that is not economically realized in second-order form is that in which the function value is determined only by the number of literals that take the value 1. Such functions are called symmetric functions and will be discussed in detail in the next section. Section 16.4 deals with the general problem of breaking Boolean functions into higher-order representations. These representations are called *decompositions*. It will be shown that symmetric functions are readily decomposable functions. A search for decompositions will constitute part of any program for optimization of higher-order (more than two-level) combinational logic functions. Any such program must be a composite of many techniques that are coordinated by a heuristic strategy, necessarily more complex than the Espresso introduced in Section 6.10. At this writing, the dominant multilevel logic optimization program is MIS [8], a program available from the Berkeley OCT tools suite.

Section 16.6 examines another special type of combinational logic circuit, the *iterative network*, made up of a cascade of identical subcircuits. We will learn that the design of an iterative network has much in common with clock-mode sequential circuit design. The iterative network provides a mechanism for reducing the problem of finding a test for a single fault in a sequential circuit to the already solved problem of finding a test for a fault in combinational logic. The concluding section of this chapter will return to the theme of Chapter 15 and develop this approach to finding a test for a sequential circuit fault.

16.2 SYMMETRIC FUNCTIONS

In Chapter 8, we encountered the expression

$$f(c, x, y) = c\bar{x}\bar{y} + \bar{c}x\bar{y} + \bar{c}\bar{x}y + cxy \tag{16.1}$$

in the generation of a sum bit in binary addition (see Fig. 8.8a). This function is an example of a *symmetric function*.

As we learned, symmetric functions may often be most economically realized with more than two levels of logic. Notice that the value of Equation 16.1 is 1 whenever an odd number of the variables are 1. The n-variable odd-parity check function is the extension of this pattern to n variables. An efficient multilevel NAND gate realization of this latter symmetric function was also given in Chapter 8.

In general, a symmetric function may be defined as follows:

Definition 16.1 A switching function $f(x_1^{j_1}, x_2^{j_2}, ..., x_n^{j_n})$ is *symmetric* with respect to the literals $x_1^{j_1}, x_2^{j_2}, ..., x_n^{j_n}$, if and only if it remains unchanged after any permutation of these literals, where the j_i may take on only the values 0 and 1, and

$$x_i^{j_i} = x_i \quad \text{if } j_i = 0 \qquad \text{and} \qquad x_i^{j_i} = \bar{x}_i \quad \text{if } j_i = 1 \qquad (16.2)$$

Consider, for example, Equation 16.1. Since it contains only three variables, we may check exhaustively all six permutations of the variables to determine if the expression remains unchanged. In Fig. 16.1, we see Equation 16.1 as it appears after each permutation of $\{c, x, y\}$. It should be clear to the reader that the function is unchanged in each case. The function

$$x\bar{y}z + \bar{x}yz + \bar{x}\bar{y}\bar{z} \qquad (16.3)$$

is also symmetric, but in this case, the literals of symmetry are x, y, and $\bar{z}$ rather than x, y, and z.

This becomes clear if we let

$$q = \bar{z}$$

Thus, Equation 16.3 becomes

$$x\bar{y}\bar{q} + \bar{x}y\bar{q} + \bar{x}\bar{y}q$$

which is clearly unchanged by a permutation of the variables.

The reader may notice that functions symmetric with respect to some complemented literals and some uncomplemented literals may not be easily identified. Symmetric functions have some very interesting properties that are expressed in the following set of theorems.

As was mentioned previously, the value of a symmetric function depends only on the number of literals of symmetry that are 1. This is set down formally in the following key theorem, which was originally stated by Shannon [7].

THEOREM 16.1 A necessary and sufficient condition that a switching function of n variables be symmetric is that it may be specified by a set of integers $\{a_k\}$, where $0 \leq a_k \leq n$, such that if exactly $a_m (m = 1, 2, ..., k)$ of the literals have the value *1*, then the function has the value *1*, and not otherwise.

	Permutation	Expression
1	Identity: $c \to c, x \to x, y \to y$	$c\bar{x}\bar{y} + \bar{c}x\bar{y} + \bar{c}\bar{x}y + cxy$
2	$c \to x, x \to y, y \to c$	$x\bar{y}\bar{c} + \bar{x}y\bar{c} + \bar{x}\bar{y}c + xyc$
3	$c \to y, y \to x, x \to c$	$y\bar{c}\bar{x} + \bar{y}c\bar{x} + \bar{y}\bar{c}x + ycx$
4	$c \to c, x \to y, y \to x$	$c\bar{y}\bar{x} + \bar{c}y\bar{x} + \bar{c}\bar{y}x + cyx$
5	$c \to x, x \to c, y \to y$	$x\bar{c}\bar{y} + \bar{x}c\bar{y} + \bar{x}\bar{c}y + xcy$
6	$c \to y, x \to x, y \to c$	$y\bar{x}\bar{c} + \bar{y}x\bar{c} + \bar{y}\bar{x}c + yxc$

Figure 16.1 Permutations of the variables of Equation 16.1.

PROOF

A. (Necessity) Suppose that a function is 1 when the first a_m literals are 1 but is 0 when some other a_m literals are 1. Suppose each of the a_m literals of the first set is permuted to one of the literals of the second set. Clearly, a different function will result. Thus, the function would not be symmetric.

B. (Sufficiency) Suppose that a function is 1 if and only if exactly a_m of the literals are 1. Following any permutation of the literals, the function will still be 1 when any a_m of the literals are 1. It is much the same for any a_m in $\{a_k\}$, so the function is symmetric. Q.E.D.

We may now express any symmetric function in the form

$$S_{\{a_k\}}^n(x_1^{j_1}, x_2^{j_2}, ..., x_n^{j_n}) \tag{16.4}$$

Thus, Equation 16.1 may be written as

$$S_{1,3}^3(c, x, y) \tag{16.5}$$

and Equation 16.3 as

$$S_1^3(x, y, \bar{z}) \tag{16.6}$$

Similarly, the n-variable even-parity check function may be written

$$S_{0,2,4,...,n}^n(x_1, x_2, ..., x_n) \tag{16.7}$$

if n is even. If n is odd, the last subscript is $n - 1$.

16.3 BOOLEAN COMBINATIONS OF SYMMETRIC FUNCTIONS

Theorem 16.1 not only permits a convenient notation for symmetric functions, but it also provides for a considerable simplification of the algebraic operations on symmetric functions.

THEOREM 16.2

$$S_{a_1, ..., a_j, b_1, ..., b_m}^n(x_1^{j_1}, x_2^{j_2}, ..., x_n^{j_n}) + S_{a_1, ..., a_j, c_1, ..., c_k}^n(x_1^{j_1}, x_2^{j_2}, ..., x_n^{j_n})$$
$$= S_{a_1, ..., a_j, b_1, ..., b_m, c_1, ..., c_k}^n(x_1^{j_1}, x_2^{j_2}, ..., x_n^{j_n}) \tag{16.8}$$

PROOF The proof follows immediately from Theorem 16.1. That is, a function that is 1 when $a_1, ..., a_j, b_1, ..., b_{m-1}$ or b_m of the literals are 1 or when $a_1, ..., a_j, c_1, ..., c_{k-1}$ or c_k of the same literals are 1 is 1 when $a_1, ..., a_j, b_1, ..., b_m, c_1, ..., c_{k-1}$ or c_k of the literals are 1.

THEOREM 16.3

$$[S_{a_1, ..., a_j, b_1, ..., b_m}^n(x_1^{j_1}, x_2^{j_2}, ..., x_n^{j_n})] \cdot [S_{a_1, ..., a_i, c_1, ..., c_k}^n(x_1^{j_1}, x_2^{j_2}, ..., x_n^{j_n})]$$
$$= S_{a_1, ..., a_i}^n(x_1^{j_1}, x_2^{j_2}, ..., x_n^{j_n}) \tag{16.9}$$

PROOF Similar to that of Theorem 16.2.

■ **EXAMPLE 16.1**

$$S_{1,2,4}^5(V, W, X, Y, Z) + S_{2,3,4}^5(V, W, X, Y, Z) = S_{1,2,3,4}^5(V, W, X, Y, Z)$$

while

$$[S_{1,2,4}^5(V, W, X, Y, Z)] \cdot [S_{2,3,4}^5(V, W, X, Y, Z)] = S_{2,4}^5(V, W, X, Y, Z)$$

It is possible to state another theorem that follows directly from Theorem 16.1. Since the general statement of the theorem is rather awkward, we will instead illustrate it by example. ∎

■ **EXAMPLE 16.2**

$$\overline{S^6_{1,2,3,6}(x_1, x_2, ..., x_6)} = S^6_{0,4,5}(x_1, x_2, ..., x_6) \tag{16.10}$$

Note that $S^6_{0,4,5}$ is 1 when 0, 4, and 5 of the literals $x_1, x_2, ..., x_6$ are 1. The function $S^6_{1,2,3,6}$ is never 1 when $S^6_{0,4,5}$ is 1, but the former function is 1 for every number of 1 laterals for which $S_{0,4,5} = 0$. The complement of any symmetric function of any number of variables may be obtained in this way. ∎

The theorem illustrated by Example 16.2 concerns the complement of a symmetric function. The next theorem concerns the function obtained by complementing the variables of a symmetric function.

THEOREM 16.4

$$S^n_{a_1, a_2, ..., a_k}(x^{j_1}_1, x^{j_2}_2, ..., x^{j_n}_n) = S^n_{n-a_1, ..., n-a_k}(x^{1-j_1}_1, x^{1-j_2}_2, ..., x^{1-j_n}_n) \tag{16.11}$$

PROOF If a_i of the literals $x^{j_1}_1, x^{j_2}_2, ..., x^{j_n}_n$ are 1, then $n - a_i$ of these literals are 0, and $n - a_i$ of the literals $x^{1-j_1}_1, x^{1-j_2}_2, ..., x^{1-j_n}_n$ are 1. Thus, the two sides of Equation 16.11 are equal for any assignment of variable values.

■ **EXAMPLE 16.3**

$$S^4_{2,3}(A,B,C,D) = AB\bar{C}\bar{D} + A\bar{B}C\bar{D} + A\bar{B}\bar{C}D + \bar{A}BC\bar{D} + \bar{A}B\bar{C}D + \bar{A}\bar{B}CD$$
$$+ ABC\bar{D} + AB\bar{C}D + A\bar{B}CD + \bar{A}BCD$$
$$= S^4_2(\bar{A},\bar{B},\bar{C},\bar{D}) + S^4_1(\bar{A},\bar{B},\bar{C},\bar{D}) = S_{4-2,4-3}(\bar{A},\bar{B},\bar{C},\bar{D}) \tag{16.12}$$
∎

THEOREM 16.5 Any symmetric function $S^n_{a_1, a_2, ..., a_k}(x^{j_1}_1, x^{j_2}_2, ..., x^{j_n}_n)$ may be represented by the conjunction of a subset of the following *elementary symmetric functions*:

$$\sigma^n_0 = x^{1-j_1}_1 \cdot x^{1-j_2}_2 \cdots x^{1-j_n}_n$$
$$\sigma^n_1 = x^{j_1}_1 \cdot x^{1-j_2}_2 \cdots x^{1-j_n}_n + x^{1-j_1}_1 \cdot x^{j_2}_2 \cdot x^{1-j_3}_3 \cdots x^{1-j_n}_n$$
$$+ \cdots + x^{1-j_1}_1 \cdot x^{1-j_2}_2 \cdots x^{1-j_{n-1}}_{n-1} \cdot x^{j_n}_n$$

.

.

.

$$\sigma^n_n = x^{j_1}_1 x^{j_2}_2 \cdots x^{j_n}_n. \tag{16.13}$$

PROOF Each elementary symmetric function σ_i is 1 if and only if exactly i of the literals are 1. Thus, by Theorems 16.1 and 16.2, any symmetric function may be formed as the conjunction of a subset of these functions.

■ **EXAMPLE 16.4**

$$S^4_{0,2,4}(A, B, C, \bar{D}) = S^4_0(A, B, C, \bar{D}) + S^4_2(A, B, C, \bar{D}) + S^4_4(A, B\,C, \bar{D})$$
$$= \bar{A}\bar{B}\bar{C}D + (AB\bar{C}D + A\bar{B}CD + A\bar{B}\bar{C}\bar{D} + \bar{A}BCD + \bar{A}B\bar{C}\bar{D}$$
$$+ \bar{A}\bar{B}C\bar{D}) + ABC\bar{D}$$
$$= \sigma^4_0 + \sigma^4_2 + \sigma^4_4$$
∎

■ **EXAMPLE 16.5**

A certain communication system utilizes 16 parallel lines to send four four-bit characters at a time. Each of these four-bit characters has a parity bit to establish even parity over the four bits of that character. Parity is to be checked at a certain relay point in the communications system, and retransmission of all four characters is to be demanded by the relay point if there is odd parity over two or more of the characters.

Write, in symmetric function notation, the expression for a combinational circuit whose output is 1 whenever retransmission is to be demanded. Let w_1, w_2, w_3, w_4; x_1, x_2, x_3, x_4; y_1, y_2, y_3, y_4; and z_1, z_2, z_3, z_4 represent the bits of the four characters.

SOLUTION   It is easily seen that the overall expression is not a symmetric function. Consider, for example, the following sets of bits, which might be received at the relay point:

w_1 w_2 w_3 w_4	x_1 x_2 x_3 x_4	y_1 y_2 y_3 y_4	z_1 z_2 z_3 z_4
1 1 1 0	1 1 1 0	1 1 0 0	0 0 0 0

w_1 w_2 z_3 w_4	x_1 x_2 z_4 x_4	y_1 y_2 y_3 y_4	z_1 z_2 w_3 x_3
1 1 0 0	1 1 0 0	1 1 0 0	0 0 1 1

For the first set of bits, the functions should be 1 since the first two characters have odd parity. The second set of bits has the same number of 1's, but the output should be 0 since all characters have correct parity.

It is, however, straightforward to write the parity check function over each set of four bits as a symmetric function. For example, odd parity over w_1, w_2, w_3, and w_4 is given by

$$P_w = S^4_{1,3}(w_1, w_2, w_3, w_4)$$

Similarly, the function that determines from the four parity checks whether two or more of these are 1 is also symmetric. Thus, overall function may be expressed as

$$f = S^4_{2,3,4}[S^4_{1,3}(w_1, w_2, w_3, w_4), S^4_{1,3}(x_1, x_2, x_3, x_4)$$
$$S^4_{1,3}(y_1, y_2, y_3, y_4), S^4_{1,3}(z_1, z_2, z_3, z_4)] \quad ■$$

We have said nothing so far about the physical implementation of symmetric functions. In terms of NAND or NOR gates, symmetric functions enjoy no advantage over other functions with respect to economic realization. In applications that will tolerate pass transistors or transmission gates, symmetric functions may be realized quite economically.

16.4   SIMPLE DISJOINT DECOMPOSITION

Simplification of circuits by factoring is a satisfactory technique in many cases, but it is largely a trial-and-error process, dependent on a good deal of experience and often just plain luck in spotting common factors in the function. It is evident that a more systematic method of finding simpler designs of higher order would be very useful. In essence, the process of simplification involves breaking a complex single function into a number of related simpler functions. The general name for this process is *functional decomposition*. The methods to be presented here, which we believe to be the most generally useful, are primarily the work of Ashenhurst [2] and Curtis [4].

The first type of decomposition we will consider is the simple disjoint decomposition.

Definition 16.2   Let $x_1, x_2, ..., x_n$ be a set of n switching variables and A and B be disjoint subsets of X such that $A \cup B = X$. If $F[\Phi(A), B]$ takes on the same functional values as $f(X)$ whenever the latter are specified, then F and Φ are said to form a *simple disjoint decomposition* of f. The set of variables A shall be known as the *bound* variables, the set B as the *free* variables. The partitioning of the variables shall be indicated by $B | A$. The basic form of circuit realization of the simple disjoint decomposition is shown in Fig. 16.2.

■ EXAMPLE 16.6
_____

Obtain a simple disjoint decomposition of
$$f_1(x_1, x_2, x_3, x_4) = x_1 \bar{x}_2 x_3 x_4 + x_1 x_2 x_3 \bar{x}_4 + \bar{x}_1 \bar{x}_2 \bar{x}_3 \bar{x}_4 + \bar{x}_1 x_2 \bar{x}_3 x_4 \quad (16.14)$$

SOLUTION

$$f_1(x_1, x_2, x_3, x_4) = x_1 x_3 (x_2 \bar{x}_4 + \bar{x}_2 x_4) + \bar{x}_1 \bar{x}_3 (\bar{x}_2 \bar{x}_4 + x_2 x_4) \quad (16.15)$$
$$= x_1 x_3 \, \Phi(x_2, x_4) + \bar{x}_1 \bar{x}_3 \, \overline{\Phi(x_2, x_4)} \quad (16.16)$$
$$= F[\Phi(x_2, x_4), x_1, x_3] \quad (16.17)$$

where
$$\Phi = x_2 \bar{x}_4 + \bar{x}_2 \, x_4 \quad (16.18)$$

This corresponds to the partitioning $x_1, x_3 | x_2, x_4$.   ■

Nothing was said in the previous example as to how it was determined that a simple disjoint partition of the form $x_1, x_3 | x_2, x_4$ could actually be obtained.

Let us explore the possibility of a general method of determining whether or not a particular decomposition exists. Consider, for example, the function
$$f(x_1, x_2, x_3, x_4) = \sum m(4, 5, 6, 7, 8, 13, 14, 15) \quad (16.19)$$

which may be decomposable into the form $F[\Phi(x_1, x_4), x_2, x_3]$. As a first step, we can collect the minterms containing each of the four combinations of $x_2 x_3$ and then factor these terms out to give the form
$$f(x_1, x_2, x_3, x_4) = \bar{x}_2 \bar{x}_3 \, \alpha(x_1, x_4) + \bar{x}_2 x_3 \, \beta(x_1, x_4)$$
$$+ x_2 x_3 \lambda(x_1, x_4) + x_2 \bar{x}_3 \, \delta(x_1, x_4) \quad (16.20)$$

This factoring can be done for any function whether it is decomposable or not. Now we will make a special map of the function, called a *partition map* (Fig. 16.3).

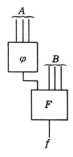

Figure 16.2   Simple disjoint decomposition.

$x_2x_3 \backslash x_1x_4$	00	01	11	10
00	0 [0]	0 [1]	0 [9]	1 [8]
01	0 [2]	0 [3]	0 [11]	0 [10]
11	1 [6]	1 [7]	1 [15]	1 [14]
10	1 [4]	1 [5]	1 [13]	0 [12]

Figure 16.3 Partition map.

This is simply a K-map with the free variables running down the side and the bound variables across the top. Note that the minterm numbers are still those based on the original ordering of the variables specified in Equation 16.19.

Comparing the partition map with Equation 16.20, we can see that the first row must correspond to the function $\alpha(x_1, x_4)$, the second row to $\beta(x_1, x_4)$, the third row to $\lambda(x_1, x_4)$, and the fourth row to $\delta(x_1, x_4)$. Again, this is still completely general. Now recall that for a simple disjoint function, F must be a function of x_2, x_3 and a *single* function $\Phi(x_1, x_4)$. In the above, let the first row function α be the function Φ. Then, for the condition of a *single* function to be met, the remaining functions must be either α, $\bar{\alpha}$, 0, or 1. This, in turn, means that the remaining rows must be either identical, complementary, all 0's, or all 1's. In this example, the conditions are met and the function can be read off the partition matrix as

$$\begin{aligned} f(x_1, x_2, x_3, x_4) &= \bar{x}_2\bar{x}_3\Phi(x_1, x_4) + x_2\bar{x}_3\overline{\Phi(x_1, x_4)} + x_2x_3 \\ &= F[\Phi(x_1, x_4), x_2, x_3] \end{aligned} \tag{16.21}$$

where $\Phi(x_1, x_4) = x_1\bar{x}_4$.

We now have a rule for detecting a simple disjoint decomposition on the partition map. However, this rule may be rather difficult to apply in large partition maps. As a further aid, let us define *column multiplicity* as the number of different column patterns of 1's and 0's. For example, in the partition of Fig. 16.3, there are two distinct column patterns

$$\begin{bmatrix} 0 \\ 0 \\ 1 \\ 1 \end{bmatrix} \quad \text{and} \quad \begin{bmatrix} 1 \\ 0 \\ 1 \\ 0 \end{bmatrix}$$

The following theorem offers a simple method by which it can be determined if a partition map corresponds to a simple disjoint decomposition.

THEOREM 16.6 A partition map corresponds to a simple disjoint decomposition if and only if the column multiplicity is less than or equal to 2.

PROOF If a function is decomposable into row functions, the functional values of each row of the Karnaugh map must form one of the four functions, 0, 1, Φ, or $\bar{\Phi}$. Rows corresponding to the functions 0 and 1 have the same entry in every column and thus have no effect on column multiplicity. If a column has a 1 in some Φ row, it must have a 0 in every $\bar{\Phi}$ row and a 1 in every Φ row. Similarly, a 0 in a Φ row implies a 0 in every Φ row and a 1 in every $\bar{\Phi}$ row. Thus, there are only two distinct types of columns possible. If we assume a column multiplicity of 2, we may similarly deduce the existence of a simple disjoint decomposition.

Q.E.D.

With the above, we have the basic rule needed to find decompositions. However, the amount of work is still rather extensive, since there are many possible partitions of the variables into bound and free variables. To reduce the amount of labor involved, we introduce *decomposition charts*, which are simply K-maps showing the position of the minterms for all possible combinations of bound and free variables. The decomposition charts for four-variable functions are shown in Fig. 16.4. To use the charts, we simply circle the minterm numbers of the function on all charts and then look for column multiplicities of 2 or 1. Note that the 4 × 4 charts should be viewed sideways to reverse the order of bound and free variables. Thus, for example, the leftmost 4 × 4 chart will indicate decompositions of the form $F(\Phi(x_3, x_4), x_1, x_2)$ when viewed normally and decompositions of the form $F(\Phi(x_1, x_2), x_3, x_4)$ when viewed sideways. The 2 × 8 charts should only be viewed normally, since decompositions with only one bound variable are trivial.

■ EXAMPLE 16.7

Use decomposition charts to find all simple disjoint decompositions of

$$f(x_1, x_2, x_3, x_4) = \sum m(4, 5, 6, 7, 8, 13, 14, 15) \tag{16.22}$$

SOLUTION The decomposition charts for this function are shown in Fig. 16.5. The maps of Figs. 16.5b and 16.5e when viewed normally indicate the partitionings $x_2 | x_1 x_3 x_4$ and $x_1 x_2 | x_3 x_4$. The maps of Figs. 16.5f and 16.5g viewed sideways indicate the partitionings $x_2 x_4 | x_1 x_3$ and $x_2 x_3 | x_1 x_4$. The resulting decompositions can then be read from the charts as

$$f(x_2 | x_1 x_3 x_4) = \bar{x}_2 \Phi(x_1, x_3, x_4) + x_2 \overline{\Phi(x_1, x_3, x_4)}$$
$$= \bar{x}_2(x_1 \bar{x}_3 \bar{x}_4) + x_2(\overline{x_1 \bar{x}_3 \bar{x}_4}) \tag{16.23}$$

where $\Phi(x_1, x_3, x_4) = x_1 \bar{x}_3 \bar{x}_4$.

x_2	0				1				x_1	0				1			
$x_3 x_4$	00	01	11	10	00	01	11	10	$x_3 x_4$	00	01	11	10	00	01	11	10
x_1 0	0	1	3	2	4	5	7	6	0	0	1	3	2	8	9	11	10
1	8	9	11	10	12	13	15	14	x_2 1	4	5	7	6	12	13	15	14

x_1	0				1				x_1	0				1			
$x_2 x_4$	00	01	11	10	00	01	11	10	$x_2 x_3$	00	01	11	10	00	01	11	10
x_3 0	0	1	5	4	8	9	13	12	0	0	2	6	4	8	10	14	12
1	2	3	7	6	10	11	15	14	x_4 1	1	3	7	5	9	11	15	13

$x_3 x_4$	00	01	11	10		$x_2 x_4$	00	01	11	10		$x_2 x_3$	00	01	11	10
00	0	1	3	2		00	0	1	5	4		00	0	2	6	4
01	4	5	7	6		01	2	3	7	6		01	1	3	7	5
$x_1 x_2$ 11	12	13	15	14		$x_1 x_3$ 11	10	11	15	14		$x_1 x_4$ 11	9	11	15	13
10	8	9	11	10		10	8	9	13	12		10	8	10	14	12

Figure 16.4 Four-variable decomposition charts.

x_2	0				1			
x_3x_4	00	01	11	10	00	01	11	10
x_1 0	0	1	3	2	④	⑤	⑦	⑥
x_1 1	⑧	9	11	10	12	⑬	⑮	⑭

(a)

x_1	0				1			
x_3x_4	00	01	11	10	00	01	11	10
x_2 0	0	1	3	2	⑧	9	11	10
x_2 1	④	⑤	⑦	⑥	12	⑬	⑮	⑭

(b)

x_1	0				1			
x_2x_4	00	01	11	10	00	01	11	10
x_3 0	0	1	⑤	④	⑧	9	⑬	12
x_3 1	2	3	⑦	⑥	10	11	⑮	⑭

(c)

x_1	0				1			
x_2x_3	00	01	11	10	00	01	11	10
x_4 0	0	2	⑥	④	⑧	10	⑭	12
x_4 1	1	3	⑦	⑤	9	11	⑮	⑬

(d)

x_3x_4	00	01	11	10
x_1x_2 00	0	1	3	2
01	④	⑤	⑦	⑥
11	12	⑬	⑮	⑭
10	⑧	9	11	10

(e)

x_2x_4	00	01	11	10
x_1x_3 00	0	1	⑤	④
01	2	3	⑦	⑥
11	10	11	⑮	⑭
10	⑧	9	⑬	12

(f)

x_2x_3	00	01	11	10
x_1x_4 00	0	2	⑥	④
01	1	3	⑦	⑤
11	9	11	⑮	⑬
10	⑧	10	⑭	12

(g)

Figure 16.5 Decomposition charts, Example 16.5.

$$f(x_1 x_2 \mid x_3 x_4) = \bar{x}_1 x_2 + x_1 x_2 \, \Phi(x_3, x_4) + x_1 \bar{x}_2 \, \overline{\Phi(x_3, x_4)}$$
$$= \bar{x}_1 x_2 + x_1 x_2 (x_3 + x_4) + x_1 \bar{x}_2 \overline{(x_3 + x_4)} \tag{16.24}$$

where $\Phi(x_3, x_4) = x_3 + x_4$.

$$f(x_2 x_4 \mid x_1 x_3) = x_2 x_4 + x_2 \bar{x}_4 \, \Phi(x_1, x_3) + \bar{x}_2 \bar{x}_4 \overline{\Phi(x_1, x_3)}$$
$$= x_2 x_4 + x_2 \bar{x}_4 (\bar{x}_1 + x_3) + \bar{x}_2 \bar{x}_4 \overline{(\bar{x}_1 + x_3)} \tag{16.25}$$

where $\Phi(x_1, x_3) = \bar{x}_1 + x_3$.

$$f(x_2 x_3 \mid x_1 x_4) = x_2 x_3 + x_2 \bar{x}_3 \, \Phi(x_1, x_4) + \bar{x}_2 \bar{x}_3 \overline{\Phi(x_1, x_4)}$$
$$= x_2 x_3 + x_2 \bar{x}_3 (\bar{x}_1 + x_4) + \bar{x}_2 \bar{x}_3 \overline{(\bar{x}_1 + x_4)} \tag{16.26}$$

where $\Phi(x_1, x_4) = \bar{x}_1 + x_4$. ∎

The same procedure can be applied to functions of more variables. The decomposition charts for five variables are shown in Fig. 16.6. The method can, in theory, be extended to any number of variables, but the number and size of the charts rapidly become impractical beyond five variables. Curtis [4] gives the charts for six variables but there are 29 of them, making the practicality of the method for six variables rather doubtful. However, the generality of the method provides the basis for computer mechanization of the process for large numbers of variables.

Functions with don't cares can be handled very simply on the decomposition charts. Circle the 1's of the function in the usual manner and indicate the don't-cares by lines through the corresponding minterms. Then assign the don't-cares to achieve a column multiplicity of 2 or less.

■ **EXAMPLE 16.8**

Determine a simple disjoint decomposition of

$$f(x_1, x_2, x_3, x_4, x_5) = \sum m(1, 2, 7, 9, 10, 18, 19, 25, 31) + d(0, 15, 20, 23, 26)$$

SOLUTION   The decomposition chart for the partition $x_3 x_4 | x_1 x_2 x_5$ is shown in Fig. 16.7a, with the minterms circled and don't-cares lined out. We see that the chart has a column multiplicity of 4 in both directions. If the don't-cares at 15 and 26 are assigned as 1's, a column multiplicity of 2 is achieved with the chart viewed in the normal direction, as shown in Fig. 16.7b. From this chart, interpreted as a K-map, we can read off the realization

$$f(x_3 x_4 | x_1 x_2 x_5) = (\bar{x}_3 \bar{x}_4 + x_3 x_4) \Phi(x_1, x_2, x_5) + \bar{x}_3 x_4 \overline{\Phi(x_1, x_2, x_5)} \quad (16.27)$$

where $\Phi(x_1, x_2, x_5) = x_2 x_5 + \bar{x}_1 x_5$.   ■

16.5   COMPLEX DISJOINT DECOMPOSITION

It is apparent that there must be many forms of decomposition other than simple disjoint. In this section, we consider decompositions resulting from the partitioning of the variables into three or more disjoint subsets.

Definition 16.3   Let $X = x_1, x_2, ..., x_n$ be a set of n switching variables and $A_1, A_2, ..., A_m$ be disjoint subsets of X such that $A_1 \cup A_2 \cup \cdots \cup A_m = X$. Decompositions of the form

$$f(X) = F[\Phi_1(A_1), \Phi_2(A_2), ..., \Phi_m(A_m)]$$

or

$$f(X) = F[\Phi_1(A_1), \Phi_2(A_2), ..., \Phi_{m-1}(A_{m-1}), A_m] \quad (16.28)$$

shall be known as multiple disjoint decompositions. Decompositions of the form

$$f(X) = F\{\lambda[\Phi(A), B], C\} \quad (16.29)$$

or, in general, of the form

$$f(X) = F\{\Phi_{m-1}[\Phi_{m-2}(...\{\Phi_1[\Phi_0(A_0), A_1], A_2\}...A_{m-2}), A_{m-1}], A_m\} \quad (16.30)$$

shall be known as iterative disjoint decompositions. Finally, combinations of these forms, such as

$$f(X) = F\{\gamma[\Phi(A), B], \lambda(C), D\} \quad (16.31)$$

shall be known as *complex disjoint* decompositions. In all cases, the decompositions are nontrivial only if all the bound sets contain at least two variables, since functions of a single variable are trivial.

Ashenhurst [2] has given a set of theorems relating the above types of decompositions to simple disjoint decompositions. These theorems enable us to extend the use of the decomposition chart to the determination of complex decompositions.

Rather than present the somewhat tedious proofs of these theorems, we will content ourselves with an example of the application of each theorem.

x_3	0				1			
x_4x_5	00	01	11	10	00	01	11	10
x_1x_2 00	0	1	3	2	4	5	7	6
01	8	9	11	10	12	13	15	14
11	24	25	27	26	28	29	31	30
10	16	17	19	18	20	21	23	22

x_2	0				1			
x_4x_5	00	01	11	10	00	01	11	10
x_1x_3 00	0	1	3	2	8	9	11	10
01	4	5	7	6	12	13	15	14
11	20	21	23	22	28	29	31	30
10	16	17	19	18	24	25	27	26

x_2	0				1			
x_3x_5	00	01	11	10	00	01	11	10
x_1x_4 00	0	1	5	4	8	9	13	12
01	2	3	7	6	10	11	15	14
11	18	19	23	22	26	27	31	30
10	16	17	21	20	24	25	29	28

x_2	0				1			
x_3x_4	00	01	11	10	00	01	11	10
x_1x_5 00	0	2	6	4	8	10	14	12
01	1	3	7	5	9	11	15	13
11	17	19	23	21	25	27	31	29
10	16	18	22	20	24	26	30	28

x_1	0				1			
x_4x_5	00	01	11	10	00	01	11	10
x_2x_3 00	0	1	3	2	16	17	19	18
01	4	5	7	6	20	21	23	22
11	12	13	15	14	28	29	31	30
10	8	9	11	10	24	25	27	26

x_1	0				1			
x_3x_5	00	01	11	10	00	01	11	10
x_2x_4 00	0	1	5	4	16	17	21	20
01	2	3	7	6	18	19	23	22
11	10	11	15	14	26	27	31	30
10	8	9	13	12	24	25	29	28

x_1	0				1			
x_3x_4	00	01	11	10	00	01	11	10
x_2x_5 00	0	2	6	4	16	18	22	20
01	1	3	7	5	17	19	23	21
11	9	11	15	13	25	27	31	29
10	8	10	14	12	24	26	30	28

x_1	0				1			
x_2x_5	00	01	11	10	00	01	11	10
x_3x_4 00	0	1	9	8	16	17	25	24
01	2	3	11	10	18	19	27	26
11	6	7	15	14	22	23	31	30
10	4	5	13	12	20	21	29	28

x_1	0				1			
x_2x_4	00	01	11	10	00	01	11	10
x_3x_5 00	0	2	10	8	16	18	26	24
01	1	3	11	9	17	19	27	25
11	5	7	15	13	21	23	31	29
10	4	6	14	12	20	22	30	28

x_1	0				1			
x_2x_3	00	01	11	10	00	01	11	10
x_4x_5 00	0	4	12	8	16	20	28	24
01	1	5	13	9	17	21	29	25
11	3	7	15	11	19	23	31	27
10	2	6	14	10	18	22	30	26

Figure 16.6 Decomposition charts for five-variable functions.

THEOREM 16.7 Let $f(X)$ be a function for which there exist two simple disjoint decompositions:

$$f(X) = F[\lambda(A, B), C]$$
$$= G[\Phi(A), B, C] \tag{16.32}$$

Then there exists an iterative disjoint decomposition

$$f(X) = F[\rho[\Phi(A), B], C] \tag{16.33}$$

x_2x_3	00				01				11				10			
x_4x_5	00	01	11	10	00	01	11	10	00	01	11	10	00	01	11	10
x_1 0	0	1	3	2	4	5	7	6	12	13	15	14	8	9	11	10
1	16	17	19	18	20	21	23	22	28	29	31	30	24	25	27	26

x_1x_3	00				01				11				10			
x_4x_5	00	01	11	10	00	01	11	10	00	01	11	10	00	01	11	10
x_2 0	0	1	3	2	4	5	7	6	20	21	23	22	16	17	19	18
1	8	9	11	10	12	13	15	14	28	29	31	30	24	25	27	26

x_1x_2	00				01				11				10			
x_4x_5	00	01	11	10	00	01	11	10	00	01	11	10	00	01	11	10
x_3 0	0	1	3	2	8	9	11	10	24	25	27	26	16	17	19	18
1	4	5	7	6	12	13	15	14	28	29	31	30	20	21	23	22

x_1x_2	00				01				11				10			
x_3x_5	00	01	11	10	00	01	11	10	00	01	11	10	00	01	11	10
x_4 0	0	1	5	4	8	9	13	12	24	25	29	28	16	17	21	20
1	2	3	7	6	10	11	15	14	26	27	31	30	18	19	23	22

x_1x_2	00				01				11				10			
x_3x_4	00	01	11	10	00	01	11	10	00	01	11	10	00	01	11	10
x_5 0	0	2	6	4	8	10	14	12	24	26	30	28	16	18	22	20
1	1	3	7	5	9	11	15	13	25	27	31	29	17	19	23	21

Figure 16.6 (continued)

where

$$\rho[\Phi(A), B] = \lambda(A, B)$$

As an example of the application of this theorem, consider the function

$$f(x_1, x_2, x_3, x_4, x_5) = \sum m(5, 10, 11, 14, 17, 21, 26, 30)$$

Two decomposition charts of this function are shown in Fig. 16.8. The first indicates the existence of a decomposition

$$f(X) = F[\lambda(x_1, x_3, x_5), x_2, x_4] \tag{16.34}$$

while the second, viewed sideways, shows a decomposition

$$f(X) = G[\Phi(x_1, x_3), x_2, x_4, x_5] \tag{16.35}$$

Thus, the conditions of Theorem 16.7 are met with $A = (x_1\ x_3)$, $B = (x_5)$, and $C = (x_2, x_4)$. From Fig. 16.8a, we read off the realization

x_1	0				1			
$x_2 x_5$	00	01	11	10	00	01	11	10
$x_3 x_4$ 00	~~0~~	(1)	(9)	8	16	17	(25)	24
01	(2)	3	11	(10)	(18)	(19)	27	~~26~~
11	6	(7)	~~15~~	14	22	~~23~~	(31)	30
10	4	5	13	12	~~20~~	21	29	28

(a)

x_1	0				1			
$x_2 x_5$	00	01	11	10	00	01	11	10
$x_3 x_4$ 00	~~0~~	(1)	(9)	8	16	17	(25)	24
01	(2)	3	11	(10)	(18)	(19)	27	(26)
11	6	(7)	(15)	14	22	~~23~~	(31)	30
10	4	5	13	12	~~28~~	21	29	28

(b)

Figure 16.7 Decomposition charts, Example 16.7.

x_1	0				1			
$x_3 x_5$	00	01	11	10	00	01	11	10
00	0	1	(5)	4	16	(17)	(21)	20
01	2	3	7	6	18	19	23	22
$x_2 x_4$ 11	(10)	(11)	15	(14)	(26)	27	31	(30)
10	8	9	13	12	24	25	29	28

(a)

x_2	0				1			
$x_4 x_5$	00	01	11	10	00	01	11	10
00	0	1	3	2	8	9	(11)	(10)
01	4	(5)	7	6	12	13	15	(14)
$x_1 x_3$ 11	20	(21)	23	22	28	29	31	(30)
10	16	(17)	19	18	24	25	27	(26)

(b)

Figure 16.8 Decomposition charts indicating iterative decomposition.

$$f(x) = \bar{x}_2 \bar{x}_4 \lambda(x_1, x_3, x_5) + x_2 x_4 \overline{\lambda(x_1, x_3, x_5)} \qquad (16.36)$$

where $\lambda(x_1, x_3, x_5) = x_3 x_5 + x_1 x_5$.

Thus, from Theorem 16.7, $\lambda(x_1, x_3, x_5)$ can be decomposed according to the partition $x_1 x_3 | x_5$. By inspection of the equation for λ, we see that the decomposition is given by

$$\rho[\Phi(A), B] = x_5(x_3 + x_1)$$

where $\Phi(A) = x_3 + x_1$.

This example is trivial in the sense that we did not need Theorem 16.7 to tell us that $\lambda(x_1, x_3, x_5)$ could be factored in the above manner. However, the theorem tells us that no matter how large the number of variables, the charts will enable us to locate decompositions in which the subfunctions can, in turn, be further decomposed.

THEOREM 16.8   Let $f(X)$ be a function for which there exist two simple disjoint decompositions

$$f(X) = F[\lambda(A), B] \tag{16.37}$$

$$= G[\Phi(B), A] \tag{16.38}$$

Then there exists a multiple disjoint decomposition

$$f(X) = H[\lambda(A), \Phi(B)] \tag{16.39}$$

The conditions of Equations 16.37 and 16.38 will be indicated by a single decomposition chart, which satisfies the column multiplicity requirements when viewed both normally and sideways. An example is the parity function considered earlier. Consider the map shown in Fig. 16.9a. Clearly, we have a column multiplicity of 2 in both directions. Viewed normally, the chart indicates the decomposition

$$f(x_1 x_2 | x_3 x_4) = (\bar{x}_1 \bar{x}_2 + x_1 x_2)\overline{\Phi(x_3 x_4)} + (\bar{x}_1 x_2 + x_1 \bar{x}_2)\Phi(x_3, x_4) \tag{16.40}$$

where $\Phi(x_3 x_4) = x_3 \bar{x}_4 + \bar{x}_3 x_4$.

Viewed sideways, the chart indicates

$$f(x_3 x_4 | x_1 x_2) = (\bar{x}_3 \bar{x}_4 + x_3 x_4)\overline{\lambda(x_1, x_2)} + (\bar{x}_3 x_4 + x_3 \bar{x}_4)\lambda(x_1, x_2) \tag{16.41}$$

where $\lambda(x_1 x_2) = \bar{x}_1 x_2 + x_1 \bar{x}_2$.

The theorem indicates the existence of a multiple decomposition of the form $H[\Phi(x_3, x_4), \lambda(x_1, x_2)]$. To determine H, we make a K-map of $F(\Phi, x_1, x_2)$ as given by Equation 16.40 (Fig. 16.9b). The values of $\lambda(x_1, x_2)$ are indicated on this map, immediately enabling us to construct the map of $H(\Phi, \lambda)$ as shown in Fig. 16.9c. From this map, we read off

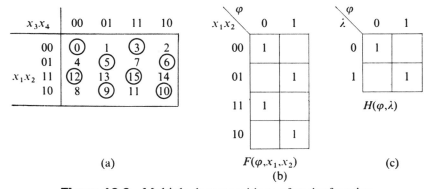

Figure 16.9   Multiple decompositions of parity function.

$$H(\Phi, \lambda) = \bar{\Phi}\bar{\lambda} + \Phi\lambda \tag{16.42}$$

where $\Phi = x_3 \oplus x_4$ and $\lambda = x_1 \oplus x_2$.

This will be recognized as the same result as obtained by factoring. Again the example is perhaps trivial, but the method extends to functions of any number of variables.

THEOREM 16.9 Let $f(X)$ be a function for which there exist two simple disjoint decompositions

$$f(X) = F[\lambda(A), B, C] \tag{16.43}$$
$$= G[\Phi(B), A, C] \tag{16.44}$$

Then there exists a multiple disjoint decomposition

$$f(X) = H[\lambda(A), \Phi(B), C] \tag{16.45}$$

As an example, consider the parity function of five variables $S_{1,3,5}^5(x_1 x_2 x_3 x_4 x_5)$, for which two decomposition charts are shown in Fig. 16.10. Chart (a), viewed sideways, gives the decomposition

$$f(x_3 x_4 x_5 | x_1 x_2) = (\bar{x}_3\bar{x}_4\bar{x}_5 + \bar{x}_3 x_4 x_5 + x_3\bar{x}_4 x_5 + x_3 x_4\bar{x}_5)\lambda(x_1, x_2)$$
$$+ (\bar{x}_3\bar{x}_4 x_5 + \bar{x}_3 x_4\bar{x}_5 + x_3\bar{x}_4\bar{x}_5 + x_3 x_4 x_5)\overline{\lambda(x_1, x_2)} \tag{16.46}$$

where $\lambda(x_1, x_2) = \bar{x}_1 x_2 + x_1\bar{x}_2$.

Chart (b), viewed sideways, gives the decomposition

$$f(x_1 x_2 x_5 | x_3 x_4) = (\bar{x}_1\bar{x}_2\bar{x}_5 + \bar{x}_1 x_2 x_5 + x_1\bar{x}_2 x_5 + x_1 x_2\bar{x}_5)\Phi(x_3, x_4)$$
$$+ (\bar{x}_1\bar{x}_2 x_5 + \bar{x}_1 x_2\bar{x}_5 + x_1\bar{x}_2\bar{x}_5 + x_1 x_2 x_5)\overline{\Phi(x_3, x_4)} \tag{16.47}$$

where $\Phi(x_3, x_4) = \bar{x}_3 x_4 + x_3\bar{x}_4$.

x_3			0				1	
$x_4 x_5$	00	01	11	10	00	01	11	10
$x_1 x_2$ 00	0	①	3	②	④	5	⑦	6
01	⑧	9	⑪	10	12	⑬	15	⑭
11	24	㉕	27	㉖	㉘	29	㉛	30
10	⑯	17	⑲	18	20	㉑	23	㉒

(a)

x_1			0				1	
$x_2 x_5$	00	01	11	10	00	01	11	10
$x_3 x_4$ 00	0	①	9	⑧	⑯	17	㉕	24
01	②	3	⑪	10	18	⑲	27	㉖
11	6	⑦	15	⑭	㉒	23	㉛	30
10	④	5	⑬	12	20	㉑	29	㉘

(b)

Figure 16.10 Decomposition chart for $S_{1,3,5}^5(x_1, x_2, x_3, x_4, x_5)$.

F.16.10

The conditions of Theorem 16.9 are thus satisfied with $A = \{x_1, x_2\}$, $B = \{x_3, x_4\}$, and $C = \{x_5\}$. To determine the composite function, we construct a K-map of Equation 16.46, as shown in Fig. 16.11a, indicating the corresponding values of $\Phi(x_3, x_4)$. From these, the map of $H[\lambda(x_1, x_2), \Phi(x_3 x_4), x_5]$ may be constructed as shown in Fig. 16.11b. From the map, we read off the function

$$H(\lambda, \Phi, x_5) = (\bar{\lambda}\bar{x}_5 + \lambda x_5)\Phi + (\bar{\lambda}x_5 + \lambda\bar{x}_5)\bar{\Phi} \qquad (16.48)$$

This completes the decomposition corresponding to Theorem 16.9, but we note that the final map of H (Fig. 16.10b) is also decomposable. Thus, we can define

$$\rho(\lambda, x_5) = \bar{\lambda}x_5 + \lambda\bar{x}_5$$
$$= \lambda \oplus x_5$$

Then the function becomes

$$f(X) = \bar{\rho}\Phi + \rho\bar{\Phi} \qquad (16.49)$$

where $\rho = \lambda \oplus x_5$, $\Phi = x_3 \oplus x_4$, and $\lambda = x_1 \oplus x_2$.

It is apparent that functions might also be decomposed in a nondisjoint fashion, and Curtis [4] has developed the theory quite extensively. However, the methods involved are so complex as to make them of doubtful value. It is apparent that there is little point trying to decompose a function unless the possibility of significant economy exists. As we indicated earlier, many of the functions whose two-level realizations are most expensive are subject to disjoint decomposition.

The reader may have noted in the example illustrating Theorem 16.9 that the two decomposition charts (Fig. 16.10) are identical. Since the value of a symmetric function is dependent only on the *number* of variables taking on the value 1, it is apparent that this condition (identical decomposition charts) will always hold for symmetric functions. Thus, we may conclude that one or more of the theorems given here will be applicable to most (if not all) symmetric functions. This observation, in turn, suggests that the methods of disjoint decomposition will generally be adequate for the simplification of symmetric functions.

Identification of decompositions may be added to the techniques discussed in Section 8.8 as components of any multilevel combinational logic optimization program. To identify decompositions, MIS [8, 13], the most popular multilevel optimization program, reduces the original multilevel network to sum-of-products

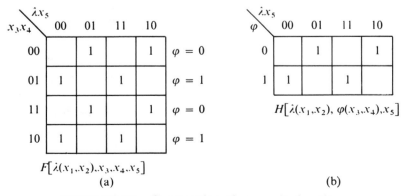

Figure 16.11 Construction of composite function.

form. If the original realization was a particularly efficient structure, such as a ripple carry adder, the optimization program may not be able to reconstruct a network as efficient as the original.

16.6 ITERATIVE NETWORKS

An iterative network is a highly repetitive form of a combinational logic network. This repetitive structure makes it possible to describe iterative networks utilizing techniques already developed for sequential circuits. We will limit our discussion to one-dimensional iterative networks represented by the cascade of identical cells given in Fig. 16.12a. A typical cell with appropriate input and output notation is given in Fig. 16.12b. Note the two distinct types of inputs: primary inputs from the outside world and secondary inputs from the previous cell in the cascade. Similarly, there are two types of outputs: primary to the outside world and secondary to the next cell in the cascade. At the left of the cascade are a set of boundary inputs that we will denote in the same manner as secondary inputs. In some cases, these inputs will be constant values. A set of boundary outputs emerges from the rightmost cell in the cascade. Although these outputs are to the outside world, they will be labeled in the same manner as secondary outputs. In a few cases, the boundary outputs will be the only outputs of the iterative network.

The next three examples will serve to illustrate the analysis and synthesis of a few practical iterative networks. We will find that it is possible to use all the techniques associated with the synthesis of clock-mode sequential circuits to synthesize iterative networks. As with sequential circuits, we will first analyze a network already laid out in iterative form. By first translating the network of Example 16.9 to a form very similar to a state table, we hope to make the reverse process more meaningful.

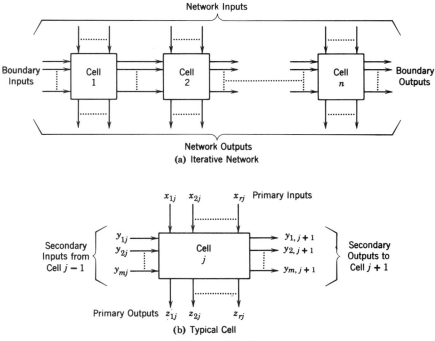

(a) Iterative Network

(b) Typical Cell

Figure 16.12

■ **EXAMPLE 16.9**

Although the reader is accustomed to thinking of counters as self-contained sequential circuits, such need not be the case. Very often, it is desirable to include a counting capability in a general-purpose arithmetic register in a computer. Going one step further to include the possible counting capability in a bank of several arithmetic registers provides a rationale for the network of Fig. 16.13.

The inputs $x_1 \cdots x_n$ in Fig. 16.13 come from the output of a data bus. Depending on control inputs to the bus, this information vector might be the contents of any one of several data registers. The outputs of the iterative network $z_1, z_2, ..., z_n$ represent the new contents of the same register after incrementing. A control pulse will cause the new information to be triggered back into the same register. The arrangement in Fig. 16.13 makes it possible to share the incrementing logic over the entire bank of registers rather than including it separately in each register. Admittedly, in an MSI environment, the economic advantage would be important only if the number of registers in the bank was quite large. Another criticism of the network is the possible need for the effect of x_1 to propagate through $n - 1$ levels of AND gates. A practical realization would no doubt use multiinput AND gates. Finally, the AND gate and exclusive-OR network for bit 1 and AND gate for bit n are unnecessary. Nevertheless, we will consider the format of Fig. 16.13, which is an example of a simple iterative network. The reader will easily find that the number representing $x_1 \cdots x_n$ (x_1 is the least significant bit) is actually incremented by noting that a bit is complemented whenever all less significant bits are 1.

We base our analysis on the typical cell, cell j, given in Fig. 16.14a. The functional values of y_{j+1} and z_j are tabulated in Fig. 16.14b for all combinations of values of inputs x_j and y_j. Note that this tabulation very closely resembles the *transition table* of clock-mode sequential circuits. The primary inputs are listed horizontally as are sequential-circuit inputs. The secondary inputs $y_{1_j} \cdots y_{m_j}$ (in this case, only y_j) take the place of the present states of memory elements in the table. The secondary outputs take the place of the memory element next states. By defining $y = 0$ as state A and $y = 1$ as state B, we translate the transition table to the *state table* form of Fig. 16.14c. ■

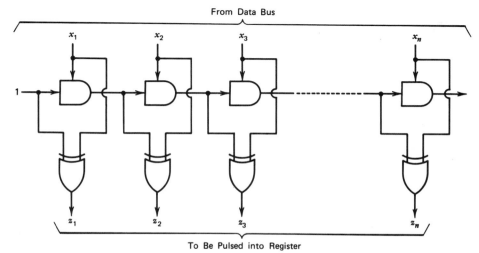

From Data Bus

To Be Pulsed into Register

Figure 16.13 Increment network.

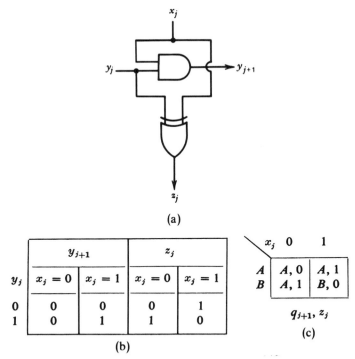

Figure 16.14 Development of state table.

As was the case with sequential circuits, the first step in the synthesis of an iterative network will be to reduce an English-language description to state table form. From there, all the techniques of Chapter 10 may be used to obtain an iterative network from a state table. The following two examples will serve to illustrate the synthesis process.

■ **EXAMPLE 16.10**

A set of bits $x_1 \cdots x_n$ may be considered as received information in a communications system. A particular decoding scheme requires a check for odd parity over the first bit, the first two bits, the first three bits, etc. Design an iterative network that will accomplish this function.

SOLUTION The reader will notice that the desired outputs are the symmetric functions S_1^1, S_1^2, $S_{1,3}^3$, $S_{1,3}^4$, etc. In general, sets of symmetric functions can quite easily be implemented as iterative networks.

A state table may be constructed directly by letting the *secondary input state* to block j be A if parity over primary inputs $x_1, x_2, ..., x_{j-1}$ is odd. If parity over these bits is even, the secondary input state will be B. The output z_j is to be 1 if there is odd parity over bits $x_1, x_2, ..., x_j$. Thus, $z_j = 1$ when $x_j = 0$ and the input state is A and when $x_j = 1$ and the input state is B. The secondary output will be A precisely when $z_j = 1$ and will be B when $z_j = 0$. Thus, we have the state table of Fig. 16.15. ■

The next example will be the first to require more than two secondary input states. We will continue the design process in this case to develop the network realization of a single cell from the state table.

x_j	0	1
B	$B, 0$	$A, 1$
A	$A, 1$	$B, 0$

q_{j+1}, z_j

Figure 16.15 State table for iterative parity checker.

■ EXAMPLE 16.11

Bits $x_1, x_2, ..., x_n$ are interrupt signals sent to a computer central processor from n peripheral equipments to request service from the processor. At any given time, the computer can communicate with no more than two of these peripherals. Peripheral 1 (corresponding to interrupt x_1) has the highest priority, with peripheral 2 second highest, etc. It is, therefore, desired to construct a network with outputs $z_1, z_2, ..., z_n$, such that no more than two of the outputs will be 1 at a given time. The 1 outputs will correspond to the two highest-priority inputs with active interrupts. (Interrupt i is active if $x_i = 1$.) If only one interrupt is active, only one output line will be 1 and no output will be 1 in the case of no active interrupts.

SOLUTION   We define three secondary input states for cell j. The secondary input state will be A if no interrupts of higher priority than peripheral j are active. The state is B if one higher-priority interrupt is active and the state is C if two or more higher-priority interrupts are active. Thus, the output $z_j = 1$ only if $x_j = 1$ and the input state is A or B. If $x_j = 0$, the secondary output state is the same as the input state. If $x_j = 1$, an input state B must generate an output state C and an input state A must generate an output state B. This information is tabulated in Fig. 16.16a.

A convenient state assignment defines $y_2 y_1 = 00$ as A, $y_2 y_1 = 01$ as B, and $y_2 y_1 = 11$ as C. This leads to the transition table of Fig. 16.16b. The equations for the secondary outputs or next-state equations (16.50, 16.51) may be obtained directly by treating the transition table as a K map.

$$y_{2,j+1} = y_{2j} + x_j \cdot y_{1j} \tag{16.50}$$

$$y_{1,j+1} = y_{1j} + x_j \tag{16.51}$$

$$z_j = x_j \cdot \bar{y}_{2j} \tag{16.52}$$

The primary output z_j is similarly obtained from the transition table. The network for a single cell of the priority network may be obtained by implementing these equations. ■

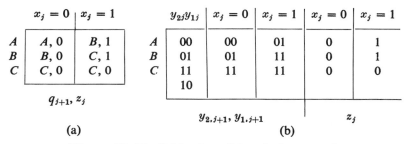

	$x_j = 0$	$x_j = 1$
A	$A, 0$	$B, 1$
B	$B, 0$	$C, 1$
C	$C, 0$	$C, 0$

q_{j+1}, z_j

$y_{2j}y_{1j}$	$x_j = 0$	$x_j = 1$	$x_j = 0$	$x_j = 1$
A	00	01	0	1
B	01	11	0	1
C	11	11	0	0
	10			

$y_{2,j+1}, y_{1,j+1}$ z_j

(a)   (b)

Figure 16.16 Tables for cell in priority network.

Two additional examples of iterative circuits of considerable interest are the adder and magnitude comparator. (See Problems 16.10 and 16.11.) Both these networks have two inputs per cell, but are otherwise of the same order of difficulty as the above example. The comparator is a special case in that there are no primary outputs. The only outputs of the overall network are the secondary outputs for the rightmost cell. As indicated in Fig. 16.12a, these outputs are called boundary outputs.

16.7   TEST GENERATION USING THE ANALOGOUS ITERATIVE NETWORK

We observed in the previous section the one-to-one correspondence between a cell table in an iterative network and the transition table of an analogous clock-mode sequential circuit. A series of inputs applied to the cells of an iterative network in order from left to right will produce a series of outputs from these cells. The same series of inputs now applied as a sequence in clock-period-by-clock-period order to the analogous sequential circuit will generate the same series of outputs in sequence. There is a one-to-one correspondence between the inputs and outputs of the iterative network over cell space to the input and output sequences of the sequential circuit over time.

Suppose that it is necessary to discover an input sequence (and to determine the resulting output sequence) to accomplish some task in a sequential circuit. It may be more attractive to find the analogous series of inputs and outputs for the iterative network and to then apply the input sequence to the sequential circuit. An obvious application of this strategy is finding a test sequence for a single stuck-at fault in a sequential circuit. In this section, we shall inject the fault in the analogous iterative network and find the input sequence that will excite the fault and drive the excited fault to a primary output. In the process, the output series for both the good and faulty networks will be determined. The analogous sequential circuit may be tested for this fault by applying this same sequence of inputs and observing the resulting output sequence.

Recall that the two memory element sequential circuits used to illustrate fault simulation in Example 15.10 had no synchronizing sequence. The fault-free version of this circuit cannot be driven to a known state, if both memory elements are initialized to the unknown value U. This implies that no test sequence, which will be independent of the initial state, can be found for a single stuck-at fault. Adding an AND gate at the input to flip-flop y_1, as shown in Fig. 16.17, will make every sequence of two or more inputs a synchronizing sequence.

■ EXAMPLE 16.12

Use the Wang algorithm of Section 15.5 on the analogous iterative network to find a test for gate input p SA1 in the sequential circuit of Fig. 16.17.

SOLUTION   To construct the iterative network, we cut the feedback loops at both the inputs and outputs of the memory elements as shown. The detached combinational logic is the cell realization in the iterative network. The excitations Y_0 and Y_1 are connected to the secondary inputs of the next cell. The result is given in Fig. 16.18. The number of cells to the left and right of the starting point of the search is unknown at the outset. The cell-numbering scheme was chosen to avoid negatively numbered cells in this small example.

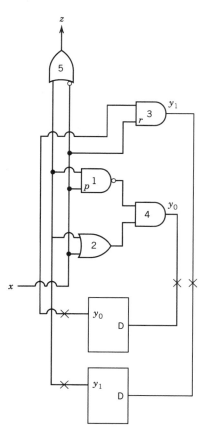

Figure 16.17   Sequential circuit with synchronizing sequence.

Arbitrarily, the fault is excited in cell 5. The values of primary input x and secondary input y_{15}, required to excite the fault, are double-underlined in Fig. 16.18. The value 1/X is required at the output of gate 2 to propagate the excited fault through gate 4. This is consistent with the already determined inputs to gate 2. A 1 (single underline) on x_6 is required to propagate the fault through gate 3 of cell 6. A 0 on x_7 will drive the fault to the primary output of cell 7.

The final step is to search for a set of primary input values consistent with the double-underlined $y_{15} = X/1$. Usually, the search will expand nodes for primary inputs in reverse cell order, beginning at the left of the cell in which the fault was excited. Figure 16.19 depicts the search process. $x_4 = 0$ is tried first. This is inconsistent with $y_{15} = X/1$, so node 0XXXX is terminal. From node 1XXXX, we try first $x_3 = 0$. Because point p in cell 3 is SA1, the outputs of both gates 1 and 2 in that cell remain undetermined. From node 10XXX, we let $x_2 = 0$. This will cause the outputs of gate 3 in cell 2 and gate 2 in cell 3 to be 0. This is inconsistent with $y_{04} = X/1$. Backtracking to let $x_2 = 1$ yields node 101XX. Continuing from this point will lead to a solution only if a set of values is found for which $y_{02} = 1$. This is no easier than establishing y_{04}, which has not yet been accomplished. The unmodified Wang algorithm would continue back from cell to cell without finding a solution. A simple modification would be to expand nonterminal nodes with all combinations of inputs in a given cell before moving back to the next cell. This would call for expanding node 11XX before node 101X, thereby finding success node 110X. The network values for this node are given in Fig. 16.18. The values of y_{12}

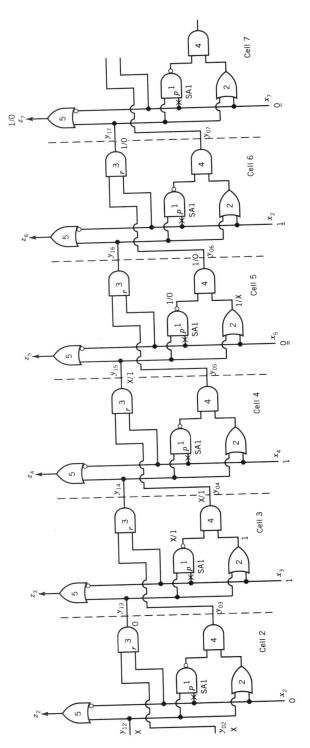

Figure 16.18 Test generation in an iterative network.

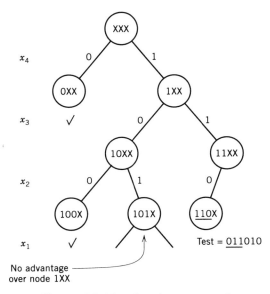

Figure 16.19   Consistency search.

and y_{02} remain unspecified. Thus, the input sequence 011010, applied clock period by clock period, will test for p SA1, regardless of the initial state of the sequential circuit.   ∎

One advantage of the iterative network is that the single fault can be imposed in each cell to guarantee the accuracy of the results. However, the computations involved in searching a network are more time-consuming than searching a state table or a RTL description of a larger sequential circuit in the time domain. The complexity of digital systems to be tested, particularly the large number of possible input combinations, dictates that any time-based test search system must depend on heuristic guidance. No single approach has achieved sufficient acceptance to warrant treatment here. SCIRTSS [10, 11], a sequential circuit test search system developed by the author has been demonstrated to be effective for a broad class of RTL descriptions.

PROBLEMS

16.1   Simplify the following symmetric function expressions:

(a) $[S_{1,3,5,7}^{7}(x_1, ..., x_7)] \cdot [S_{4,5,6,7}^{7}(x_1, ..., x_7)]$

(b) $\overline{S_{2,3,5}^{5}(x_1, ..., x_5)} + S_{3}^{5}(x_1, ..., x_5)$

(c) $[S_{1,3,5}^{5}(\bar{x}_1, \bar{x}_2, ..., \bar{x}_5)] \cdot [S_{2,3,4,5}^{5}(x_1, x_2, ..., x_5)]$

16.2   How many different functions of the form $f(x_1, x_2, ..., x_6)$ are there that are symmetric with respect to the six literals $x_1, x_2, ..., x_6$?

16.3   How many of the 16 functions of two variables are symmetric (with respect to any of the four possible pairs of literals)?

16.4   Consider the 10 bits $x_1, x_2, ..., x_{10}$ arranged in ascending order. There are eight possible ways of selecting three successive bits from this arrangement. The function that is 1 when a majority of some such set of three bits is 1 is, of course, a symmetric function.

(a) Write, as a function of symmetric functions, an expression that will be 1 whenever the majorities of bits in an odd number of the sets of three bits are 1.

(b) Is the resulting expression a symmetric function? Why?

(c) Repeat part (a), this time considering all $_{10}C_3 = 120$ possible combinations of 3 of the 10 bits. Simplify.

(d) Is the function in part (c) symmetric? Why?

16.5 Determine a simple disjoint decomposition for each of the following functions for which such a decomposition exists.

(a) $f(x_1, x_2, x_3, x_4) = \sum m(0, 5, 6, 7, 9, 10, 11, 12)$

(b) $f(x_1, x_2, x_3, x_4) = \sum m(0, 3, 4, 7, 12, 15)$

(c) $f(x_1, x_2, x_3, x_4) = \sum m(0, 2, 4, 9, 6, 13)$

(d) $f(x_1, x_2, x_3, x_4, x_5) = \sum m(3, 10, 14, 17, 18, 22, 23, 24, 27, 28, 31)$

16.6 Repeat Problem 16.5 for the following two functions.

(a) $f(x_1, x_2, x_3, x_4, x_5) = \sum m(1, 3, 5, 7, 9, 10, 13, 14, 21, 23, 26, 27, 29, 30)$

(b) $f(x_1, x_2, x_3, x_4, x_5) = \sum m(1, 3, 9, 10, 17, 19, 24, 27)$

16.7 There exists an iterative disjoint decomposition for the following function of five variables. Determine this decomposition and express it in the form of Equation 16.33.

$$f(x_1, x_2, x_3, x_4, x_5) = \sum m(2, 6, 8, 12, 16, 20, 26, 30)$$

16.8 Determine a multiple disjoint decomposition of the four-variable function.

$$f(x_1, x_2, x_3, x_3, x_4) = \sum m(0, 3, 4, 7, 9, 10, 13, 14)$$

16.9 Develop an original five-variable example of the multiple disjoint decomposition given by Equation 16.45. Let the variables be partitioned as given by

$$f(x_1, x_2, x_3, x_4, x_5) = H[\lambda(x_1, x_2), \Phi(x_3, x_4), x_5]$$

16.10 Each cell of an iterative network is to have one primary input and one primary output. Let the primary input of cell i be x_i and the primary output be z_i. Determine the cell table for cell i such that $z_i = 1$ if and only if $x_{i-2} = x_{i-1} = x_1$. Determine a network realization of the typical cell i.

16.11 The n-bit binary adder of Fig. 8.13 may be regarded as an iterative network. A single cell of the network is given in circuit form as Fig. 8.9. Determine the state table of a single cell of the adder.

16.12 A circuit that will compare the magnitudes of two 18-bit numbers is to be formulated as an iterative network. Each cell is to have two secondary outputs and no primary outputs. The input numbers are given as vectors **X1** and **X2**. The secondary outputs of cell j should indicate whether the binary number represented by $X1_{0:j-1}$ is greater than, equal to, or less than the binary number represented by $X2_{0:j-1}$. Determine the state table of a single cell and a circuit realization of that cell. Sketch the complete network. What will be the output of the network?

16.13 Use the analogous iterative network to find a test sequence that will detect the fault, input point p of gate AND2 SA1, in the sequential circuit shown in Fig. P16.13. Use the Wang algorithm and assume that the flip-flop is

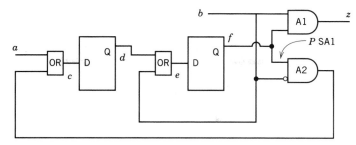

Figure P16.13

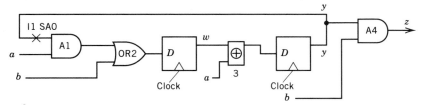

Figure P16.14

initially in an "unknown state." Show your work, including at least three copies of the combinational logic. Indicate on your iterative network diagram which inputs are established by the forward drive and which by the consistency search. Show clearly your final input sequence as a time sequence leaving "don't care" inputs whenever possible. How many clock periods will be required?

16.14 Use the analogous iterative network to find a test sequence that will detect the fault, input I1 of gate 1 SA0, in the sequential circuit shown in Fig. P16.14. Use the same process as described in Problem 16.13.

16.15 Use the analogous iterative network and the Wang algorithm to find a test for memory element output y_0, SA0 in the finite memory circuit of Fig. 10.10.

BIBLIOGRAPHY

1. Marcus, M. P. "The Detection and Identification of Symmetric Switching Functions with the Use of Tables of Combinations," *IRE Trans. Electronic Computers*, **EC-5:** 4, 237–239 (Dec. 1956).

2. Ashenhurst, R. L. "The Decomposition of Switching Functions," in *Proceedings of the Symposium on the Theory of Switching*, April 2–5, 1957, Vol. 29 of *Annals of Computers*. Harvard University, Cambridge, Mass., 1959, pp. 74–116.

3. Maley, G. A. and J. Earle. *The Logic Design of Transistor Digital Computers*. Prentice-Hall, Englewood Cliffs, N.J., 1963.

4. Curtis, H. A. *Design of Switching Circuits*. Van Nostrand, New York, 1962.

5. Harrison, M. *Introduction to Switching and Automata Theory*. McGraw-Hill, New York, 1965.

6. Birkhoff, G. and S. MacLane. *Survey of Modern Algebra*, 3rd ed. Macmillan, New York, 1965.

7. Shannon, C. E. "A Symbolic Analysis of Relay and Switching Circuits," *Trans. AIEE*, **57,** 713–723 (1938).

8. Brayton, R., R. Rudell, A. Sangiovanni-Vincentelli, and A. Wang. "MIS: A Multiple Level Logic Optimization System," *IEEE Trans. Computer Aided Design of Integrated Circuits and Systems*: Vol. CAD-6, 1062–1081 (Nov. 1987).

9. Kubo, H. "A Procedure for Generating Test Sequences to Detect Sequential Circuit Failures," *NEC J. Research and Development*, **12:** 59–78 (Oct. 1968).

10. Mohsenni Behbahani, A., F. J. Hill, and M. Y. Patel. "Intelligence Driven Test Sequence Generator for VLSI Design," in *Microprocessing and Microprogramming*, Vol. 18. North-Holland, Amsterdam, 1986, pp. 355–362.

11. Hill, F. J. and B. Huey. "SCIRTSS: A Search System for Sequential Circuit Test Sequences," *IEEE Trans. Computers*, **C-26:** 490–502 (May 1977).

12. Miczo, A. *Digital Logic Testing and Simulation*. Harper and Row, New York, 1986.

13. Spickelmeier, R. (ed.). *OCT Release 3.0 Reference Manual*. University of California, Berkeley, Calif., 1990.

APPENDIX A

SELECTION OF MINIMAL CLOSED COVERS

In Chapter 13, after finding the set of all maximal compatibles, the selection of a minimal cover was carried out by an essentially trial-and-error process. This was practical since the number of maximal compatibles was reasonable and there were few implications. In very complex cases, with many compatibles and interacting implications, a more systematic procedure would be highly desirable. Unfortunately, no procedure exists, other than complete enumeration, that guarantees finding the minimal equivalent of a state table in every case. However, a paper published by Grasselli and Lucio [1] provides the basis for a straightforward procedure that will reduce the number of alternatives considerably in most problems.

As a starting point, it is necessary to formalize the procedure of combining sets of compatibility pairs into larger compatibility classes. We have already seen that this can be done, but there is some question as to the effect of this procedure on closure. Theorem A.1 provides the answer to this question.

THEOREM A.1 Let $\{q_1, q_2, ..., q_n\} = \{q_i\}$ be a set of states that are pairwise compatible in a single closed collection. Then $\{q_i\}$ is a compatibility class within some closed collection.

PROOF Since the initial collection is closed, the next states $\delta(q_1, X_j)$ and $\delta(q_2, X_j)$ for any input X_j are a compatible pair in the same collection. The same holds true for the next states for any other pair in $\{q_i\}$. Thus, all the next states of $\{q_i\}$ for a given input X_j are pairwise compatible and can form a single compatibility class $\{\delta[q_i, X_j]\}$. Therefore, the requirements of Definition 13.7 are satisfied with respect to $\{q_i\}$. Similarly, all the next states of $\{\delta[q_1 X_j]\}$ are pairwise compatible in the original closed collection and can form a single compatibility class, satisfying Definition 13.7 for $\{\delta[q_i, X_j]\}$. This process continues until a complete closed collection is formed. Since only a finite number of distinct classes can be formed from a finite set of states, the process must terminate. Q.E.D.

THEOREM A.2 The union of all distinct maximal compatibles formed from the states of a circuit S is a closed collection of compatibility classes.

PROOF Starting with any maximal compatible, generate a closed collection by the process described in the proof of Theorem A.1. Every class in this collection must either be a maximal compatible or else be included in a maximal compatible. Thus, Definition 13.7 is satisfied. Q.E.D.

The reader can easily verify that the collections of maximal compatibles found in the examples of Chapter 13 are indeed closed collections. These collections of maximal compatibles may be used as a starting point in the process of finding a minimal cover. In these examples, we were able to find minimal covers without great difficulty, simply by inspection. In general, this may not be possible. First, in a problem of any complexity, there will be a large number of maximal compatibles. Second, all possible subsets of the maximal compatibles are also candidates for inclusion in a minimal cover. The first step then is to reduce the number of candidates.

If C_α and C_β are compatibility classes and P is a collection of classes, we mean by $C_\alpha \subseteq C_\beta$ that all the states in C_α are included in C_β, and by $C_\beta \in P$ that C_β is a member of P.

Definition A.1   The *class set* P_α implied by a compatibility class C_α is the set $\{D_{\alpha_1}, D_{\alpha_2}, ..., D_{\alpha_n}\} = \{D_{\alpha_i}\}$, of all classes implied by C_α, such that

1. D_{α_i} has more than one element.
2. $D_{\alpha_i} \nsubseteq C_\alpha$.
3. $D_{\alpha_i} \nsubseteq D_{\alpha_j}$ if $D_{\alpha_j} \in P_\alpha$.

Less formally, the class set is made up of the compatibility classes, other than singletons, implied by a given set and not included in the given set. Figure A.1 shows a small implication table, the determination of the maximal compatibles, and the class sets implied by the maximal compatibles. Note that the class (*abde*) does not imply a class set since all the pairs implied by this class are subsets of the class. Class (*bcd*), however, implies three pairs that are not included in the class.

Condition 3 of Definition A.1 may be applicable when some class of three or more states implies all possible pairs of states from another set of three or more states. Consider the two partial state tables shown in Fig. A.2 and assume that in both cases, (*abc*) is a compatibility class. In both cases, (*abc*) implies the classes (*de*), (*df*), and (*ef*). For the table of Fig. A.2, the three classes are implied for different inputs; hence, they do not combine into a single implied class. For the table of Fig. A.2b, however, all three classes are implied by a single input, so they combine into a single implied class (*def*). By Condition 3, only this larger set need be included in the class set. Note that this situation cannot be found with the implication table alone, but requires reference to the state table. In the case of Fig. A.1, it was not necessary to refer to the state table, since the pairs appearing did not make up all possible combinations of some larger set of states.

Definition A.2   A compatibility class C_α excludes a compatibility class C_β if and only if

1. $C_\beta \subseteq C_\alpha$.
2. $P_\alpha \subseteq P_\beta$.

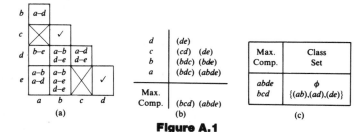

Figure A.1

q^v	q^{v+1}				q^v	q^{v+1}	
	$X = 0$	$X = 1$	$X = 2$			$X = 0$	$X = 1$
a	d	f	$-$		a	d	f
b	e	$-$	e		b	e	$-$
c	$-$	d	f		c	f	d
	(a)					(b)	

Figure A.2

where P_α and P_β are the class sets implied by C_α and C_β, respectively.

A class that is not excluded will be called a *prime* compatibility class.

Suppose a minimal set of compatibility classes covering a given state table has been found, including some class C_β that is not prime. Let C_α be a class that excludes C_β. Since C_α includes all the states covered by C_β and does not imply any classes not implied by C_β, both cover and closure will be maintained if C_α replaces C_β. We have thus proved the following theorem:

THEOREM A.3 At least one minimal covering will consist solely of prime compatibility classes.

The concept of "exclusion" is used to eliminate classes from consideration in much the same manner that dominance is used to eliminate prime implicants. An excluded class covers only states that are included in some larger class, as specified by condition 1 of Definition A.2. Condition 2 ensures that closure is not violated by the substitution of the larger class for the smaller one. It is important to note that not all minimal covers consist of prime classes and that not all covers consisting only of prime classes are minimal. Theorem A.3 simply allows us to restrict our attention to prime classes, with the assurance that a minimal cover can still be found.

■ EXAMPLE A.1

Determine the prime compatibility classes for the circuit partially described by the implication table of Fig. A.3.

SOLUTION The first step is the determination of the maximal compatibles, as shown in Fig. A.4.

The process of determining the prime compatibles starts with the maximal compatibles, which are themselves prime. These are listed in Fig. A.5a, together with the implied class sets.

The remaining prime classes are found by deleting one state at a time from each of the maximal compatibles. If the implied class set is smaller than that of the maximal compatible or any other prime class containing the subset, then it is prime. After this has been done for all of the maximal compatibles, the procedure is repeated for the resultant prime classes, until no further decomposition takes place.

For example, the maximal compatible (134) calls for a check of (13), (14), and (34). We see that (13) and (14) imply (56), just as does (134), so they are not prime. Class (34), however, implies the empty set and must be included in the list of Fig. 11.15*b*. Note that a subset must be checked against *all* prime classes that contain

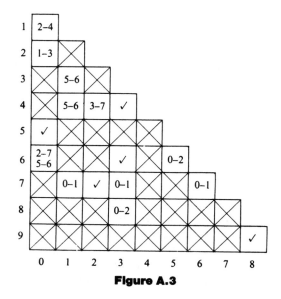

Figure A.3

8	(89)							
7	(89)							
6	(89)	(67)						
5	(89)	(67)	(56)					
4	(89)	(67)	(56)					
3	(89)	(367)	(56)	(38)	(34)			
2	(89)	(367)	(56)	(38)	(34)	(27)	(24)	
1	(89)	(367)	(56)	(38)	(134)	(27)	(24)	(137)
0	(89)	(367)	(056)	(38)	(134)	(27)	(24)	(137)	(02)	(01)

M.C. | (056) (134) (137) (367) (01) (02) (24) (27) (38) (89)

Figure A.4

Max Comp.	056	134	137	367	01	02	24	27	38	89
Class Sets	27 02	56	56 01	01	24	13	37	ϕ	02	ϕ

(a)

Other Prime Classes	05	06	34	17	36	1	56
Class Sets	ϕ	27 56	ϕ	01	ϕ	ϕ	02

(b)

Figure A.5 Determination of prime classes, Example A.1.

it. For example, subset (37), obtained from maximal compatible (137), is not excluded by (137) but is excluded by (367).

Also note that a prime class that implies the empty set excludes all its subsets and thus need not be checked for further prime classes. In this case, only (06), (17), and (56) need be checked further. Class (17) calls for a check of (1) and (7). We see that (7) is excluded by (27), but (1) is included in no prime class implying the empty set, so (1) must be added to the list of prime classes. ∎

We are now at much the same position as in the Quine–McCluskey method when the determination of the prime implicants has been completed. Just as the prime implicants are the only products that need be considered in realizing a combinational function, so only the prime classes need be considered in finding a cover for a state table. The first step in selecting a cover from the prime classes is to determine any essential prime classes, that is, prime classes that are the only ones including one or more states. Again, the parallel to the Quine–McCluskey method is noted.

Next, we delete the covered states from the unselected classes. If the resulting reduced classes are then excluded, they may be removed from further consideration. This step is roughly comparable to a reduction of the prime implicant table. We then check again for classes made essential by this second step. This process may be continued until no further reduction is possible. We will see in the completion of this example how the choice of prime compatibles can often be further narrowed through the intuitive treatment of a pictorial model.

■ EXAMPLE A.1 (continued)

Find a minimal cover from the set of prime compatibility classes.

We first note that (89) is essential to cover (9). We then delete (8) from (38), leaving (3), which is excluded by (34). So (38) is discarded. This does not make any other sets essential, so we set up a pictorial representation of the implications of the remaining classes (Fig. A.6), which we will refer to as the *implication graph*. Each class that has been selected or is still under consideration (not excluded) is represented by a circle. Selected classes are indicated by an asterisk. Every class implied by a class C_α is indicated by an arrow from C_α to that class. If only one class can satisfy an implication, we indicate a *strict implication* by a solid arrow. If an implication can be satisfied by any of two or more classes, we indicate an *optional implication* by dashed arrows.

A good place to start the reduction process is with classes that imply other classes but satisfy no implications themselves—for example, class (17). Note that the chain of implications set up by the selection of (17)

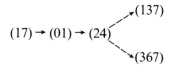

includes classes that cover both (1) and (7). Therefore, (17) may be discarded. By the same argument, (06) may be discarded. Each time a class is discarded, we should check to see if any of the remaining classes have been made essential as a result. This has not yet occurred, so we continue.

Next, let us compare classes that overlap, such as (056) and (56). Both satisfy the same implications, and both imply (02), thus providing a cover for (0). Thus,

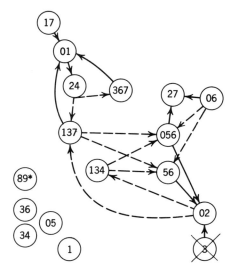

Figure A.6   Implication graph for Example A.1.

(056) can accomplish nothing that (56) cannot, and it implies more classes. Therefore, (056) may be discarded. Even after discarding (056), no classes have become essential. To facilitate the discussion, we redraw the graph, incorporating the above simplifications as shown in Fig. A.7.

Let us make a tentative selection of (137) to see what sort of cover results. The strict implication of (137) requires that (01), (24), (56), and (02) also be selected. With the addition of the essential class (89), we have a six-state closed cover

$$(137)(01)(02)(24)(56)(89)$$

Because of the strict implication, it is evident that this six-state cover is the smallest that can be found by using (137). Next, let us eliminate (137) and see what possibilities remain. The implication graph with (137) removed is shown in Fig. A.8. Here we see two *cycles*, sets of classes that must be selected as a group because of the cyclic pattern of implication. If we select cycle A, the addition of (05) and (89) will provide a five-state closed cover

$$(01)(24)(367)(05)(89) \tag{A.1}$$

If we select cycle B, we also get a five-state closed cover

$$(02)(134)(56)(27)(89) \tag{A.2}$$

The reader can quickly verify that any cover not including one of the cycles must have at least six states. Thus, the covers of (A.1) and (A.2) are minimal.   ∎

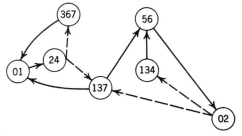

Figure A.7   Simplified implication graph for Example A.1.

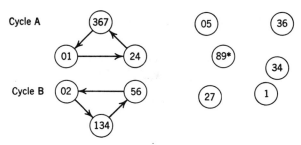

Figure A.8   Final implication table for Example A.1.

It should be emphasized that the above procedure is not a complete or exact algorithm. Rather, it consists of the application of a few "common-sense" rules in order to reduce the number of candidates for inclusion in a cover to a point where a trial-and-error check of all possibilities is practical. Also, the implication graphs are not an essential feature of the method. The same information is contained in the list of prime classes and class sets. However, the graphs, like Karnaugh maps, make it easier to recognize significant patterns and relationships. The reader who desires practice with this method should rework some of the more difficult exercises of Chapter 13.

BIBLIOGRAPHY

1. Grasselli, A. and F. Lucio. "A Method for Minimizing the Number of Internal States in Incompletely Specified Sequential Networks," *IEEE Trans. Electronic Computers*, **EC-14:** 3, 330–359 (June 1965).

APPENDIX B

SUMMARY OF AHPL SYNTAX

B.1 OPERAND CONVENTION

In AHPL as in other hardware description languages literal values are assigned to physical signals such as registers and buses. Signals may be scalars or may be grouped as vectors or two dimensional arrays. A vector is simply a collection of operands arranged in a one-dimensional array. If a register, PC has 18 bits, then $\rho PC = 18$; and the individual bit positions are denoted $PC[0]$, $PC[1]$, $PC[2]$ ... $PC[17]$. Note that the first position of the vector is denoted $PC[0]$, a procedure that is known as a 0-origin indexing.

A matrix is a two-dimensional array of operands as illustrated in Fig. B.1. The vector

$$(\mathbf{M}{<}i{>}[0],\ \mathbf{M}{<}i{>}[1],\ \ldots\ldots\ \mathbf{M}{<}i{>}[\rho_1-1])$$

is known as the ith row of $\mathbf{M}$ denoted as $\mathbf{M}{<}i{>}$. Similarly the vector

$$(\mathbf{M}{<}0{>}[j],\ \mathbf{M}{<}1{>}[j],\ \ldots\ldots\ \mathbf{M}{<}\rho_2-1{>}[j])$$

is the jth column of $\mathbf{M}$.

The most commonly encountered two dimensional arrays in AHPL are random access memories. In this case each row of $\mathbf{M}$ is a memory word. A stack of registers can also be treated as a two dimensional array. The notation $\mathbf{M}{<}i{:}k{>}$ can be used to form a new array consisting of rows i through k of $\mathbf{M}$. $\mathbf{M}[r{:}n]$ is an array consisting of columns r through n of $\mathbf{M}$.

B.2 AHPL OPERATORS

In the hardware translation of AHPL, the signal variables (e.g., buses and registers) are capable of assuming only values of one "type", Boolean (0's and 1's). Similarly BOOLEAN is the only type and 0 and 1 are the only values processed by the most common simulator of AHPL. A special syntax is needed for describing complex combinational logic networks. To facilitate the description of such networsk and for use in the process of translating them to a hardware realization a second type, INTEGER (scalar only), is allowed. The use of integers and operators

$$\begin{bmatrix} \mathbf{M}{<}0{>}\,[0] & \mathbf{M}{<}0{>}\,[1] & \mathbf{M}{<}0{>}\,[2] & \ldots\ldots\ldots\ldots\ldots\ldots\ldots\ldots\ & \mathbf{M}{<}0{>}\,[\rho_1-1] \\ \mathbf{M}{<}1{>}\,[0] & \mathbf{M}{<}1{>}\,[1] & \ldots\ldots\ldots\ldots\ldots\ldots\ldots\ldots\ldots\ & \mathbf{M}{<}1{>}\,[\rho_1-1] \\ \mathbf{M}{<}\rho_2-1{>}\,[0] & \ldots\ldots\ldots\ldots\ldots\ldots\ldots\ldots\ldots\ldots\ & \mathbf{M}{<}\rho_2-1{>}[\rho_1-1] \end{bmatrix}$$

Figure B.1 Array

thereon is only a descriptive mechanism. Once translated to hardware (or when simulated) combinational logic networks process only Boolean values.

The binary and unary operators of AHPL are summarized in Fig. B.2. Vectors and arrays of Boolean values and signals are common in AHPL. However, the use of integers will be limited to scalar values. Binary bit by bit Boolean operations are valid on arrays and vectors where operands have the same dimensions. The only valid exception is the case where one of the operands is a scalar Boolean value. In this case the operation is treated as if the scalar value is replaced with an array or vector of the same dimensions as the other operand in which every element has the same value as the scalar operand.

Among the integer operations applicable in combinational logic unit description is the relational operator $(x \mathcal{R} y)$, which evaluates to 1 if the relation $\mathcal{R}$, is satisfied and to 0 if the relation is not satisfied. The relation $\mathcal{R}$, may be any of the relations that typically compare numbers such $=, >, \geq, <,$ or $\leq$. For example

$$(4 > 3) = 1$$

since 4 is indeed greater than 3. Similarly,

$$(4 < 3) = 0.$$

Much of the power of APL derives from the mixed operators, which operate on various combinations of scalars, vectors, and matrices. A few of these which have been incorporated into AHPL are listed in Fig. B.3.

Operation	Name	Value Type	Restrictions
$\bar{x}$	NOT	Boolean	Applicable to either
$x \wedge y$	AND		signals or
$x \vee y$	OR		variables
$x \oplus y$	Exclusive OR		
$x + y$	Addition	Integer	Applicable to index
$x - y$	Subtraction		variables in
$x * y$	Multiplication		combinational
x / y	Division		logic unit
$(x \mathcal{R} y)$	Relational		descriptions only
$2 \uparrow n$	Power of 2		

Figure B.2 Unary and Binary Operators in AHPL

Operator	Name	Meaning
X, Y	Catenate	$X[0] \ldots X[\rho_1 - 1], Y[0] \ldots Y[\rho_1 - 1]$
$M ! N$	Row catenate	The result is an array with the rows of **M** above the rows of **N**.
$n \mathsf{T} p$	Binary encode	An n element vector whose rightmost elements are the bits in a binary representation of p
$\odot / X$	Reduction	$X[0] \odot X[1] \odot X[2] \ldots \odot X[\rho_1 - 1]$
$Z = \odot / M$	Row reduction	$Z[i] = \odot / M<i>$
$Z = \odot // M$	Column reduction	$Z[i] = M<0>[i] \odot M<1>[i] \ldots M<\rho_2 - 1>[i]$

Figure B.3 Mixed Operators in AHPL

The *concatenate* operator simply joins vectors together to form larger vectors. Thus, if $X = (1, 1, 0)$ and $Y = (1, 0, 1)$ then specifies that $X, Y = (1, 1, 0, 1, 0, 1)$. The notation **M ! N** indicating row catenation is a unique feature of AHPL. This operation is valid only if **M** and **N** have the same number of columns. The result is a matrix whose first rows are the rows of **M** and whose last rows are the rows of **N**. One or more of the arguments of row catenation may be a vector.

The binary encode operation of Fig. B.3 is a special case of the standard APL encode. That is, n T p expresses the binary equivalent of the decimal number p as a n-element vector. The least significant bit of the binary number is the rightmost element of n T p. This notation provides a convenient method of expressing constants in AHPL. For example,

$$8 \text{ T } 100 = (0, 1, 1, 0, 0, 1, 0, 0)$$

and

$$8 \text{ T } 0 \quad = (0, 0, 0, 0, 0, 0, 0, 0)$$

An operation that is applied to all elements of a vector or matrix is called a reduction. The AND reduction of a vector is denoted by

$$z = \wedge/X$$

and signifies that

$$z = ((X[0] \wedge X[1]) \wedge X[2]) \wedge \ldots) \wedge X[\rho-1]).$$

B.3 AHPL MODULES

The fundamental stand-alone descriptive element of AHPL is called a module. Usually a module will be an independent sequential circuit with internal sequencing of control. Although an AHPL description may consist of a single module, in a complex system AHPL will also define the relationship between modules. There will be wires interconnecting modules that must be identified as outputs in one module and inputs in another. The syntax which partitions an overall digital system into modules is given in Fig. B.4.

The syntax of an individual module description is also given in Fig. B.4. An AHPL module description begins with the keyword MODULE followed by the

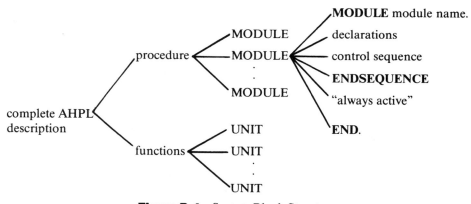

Figure B.4 System Block Structure

module name and ends with the word END. In between are the declarations followed by the control sequence. The keyword ENDSEQUENCE may follow the control sequence. If so, the list of statements between ENDSEQUENCE and END is active every clock period, regardless of the current control state. Any number of transfer and connection statements (in any order), separated by semicolons and terminated by a period, may be included after ENDSE-QUENCE. Each step in the control sequence is, likewise, a list of any number of transfer or connection statements (in any order) separated by semicolons and terminated by a period. The only distinction is that the last statement in the step may or may not be a branch.

Input and output signals, memory elements, buses and combinational logic units must be declared. A declared bus bit is a wire to which a connection may be established in more than one control state and whose output may fan out to multiple connections. In general a bus is a vector of bus bits. The keyword BUSES identifies the list of declarations of buses internal to the module. Buses to which connections can be made in more than one module must be separately declared after the keyword COMBUSES. Declaration of a bus as a COMBUS may signal a special hardware realization in the synthesis process. Only the keywords COM-BUSES and CLUNITS are absent from the AHPL module template of Section 11.8.

A file submitted to AHPL software consists of the set of module descriptions followed by a section labeled *functions* in Fig. B.4. This section is a collection of the descriptions of those combinational logic units invoked in the modules.

B.4 AHPL TRANSFER AND CONNECTION STATEMENTS

An individual clocked transfer statement may take on any of the four forms listed in Fig. B.5a. *DV* is an individual target vector. Only row selection $<r{:}s>$, column selection $[r{:}s]$, and concatenation operators may be used in assembling the target vector. **DM** is an array whose rows are individual target vectors. The bits of the vector *F* select those rows of the target vector to be clocked. *OCLV* is a vector source expression (any of the Boolean operators in Fig. B.2 and B.3) of the same dimension as *DV*. *OCLM* is an array whose rows are vector source expressions. The vector *G*, containing only one non zero element selects the active source expression. The syntax for bus and output connection statements, given in Fig. B.5b are similar but involve only right hand side conditions.

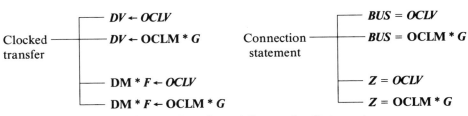

Figure B.5 Transfer and Connection Statements

INDEX